Water Chemistry

Green Science and Technology of Nature's Most Renewable Resource

Water Chemistry

Green Science and Technology of Nature's Most Renewable Resource

Stanley E. Manahan

CRC Press
Taylor & Francis Group
Boca Raton London New York

CRC Press is an imprint of the
Taylor & Francis Group, an **informa** business

CRC Press
Taylor & Francis Group
6000 Broken Sound Parkway NW, Suite 300
Boca Raton, FL 33487-2742

International Standard Book Number: 978-1-4398-3068-0 (Paperback)

Library of Congress Cataloging-in-Publication Data

Manahan, Stanley E.
Water chemistry : green science and technology of nature's most renewable resource / by Stanley E. Manahan.
p. cm.
Includes bibliographical references and index.
ISBN 978-1-4398-3068-0
1. Environmental chemistry. 2. Water chemistry. 3. Water--Pollution. I. Title.

TD193.M362 2010
628--dc22 2010005215

Visit the Taylor & Francis Web site at
http://www.taylorandfrancis.com

and the CRC Press Web site at
http://www.crcpress.com

Contents

Preface

Water Chemistry: Green Science and Technology of Nature's Most Renewable Resource is designed to provide a comprehensive overview of the chemistry of water in the environment within the broad framework of sustainability, an issue of great concern as the demands of Earth's increasing human population threaten to overwhelm our planet's carrying capacity. With this goal in mind, the book is more than just a basic coverage of the chemistry of water and relates the science and technology of this amazing substance to areas essential to sustainability science, including environmental chemistry, green chemistry, industrial ecology, and green (sustainable) science and technology. Such an approach views planet Earth as consisting of five mutually interacting spheres: the hydrosphere, which is the main area addressed by this book, the atmosphere, the geosphere, the biosphere, and the anthrosphere. Because of its overwhelming influence on the environment, it is essential to give attention to the anthrosphere, that part of the environment constructed and operated by humans. The anthrosphere has pervasive effects on the hydrosphere in areas such as water consumption and pollution. Furthermore, water plays a key role in ensuring the sustainable operation of the anthrosphere and its maintenance in a manner that will enable it to operate in harmony with the environment for generations to come.

Those with a need to study and understand water chemistry come from very diverse backgrounds, especially with respect to their understanding of chemistry. Many, such as students of engineering, will have taken university-level courses in chemistry. But there are others who may have had only a single course in general chemistry or no university-level chemistry at all. To accommodate this diversity, the chemistry in *Water Chemistry* is kept as simple as is consistent with thorough coverage of the topic. In addition, two chapters are provided at the end of the book to provide readers with the basic chemistry they may need to understand water chemistry. Chapter 11, "Fundamentals of Chemistry," defines chemistry and introduces the essential concepts of elements, chemical bonding, chemical reactions and equations, and solutions. Chapter 12, "Organic Chemistry," is a basic coverage of this discipline, which is essential to understand organic water pollutants, the organic chemicals used to treat water, and the organic polymers used to make filters and membranes used in water treatment, water desalination, and water analysis.

Chapter 1 of this book, "Water and the Five Spheres of the Environment," sets the framework of the book within the broader context of the environment as a whole. It explains how the hydrosphere relates to the geosphere, atmosphere, biosphere, and anthrosphere. It covers biogeochemical cycles and the essential role of water in these basic cycles of materials. Environmental chemistry is defined and the role of water in environmental chemistry is explained. Emerging as a major subdiscipline in chemistry since the 1990s, green chemistry is explained with emphasis on the part played by water in the modern practice of sustainable chemistry.

Chapter 2, "Properties of Water and the Hydrosphere," is an overview of water's properties and the hydrosphere's place in the environment. Although this chapter deals mostly with the nonchemical aspects of water, it begins with an explanation of "the fantastic water molecule" and how the chemical nature of water determines its physical behavior. The hydrological cycle is illustrated along with the various places where usable water is found, including standing bodies of water, flowing water, and groundwater. This chapter also covers water as an essential resource, water utilization in the anthrosphere, and impoundment and transfer of water. A summary is given of the various forms of aquatic life, which are strongly influenced by and have a strong influence on the chemistry of water. Another section of the chapter explains the relationship of global warming to water supply and quality.

Chapters 3 through 9 constitute the core of water chemistry. Chapter 3, "Fundamentals of Aquatic Chemistry," covers water acidity and alkalinity, gases in water, metals in water, and complexation and chelation of metal ions. Chapter 4, "Oxidation–Reduction," explains oxidation and reduction in water, the transfer of electrons between species, and the key concept of electron activity as expressed by pE, conceptually the negative log of electron activity. Chapter 5, "Phase Interactions," deals with the interactions of species between solution in water and gas or solid phases. The chapter also covers sediments in bodies of water and streams and the unique importance of colloids in water. Microorganisms, especially bacteria and algae, have a strong influence on water chemistry and are discussed in Chapter 6, "Aquatic Microbial Biochemistry." Chapter 7, "Water Pollution," includes a discussion of pollutant heavy metals, inorganic pollutants, organic pollutants, algal nutrients, and other water pollutants. The treatment of water for municipal and industrial uses, the treatment of wastewater, and water reuse and recycling are included in Chapter 8, "Water Treatment." An overview of water analysis is provided in Chapter 9, "Chemical Analysis of Water and Wastewater."

Water and energy are very closely related. One of the greater uses of water is for cooling electric power plants. Greenhouse warming largely results from carbon dioxide emitted to the atmosphere in the combustion of fossil fuels. Sustainable biomass energy sources are dependent on water to grow plants; algae can be grown in water as a renewable biomass energy resource. The availability of abundant sustainable energy enables substantial extension of available water supplies. Means by which this may be done, all of which require energy, include purification of municipal wastewater to drinking water standards, desalination of brackish groundwater, and even desalination of seawater. With such relationships in mind, Chapter 10, "Sustainable Energy: The Key to Everything," provides an overview of energy sources, including how they relate to water resources. As part of this chapter, a system of industrial ecology is proposed based on the thermochemical gasification of biomass and using renewable wind energy to produce synthetic natural gas as well as by-products, including ammonia used as fertilizer.

The author welcomes questions and input from readers. He can be contacted by e-mail at the following address: manahans@missouri.edu.

Stanley E. Manahan

Author

Stanley E. Manahan is professor emeritus of chemistry at the University of Missouri–Columbia, where he has been on the faculty since 1965. He received his AB in chemistry from Emporia State University in Kansas in 1960 and his PhD in analytical chemistry from the University of Kansas in 1965. Since 1968, his primary research and professional activities have been in environmental chemistry, with recent emphasis on hazardous waste treatment. His latest research has involved gasification of wastes and gasification of sewage sludge and crop by-product biomass for energy production. Professor Manahan has taught courses on environmental chemistry, hazardous wastes, toxicological chemistry, and analytical chemistry, and has lectured on these topics throughout the United States as an American Chemical Society Local Sections Tour Speaker and in a number of countries including France, Italy, Austria, Japan, Mexico, and Venezuela. Professor Manahan has written books on environmental chemistry (*Environmental Chemistry*, 9th ed., 2010, Taylor & Francis/CRC Press, and *Fundamentals of Environmental Chemistry*, 3rd ed., 2009, Taylor & Francis/CRC Press), green chemistry (*Green Chemistry and the Ten Commandments of Sustainability*, 2nd ed., 2005, ChemChar Research, Inc.), general chemistry (*Fundamentals of Sustainable Chemical Science*, 2009, Taylor & Francis/CRC Press), hazardous wastes and industrial ecology (*Industrial Ecology: Environmental Chemistry and Hazardous Waste*, 1999, Lewis Publishers/CRC Press), toxicological chemistry (*Toxicological Chemistry and Biochemistry*, 3rd ed., 2002, Lewis Publishers/CRC Press), applied chemistry, and quantitative chemical analysis. He is the author or coauthor of approximately 90 research articles.

1 Water and the Five Spheres of the Environment

1.1 WATER

A photograph of the Earth from outer space shows a unique and marvelous planet, a blue globe with areas of red from desert lands and large white masses of cloud. Such a picture speaks of a unique speck of the universe, far different from the inferno of the Sun, the sterile rock of Earth's moon, or the red, rocky surface of Mars. Earth is the mother of life including our own and all life forms currently known to humankind. She is nurturing and remarkably forgiving to the abuses of humankind, although there are limits to her bounty that humans are almost certainly approaching now. There are limits to the number of people that she can support and to the demands for materials and energy that each places on her. There is a limit to the amount of global warming carbon dioxide that her atmosphere can tolerate, a level that may be in the process of being exceeded even as these words are written. There are limits to the amounts of food that can be grown on her surface. Ominously for world food supplies, the once-bountiful food resource of her oceans is now declining. Humankind faces the enormous challenge of sustainability, of keeping the Earth as a living planet essential to all humans.

If one were to name one feature of the Earth among the many that make it a hospitable place to live, it would be water. Water, H_2O, a molecule with a deceptively simple formula, but one with remarkably complex properties that make it the medium in which life can exist. As we deal with the sustainability challenge, it is essential to know about water, its properties, where it comes from, how it is used, how it is purified, and how it may be treated to keep it pure. Fortunately, properly used, water is nature's most sustainable and renewable resource. The Earth herself does most of the heavy lifting in sustaining the water resource. This is done through the mechanism of the hydrological cycle in which heat from the Sun evaporates water from the oceans into the atmosphere, winds powered by solar heating of the atmosphere carry water vapor over land, and purified water falls to the surface, eventually (often with a load of contaminants picked up from land) flowing into the oceans to start the whole cycle over again. Some of the water percolates below Earth's surface where it accumulates as groundwater, much of which is pumped out by humans for municipal, agricultural, and industrial uses.

It is essential for humans to protect and nurture the precious bounty of fresh water provided to them by the hydrological cycle. The most direct way in which this is done is through water conservation, using as little water as possible and minimizing the impurities placed into it. In many areas with abundant water resources, little thought is given to conserving water because it is so abundant and cheap. In other parts of the world, getting enough water is an overriding concern. In some of these water-deprived areas, the amount of water available for each person to use is only about the quantity that one person leaves running down the drain in brushing teeth in more water-rich areas! Moreover, even the small amounts of water used in many parts of the world may contain microorganisms that cause debilitating disease or toxic substances such as arsenic in some groundwater sources.

The need to know about water is self-evident. In addition, a knowledge of water requires understanding of its chemistry—*water chemistry* or *aquatic chemistry*. This book is about water chemistry. To understand water chemistry it is essential to know more than just the chemical properties of H_2O and how this molecule interacts with solutes dissolved in water. Therefore, the material in this book relates water chemistry to the places on, above, and below Earth's surface where water may be found and the book begins with a discussion of water's interactions with these places. It is convenient and logical to divide Earth's environment among five closely related spheres; such a division serves as an organizational framework for this book (see Figure 1.1). These spheres are the *hydrosphere*, that part of the environment consisting of water; the *geosphere*, made of solid rock and soil as well as molten rock at greater depths; the *atmosphere*, composed of a thin layer of air; the *biosphere*, consisting of living organisms; and the *anthrosphere*, made up of the things that humans make and operate. The relationships of each of these five environmental spheres to water and the hydrosphere are discussed in this chapter.

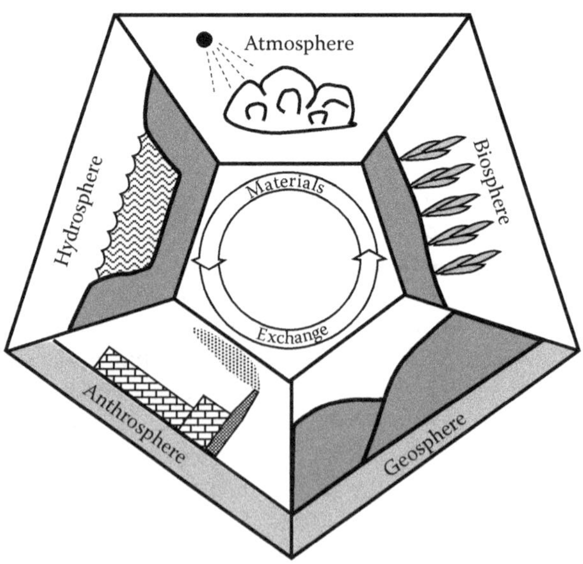

FIGURE 1.1 It is useful to consider Earth's environment from the viewpoint of five closely interrelated spheres. Water is found in the hydrosphere, but it strongly influences and is strongly influenced by the other four environmental spheres.

1.2 THE HYDROSPHERE

Water, chemical formula H_2O, comprises the *hydrosphere*. The role of water in the environment as a whole is best understood through the hydrological cycle mentioned at the beginning of this chapter and illustrated and discussed in more detail in Chapter 2.

Water is a key resource in the maintenance of sustainability. Shortages of water from climate-induced droughts have been responsible for the declines of major civilizations. Variations in water supply cause severe problems for humans and other life forms. Devastating floods displace and even kill large numbers of people throughout the world and destroy homes and other structures. Severe droughts curtail plant productivity resulting in food shortages for humans and animals in natural ecosystems and often necessitating slaughter of farm animals during catastrophic droughts.

Water supply, quality of water, and the nature of the hydrosphere are strongly affected by the interaction of the hydrosphere with the other environmental spheres. These interactions are discussed in the remainder of this chapter.

1.3 THE GEOSPHERE AND THE HYDROSPHERE

The *geosphere* (Figure 1.2) is the solid Earth. It is the medium upon which most food is grown and is a source of metals, plant fertilizers, construction materials, and fossil fuels. Large quantities of consumer and industrial wastes have been discarded to the geosphere, a practice that continues. Managing and preserving the geosphere is of utmost importance to achieve sustainability.

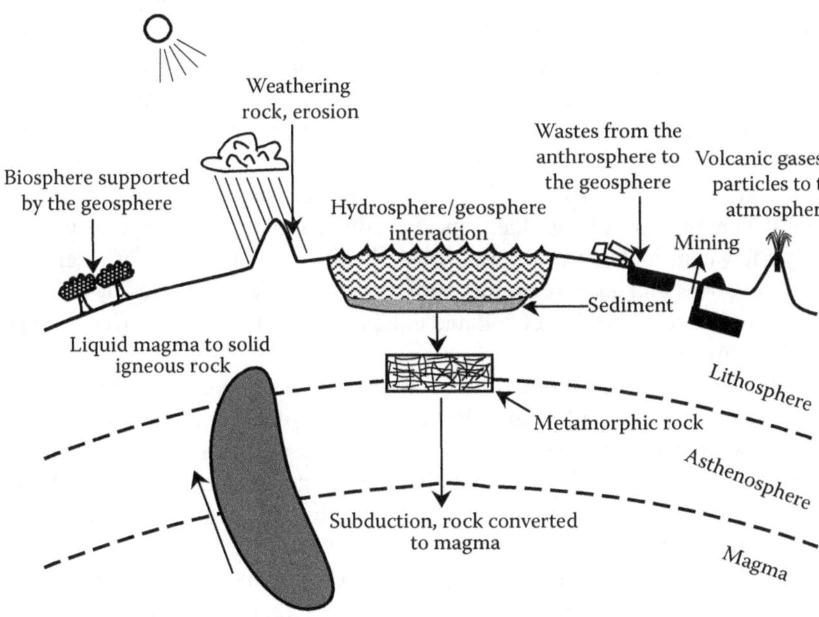

FIGURE 1.2 Representation of the geosphere showing its strong interactions with other spheres of the environment.

The geosphere consists of rocks, the minerals that compose rocks, soil, sediments, the molten rock that constitutes Earth's outer core or mantle, and the highly compressed iron-rich solid material that makes up Earth's inner core. The outermost solid layer of the Earth is the *lithosphere* composed of relatively strong solid rock, varying in thickness from just a few to as much as 400 km, averaging about 100 km. Earth's *crust* is the outer layer of the lithosphere, which is only 5–40 km thick and composed of rocks that are somewhat less dense than the rocks deeper in the lithosphere. Below the lithosphere to a depth of approximately 250 km and resting on the viscous liquid rock mantle is the *athenosphere* composed of hot rocks that are relatively weak and plastic. Essentially all interactions of the hydrosphere with the geosphere are with the crust relatively near Earth's surface.

There are about 2000 known minerals found in the lithosphere characterized by definite chemical composition and crystal structure. Most rocks in the geosphere are composed of only about 25 common minerals. The crust is 49.5% oxygen and 25.7% silicon, and this is reflected in the chemical composition of rocks, which are mostly various chemical combinations of oxygen and silicon with smaller amounts of aluminum, iron, carbon, sulfur, and other elements. The chemical compositions and solubilities of minerals have a very strong influence on water that contacts them and on the hydrosphere. For example, groundwater in contact with mineral NaCl (halite) becomes too saline for irrigation or drinking.

Molten rock that penetrates to near the top of Earth's crust cools and solidifies to form *igneous rock*. Exposed to the water and atmosphere, igneous rock undergoes physical and chemical changes in a process called *weathering*. Weathered rock material carried by water and deposited as sediment layers may be compressed to produce secondary minerals, of which clays are an important example.

A crucial part of the crust is the thin layer of weathered rock, partially decayed organic matter, air spaces, and water composing *soil*, obviously of great importance because it supports plant life that provides food upon which humans and other animals are reliant for their existence. This essential layer of soil is extremely thin compared to Earth's diameter. If the Earth were the size of a classroom globe, the average thickness of soil on its surface would only be about the dimensions of a human cell! The top layer of soil that is most productive of plants is *topsoil*, which is often only a few centimeters thick in many locations, or even nonexistent where poor cultivation practices and adverse climatic conditions have led to its loss by wind and water erosion. The conservation of soil and enhancement of soil productivity are key aspects of sustainability.

The interactions between the geosphere and the hydrosphere are many and very important, as noted above with respect to the influence of minerals on water composition. Major segments of the hydrosphere including the oceans, rivers, and lakes rest on the geosphere, and groundwater exists in aquifers underground. The weathering of rock that converts exposed igneous rock to sediments, sedimentary rock, and eventually soil is largely through the action of water. Vast deposits of limestone formed in the geosphere millions of years ago were produced as a by-product of photosynthesis of algae in water that raised the pH such that the calcium and magnesium carbonate salts that compose limestone precipitated from water. Even earlier, photosynthetic cyanobacteria in water produced elemental oxygen that oxidized soluble Fe^{2+} ion in water, initially precipitating $Fe(OH)_3$ that produced the deposits of iron oxides that now serve as iron ores.

The nature and quality of water available for human use are largely the result of the interaction of the hydrosphere with the geosphere. Two of the key water quality parameters are water hardness, generally in the form of dissolved Ca^{2+} ion, and water alkalinity as dissolved bicarbonate, HCO_3^-; both are discussed in detail in Chapter 3. These two ions get into water that is in contact with geospheric limestone, $CaCO_3$, as the result of the reaction of this mineral with CO_2 dissolved in water. A variety of trace elements present in minute quantities are picked up by groundwater in contact with geospheric minerals. In some cases these are not desirable as is the case with toxic levels of arsenic in water from some wells in Bangladesh.

An important hydrosphere/geosphere interaction occurs in soil where water acts as a transfer medium to move essential nutrients from soil to plant roots. This soil solution, as it is called, dissolves nutrients such as Ca^{2+} ion, K^+ ion, NO_3^- (a source of plant nitrogen), $H_2PO_4^-$ ion (a source of fertilizer phosphorus), and SO_4^{2-} ion (used by plants to synthesize sulfur-containing proteins) from soil and into the plant vascular system where they are involved in plant metabolism. Water itself is an essential constituent of soils for plant growth and an adequate, but not excessive, level of water is required in soil in order for plants to thrive.

An interesting, potentially deadly, interaction between the hydrosphere and the geosphere is posed by gases of geospheric origin trapped beneath lakes that can erupt suddenly and asphyxiate people in the vicinity. This happened in 1986 in the African country of Cameroon when an estimated 1.6 million metric tons of volcanic carbon dioxide evolved from supersaturated dissolved CO_2 in the waters of Lake Nyos that asphyxiated 1700 people. An even greater potential problem is posed by much larger Lake Kivu located in a volcanically active region between the African countries of Congo and Rwanda. This freshwater lake is underlain by vast deposits of an estimated 392 billion cubic yards of carbon dioxide and 78 billion cubic yards of methane, enough methane to be of commercial interest as a source of fuel. Huge bubbles of these gases, known in Swahili as *mazuku* or "evil wind" kill about 100 people each year along Lake Kivu's northern shore. There is great concern that the levels of trapped gases are reaching a saturation point that could result in a sudden, massive release of gas that could spread throughout the area and kill many more people than those who died in the Lake Nyos incident.

1.4 THE ATMOSPHERE AND THE HYDROSPHERE

The *atmosphere* is a layer of gases with a total mass of about 5.15×10^{15} tons that surrounds the solid Earth. On a dry basis (excluding water vapor), the air that composes the atmosphere is 78.1% by volume of nitrogen gas (N_2), 21.0% O_2, 0.9% noble gas argon, and 0.039% CO_2, a level that keeps increasing by almost 0.002% per year. In addition, there are numerous trace gases in the atmosphere at levels below 0.002% including ammonia, carbon monoxide, helium, hydrogen, krypton, methane, neon, nitrogen dioxide, nitrous oxide, ozone, sulfur dioxide, and xenon. Relative to Earth's diameter, the atmosphere is really quite thin. People who travel to the tops of higher mountains readily notice the lower density of the air in reaching altitudes of only 3000 or 4000 meters (m). More than 99% of the atmosphere's mass is within 40 kilometers (km) of Earth's surface, with the majority of the air below 10 km altitude (compared to Earth's diameter of almost 13,000 km). Again, to use the analogy of a classroom globe, if the Earth were of the size of such a globe, virtually the whole atmosphere would be contained within a layer of the thickness of varnish on it!

Humans think first of the atmosphere as a source of oxygen essential to their existence; deprived of oxygen, they expire within a few minutes. In addition to providing oxygen, the atmosphere provides many other support functions essential for life. One of these is to act as a blanket to keep Earth's surface at an average temperature of about 15°C at sea level and within a temperature range that enables life to exist. As well as maintaining temperature levels conducive to life, Earth's atmosphere absorbs very short-wavelength ultraviolet radiation from the Sun, which, if it reached organisms on Earth's surface, would tear apart the complex biomolecules essential for life. In addition to providing oxygen required by animals, the atmosphere is the source of carbon dioxide, which green plants absorb from the atmosphere and convert into biomass. The atmosphere's carbon dioxide (present at only about 0.039%) is the source of carbon for the biomass in all food and biomass-based raw materials. The chemically bound nitrogen that all organisms require for their protein is taken from the atmosphere, which consists mostly of elemental nitrogen gas. Bacteria growing on the roots of some plants accomplish this task as do nitrogen-fixing industrial processes operating under rather severe conditions.

The atmosphere (Figure 1.3) is divided into layers based on temperature. Living organisms experience the lowest layer called the *troposphere* characterized with decreasing temperature and density with increasing altitude and extending from surface level, where the average temperature is 15°C, to about 11 km (the approximate cruising altitude of commercial jet aircraft) where the average temperature is –56°C. The very cold region at the upper limits of the troposphere is called the *tropopause.* It has special significance for sustaining life on Earth because water vapor reaching it becomes frozen into ice crystals that tend to settle. Were this is not the case, over the eons the intense ultraviolet radiation above the tropopause would long ago have broken apart the water molecules in the atmosphere and the very light elemental hydrogen (H_2) produced would have escaped to outer space leading to a dry planet that could not support life.

Above the troposphere is the *stratosphere* in which the average temperature increases from about –56°C at its lower boundary to about –2°C at its upper limit. The heating effect in the stratosphere is due to energy absorbed by air molecules from the intense solar radiation that impinges on it.

A key aspect of atmospheric chemistry is the occurrence of *photochemical reactions* that occur in it. These take place when quanta of electromagnetic radiation from the Sun (basically small packets of energy represented as $h\nu$) are absorbed by molecules often causing the molecules to fall apart and producing highly reactive species with unpaired electrons called *free radicals* that can undergo series of chain reactions forming a variety of products such as the noxious oxidant species characteristic of photochemical smog.

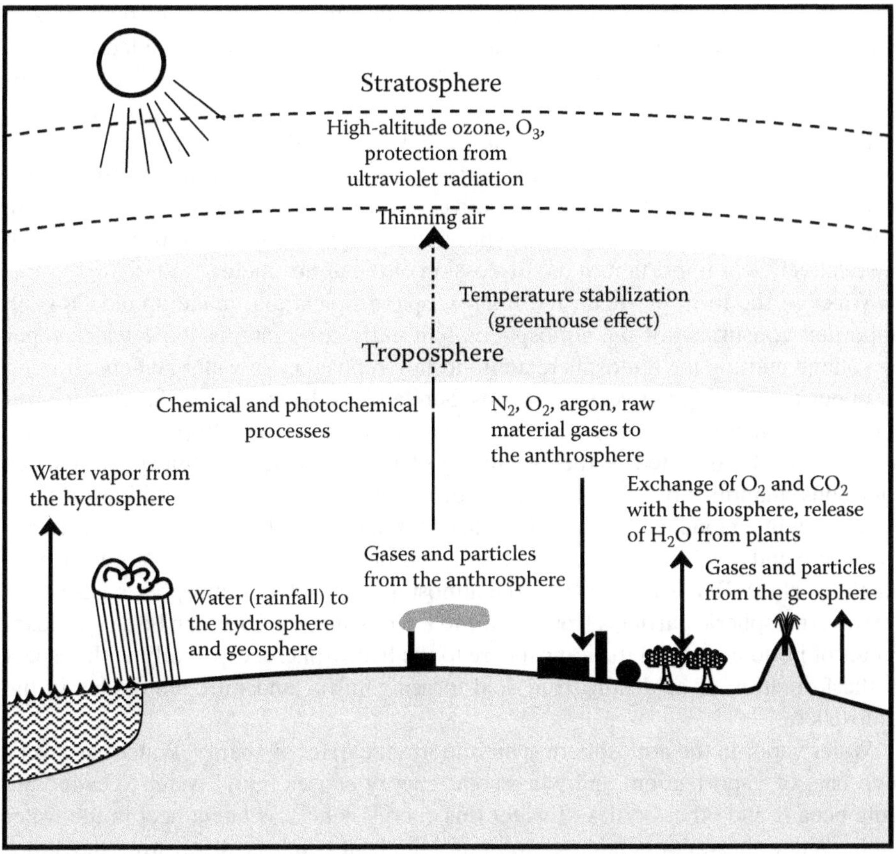

FIGURE 1.3 Some major aspects of the atmosphere. The troposphere characterized by decreasing temperature with increasing altitude is the atmospheric layer closest to the Earth. Most of the atmosphere's air and water vapor are in the troposphere. Temperature increases with increasing altitude in the stratosphere, which contains life-sustaining ozone that acts as an essential filter to damaging ultraviolet radiation.

One of the most significant photochemical reactions occurs at altitudes above approximately 11 km in the stratosphere where ultraviolet solar radiation penetrates far enough to cause photochemical dissociation of molecular oxygen:

$$O_2 + h\nu \rightarrow O + O \tag{1.1}$$

Especially at higher altitudes in the stratosphere the oxygen atoms can persist for some time but they eventually react with molecules of O_2 to produce *ozone*, O_3:

$$O_2 + O \rightarrow O_3 \tag{1.2}$$

The result of this reaction is a very dilute zone of ozone spread over many kilometers of altitude in the stratosphere. This ozone zone layer (which in a pure form at sea level would be only 3 mm thick!) is essential for the existence of humans and much other life on Earth's surface because it absorbs a portion of ultraviolet radiation, which otherwise would penetrate to Earth's surface and destroy essential life biomolecules in exposed plants and animals. Some classes of chemical species, especially the chlorofluorocarbons or Freons once widely used as refrigerants, are known to react in ways that destroy stratospheric ozone, and their elimination from commerce has been one of the major objectives of efforts to sustain life on the Earth. Above the stratosphere are the atmospheric mesosphere and thermosphere, which are relatively less important in the discussion of the atmosphere.

Water in the form of vapor and microscopic droplets that make up clouds is an important constituent of the atmosphere. Normally, air contains 1–3% water vapor by volume making the atmosphere an important repository of water and constituting an important overlap between the atmosphere and the hydrosphere. As air becomes cooler with increasing altitude, water vapor condenses into droplets and the air becomes drier. As noted at the beginning of this chapter, the atmosphere carries enormous quantities of water over great distances. Atmospheric conditions of temperature, wind, cloud cover, and other variables determine how far this water travels and where and in which form it is discharged as precipitation from the atmosphere to Earth's surface. Water falling from the atmosphere dissolves atmospheric gases and carries atmospheric particles from the air to Earth's surface. This can be a significant vector of pollutants from the atmosphere to the hydrosphere, especially with respect to the formation of acid rain from acid-forming sulfur and nitrogen oxides in the atmosphere.

Water vapor in the atmosphere is an important carrier of energy. Water has a very high heat of vaporization, and when solar energy causes liquid water to evaporate from oceans and other bodies of water this energy is held as latent heat in the water vapor. When the vapor condenses to liquid, the heat is released warming the atmosphere and producing warmer air that rises. This effect can be powerful and, for instance, is the driving force behind the enormous energy from wind and air currents in hurricanes. Because of the angle in which the Earth receives energy from the Sun, solar heating is most intense at the equator. Water evaporated by this energy carries latent heat into the atmosphere and moisture-laden air moves away from the equator carrying energy with it. This energy is released when rain forms far from the equator and this phenomenon is responsible for much of the redistribution of energy around the globe.

1.5 THE BIOSPHERE AND THE HYDROSPHERE

The *biosphere* (Figure 1.4) includes living organisms and the materials and structures that they produce. *Biology* is the science of the living organisms in the biosphere. Living organisms are characterized by particular classes of biochemical substances (biomolecules). These include proteins that are the basic building blocks of organisms, carbohydrates made by photosynthesis and metabolized by organisms for energy, lipids (fats and oils), and the all-important nucleic acids (DNA, RNA), genetic material that defines the essence of each individual organism and acts as codes to direct protein biosynthesis and reproduction. In the biosphere a hierarchical

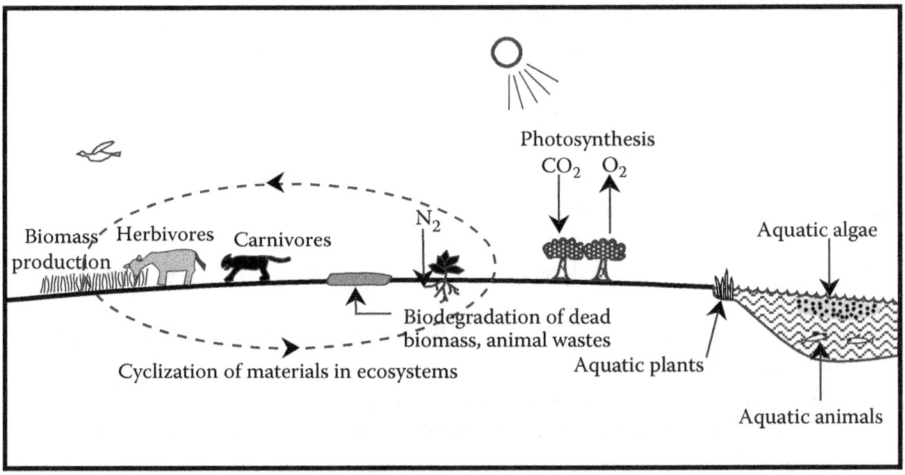

FIGURE 1.4 The biosphere. Organisms in the biosphere occupy ecosystems in which materials undergo complete recycling. These systems are based on photosynthetic plants and (in water) photosynthetic algae that form the basis of the food chain in different ecosystems. Photosynthesis removes carbon dioxide from the atmosphere and returns elemental oxygen. Elemental nitrogen from the atmosphere is chemically combined into biomolecules by the action of nitrogen-fixing bacteria, such as those that grow on legume plant roots. Dead biomass is broken down to simple inorganic materials by fungi and bacteria.

organization prevails, beginning at the molecular level and progressing through living cells, organs, organisms, and the biosphere, itself. Organisms are capable of carrying out metabolic processes in which they chemically alter substances. They reproduce and their young ones undergo various stages of development. They also demonstrate heredity and may be altered by mutation processes resulting from modifications of DNA.

Organisms comprising the biosphere may be as simple as individual cells of bacteria and photosynthetic algae, so small that a microscope is required to view them or as complex as human beings consisting of a variety of organs and with brains that can carry out thought and reasoning processes. Biologists recognize six kingdoms of organisms. Three of these may exist as single cells. *Archaebacteria* and *eubacteria* do not have defined cell nuclei. *Protists* are single-celled organisms but do have defined nuclei in which cellular DNA is located and may have distinct structures, such as mouth-like structures for ingesting food and light-sensitive structures that act as primitive "eyes." At more complex levels are generally multicelled *plantae* (plants) and *animalia* (animals), as well as *fungi* including yeasts, molds, and mushrooms.

Organisms exist in *ecosystems* made up of interacting communities in a defined environment with which they interact. The entire biosphere is composed of all the ecosystems on the Earth. A key aspect of a biological ecosystem is its *productivity*, that is, its ability to produce biomass by photosynthesis. The organisms that photosynthetically remove carbon dioxide from the atmosphere and fix it in the form of organic matter that is further converted by biochemical processes into proteins, fats, DNA, and other life molecules constitute the basis of the whole ecosystem food web upon which the remainder of the organisms in the food web depend for their existence. The major photosynthetic organisms in terrestrial ecosystems are plants growing in soil. In water, the major contribution is made by algae, a form of plants growing in water. Some protozoa and some bacteria also have photosynthetic capabilities.

The biosphere is strongly connected to the other environmental spheres. It has strong effects on these spheres and, in turn, is strongly influenced by them. Some important examples of these are given below:

- Organisms in the biosphere generate a variety of materials that humans use.
- Conditions in the geosphere, especially of soil, are crucial in determining biospheric productivity.
- The biosphere is strongly influenced by large quantities of pesticides and fertilizers produced by anthrospheric processes to enhance the health and productivity of organisms, particularly farm crops, in the biosphere.
- Most of the major efforts to control pollutants, wastes, and toxic substances from the anthrosphere are designed to protect organisms, including humans, in the biosphere.
- The nature of the anthrosphere and the ways in which it is operated strongly influence the nature of the biosphere and its productivity.

The biosphere has especially strong interactions with the hydrosphere. Water is the medium of life; were it not for water there would be no biosphere. The availability of water largely determines the kinds and numbers of organisms in the biosphere. A rain forest, for example, supports a high mass of living organisms of many species that are generally prolific users of the abundant water that the rain forest climate provides. An arid desert in contrast may have relatively small populations of only a limited number of species that are particularly well adapted to arid conditions. The organisms in regions that alternate between monsoonal wet seasons and dry seasons must have mechanisms for adjusting to variations in water supply. In all living organisms, water is the milieu in which proteins, enzymes, nucleic acids, and other life molecules exist within cells. The blood in humans and other animals is an aqueous solution of salts in which is suspended blood cells and other bodies. Through the process of transpiration (Figure 1.5) water carries essential nutrients from soil into roots, through plant tissues, and to leaf surfaces where it evaporates to the atmosphere.

As well as being strongly influenced by the other environmental spheres, the biosphere has a strong influence on the other spheres. For example, the oxygen that all humans breathe was put into the atmosphere by the actions of photosynthetic bacteria eons ago. Lichens, mixed communities of algae and fungi growing synergistically (to mutual advantage), act upon geospheric rock to break it down to soil that is hospitable to plant life. In less industrialized societies, the nature of the anthrosphere made and operated by humans depends strongly on the surrounding biosphere. Abundant herds of bison provided the sturdy hides used by Native Americans for their mobile-home TPs. The availability and kinds of wood have largely determined the nature of dwellings constructed in many societies. More than 2000 years ago,

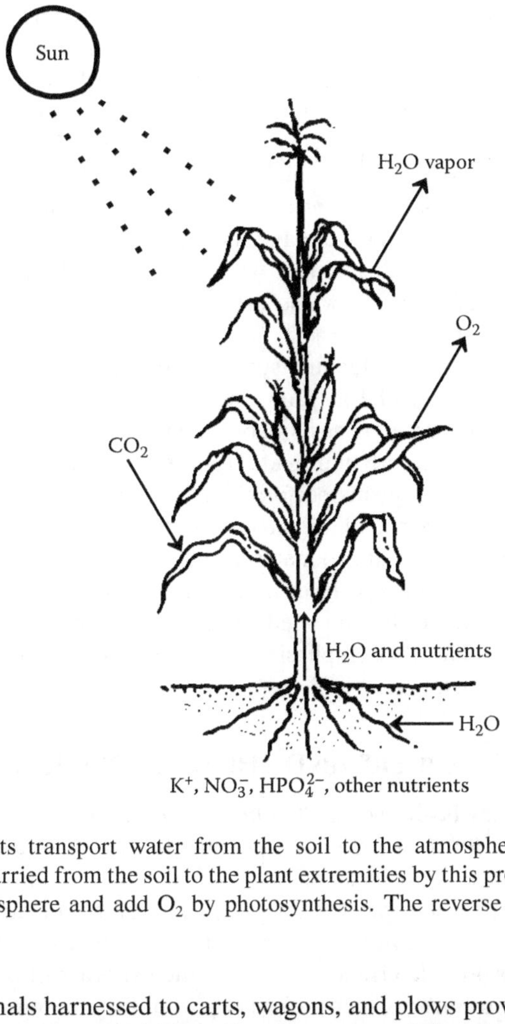

FIGURE 1.5 Plants transport water from the soil to the atmosphere by transpiration. Nutrients are also carried from the soil to the plant extremities by this process. Plants remove CO_2 from the atmosphere and add O_2 by photosynthesis. The reverse occurs during plant respiration.

domesticated animals harnessed to carts, wagons, and plows provided humans with mobility and the means to cultivate soil, and many societies, including the Amish farmers in the United States, still use horses, donkeys, mules, oxen, and water buffalo for such purposes. As petroleum dwindles as a source of raw materials, such as plastics, it is likely that materials produced in the biosphere will play a stronger future role in determining the kinds of materials used in the anthrosphere.

Over eons of evolution, the biosphere has developed as an exemplary survivor that can teach humans important lessons in achieving sustainability. One important factor is that organisms carry out their activities under very mild conditions, which is one of the goals of sustainable green technology. Even the conditions of boiling water in which some thermophilic bacteria thrive in locations such as the thermal hot springs of Yellowstone National Park are mild compared to the high-temperature, high-pressure conditions required in many chemical syntheses. Furthermore, the intolerance of living organisms to toxic substances demonstrates which of these substances should be avoided in performing chemical processes in the anthrosphere.

Over hundreds of millions of years of evolution, organisms have had to evolve sustainable ecosystems for their own survival, completely recycling materials and enhancing their environment. In contrast, humans have behaved in ways that are unsustainable with respect to their own existence, exploiting nonrenewable resources and polluting the environment on which they depend for their own survival. The complex, sustainable ecosystems in which organisms live sustainably in relationship with each other and their surroundings serve as models for anthropogenic systems. By taking lessons from the biosphere and its long-established ecosystems, humans can develop much more sustainable industrial systems. Such systems have come to be known as *industrial ecosystems* and are studied in the discipline of *industrial ecology*.[1]

The biosphere is a key to achieving sustainability, especially with respect to its ability to perform photosynthesis. Biogenic materials have the potential to be produced in a much greener, more sustainable manner than do materials produced by humans because plants use carbon dioxide from the atmosphere as a carbon source and are solar powered. Organisms perform syntheses under inherently mild, therefore safe, conditions. Furthermore, organisms are particularly well adapted to make a variety of complex and specialized materials that are very difficult or impossible to make by purely chemical means. In order for organisms in the biosphere to fulfill their potential to provide for human needs, it is crucial to have adequate supplies of high quality water, a fact that emphasizes the important connection between the biosphere and the hydrosphere.

1.6 THE ANTHROSPHERE AND THE HYDROSPHERE

The *anthrosphere* may be defined as *that part of the environment made or modified by humans and used for their activities.* Although some environmentalists object to recognizing the anthrosphere as an environmental sphere, examination of any area in which humans live reveals that it is definitely a part of the surroundings. Buildings, roads, airports, factories, power lines, and numerous other things constructed and operated by humans provide visible evidence of the existence of the anthrosphere on the Earth. The anthrosphere is such a distinct part of the environment and its potential effects so profound that many scientists agree with the contention of Nobel Prize-winning atmospheric chemist Paul Crutzen that the Earth is undergoing a transition from the Halocene geological epoch to a new one, the *anthropocene.* This is occurring because human activities are now quite significant compared to nature in their impact on Earth's environment and are changing Earth's fundamental physics, chemistry, and biology. Some scientists fear that, especially through changes in global climate, activity in the anthrosphere will detrimentally alter Earth's relatively stable, nurturing environment and produce one that is much more challenging to human existence.

Preindustrial humans had some influence on the environment. For example, in areas now occupied by the present-day United States, native Americans employed fire to destroy forests and create grasslands that supported game animals. Compared to the modern industrial era, however, the impact of humans until just a few centuries ago was comparatively benign. Particularly with the development of fossil fuel resources and large machines powered by these fuels, during the past approximately 200 years, humans have built a pervasive anthrosphere that has a massive environmental impact.

There are many distinct segments of the anthrosphere as determined by factors such as the following:

- Where humans dwell
- How humans and their goods move
- How commercial goods and services are provided
- How renewable food, fiber, and wood are provided
- How energy is obtained, converted into other forms, and distributed
- How wastes are collected, treated, and disposed

Considering these aspects, it is possible to list a number of specific things that are parts of the anthrosphere as shown in Figure 1.6. These include dwellings as well as other structures used for manufacturing, commerce, education, and government functions. Utilities include facilities for the distribution of water, electricity, and fuel as well as systems for the collection and disposal of municipal wastes and wastewater (sewers), and—of particular importance to sustainability—systems for materials recycle. Transportation systems include roads, railroads, airports, and waterways constructed or modified for transport on water. The anthrospheric segments used in food production include cultivated fields for growing crops and water systems for irrigation.

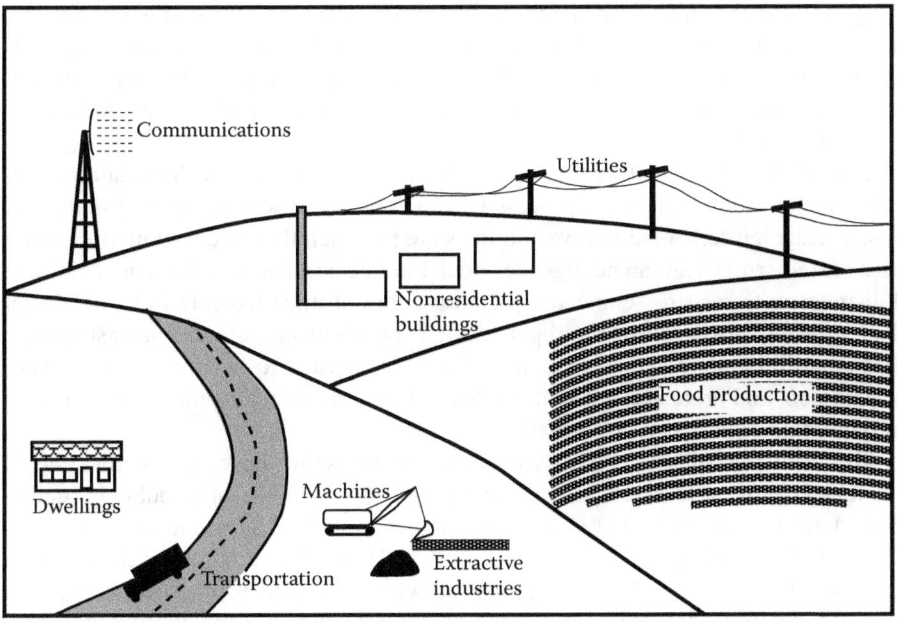

FIGURE 1.6 The anthrosphere consists of machines, structures, modifications to the geosphere, and other things made and operated by humans.

A variety of machines, including automobiles, trains, construction machinery, and airplanes are part of the anthrosphere. The communications sector of the anthrosphere includes radio transmitter towers, satellite dishes, and fiber optics networks. Oil and gas wells are employed for extracting fuels from the geosphere and mines for removing coal and minerals.

Much of the anthrosphere may be classified as *infrastructure* made up of utilities, facilities, and systems that large numbers of people must use in common and that are essential for a society to operate properly. A large portion of the infrastructure consists of physical components, including electrical power-generating facilities and distribution grids, communications systems, roads, railroads, air transport systems, airports, buildings, water supply and distribution systems, and waste collection and disposal systems. Another essential part of the infrastructure consists of laws, regulations, instructions, and operational procedures. Components of the infrastructure may be in the public sector, such as U.S. highways or some European railroads, or privately owned and operated, such as the trucks that use the highways or airlines that use publicly owned airports.

Infrastructure has been compared to a computer operating system, which enables a computer to operate programs for word processing, record keeping, calculation, drawing, communication, and other common computer operations, and to properly record, store, correlate, and output the products of such programs. The infrastructure of the anthrosphere enables acquisition and processing of materials, the conversion of materials into manufactured items, and the distribution of such items and most other activities that occur in the anthrosphere. Many computer users are all too familiar with the loss of productivity that can occur with outdated and poorly designed operating systems that have a tendency to crash. Similarly, an outdated, cumbersome, poorly designed, worn-out anthrospheric infrastructure causes economic systems and societies to operate in a very inefficient manner that is inconsistent with sustainability. Catastrophic failure can result, as occurred with several massive failures of electrical power grids.

Parts of the infrastructure tend to deteriorate with age. One of the greatest problems is corrosion, a chemical process in which metals, such as the steel that composes bridge girders, tend to revert to the state in which they occur in nature (in the case of steel, rust). Human negligence and deliberate damage—misuse, neglect, vandalism, even terrorism—can cause premature loss of infrastructure function. Often the problem starts with improper design of elements of the infrastructure. Sustainability requires that elements of the infrastructure be properly designed, maintained, and protected to avoid the expense and material and energy use entailed in rebuilding infrastructure prematurely.

To date much of the resources put into infrastructure have been dedicating to "conquering," or at least temporarily subduing nature. For example, publicly funded dikes have been constructed along major rivers to protect surrounding areas from flooding. These measures may succeed for decades until a "100-year" deluge causes the dike system to fail. Now the emphasis with respect to infrastructure must be directed toward sustainability and the maintenance of environmental quality. Such elements as highly effective waste treatment systems with recovery of materials and energy from wastes, high-speed rail systems to replace inefficient movement of people and freight by private carriers, and electrical systems that use wind power to the maximum extent possible are examples of sustainable infrastructure.

The anthrosphere and the hydrosphere are intimately connected and each has profound effects on the other. Water has many uses in the anthrosphere, much of which is devoted to the collection, treatment, and storage of water. The energy of flowing water is harnessed to provide hydroelectric power. Water is used for cooling powerplants. High-pressure steam is capable of transferring large amounts of energy and chilled water is used to cool buildings. Enormous quantities of water are employed in manufacturing processes. Water is pumped from the ground and diverted from surface sources to irrigate crops.

Water in the hydrosphere has profound, often destructive, effects on the anthrosphere. Bridges and roadways are often damaged by floodwaters. Water erosion in agricultural lands is a long-standing problem. Water is a key agent in corrosion of metals in the anthrosphere.

Virtually all water pollution problems are the result of anthrospheric activities. Most of the early efforts to control pollution were dedicated to water pollution, which is still an important issue. Leachate produced by water in contact with improperly disposed industrial wastes can be a serious water pollution problem. Residues of herbicides used on land for agricultural production are frequently encountered in water supplies. Water pollution is an important area of water chemistry and Chapter 7 of this book, "Water Pollution," is dedicated to it.

1.7 WATER AND BIOGEOCHEMICAL CYCLES

There is a constant exchange of matter among the five major environmental spheres as described by cycles of matter. It is convenient to classify the various cycles by the chemical species that they involve, most commonly elements such as carbon and nitrogen. One particularly important cycle is that of a chemical compound, water, shown in the hydrological cycle in Chapter 2, Figure 2.2. Another important cycle is the rock cycle in which molten rock solidifies, undergoes weathering, may be carried by water and deposited as sedimentary rock, is converted into metamorphic rock by heat and pressure, and is eventually buried at great depths and melted to produce molten rock again. Since natural cycles of matter usually involve geochemical and biological processes, they are frequently called biogeochemical cycles, a term that particularly describes plant and animal nutrients including carbon and nitrogen. It should be kept in mind that anthrospheric processes are also very much involved in the important cycles of matter.

Cycles of matter involve various reservoirs of matter, such as the ocean, the atmosphere, and parts of the geosphere, as well as conduits through which matter moves between these reservoirs. Two main classifications of cycles of matter are endogenic cycles that occur below or directly on the surface of the geosphere and exogenic cycles, which involve surface phenomena and an atmospheric component. The rock cycle is an endogenic cycle and the hydrological cycle is an exogenic cycle.

Figure 1.7 illustrates one of the key biogeochemical cycles, the carbon cycle. An important reservoir of carbon is in the atmosphere as carbon dioxide. Photosynthetic processes by plants extract significant amounts of carbon from the atmosphere and

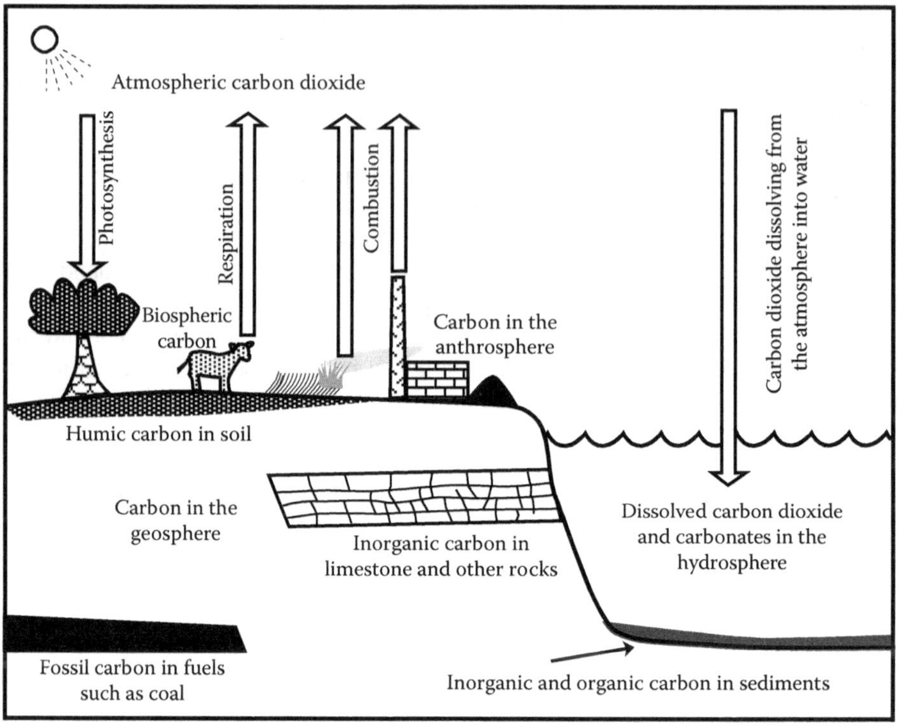

FIGURE 1.7 The carbon cycle describes locations in various environmental spheres in which carbon is found and the interchanges of carbon in various chemical forms among these spheres. Consideration of the carbon cycle is of utmost importance in maintaining sustainability.

fix it as biological carbon in the biosphere. In turn, animals and other organisms in the biosphere release carbon dioxide back to the atmosphere through the respiration processes by which they utilize oxygen and food for energy production. More carbon dioxide is released to the atmosphere by the combustion of biological materials such as wood and combustion of fossil fuels including coal and petroleum. Carbon dioxide from the atmosphere dissolves in water to produce dissolved carbon dioxide and inorganic carbonates. Solid carbonates, particularly limestone, dissolve in bodies of water to also produce dissolved inorganic carbon. The biodegradation of organic matter by organisms in water and sediments also produces dissolved inorganic carbon species in water. Large amounts of carbon are held in the geosphere. The major forms of this carbon are fossil carbon of ancient plant origin held by fossil fuels, such as coal and petroleum, and inorganic carbon in carbonate rocks, especially limestone. Organic carbon from partially degraded plant material is present as humic material in soil, a source that is added to by organisms breaking down plant biomass from the biosphere.

The carbon cycle is extraordinarily important in maintaining sustainability because a major part of it is the fixation of carbon from highly dilute atmospheric carbon dioxide into biomass by photosynthesis carried out by green plants. Biomass is a source of food, chemical energy, and raw materials and the carbon cycle contains the main pathway by which solar energy is captured and converted into a form of energy that can be utilized by organisms and as fuel.

Other important cycles of matter are linked to the carbon cycle. The oxygen cycle describes movement of oxygen in various chemical forms through the five environmental spheres. At 21% elemental oxygen by volume, the atmosphere is a vast reservoir of this element. This oxygen becomes chemically bound as carbon dioxide by respiration processes of organisms and by combustion. Oxygen is added to the reservoir of atmospheric oxygen by photosynthesis. Oxygen is a component of biomass in the biosphere and most rocks in Earth's crust are composed of oxygen-containing compounds. With its chemical formula of H_2O, water in the hydrosphere is predominantly oxygen.

Another important life-element cycle is the nitrogen cycle. The atmosphere is almost 80% by volume elemental nitrogen. Chemically bound nitrogen is essential for life-molecule proteins, but the N_2 molecule of elemental nitrogen is such a stable species that binding atmospheric nitrogen into forms that can be utilized in biomolecules is a difficult thing to do. Humans do it in the anthrosphere using precious-metal catalysts at high temperatures and extremely high pressures, expending a great deal of energy in the process. Nitrogen becomes bound to oxygen under the harsh conditions inside an internal combustion engine and is released back to the atmosphere as pollutant nitrogen oxide compounds. In contrast to the harsh conditions under which elemental nitrogen becomes chemically bound in the anthrosphere, bacteria, such as the *Rhizobium* bacteria that grow on the roots of legume plants accomplish the same feat in a very energy-efficient manner at the mild temperature and pressure conditions of their surroundings. Nitrogen occurs in chemical compounds (ammonia and nitrate) dissolved in water in the hydrosphere and as the same chemical species in soil, to which it is added as fertilizer. Microorganisms in soil and water convert chemically bound nitrogen back into elemental nitrogen, which returns to the atmosphere.

Two other biogeochemical cycles involving important life elements are the phosphorous cycle and the sulfur cycle. Sulfur occurs abundantly in the geosphere and is released to the atmosphere as organic sulfur by organisms in the oceans and as pollutant sulfur oxides by anthrospheric processes, such as combustion of sulfur-containing coal. Phosphorus is an essential life element in DNA and other nucleic acids and in species designated ATP and ADP required for energy transfer in living systems. Unlike the other cycles mentioned here, the phosphorus cycle is exclusively endogenous without a significant atmospheric reservoir.

1.8 WATER AND ENVIRONMENTAL CHEMISTRY

At the beginning of this chapter, the environment was defined as consisting of five spheres: The hydrosphere, the atmosphere, the geosphere, the biosphere, and the anthrosphere, that is, water, air, earth, life, and those parts of the environment consisting of human constructs and activities. The chemistry of the environment, *environmental chemistry*, may be defined as *the study of the sources, reactions, transport, effects, and fates of chemical species in the hydrosphere, the atmosphere, the geosphere, and the biosphere and the effects of human activities thereon.*[2] This definition is illustrated for a typical environmental pollutant in Figure 1.8. Pollutant sulfur dioxide is generated in the combustion of sulfur in coal, transported to the atmosphere with flue gas, and oxidized by chemical and photochemical processes to sulfuric acid. The sulfuric acid, in turn, falls as acidic precipitation, where it may have detrimental effects such as toxic effects on trees and other plants. Eventually, the sulfuric acid is carried by stream runoff to a lake or ocean where its ultimate fate is to be stored in solution in the water or precipitated as solid sulfates.

Environmental chemistry is complicated by the continuous and variable interchange of chemical species among various environmental spheres. This complexity is illustrated for sulfur species in Figure 1.8. The sulfur in coal is taken from the geosphere, converted into gaseous sulfur dioxide by an anthrospheric process (combustion), transported and undergoes chemical reactions in the atmosphere, may affect plants in the biosphere, and ends up in a sink in the hydrosphere or back in the geosphere. Throughout this sequence, the sulfur takes on several forms including organically bound sulfur or pyrite (FeS_2) in coal, sulfur dioxide produced in the combustion of coal, sulfuric acid produced by oxidation of sulfur dioxide in the atmosphere, and sulfate salts produced from sulfuric acid when it reaches the geosphere. Throughout an environmental system there are variations in temperature, mixing, intensity of solar radiation, input of materials, and various other factors that strongly influence chemical conditions and behavior. Because of its complexity, environmental chemistry must be approached with simplified models.

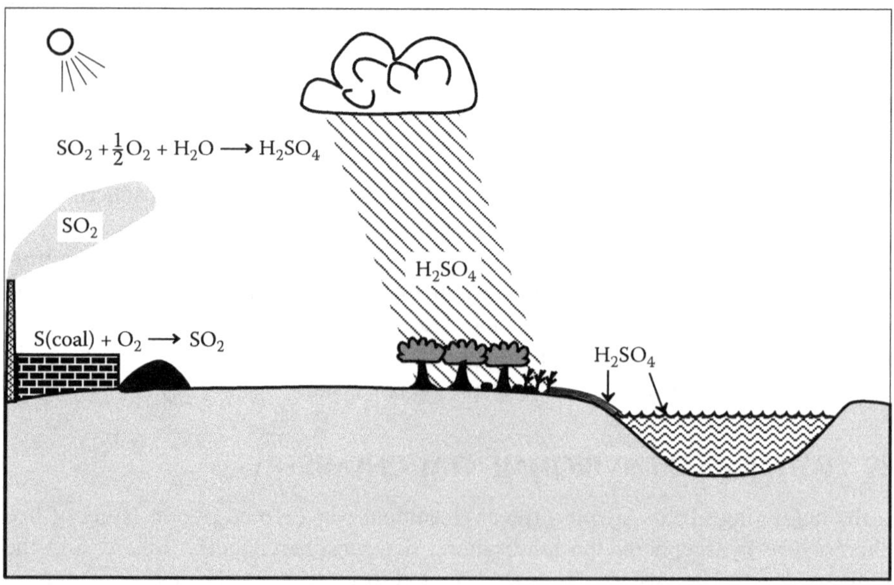

FIGURE 1.8 Illustration of the definition of environmental chemistry by the example of pollutant sulfuric acid formed by the oxidation of sulfur dioxide generated during the combustion of coal.

Environmental chemistry can be divided into several categories based on the five spheres of the environment. This book deals with one of these, aquatic chemistry, the chemistry of the hydrosphere. The other categories of environmental chemistry are addressed briefly and also come up in later portions of the book because of the overlap and interactions of the hydrosphere with the other environmental spheres.

The chemistry of the geosphere is addressed by the discipline of *geochemistry*.[3] Included among the many aspects of geochemistry are chemical composition of the geosphere, oxidation/reduction phenomena of minerals, thermodynamics, kinetics, weathering of minerals, chemistry of trace elements (those present at very low levels), and chemistry of trace substances (those present at very low concentrations). There exists an especially strong correlation between geochemistry and aquatic chemistry because so much of what happens to solids in the geosphere depends on their interactions with water and solutes in water. Included are phase interactions between water and geospheric solids, dissolution and precipitation phenomena, adsorption of solutes from water onto solids, oxidation/reduction involving species dissolved in water, behavior of colloidal particles, and complexation. Some of these topics are addressed later in this book. Soil is a specific and very important part of the geosphere and there exists a large body of knowledge on *soil chemistry*.[4]

Atmospheric chemistry is the science of chemical processes in the atmosphere. Several characteristics of the atmosphere make atmospheric chemistry unique. One of these is the extreme dilution of chemical species in the atmosphere compared to the hydrosphere or geosphere. Another is the possibility of photochemical reactions such as the one involved in the formation of stratospheric ozone (Reaction 1.2). Although photochemical reactions are most likely to occur above the troposphere where energetic ultraviolet radiation penetrates, an especially important one takes place near Earth's surface with the photochemical dissociation of nitrogen dioxide:

$$NO_2 + h\nu \rightarrow NO + O \tag{1.3}$$

The atomic oxygen product of this reaction is very reactive and reacts with organic pollutants in air to produce noxious photochemical smog.

There exists a connection between atmospheric chemistry and aquatic chemistry. A number of atmospheric chemical reactions take place within suspended water droplets in the atmosphere and in a sense are aquatic chemical reactions. Chemical species produced in or carried by the atmosphere can get into the hydrosphere and cause water pollution problems. The main one of these is sulfuric acid, H_2SO_4, generated when atmospheric pollutant sulfur dioxide, SO_2, is oxidized. In some regions, airborne sulfuric acid makes lake water so acidic that it kills fish fingerlings. Acidic rainwater can leach so much Al^{3+} ions from the geosphere that it becomes toxic to trees and other plants.

The biosphere's connection to environmental chemistry is mentioned in various contexts throughout the book as it relates to environmental chemical processes in water. *Biochemistry* deals with chemical phenomena in the biosphere. A significant concern in environmental science is the effect of environmental pollutants on organisms. This area is covered by *toxicological chemistry* defined as *the chemistry of toxic substances with emphasis upon their interactions with biologic tissue and living organisms.*[5]

1.9 WATER AND GREEN CHEMISTRY

In its earlier stages of development dating from around the 1960s, the emphasis in environmental chemistry was upon detection of pollution, cataloging its adverse effects, and controlling pollutants once they were produced. This resulted in an abundance of laws and regulations in countries throughout the world that, in many areas, prevented pollution from becoming worse and in a number of cases improved air and water quality and controlled toxic and hazardous substances improperly disposed on or buried beneath Earth's surface. Even as these efforts gained success, it became obvious that such measures were increasingly expensive to implement as allowable pollutant levels became lower. The "command-and-control" approach was hard to monitor and potential polluters were always tempted to avoid controls to save costs. A better way was needed. Insofar as chemistry was concerned, a better way emerged in the practice of green chemistry.

Formulated in the late 1990s by Paul Anastas[6] and others, green chemistry is the practice of chemical science and engineering in ways that minimize consumption of materials and energy, produce the least possible amount of waste, and is inherently safe.[7] Figure 1.9 illustrates this definition of green chemistry. Ideally, green chemistry is practiced within a framework of *industrial ecology*, which is a comprehensive approach to production, distribution, utilization, and termination of goods and services in a manner that maximizes mutually beneficial utilization of materials and energy while preventing the production of wastes and pollutants.[8] As the area of endeavor charged with using science to implement production of goods and provision

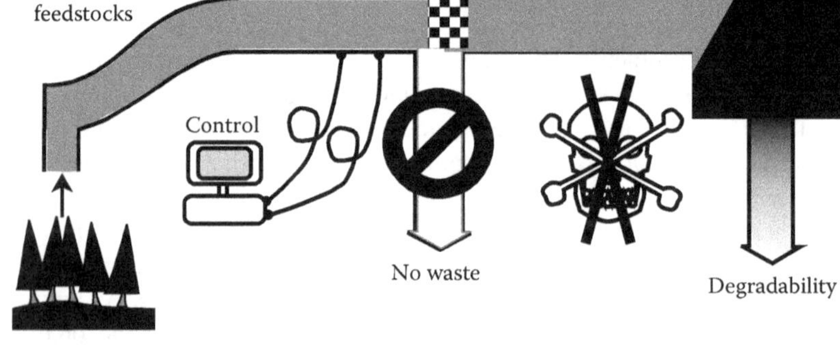

FIGURE 1.9 Illustration of the definition of green chemistry.

of services, engineering obviously has strong implications for environment and sustainability. Based largely on chemical engineering, *green engineering* is the environmentally conscious design of chemical and other processes.[9] Within the short time that it has been recognized as a distinct area of endeavor, green chemistry has become an integral, important part of the practice of chemical science.

Much of green chemistry has to do with water. Green chemistry emphasizes renewable materials of which water is the most renewable of all. Although many pollution and waste problems in the past resulted from the use of hydrocarbon and halocarbon solvents and reaction media, green chemistry utilizes water instead to the maximum extent possible. Green chemistry seeks to use only minimum quantities of material, a practice that reduces water consumption. In order to best utilize water in the practice of green chemistry it is essential to have an understanding of aquatic chemistry, the topic of this book.

LITERATURE CITED

1. Manahan, S. E., Toward a greener anthrosphere through industrial ecology, Chapter 11 in *Green Chemistry and the Ten Commandments of Sustainability*, 2nd ed., ChemChar Research, Inc., Columbia, MO, 2005, pp. 279–303.
2. Manahan, S. E., *Environmental Chemistry*, 9th ed., Taylor & Francis/CRC Press, Boca Raton, FL, 2010.
3. Albarède, F., *Geochemistry: An Introduction*, 2nd ed., Cambridge University Press, Cambridge, UK, 2009.
4. Sposito, G., *The Chemistry of Soils*, Oxford University Press, Oxford, UK, 2008.
5. Manahan, S. E., *Toxicological Chemistry and Biochemistry*, 3rd ed., Taylor & Francis/CRC Press, Boca Raton, FL, 2002.
6. Anastas, P. T. and J. C. Warner, *Green Chemistry: Theory and Practice*, Oxford University Press, New York, 2000.
7. Manahan, S. E., *Green Chemistry and the Ten Commandments of Sustainability*, 2nd ed., ChemChar Research, Columbia, MO, 2006.
8. Gallopoulos, N. E., Industrial ecology: An overview, *Progress in Industrial Ecology*, 3, 10–27 (2006).
9. Allen, D. T. and D. R. Shonnard, *Green Engineering: Environmentally Conscious Design of Chemical Processes*, Prentice-Hall, Upper Saddle River, NJ, 2002.

SUPPLEMENTARY REFERENCES

Ahluwalia, V. K. and M. Kidwai, *New Trends in Green Chemistry*, Kluwer Academic, Boston, 2004.

Anastas, P., Ed., *Handbook of Green Chemistry*, Wiley-VCH, New York, 2010.

Ayres, R. U. and L. W. Ayres, Eds., *A Handbook of Industrial Ecology*, Edward Elgar Publishing, Cheltenham, UK, 2002.

Baird, C. and M. Cann, *Environmental Chemistry*, 4th ed., W. H. Freeman, New York, 2008.

Botkin, D. B. and E. A. Keller, *Environmental Science: Earth as a Living Planet*, 6th ed., Wiley, Hoboken, NJ, 2007.

Connell, D. W., *Basic Concepts of Environmental Chemistry*, 2nd ed., Taylor & Francis/CRC Press, Boca Raton, FL, 2005.

El-Haggar, S. M., *Sustainable Industrial Design and Waste Management: Cradle-to-Cradle for Sustainable Development*, Elsevier Academic Press, Amsterdam, 2007.

Graedel, T. E. and B. R. Allenby, *Industrial Ecology*, 2nd ed., Prentice-Hall, Upper Saddle River, NJ, 2003.
Hites, R. A., *Elements of Environmental Chemistry*, Wiley, Hoboken, NJ, 2007.
Holtzapple, M. T. and W. Dan Reece, *Concepts in Engineering*, 2nd ed., McGraw-Hill, Dubuque, IA, New York, 2008.
Ibanez, J. G., *Environmental Chemistry: Fundamentals*, Springer, Berlin, 2007.
Lancaster, M., *Green Chemistry*, Springer-Verlag, Berlin, 2002.
Raven, P. H., L. R. Berg, and D. M. Hassenzahl, *Environment*, 6th ed., Wiley, Hoboken, NJ, 2008.
Worldwatch Institute, *State of the World 2009: Into a Warming World*, Worldwatch Institute, Williamsport, PA, 2003.

QUESTIONS AND PROBLEMS

The use of Internet resources is assumed in answering any of the questions. These would include such things as constants and conversion factors as well as additional information needed to complete an answer.

1. If water vapor were to penetrate in significant quantities to very high altitudes (tens of kilometers) in the atmosphere, it would be subjected to high-energy ultraviolet radiation that would split off H atoms resulting in loss of light, mobile H atoms, and H_2 molecules to outer space and eventually loss of all Earth's water. Look up the tropopause on the internet and suggest how this layer of the atmosphere has prevented loss of Earth's water.
2. What are some of the major waterborne diseases and where do they most likely occur? Although not a waterborne disease as such, why is malaria prevalent in warmer areas where there are many bodies of water?
3. Look up the pre-Columbian settlement of Mesa Verde and suggest how its fate was tied to changes in the hydrosphere.
4. Suggest with chemical and biochemical reactions how primitive cyanobacteria produced deposits of iron oxide minerals.
5. Look up how a well-intentioned program to provide pathogen-free drinking water in Bangladesh resulted in arsenic poisoning to a significant number of people.
6. Much of the water vapor in Earth's stratosphere comes from methane that has penetrated from the troposphere. After doing a little research on photochemistry and with the knowledge that short-wavelength, high-energy ultraviolet radiation from the Sun reaches the stratosphere. Suggest the photochemical process by which this water vapor is produced.
7. Suggest how transpiration in the biosphere can contribute to the hydrological cycle, especially in heavily vegetated areas.
8. Representing biomass by the simple approximate formula $\{CH_2O\}$, represent the productivity of a biological ecosystem with a chemical (biochemical) reaction.
9. What is the anthropocene? What are the arguments that we may be entering the anthropocene epoch?

10. Recently some groups have expressed extreme concern about infrastructure in the United States. What are some of these concerns as they relate to water and the hydrosphere?
11. How does corrosion relate to deterioration of the anthrosphere and its infrastructure? What is the role of water in corrosion?
12. How does water function for heat transfer in the atmosphere and the anthrosphere? What property of the water molecule makes it especially effective for heat transfer?
13. Why is it particularly important to label the carbon cycle as a biogeochemical cycle, not strictly a geochemical cycle? What are some specific "bio" aspects of the carbon cycle.
14. Suggest two ways involving biological processes in which the carbon cycle and oxygen cycle are linked.
15. How are the *Rhizobium* bacteria linked to the nitrogen cycle? What is particularly remarkable about these bacteria as it relates to the nature of the N_2 molecule?
16. Designate each of the five spheres of the environment in Figure 1.8 illustrating the definition of environmental chemistry.
17. In what respects are geochemistry and aquatic chemistry very intimately related?
18. Although atmospheric photochemical reactions are especially likely above the troposphere, one is especially significant in the troposphere. What is this reaction and why is it particularly significant?
19. Suggest ways in which toxicological chemistry may be related to aquatic chemistry.
20. Large amounts of material on green chemistry are available on the internet. Using these resources suggest specific ways in which water is involved with green chemistry.

2 Properties of Water and the Hydrosphere

2.1 THE FANTASTIC WATER MOLECULE

All of matter and all of life are all about the vast variety of molecules composing matter and living organisms. We are rightfully highly impressed by the remarkably complex molecules of DNA composed of billions of atoms that make up the genetic code of organisms. Chemists are constantly trying to synthesize new molecules, such as those in pharmaceuticals that may be extremely complex and fiendishly difficult to make. But among all molecules, one stands out for its diversity, its occurrence throughout the environment, its vast variety of uses, and its role as a medium for life. This is the molecule of water, which has the very simple formula of H_2O. Essential for life, useful for a vast variety of applications, and totally recyclable, water is the ultimate green chemical compound.

Although the chemical formula of water is simple and the H_2O molecule is small, the behavior of this unique substance is unusual and complex. The structure of the water molecule is not simply one in which all the atoms in H–O–H are in a straight line but rather the two bonds between H and O atoms form an approximate right angle such that the two H atoms are on the same "side" of the central O atom. The reason for this structure is that the outer shell of eight electrons in the O atom of H_2O is composed of four pairs of electrons, two pairs of which are in the covalent bonds between the H and O atoms and two pairs of which are not involved in bonds. It turns out that such pairs of electrons tend to be arranged in the imaginary sphere around the O atom to be as far from each other as possible. This is illustrated in Figure 2.1.

The structure of the water molecule shown in Figure 2.1 does two things that determine water's properties. The first of these is that the end of the water molecule with the two unshared pairs of electrons is relatively negative compared to the end with the two H atoms, so that the water molecule is electrically *polar*. Second, an unshared pair of electrons on the oxygen atom of one molecule can bond to an H atom on another water molecule in a special kind of bond called the *hydrogen bond*. These two characteristics, especially hydrogen bonding, mean that water molecules are strongly attracted to each other. Therefore, when water is heated, which means that the water molecules move more rapidly, a large amount of heat energy must be put into a mass of the substance to raise its temperature. Furthermore, when solid ice is melted, a very large amount of heat energy is required (to enable the molecules of H_2O to move from their relatively fixed positions in the solid to their mobile state in the liquid) compared to that needed to melt other substances, and an even larger amount of heat energy is required to convert liquid water into vapor (steam). Similarly large amounts of heat energy are *released* when steam condenses to liquid water, as liquid water cools, and when liquid water freezes.

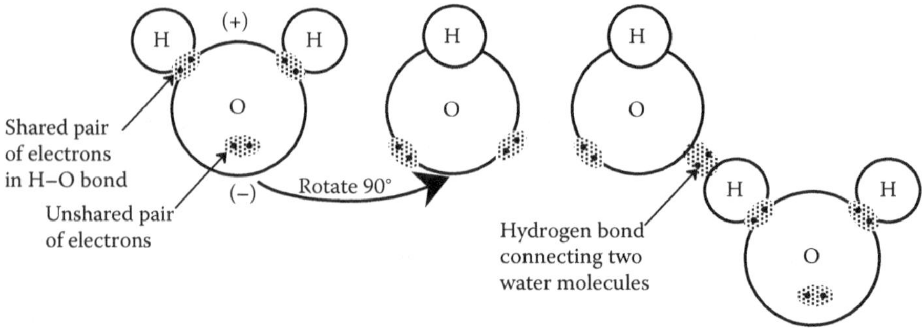

FIGURE 2.1 Because of the arrangements of the four pairs of electrons making up the outer electron shell of the O atom in the water molecule, the molecule is electrically polar and can form special bonds called hydrogen bonds, with other water molecules. These characteristics determine the chemical and physical diversity of water.

The heat storing and transfer capabilities of water are very important in determining its practical uses and influence on the environment. Steam produced by heating water in a boiler can be moved in steam lines to buildings where it is condensed in radiators releasing heat to heat the buildings. Some modern "green" buildings have cooling systems in which water is frozen at night by heat pumps when cooler temperatures and lower electricity demand make it relatively efficient to do so; the ice is then melted during hot daytime periods absorbing heat and cooling the buildings. Europe owes its relatively mild winters, despite its more northern latitudes, to the Gulf Stream, consisting of water warmed in the Gulf of Mexico region that flows near the surface of the Atlantic Ocean to European shores, cools as it releases its heat to the air in European regions, then sinks as colder, denser water and flows back to the region from which it came.

Water has several important properties that are very important in its environmental influence and practical uses. These are listed in Table 2.1 and discussed further in this chapter.

2.2 WATER AS AN ESSENTIAL RESOURCE

Throughout history, the quality and quantity of water available to humans have been vital factors in determining their well-being. Whole civilizations have disappeared because their supplies of essential water became unsustainable due to changes in climate, abuse of soil, deterioration of the watershed land upon which water was collected, and exhaustion of underground water aquifers. Even in temperate climates, fluctuations in precipitation cause problems. Devastating droughts and destructive floods are problems in many areas of the world.

Waterborne diseases such as cholera, typhoid, and dysentery killed millions of people in the past and still cause great misery in less developed countries. Ambitious programs of dam and dike construction have reduced flood damage, but they have had a number of undesirable side effects in some areas, such as inundation of farmland by reservoirs, failure of unsafe dams and destruction of fisheries. Globally,

TABLE 2.1
Important Properties of Water

Property	Effect or Significance
Excellent solvent, especially for ionic substances	Transport of nutrients and wastes in biological systems, makes nutrients available to plant roots, dissolves, transports, and deposits minerals
High surface tension	Controlling factor in physiology; governs drop and surface phenomena such as in rainfall formation
Transparent	Enables light to penetrate water to some depth enabling algae and other plants in water to carry out photosynthesis
Maximum density as a liquid at 4°C	Ice floats, water in bodies of water is stratified into layers during summer months
High heat of evaporation	Large quantities of heat are absorbed when water evaporates and are released when water vapor condenses enabling use of steam for heat transfer and strongly influencing weather phenomena.
High latent heat of fusion	Temperature stabilized at the freezing point of water
High heat capacity	Stabilizes temperatures of organisms and geographical areas

problems with quantity and quality of water supply remain and in some respects are becoming more serious. These problems include increased water use due to population growth, contamination of drinking water by improperly discarded hazardous wastes and destruction of wildlife by water pollution.

Discussions of water chemistry and biology later in this book require some understanding of the sources, transport, characteristics, and composition of water. The chemical reactions that occur in water and the chemical species found in it are strongly influenced by the environment in which the water is found and the interactions between the hydrosphere and the other environmental spheres. The chemistry of water exposed to the atmosphere is quite different from that of water at the bottom of a lake. Groundwater characteristics are strongly affected by the geochemical characteristics of the underground aquifers that contain the water. Microorganisms play an essential role in determining the chemical composition of water. Thus, in discussing water chemistry, it is necessary to consider the many general factors that influence this chemistry.

The study of water is known as *hydrology*, which has several subcategories. *Limnology* is the branch of the science dealing with the characteristics of fresh water including biological properties, as well as chemical and physical properties.[1] *Oceanography* is the science of the ocean and its physical and chemical characteristics.[2] The chemistry and biology of the Earth's vast oceans are unique because of the oceans' high salt content, great depth, and other factors.

2.3 THE HYDROLOGICAL CYCLE

Water circulates throughout Earth's environment by means of the solar-powered *hydrological cycle* (Figure 2.2).[3] Excluding chemically bound water, the total amount of water on the Earth is about 1.4 billion cubic kilometers (1.4×10^9 km^3). Of this

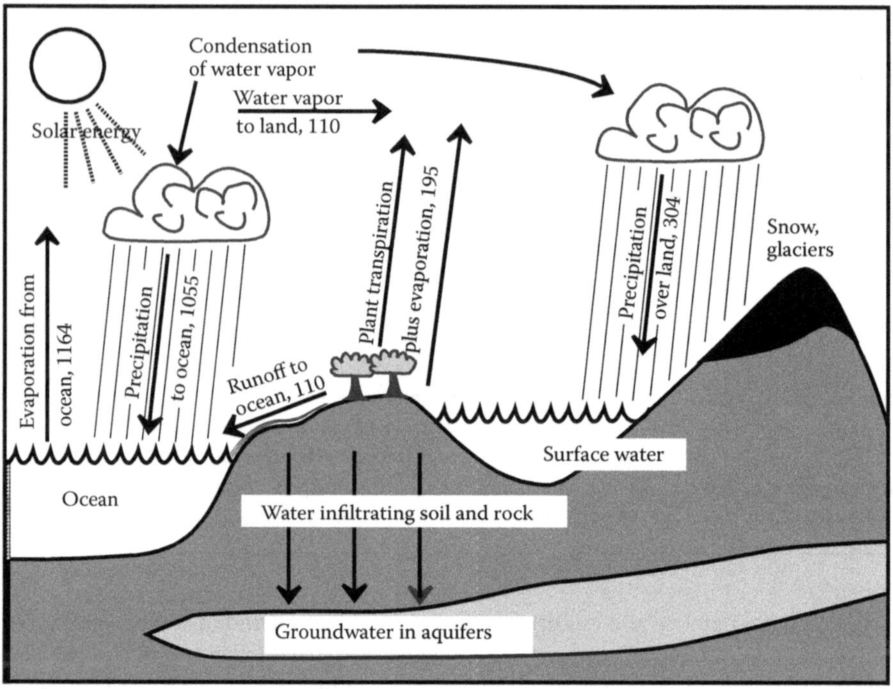

FIGURE 2.2 The hydrological cycle, quantities of water in trillions of liters per day.

amount, about 97.6% is present as salt water in Earth's oceans. About 33 million km^3 remains to be distributed elsewhere on the Earth as shown in Table 2.2. Even out of this amount about 87% is present in solid form, predominantly as polar snowcap and another 12% as groundwater. Therefore, just slightly more than 1% of all Earth's fresh water is distributed among surface water, atmospheric water, and biospheric water. This very small fraction comprises water in lakes, including water in the Great

TABLE 2.2
Distribution of Earth's Water Other than Ocean Water

Location	Quantity (L)	Percent Total Fresh Water
Snow, snowpack, ice, glaciers	2.90×10^{19}	86.9
Accessible groundwater	4.00×10^{18}	12.0
Lakes, reservoirs, ponds	1.25×10^{17}	0.37
Saline lakes	1.04×10^{17}	0.31
Soil moisture	6.50×10^{16}	0.19
Moisture in living organisms	6.50×10^{16}	0.19
Atmosphere	1.30×10^{16}	0.039
Wetlands	3.60×10^{15}	0.011
Rivers, streams, canals	1.70×10^{15}	0.0051

Lakes of North America, water in all the Earth's vast rivers (Mississippi, Congo, Amazon), groundwater to a depth of 1 km, water in the atmosphere, and water in the biosphere.

Earth's water can be considered in several *compartments*. The amounts of water and the *residence times* of water in these compartments vary greatly. The largest of these compartments consists of the *oceans*, containing more than 97% of all Earth's chemically unbound water with a residence time of about 3000 years. Oceans serve as a huge reservoir for water and as the source of most water vapor that enters the hydrological cycle. As vast heat sinks, oceans have a tremendous moderating effect on the climate.

The majority of Earth's water not held in the oceans is bound as snow and ice. Antarctica contains about 85% of all the ice in the world. Most of the rest is contained in the permanent ice pack in the Arctic Ocean and in the Greenland ice pack; a small fraction is present in mountain glaciers and snowpack. A major concern with respect to global warming is the melting of ice pack around the world and the accompanying rise in sea levels that may flood low-lying areas of the world.

Groundwater is water held below the surface in porous rock formations called *aquifers*. It influences, and is strongly influenced by the mineral matter with which it is in contact. It dissolves minerals from and deposits them on rock surfaces with which it is in contact. Groundwater is replenished by water flowing in from the surface, and it discharges into bodies of water that are below its level.

Bodies of fresh water include *lakes*, *ponds*, and *reservoirs*. Water flows from higher elevations back to the ocean through *rivers* and *streams*. The rate at which water flows in a stream is called its *discharge*. For the Mississippi River, the average discharge is 50 billion liters per hour. Collectively, water in lakes, ponds, reservoirs, rivers, and streams is called *surface water*. Groundwater and surface water have appreciably different characteristics. Many substances either dissolve in surface water or become suspended in it on its way to the ocean. Surface water in a lake or reservoir that contains the mineral nutrients essential for algal growth may support a heavy growth of algae. Surface water with a high level of biodegradable organic material, used as food by bacteria, normally contains a large population of bacteria. All these factors have a profound effect upon the quality of surface water.

There is a strong connection between the *hydrosphere*, where water is found, and the *lithosphere*, which is that part of the geosphere accessible to water. Human activities affect both. For example, disturbance of land by conversion of grasslands or forests into agricultural land or intensification of agricultural production may reduce vegetation cover, decreasing *transpiration* (loss of water vapor by plants, see Figure 1.5) and affecting the microclimate. The result is increased rain runoff, erosion, and accumulation of silt in bodies of water. The nutrient cycles may be accelerated, leading to nutrient enrichment of surface waters. This, in turn, can profoundly affect the chemical and biological characteristics of bodies of water.

An environmentally important compartment of water consists of *wetlands*, in which the water table is essentially at surface level.[4] Wetlands consist of marshes, meadows, bogs, and swamps that usually support lush plant life and a high population of animals as well. Wetlands serve as a reservoir and stabilized supply of water. They are crucial nurseries for numerous forms of wildlife.

The *atmosphere* is the smallest compartment of water and the one with the shortest residence time of about 10 days. Atmospheric water is of utmost importance in the movement of water from the oceans inland in the hydrological cycle. The atmosphere provides the crucial precipitation that gives water, required by all land organisms, to sustain river flow, fill lakes, and replenish groundwater. Furthermore, latent heat contained in atmospheric water, and released when water vapor condenses to form rain, is a major energy transport medium and one of the main ways that solar energy is moved from the equator toward Earth's poles.

Water enters the atmosphere by *evaporation* from liquid water, *transpiration* from plants, and *sublimation* from snow and ice. Water in the atmosphere is present as water vapor and as suspended droplets of liquid and ice. Water vapor in the atmosphere is called *humidity*, and the percentage of water vapor compared to the maximum percentage that can be held at a particular temperature is called the *relative humidity. Condensation* occurs at a temperature called the *dew point* when water molecules leave the vapor state and form liquid or ice particles. This process is aided by the presence of *condensation nuclei* consisting of small particles of sea salt (produced by the evaporation of water from ocean spray), bacterial cells, ash, spores, and other matter upon which water vapor condenses. Condensation, alone, does not guarantee precipitation in the form of rain or snow because the condensed water vapor may remain suspended in *clouds*; precipitation occurs when the conditions are right for the cloud particles to coalesce to particles large enough to fall.

Topographical conditions can strongly influence the degree and distribution of precipitation. A striking example of this is provided by the effects of coastal mountain ranges upon precipitation. Moisture-laden air flowing in from the ocean is forced up the sides of coastal mountain ranges, cooling as it does so, and releasing rain as it becomes supersaturated. On the other side of the range, the air warms so that the level of water vapor becomes much less than the saturation concentration, the water stays in the vapor form, clouds disappear, and rain does not fall. The area of low rainfall on the leeward side of a coastal mountain range is called a *rain shadow* (see Figure 2.3).

2.4 STANDING BODIES OF WATER

The physical condition of a body of water strongly influences the chemical and biological processes that occur in water. *Surface water* occurs primarily in streams, lakes, and reservoirs. Lakes may be classified as oligotrophic, eutrophic, or dystrophic, an order that often parallels the life cycle of the lake. *Oligotrophic* lakes are deep, generally clear, deficient in nutrients, and without much biological activity. *Eutrophic* lakes have more nutrients, support more life, and are more turbid. *Dystrophic* lakes are shallow, clogged with plant life, and normally contain colored water with a low pH. *Wetlands* are flooded areas in which the water is shallow enough to enable growth of bottom-rooted plants.

Some constructed reservoirs are very similar to lakes, while others differ a great deal from them. Reservoirs with a large volume relative to their inflow and outflow are called *storage reservoirs*. Reservoirs with a large rate of flow-through compared to their volume are called *run-of-the-river reservoirs*. The physical, chemical, and

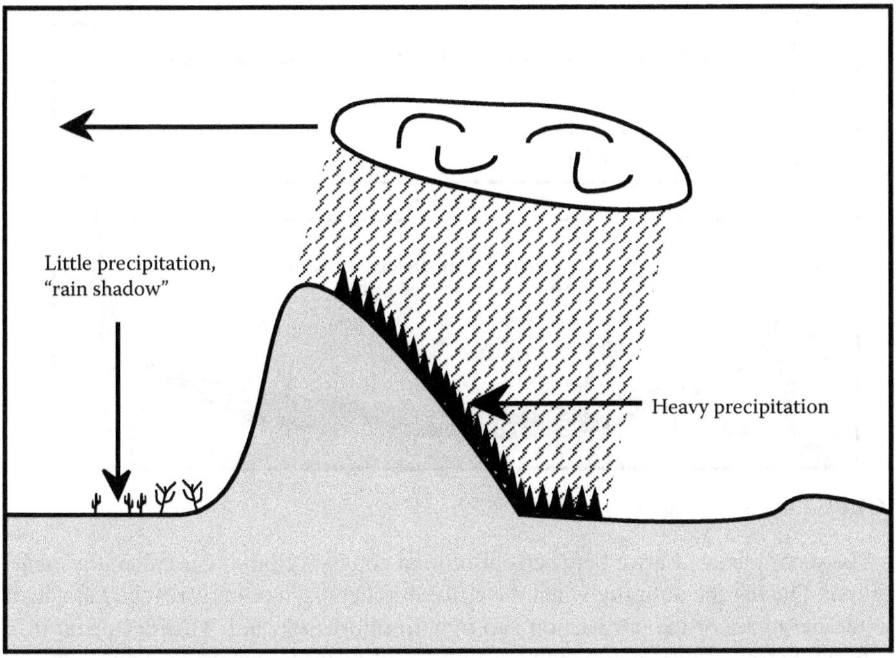

FIGURE 2.3 Illustration of a rain shadow. Vegetation is abundant on the rainy upslope side (right) and sparse in the rain shadow (left).

biological properties of water in the two types of reservoirs may vary appreciably. Water in storage reservoirs more closely resembles lake water, whereas water in run-of-the-river reservoirs is much like river water. Impounding water may have profound effects on its quality.

Estuaries constitute another type of body of water, consisting of arms of the ocean into which streams flow. The mixing of fresh and salt water gives estuaries unique chemical and biological properties. Estuaries are the breeding grounds of much marine life, which makes their preservation very important.

Water's unique temperature–density relationship results in the formation of distinct layers within nonflowing bodies of water, as shown in Figure 2.4. During the summer, a surface layer (*epilimnion*) is heated by solar radiation and, because of its lower density, floats upon the bottom layer, or hypolimnion. This phenomenon is called thermal stratification. When an appreciable temperature difference exists between the two layers, they do not mix, but behave independently and have very different chemical and biological properties. The epilimnion, which is exposed to light, may have a heavy growth of algae. As a result of exposure to the atmosphere and (during daylight hours) because of the photosynthetic activity of algae, the epilimnion contains relatively higher levels of dissolved oxygen, and is said to be aerobic or oxic. In the hypolimnion, bacterial action on biodegradable organic material consumes oxygen and may cause the water to become anaerobic or anoxic, that is, essentially free of oxygen. As a consequence, chemical species in a relatively reduced form tend to predominate in the hypolimnion.

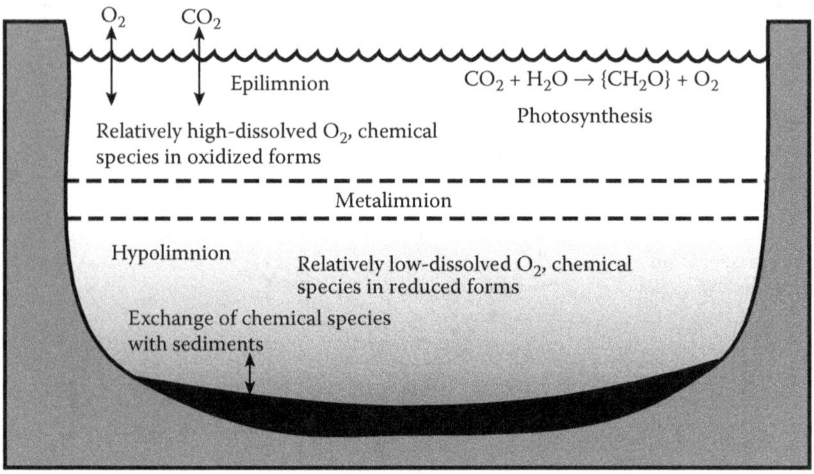

FIGURE 2.4 Stratification of a lake.

The shear plane, or layer between epilimnion and hypolimnion, is called the *metalimnion.* During the autumn, when the epilimnion cools, a point is reached at which the temperatures of the epilimnion and hypolimnion are equal. This disappearance of thermal stratification causes the entire body of water to behave as a hydrological unit, and the resultant mixing is known as *overturn.* An overturn also generally occurs in the spring. During the overturn, the chemical and physical characteristics of the body of water become much more uniform, and a number of chemical, physical, and biological changes may result. Biological activity may increase from the mixing of nutrients. Changes in water composition during overturn may cause disruption in water-treatment processes.

The chemistry and biology of the Earth's vast oceans are unique because of the ocean's high salt content, great depth, and other factors. Oceanographic chemistry is a discipline in its own right. The environmental problems of the oceans have increased greatly in recent years because of release of pollutants to oceans, oil spills (especially the horrific 2010 release of crude oil from the failed Deepwater Horizon oil well in the Gulf of Mexico), and increased utilization of natural resources from the oceans.

2.5 FLOWING WATER

Surface water that flows in streams and rivers originates from precipitation that falls initially on areas of land called the *watershed.* Watershed protection has become one of the most important aspects of water conservation and management. To a large extent, the quantity and quality of available water depends upon the nature of the watershed. An important characteristic of a good watershed is the ability to retain water for a significant length of time. This reduces flooding, allows for a steady flow of runoff water, and maximizes recharge of water into groundwater reservoirs (aquifer recharge). Runoff is slowed and stabilized by several means. One is to minimize cultivation and forest cutting on steeply sloping portions of the watershed, another is to use terraces and grass-planted waterways on cultivated land. The preservation of wetlands maximizes aquifer recharge, stabilizes runoff, and reduces turbidity of the runoff water. Small impoundments in the feeder streams of the watershed have similar beneficial effects.

2.5.1 Sedimentation by Flowing Water

The action of flowing water in streams cuts away stream banks and carries sedimentary materials for great distances. Sedimentary materials may be carried by flowing water in streams as the following:

- *Dissolved load* from sediment-forming minerals in solution
- *Suspended load* from solid sedimentary materials carried along in suspension
- *Bed load* dragged along the bottom of the stream channel

The transport of calcium carbonate as dissolved calcium bicarbonate provides a straightforward example of dissolved load. Water with a highly dissolved carbon dioxide content (usually present as the result of bacterial action) in contact with calcium carbonate formations contains Ca^{2+} and HCO_3^- ions. Flowing water containing calcium in this form may become more basic by loss of CO_2 to the atmosphere, consumption of CO_2 by algal growth, or contact with dissolved base, resulting in the deposition of insoluble $CaCO_3$:

$$Ca^{2+} + 2HCO_3^- \rightarrow CaCO_3(s) + CO_2(g) + H_2O \tag{2.1}$$

Most flowing water that contains dissolved load originates underground, where it dissolves minerals from the rock strata that it flows through.

Most sediments are transported by streams as suspended load, obvious from the appearance of "mud" in the flowing water of rivers draining agricultural areas or finely divided rock in Alpine streams fed by melting glaciers. Under normal conditions, finely divided silt, clay, or sand make up most of the suspended load, although larger particles are transported in rapidly flowing water. The degree and rate of movement of suspended sedimentary material in streams are functions of the velocity of water flow and the settling velocity of the particles in suspension.

Bed load is moved along the bottom of a stream by the action of water "pushing" particles along. Particles carried as bed load do not move continuously. The grinding action of such particles is an important factor in stream erosion.

Typically, about 2/3 of the sediment carried by a stream is transported in suspension, about 1/4 in solution, and the remaining relatively small fraction as bed load. The ability of a stream to carry water increases with both the overall rate of flow of the water (mass per unit time) and the velocity of the water. Both of these are higher under flood conditions, so floods are particularly important in the transport of sediments.

Streams mobilize sedimentary materials through *erosion*, *transport* materials along with stream flow, and release them in a solid form during *deposition*. Deposits of stream-borne sediments are called *alluvium*. As conditions such as lowered stream velocity begin to favor deposition, larger, more settleable particles are released first. This results in *sorting* such that particles of a similar size and type tend to occur together in alluvium deposits. Much sediment is deposited in flood plains where streams overflow their banks.

2.5.2 Free-Flowing Rivers

Rivers in their natural state are free-flowing. Unfortunately, the free-flowing characteristics of some of the world's finest rivers have been lost to development for power generation, water supply, and other purposes. Many beautiful river valleys have been flooded by reservoirs, and other rivers have been largely spoiled by straightening channels and other measures designed to improve navigation. One of the greater losses from dam construction has consisted of highly productive farmland in river floodplains. Esthetically, an unfortunate case was the flooding early in the 1900s of the Hetch-Hetchy Valley in Yosemite National Park by a dam designed to produce hydroelectric power and water for San Francisco. More recently, proposals have been made to drain the valley in an attempt to restore it to some of its original beauty. One of the larger dam removal projects in the United States was the removal of the Marmot Dam on Oregon's Sandy River in 2007 followed by the removal of the Little Sandy dam on the Little Sandy River in 2008. The Sandy River now flows freely from its headwaters on Mount Hood's Reid Glacier to its outlet on the Columbia River at the lower end of the Columbia River Gorge 23 km east of Portland, Oregon. From 2009, populations of salmon and steelhead trout have begun to increase on the Sandy River as a consequence of changes in the river flow resulting from dam removal.

2.6 GROUNDWATER

Most *groundwater* originates as *meteoritic* water from precipitation in the form of rain or snow and enters underground aquifers through *infiltration* (Figure 2.5). The rock and soil layer in which all pores are filled with liquid water is called the *zone of saturation*, the top of which is defined as the *water table*. Water infiltrates into aquifers in areas called *recharge zones*. Groundwater may dissolve minerals from the formations through which it passes. Most microorganisms originally present in groundwater are filtered out as it seeps through mineral formations. Occasionally, the content of undesirable salts may become excessively high in groundwater, although it is generally superior to surface water as a domestic water source. Groundwater is a vital resource in its own right; it plays a crucial role in geochemical processes such as the formation of secondary minerals. The nature, quality, and mobility of groundwater are all strongly dependent upon the rock formations in which the water is held. Physically, an important characteristic of such formations is their *porosity*, which determines the percentage of rock volume available to contain water. A second important physical characteristic is *permeability*, which describes the ease of flow of the water through the rock. High permeability is usually associated with high porosity. However, clays, which are common secondary mineral constituents of soil, tend to have low permeability even when a large percentage of the volume is filled with water.

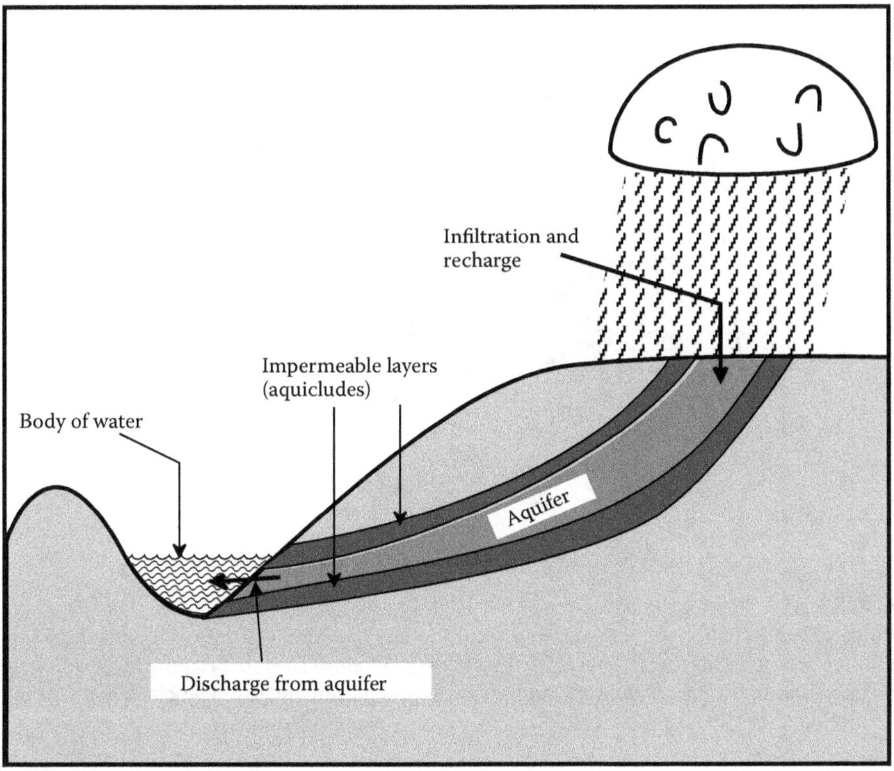

FIGURE 2.5 Groundwater in an aquifer.

Groundwater that is put to use for various purposes is usually taken from *water wells*. Poor design and mismanagement of water wells can result in problems of water pollution, land subsidence where the water is pumped out, and severely decreased production. As an example, when soluble iron or manganese are present in groundwater, exposure to air at the well wall can result in the formation of solid deposits of insoluble iron and manganese oxides produced by bacterially catalyzed oxidation processes that result from contact with oxygen in the air. The deposits fill the spaces that water must traverse to enter the well. As a result, they can seriously impede the flow of water into the well from the water-bearing aquifer. This creates major water source problems for municipalities using groundwater for water supply. As a result of this problem, chemical or mechanical cleaning, drilling of new wells, or even acquisition of new water sources may be required.

Groundwater is the major interface by which the hydrosphere and geosphere interact with each other. To enter an aquifer as groundwater, surface water normally has to percolate through many meters of soil and rock, picking up dissolved materials and altering the chemical nature of the solids that it contacts in the process. Some groundwater may spend many decades, even thousands of years, in contact with underground water formations before it again reaches the surface, such as through wells or springs.

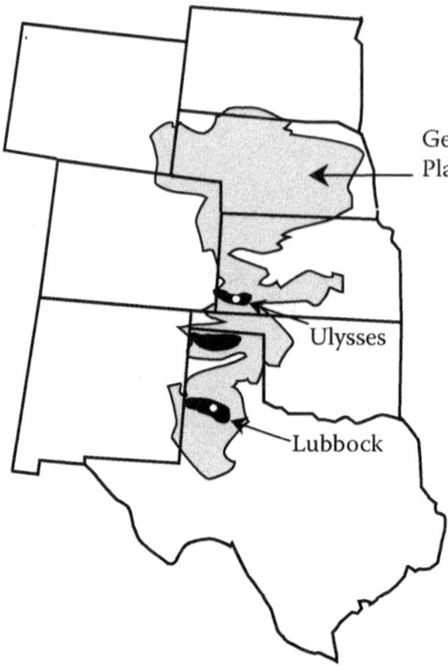

FIGURE 2.6 The Ogallala aquifer of the High Plains states of the United States is a vast source of high-quality water remaining from the last Ice Age. Unfortunately, much of this resource is being squandered to irrigate relatively low-value crops that require large amounts of water, and the water table in the aquifer is dropping dramatically. The black areas in the illustration above show areas of particularly severe water depletion.

The world abounds with examples of groundwater depletion, one of the most obvious manifestations of water over-use. Water pumped from below ground in Mexico City, which is built on an old lake bed, has caused much of the city to sink, damaging many of the structures in it. In the United States, wasteful use of groundwater is illustrated by the depletion of the High Plains Aquifer commonly called the Ogallala Aquifer. Figure 2.6 shows this formation lies beneath much of Nebraska, western Kansas, the Oklahoma and Texas panhandles, and small sections of eastern Wyoming, Colorado, and New Mexico. Although it is recharged from surface water in parts of Nebraska, it is largely composed of fossil water from the last Ice Age! Although it contains an astounding amount of water, enough to cover the entire United States to a depth of around 1.5 m, it is being rapidly depleted for irrigation water and the water table in the aquifer has dropped drastically over the last several decades.

2.7 WATER UTILIZATION

In the continental United States, an average of approximately 1.48×10^{13} L of water fall as precipitation each day, which translates to 76 cm/year. Of that amount, approximately 1.02×10^{13} L/day, or 53 cm/year, are lost by evaporation and transpiration. Thus, the water theoretically available for use is approximately 4.6×10^{12} L/day, or only 23 cm/year. At present, the United States uses 1.6×10^{12} L/day, or 8 cm of the average annual precipitation. This amounts to an almost 10-fold increase from a usage of 1.66×10^{11} L/day in 1900. Since 1900 per capita use has increased from about 40 L/day in 1900 to around 600 L/day now. Much of this increase is accounted for by high agricultural and industrial use, which each account for approximately 46% of total consumption. Municipal use consumes the remaining 8%.

Although water use in the United States increased steadily throughout most of the 1900s, surprisingly, around 1980 growth in water use in the United States leveled out, despite increasing population (Figure 2.7). This encouraging trend has been attributed to the success of efforts to conserve water, especially in the industrial (including power generation) and agricultural sectors. Conservation and recycling have accounted for much of the decreased use in the industrial sector. Irrigation water has been used much more efficiently by replacing spray irrigators, which lose large quantities of water to the action of wind and to evaporation, with irrigation systems that apply water directly to soil. Trickle irrigation systems that apply just the amount of water needed directly to plant roots are especially efficient.

A major problem with water supply is its nonuniform distribution with location and time. As shown in Figure 2.8, precipitation falls unevenly in the continental United States. This causes difficulties because people in areas with low precipitation often consume more water than people in regions with more rainfall. Rapid population growth in the more arid southwestern states of the United States during recent decades has further aggravated the problem and as of 2005, much of the western United States was suffering the effects of a severe drought classified by some experts

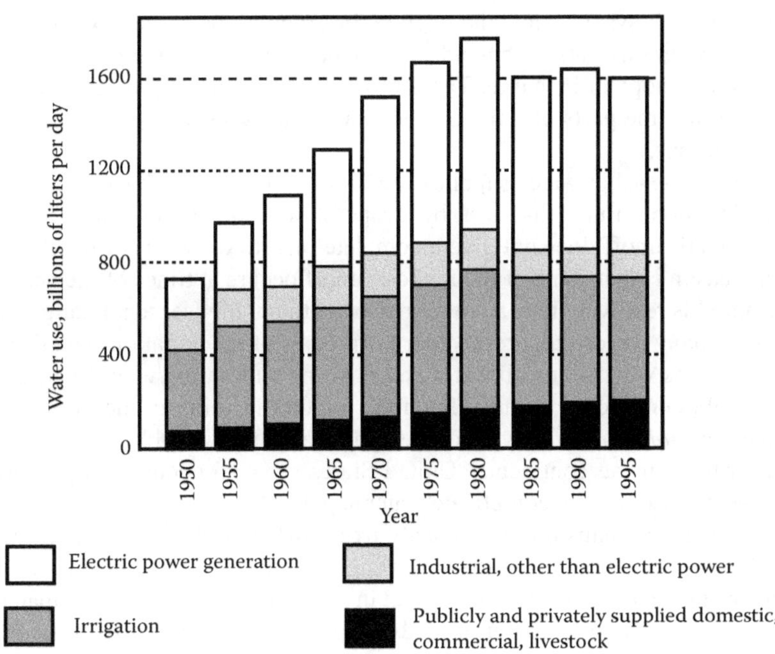

FIGURE 2.7 Trends in water use in the United States. (Data from U.S. Geological Survey, Water use in the United States, http://water.usgs.gov/watuse/.)

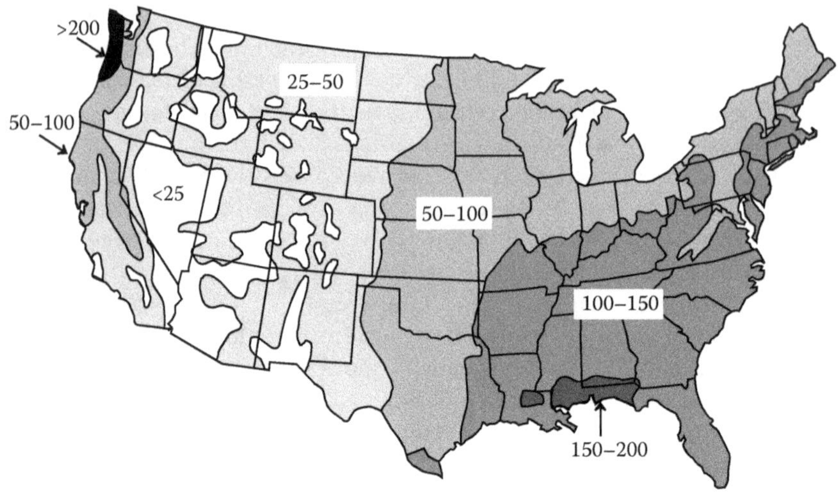

FIGURE 2.8 Distribution of precipitation in the continental United States, showing average annual rainfall in centimeters.

as a 500-year event. Water shortages are becoming more acute in this region which contains six of the nation's eleven largest cities (Los Angeles, Houston, Dallas, San Diego, Phoenix, and San Antonio). Other problem areas include Florida, where overdevelopment of coastal areas threatens Lake Okeechobee; the Northeast, plagued by deteriorating water systems; and the High Plains, ranging from the Texas panhandle to Nebraska, where irrigation demands on groundwater are dropping the water table steadily with no hope of recharge. These problems are minor, however, in comparison to those in some parts of Africa where water shortages are contributing to real famine conditions.

Available, renewable water supplies are largely determined by *runoff* equal to total precipitation minus that lost by evaporation/transpiration and infiltration. Average annual runoff does not give a complete picture of water availability, however. One reason is that some of the highest runoff occurs at times of the year when water demand is relatively low. Another reason is the highly seasonal nature of precipitation in many regions that receive large amounts of precipitation such that there are rainy seasons with too much water and flooding alternating with dry seasons in which drought conditions prevail. Therefore, a more meaningful measure of available water is *stable runoff*, which refers basically to available water. Periodic droughts, which in the continental United States seem to occur in approximately 30-year cycles, also greatly complicate water supply.

The water that humans use is primarily fresh surface water and groundwater. In arid regions, a small fraction of the water supply comes from the ocean. Saline or brackish groundwaters may also be utilized in some areas. It may be anticipated that in the future as demand for water exceeds supply, a major activity in sustainability will be in purifying water from oceans and brackish groundwater sources to a drinking water standard. An abundant, largely unutilized feedstock of contaminated water that can be purified consists of wastewater from municipal sources. With a much lower content of salt than seawater or water from some brackish groundwater sources, wastewater is relatively easy and inexpensive to purify, a possibility that is discussed in Chapter 8. Two barriers to such use are the public's revulsion to using water that has been part of sewage and potential problems from pharmaceuticals and their metabolic products in such water.

A particularly striking example of water use—and misuse—is the extraordinarily heavy pumping of water from the Oglalla aquifer which underlies much of the U.S. High Plains, ranging from the Texas panhandle to Nebraska (see Figure 2.6). The exploitation of this resource is discussed in Section 2.6.

2.8 IMPOUNDMENT AND TRANSFER OF WATER

Efforts to cope with demand for water have resulted in significant alterations in the pattern of water flow and distribution. Vast irrigation systems were known to antiquity, notably in irrigation from the Nile River in ancient Egypt and by the Incas in pre-Columbian South America. In a feat of engineering, the Romans built large aqueducts, some of which are still in use more than 20 centuries later. Typically, water is impounded in a reservoir produced by damming a river. From the reservoir the water may be moved through pipes, tunnels, or canals. One of the most massive projects utilizing all such measures is the one providing the water supply for Los Angeles and southern California. This project started with the transfer of water from the Owens Valley in northern California, a distance of about 250 miles. It now involves transfer of water from the Feather and Sacramento Rivers, as well as from the Colorado River. The huge Aswan High Dam forming Lake Nasser has stabilized the flow of Egypt's Nile River. China has undertaken a huge project to move water over 600 miles from the Yangtze River to the area around Bejing. Another huge project designed largely to provide hydroelectric power is located in Quebec.

Impounding water in reservoirs may have some profound effects upon water quality resulting from factors such as different velocities, changed detention time, and altered surface-to-volume ratios relative to the streams that were impounded. Beneficial changes due to impoundment include decreases in organic matter, turbidity, and hardness (calcium and magnesium content). Some detrimental changes are lower oxygen levels due to decreased reaeration, decreased mixing, accumulation of pollutants, lack of bottom scour produced by flowing stream water, and increased growth of algae. Algal growth may be enhanced when suspended solids settle from impounded water, causing increased exposure of the algae to sunlight. Stagnant water in the bottom of a reservoir may be of low quality. Oxygen levels frequently go to almost zero near the bottom, and odorous hydrogen sulfide is produced by the biochemical reduction of sulfur compounds in the low oxygen environment. Insoluble oxidized iron and manganese species are reduced to soluble ions, which must be removed prior to using the water. A major detrimental effect of water impoundment is simply loss of water by evaporation and by infiltration into surrounding rock strata. In some cases this can amount to 10% or more of the flow of the river impounded. Evaporative losses can cause increased water salinity. Impounded water drops its load of suspended silt, which will eventually fill a reservoir. Unforeseen detrimental effects of impoundment are illustrated vividly by Lake Nasser. The silt deposited in the lake is no longer available to provide nutrients for the fields irrigated by the river water downstream. Esthetically, irrigation canals have been found to be totally inadequate substitutes for wild, free-flowing rivers.

One of the greater environmental disasters resulting from water diversion has been the deterioration of the Aral Sea, a landlocked body of saline water in the former Soviet Union Republics of Uzbekistan and Kazakhstan in the south central part of Asia. Beginning with diversion of water flowing into the sea from the Amu Dar'ya and Syr Dar'ya rivers to irrigate cotton in 1918, the water flow into the Aral sea was severely curtailed. By 1990 the Sea had lost 2/3 of its volume, the salinity of the water had increased proportionally, and the once thriving body of water had essentially "died." Salty dust blown from its shore area has contaminated the surrounding farmland. This dead body of water serves as a stark reminder of the consequences of thoughtless technology that does not adequately consider environmental consequences.

2.9 WATER: A VERY USEFUL GREEN SUBSTANCE

Water is the ultimate green substance, essentially indestructible and totally renewable, required by all living organisms and nontoxic, except in cases of gross abuse, such as from drinking enormous quantities of it. Water is the greenest of all solvents with a vast variety of applications.

As illustrated in Figure 2.7, the major uses of water are in electrical power generation (for cooling water) and for irrigation. Withdrawal of water for industrial cooling applications is relatively sustainable because usually no more than 5% is lost to evaporation leaving the remainder available for other applications. Some manufacturing operations do require very large amounts of water—as examples, about 8000 L to produce a kilogram of aluminum and an astounding 350,000–400,000 L to manufacture an automobile.

Most agricultural use of water is for irrigation. Sometimes as little as 10% of the water withdrawn for irrigation actually reaching the crops, with the rest lost to evaporation and infiltration. Furthermore, irrigation water may become seriously contaminated, such as by salts leached from the soil and by fertilizers. However, more efficient irrigation and cropping practices have a high potential for water conservation, while still serving irrigation needs effectively. One such technique is "trickle irrigation," in which just enough water drips on the plant roots to keep them moist (see Section 2.7). Buildup of salts in soil can be a problem with low-consumption irrigation techniques.

Water is not destroyed, but it can be lost for practical use. The three ways in which this may occur are the following:

- *Evaporative losses*, such as occur during spray irrigation and when water is used for evaporative cooling.
- *Infiltration* of water into the ground, often in places and ways that preclude its later uses as groundwater.
- *Degradation* from pollutants, such as salts picked up by water used for irrigation.

The total of the three factors above is called water *consumption*. In many cases it is only a fraction of *withdrawal* consisting of the amount of water that is taken and run through a water system for some purpose. An example in which consumption is only a small fraction of withdrawal is when water is used for cooling, in which all but a small amount lost to evaporation is returned to a stream or body of water slightly warmer, but undamaged.

Some very marked changes in the pattern of water use are inevitable. The impact may be particularly severe upon agriculture, as illustrated, for example, by drastic curtailment of irrigation water to California agriculture during some recent times. In the southwestern United States, for example, agriculture accounts for the bulk of total water usage—85% in California, 90% in New Mexico and Arizona, and 68% in Texas. In some areas, industries and municipalities are willing to buy their water at prices up to 10 times that paid for irrigation water. The increased cost of water could have marked effects on food prices and availability in the United States.

Water continues to be the subject of heated disputes among land owners and governmental agencies. The state of South Dakota has protested the release of water from reservoirs in the state to maintain barge traffic on the lower Missouri River. Suggestions to transfer water from Washington, Oregon, or northern California to meet growing demand in southern California generate heated discussion. Numerous international disputes have arisen over water supplies, and these can be expected to intensify even to the point of warfare as water becomes more scarce relative to demand.

As with other scarce resources, *conservation* provides much of the answer for water shortages. Conservation can begin at the point where precipitation falls by proper management of the watershed. Proper agricultural practices that minimize and slow runoff, retain soil moisture, and maximize aquifer recharge are very helpful. Use of low tillage agriculture that retains water in plant residues on the surface is very helpful as is the older practice of summer fallowing in which fields are allowed to stand without crop growth in order to build up soil moisture. Construction of terraces, small ponds, and similar water-retaining structures are helpful conservation practices. Other water conservation measures may be applied to minimize domestic consumption. These include use of shower heads that consume minimum amounts of water, low-consumption toilets, appliances (washing machines, dishwashers) designed for low water consumption, and lawns that are planted with native grass species that require minimal water.

A limiting factor in water conservation and reuse is buildup of dissolved salt content. Furthermore, much of the water available for use is saltwater from the ocean or from brackish groundwater. Therefore, salt removal, *desalination*, can be an effective way of augmenting, recycling, or extending water supplies.

2.10 AQUATIC LIFE

Healthy aquatic systems support an abundance of living organisms that exist in complex ecosystems.[5] The living organisms (*biota*) in an aquatic ecosystem may be classified as either autotrophic or heterotrophic. *Autotrophic* organisms utilize solar or chemical energy to fix elements from simple, nonliving inorganic material into complex life molecules that compose living organisms. Algae are the most important autotrophic aquatic organisms because they are *producers* that utilize solar energy to generate biomass from CO_2 and other simple inorganic species.

Heterotrophic organisms utilize the organic substances produced by autotrophic organisms as energy sources and as the raw materials for the synthesis of their own biomass. *Decomposers* (or *reducers*) are a subclass of the heterotrophic organisms and consist of chiefly bacteria and fungi, which ultimately break down material of biological origin to the simple compounds originally fixed by the autotrophic organisms.

The ability of a body of water to produce living material is known as its *productivity*. Productivity results from a combination of physical and chemical factors. High productivity requires an adequate supply of carbon (CO_2), nitrogen (nitrate), phosphorus (orthophosphate), and trace elements such as iron. Water of low productivity generally is desirable for water supply or for swimming. Relatively high productivity is required for the support of fish and to serve as the basis of the food chain in an aquatic ecosystem. Excessive productivity results in decay of the biomass produced, consumption of dissolved oxygen, and odor production, a condition called *eutrophication*.

Life forms higher than algae and bacteria—fish, for example—comprise a comparatively small fraction of the biomass in most aquatic systems. The influence of these higher life forms upon aquatic chemistry is minimal. However, aquatic life is strongly influenced by the physical and chemical properties of the body of water in which it lives. *Temperature*, *transparency*, and *turbulence* are the three main physical properties affecting aquatic life. Very low water temperatures result in very slow biological processes, whereas very high temperatures are fatal to most organisms. The transparency of water is particularly important in determining the growth of algae. Turbulence is an important factor in mixing processes and transport of nutrients and waste products in water. Some small organisms (*plankton*) depend upon water currents for their own mobility.

Dissolved oxygen (DO) frequently is the key substance in determining the extent and kinds of life in a body of water. Oxygen deficiency is fatal to many aquatic animals such as fish. The presence of oxygen can be equally fatal to many kinds of anaerobic bacteria. *Biochemical oxygen demand (BOD)*, discussed as a water pollutant in Section 7.9, refers to the amount of oxygen utilized when the organic matter in a given volume of water is degraded biologically.

Carbon dioxide is produced by respiratory processes in water and sediments and can also enter water from the atmosphere. Carbon dioxide is required for the photosynthetic production of biomass by algae and in some cases is a limiting factor. High levels of carbon dioxide produced by the degradation of organic matter in water can cause excessive algal growth and biomass productivity.

The salinity of water also determines the kinds of life forms present. Irrigation waters may pick up harmful levels of salt. Marine life obviously requires or tolerates salt water, whereas many freshwater organisms are intolerant of salt.

2.11 GLOBAL WARMING AND WATER

Of all the human influences on the hydrosphere the one that may have the greatest effect in the future is global warming. Although there is a consensus among reputable scientists that global warming is underway, it has become a major policy, political, and economic issue that engenders heated discussion. It is a complex topic made even more so by the natural variations in climate that occur over decades of time. Recent trends in global temperature that have been measured with particular accuracy since the late 1900s using satellite instrumentation tend to support the idea that global warming is taking place. Figure 2.9 shows global temperature trends since 1880, the earliest that direct measurements of global temperature were made with any degree of accuracy, indicates a steady warming trend during recent decades. (Historically accurate records of abnormally cool years following gargantuan volcanic eruptions that resulted in reflection of solar energy back to space from high in the atmosphere go back farther. One of these occurred in 1816, "the year without a summer," following the astoundingly enormous Tambora, Indonesia volcanic explosion of 1815. Brutally cold years around 210 BC coming after a similar volcanic incident in Iceland were recorded in ancient China.) According to studies performed by the U.S. Goddard Institute for Space Studies,[6] during the time period from 1880 to the present the 10 warmest years recorded have occurred since 1997. The near record warmth of 2007 is all the more remarkable because the year was at a minimum of solar irradiance and the natural El Niño–La Niña cycle of the equatorial Pacific Ocean was in its cool phase. Although the coolest year since 2000, the year 2008 still ranks seventh to tenth of these record warm years.

Water vapor, carbon dioxide, methane, and other infrared-absorbing trace gases in the atmosphere contribute to global warming—the "greenhouse effect"—by allowing incoming solar radiant energy to penetrate to the Earth's surface while reabsorbing infrared radiation emanating from it. This phenomenon is essential for life on Earth as it is now known because it keeps surface temperature at a livable level. However, if too much of the outgoing infrared radiation is absorbed, the atmosphere

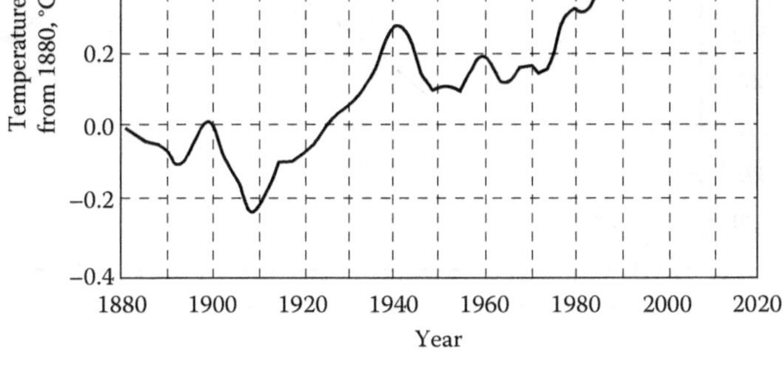

FIGURE 2.9 Global temperature trend since 1880. Earlier values are less certain because of the lack of sophisticated means of measuring temperature. More recent values are very accurate because of the use of satellite-based technologies for measuring temperature.

is heated excessively. "Greenhouse gases" have an inordinate influence on heating the atmosphere because they fill gaps in the infrared spectrum of water vapor, the substance most responsible for atmospheric warming.

As shown in Figure 2.10, atmospheric levels of "greenhouse gas" carbon dioxide have increased at a rapid rate during recent decades and are continuing to do so. The most obvious factor in increased atmospheric carbon dioxide is consumption of carbon-containing fossil fuels. In addition, release of CO_2 from the biodegradation of biomass and uptake by photosynthesis are important factors determining overall CO_2 levels in the atmosphere. The role of photosynthesis is illustrated in Figure 2.10, which shows a seasonal cycle in carbon dioxide levels in the Northern Hemisphere. Maximum values occur in April and minimum values in late September or October.

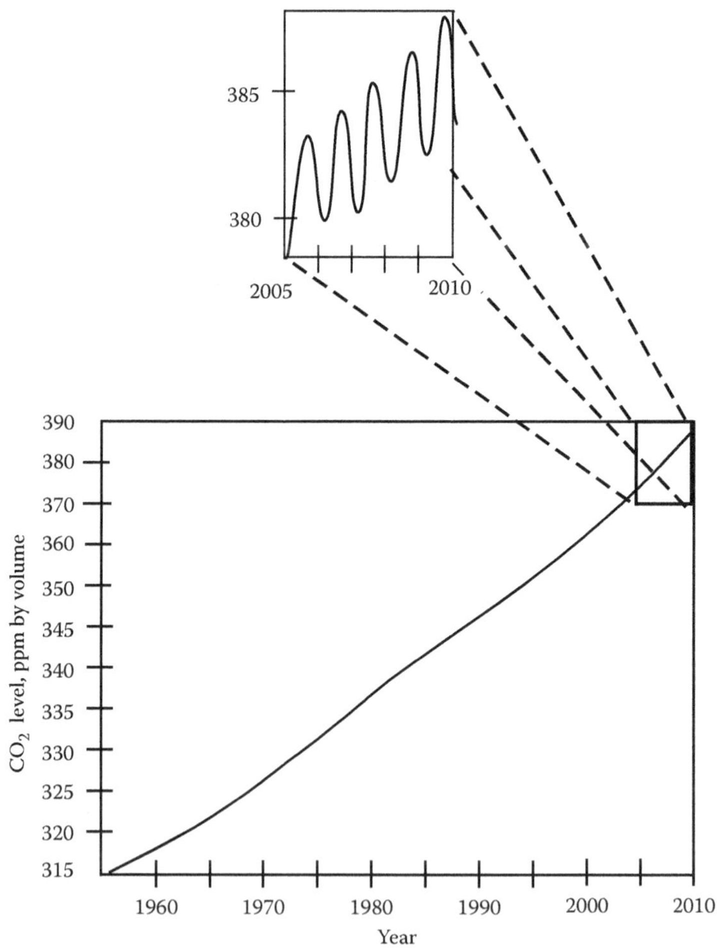

FIGURE 2.10 Increases in atmospheric CO_2 levels in recent years. The inset illustrates seasonal variations tied to photosynthetic uptake of atmospheric carbon dioxide during plant growth seasons in the Northern Hemisphere.

These oscillations are due to the "photosynthetic pulse," influenced most strongly by forests in middle latitudes and located predominantly in the Northern Hemisphere. Forests have a much greater influence than other vegetation because trees carry out more photosynthesis. Furthermore, forests store enough fixed but readily oxidizable carbon in the form of wood and humus to have a marked influence on atmospheric CO_2 content. Thus, during the summer months, forest trees carry out enough photosynthesis to reduce the atmospheric carbon dioxide content markedly. During the winter, metabolism of biota, such as bacterial decay of humus, releases a significant amount of CO_2. Therefore, the current worldwide destruction of forests and conversion of forest lands into agricultural uses contributes substantially to a greater overall increase in atmospheric CO_2 levels.

A consistent annual increase of atmospheric carbon dioxide of about one part per million per year was characteristic of the trends shown in Figure 2.10 until relatively recently, but now the annual increase is closer to 2 ppm, probably the result of many new coal-fired powerplants coming online in highly populated China and India. Although per capita carbon dioxide emissions are highest for industrialized countries, development of populous countries can be expected to add large quantities of carbon dioxide to the atmosphere in the future. Chlorofluorocarbons, which also are greenhouse gases, were not even introduced into the atmosphere until the 1930s.

Among the gases other than carbon dioxide that contribute to global warming are chlorofluorocarbons, fluorocarbons, hydrochlorofluorocarbons hydrofluorocarbons, N_2O, and, especially, methane, CH_4. Over an approximately 200-year period atmospheric methane levels approximately tripled and until the early 1990s were increasing at a rate of almost 0.02 ppm/year. The comparatively very rapid increase in methane levels has been attributed to a number of factors resulting from human activities. Among these are direct leakage of natural gas, by-product emissions from coal mining and petroleum recovery, and release from the burning of savannas and tropical forests. Biogenic sources resulting from human activities produce large amounts of atmospheric methane. These include methane from bacteria degrading organic matter such as municipal refuse in landfills; methane evolved from anaerobic biodegradation of organic matter in rice paddies; and methane emitted as the result of bacterial action in the digestive tracts of ruminant animals. For an approximately 10-year period starting around 1995, atmospheric methane levels remained essentially steady. The somewhat puzzling plateau in methane levels has been attributed to success in efforts to stop pipeline leaks, capturing methane from landfills, reduced industrial activity with the collapse of the former Soviet Union, and droughts in tropical areas. Starting around 2007, the historical rise in methane levels resumed.

Highly sophisticated computer models predict global warming from greenhouse gas emissions of at least 3.0°C and up to 5.5°C occurring over a period of just a few decades within approximately the next century. These estimates are sobering because they correspond to the approximate temperature increase since the last ice age 18,000 years past, which took place at a much slower pace of only about 1 or 2°C per 1000 years. Such warming would have profound effects on the environment.

2.11.1 Global Warming Effects on the Hydrosphere

Some of the most drastic effects of global warming can be expected on the hydrosphere and water availability. A hotter climate will likely be a dryer climate overall, severely constraining supplies of available water. Storms and rainfall events will probably be more severe when they occur causing much more rapid runoff of precipitation and reducing its availability for use. Severe drought conditions will increase demands for available water for irrigation and other uses putting additional strain on depleted water resources.

Higher atmospheric CO_2 concentrations will result in more transfer of the gas to the hydrosphere including accelerated sorption of the gas by oceans. Earth's oceans constitute a vast sink for carbon dioxide; the amount of dissolved CO_2 in the oceans is about 60 times the amount of CO_2 gas in the atmosphere. However, the times for transfer of carbon dioxide from the atmosphere to the ocean are of the order of years. Because of water stratification and low mixing rates, the times for transfer of carbon dioxide from the upper approximately 100-m layer of the oceans to ocean depths is much longer, of the order of decades. Therefore, increased absorption by oceans will lag behind the emissions of CO_2. A concern with increased levels of CO_2 in the oceans is the lowering of ocean water pH that will result. Even though such an effect will be slight, of the order of one-tenth to several tenths of a pH unit, it has the potential to strongly impact organisms that live in ocean water. Many marine organisms, including some that are key components of oceanic food chains, have shells based upon calcium carbonate. These include shellfish (clams, oysters, mussels), coral, and crustaceans including crabs, lobsters, and shrimp. At higher carbon dioxide levels in ocean water and lower pH values that result, calcium carbonate tends to dissolve:

$$CaCO_3(s) + CO_2(aq) + H_2O \rightleftarrows Ca^{2+}(aq) + 2HCO_3^-(aq) \tag{2.2}$$

This can result in some thinning of calcium carbonate-based shells, an effect that some marine biologists have reported recently.

Global warming will result in higher sea levels with estimates of 0.5–1.5 m. These will be caused by warming—therefore, expansion—of ocean water and by melting of glaciers that rest on land, particularly those in Greenland and Antarctica. As a result, many highly populated low-lying coastal areas will be more subject to flooding and some of these, including some Pacific islands that are barely above sea level, will become uninhabitable.

A very significant part of the hydrosphere consists of glacial ice that produces fresh water for large populations including inhabitants downstream from the Himalayan mountains. Some of these areas are dependent upon melting glacial ice for a steady supply of water, including irrigation water, over summer months. With increased global warming much of this glacial water supply will melt completely before summer's end and water supplies will dry up. Drastically shrinking glaciers have been documented in a number of parts of the world including Glacier National Park in North America, Mt. Kilimanjaro in Africa, and glaciers at the southern tip of South America.

REFERENCES

1. Dodson, S. I., *Introduction to Limnology*, McGraw-Hill, New York, 2005.
2. Abel, D. C. and R. L. McConnell, *Environmental Oceanography: Topics and Analysis*, Jones and Bartlett, Sudbury, MA, 2010.
3. Hubbart, J., M. Pidwirny, and A. K. Panikkar, Hydrologic cycle, in *Encyclopedia of Earth*, C. J. Cleveland, Ed., Environmental Information Coalition, National Council for Science and the Environment, Washington, DC, 2007, retrievable from http://www.eoearth.org/article/Hydrologic_cycle
4. Mitsch, W. J. and J. C. Gosselink, *Wetlands*, 4th ed., Wiley, Hoboken, NJ, 2007; Closs, G., B. Downes, and A. Boulton, *Freshwater Ecology: A Scientific Introduction*, Blackwell Publishing Co., Malden, MA, 2004.
5. Dodds, W. K., *Freshwater Ecology: Concepts and Environmental Applications*, Academic Press, San Diego, CA, 2002.
6. Goddard Institute for Space Studies, Global temperature trends: 2008 annual summation, http://data.giss.nasa.gov/gistemp/2008/

SUPPLEMENTARY REFERENCES

Brutsaert, W., *Hydrology: An Introduction*, Cambridge University Press, Cambridge, UK, 2005.

Davie, T., *Fundamentals of Hydrology*, Routledge, New York, 2008.

Desonie, D., *Hydrosphere: Freshwater Sytems and Pollution*, Chelsea House Publications, New York, 2008.

Escobar, I. C. and A. Schafer, *Sustainable Water for the Future*, Elsevier Science, Amsterdam, 2010.

Findlay, S. E. G. and R. L. Sinsabaugh, Eds., *Aquatic Ecosystems: Interactivity of Dissolved Organic Matter*, Academic Press, San Diego, CA, 2003.

Glennon, R., *Unquenchable: America's Water Crisis and What to Do About It, Island Press*, Washington, DC, 2009.

Hynes, H.B.N., *The Ecology of Running Waters*, Blackburn Press, Caldwell, NJ, 2001.

Kalff, J., *Limnology: Inland Water Ecosystems*, Prentice-Hall, Upper Saddle River, NJ, 2002.

Kaska, K., and R. J. Sager, *Water on Earth*, Holt McDougal, New York, 2007.

Kumar, A., Ed., *Aquatic Ecosystems*, A.P.H. Publishing Corp., New Delhi, 2003.

Pearce, F., *When the Rivers Run Dry: Water—The Defining Crisis of the Twenty-first Century*, Beacon Press, Boston, 2007.

Spellman, F. R., *The Science of Water: Concepts and Applications*, 2nd ed., Taylor & Francis/CRC Press, Boca Raton, FL, 2007.

Stewart, B. A. and T. A. Howell, Eds., *Encyclopedia of Water Science*, Marcel Dekker, New York, 2003.

Trimble, S., *Encyclopedia of Water Science*, 2nd ed., CRC Press, Boca Raton, FL, 2007.

Wetzel, R. G., *Limnology: Lake and River Ecosystems*, 3rd ed., Academic Press, San Diego, 2001.

QUESTIONS AND PROBLEMS

The use of Internet resources is assumed in answering any of the questions. These would include such things as constants and conversion factors as well as additional information needed to complete an answer.

1. Water has a high surface tension. With a little searching on the internet, explain what surface tension is. How does the molecular structure of water contribute to its high surface tension?

2. Describe ways in which measures taken to alleviate water supply and flooding problems might actually aggravate such problems.
3. The study of water is known as __________. __________, is the branch of the science dealing with the characteristics of fresh water, and the science that deals with about 97% of all Earth's water is called __________.
4. Consider the hydrological cycle in Figure 2.2. List or discuss the kinds or classes of environmental chemistry that might apply to each major part of this cycle.
5. Consider the unique and important properties of water. What molecular or bonding characteristics of the water molecules are largely responsible for these properties. List or describe one of each of the following unique properties of water related to (a) thermal characteristics, (b) transmission of light, (c) surface tension, (d) solvent properties.
6. Discuss how thermal stratification of a body of water may affect its chemistry.
7. Relate aquatic life to aquatic chemistry. In so doing, consider the following: autotrophic organisms, producers, heterotrophic organisms, decomposers, eutrophication, dissolved oxygen, biochemical oxygen demand.
8. Summer fallowing is practiced in more arid agricultural areas to build up soil moisture for the following year's crops. Typically, it involves leaving crop residues in place on the soil surface and avoiding the growth of plants or grass. Why is it best to avoid such growth in order for summer fallowing to be successful?
9. Western areas of the state of Oregon have a generally wet climate and the eastern sections are dryer with regions of desert. With a little research on the geography and topography of Oregon, suggest the reasons for these conditions.
10. Suggest ways in which development and urbanization may adversely affect watersheds and groundwater recharge.
11. Explain the action of streams in the sorting of sedimentary materials by size.
12. On the internet, find some pictures of stone structures, such as cathedrals, in central Mexico City dating from Spanish colonial times. Relate any structural anomalies of these structures to groundwater.
13. What is water productivity? Suggest ways in which productivity may be a self-limiting process.
14. Look up recent studies of the conditions of shells in marine shellfish and suggest how the observations may relate to atmospheric carbon dioxide levels.

3 Fundamentals of Aquatic Chemistry

3.1 INTRODUCTION TO AQUATIC CHEMISTRY

Water exhibits a rich and varied chemistry with the H_2O molecule acting as a participant in many chemical phenomena and water functioning as a unique solvent in which chemical reactions occur. Furthermore, aquatic life, especially microorganisms, play a very strong role in water chemistry (see Section 2.10 and Chapter 6). The main categories of aquatic chemical phenomena are illustrated in Figure 3.1. This chapter introduces water chemistry and discusses aquatic acid–base and complexation phenomena in water. Oxidation–reduction reactions and equilibria are discussed in Chapter 4, and details of solubility calculations and interactions between liquid water and other phases are given in Chapter 5.

Aquatic environmental chemical phenomena involve processes familiar to chemists, including acid–base, solubility, oxidation–reduction, and complexation reactions.[1] Although most aquatic chemical phenomena are discussed here from the thermodynamic (equilibrium) viewpoint, it is important to keep in mind that reaction rates (kinetics) are very important in aquatic chemistry. Biological processes play a key role in aquatic chemistry. For example, algae undergoing photosynthesis can raise the pH of water by removing aqueous CO_2, thereby converting an HCO_3^- ion into a CO_3^{2-} ion; this ion in turn reacts with Ca^{2+} in water to precipitate $CaCO_3$.

Compared to the carefully controlled conditions of the laboratory, it is much more difficult to describe chemical phenomena in natural water systems. Such systems are very complex and a description of their chemistry must take many variables into consideration. In addition to water, these systems contain mineral phases, gas phases, and organisms. As open, dynamic systems, they have variable inputs and outputs of energy and mass. Therefore, except under unusual circumstances, a true equilibrium condition is not obtained, although an approximately steady-state aquatic system frequently exists. Most metals found in natural waters do not exist as simple hydrated cations in the water, and oxyanions often occur as polynuclear species, rather than as simple monomers. The nature of chemical species in water containing bacteria or algae is strongly influenced by the action of these organisms. Thus, an exact description of the chemistry of a natural water system based upon acid–base, solubility, and complexation equilibrium constants, redox potential, pH, and other chemical parameters is not possible. Therefore, the systems must be described by simplified *models*, often based around equilibrium chemical concepts. Though not exact, nor entirely realistic, such models can yield useful generalizations and insights pertaining to the nature of aquatic chemical processes, and provide guidelines for the description and

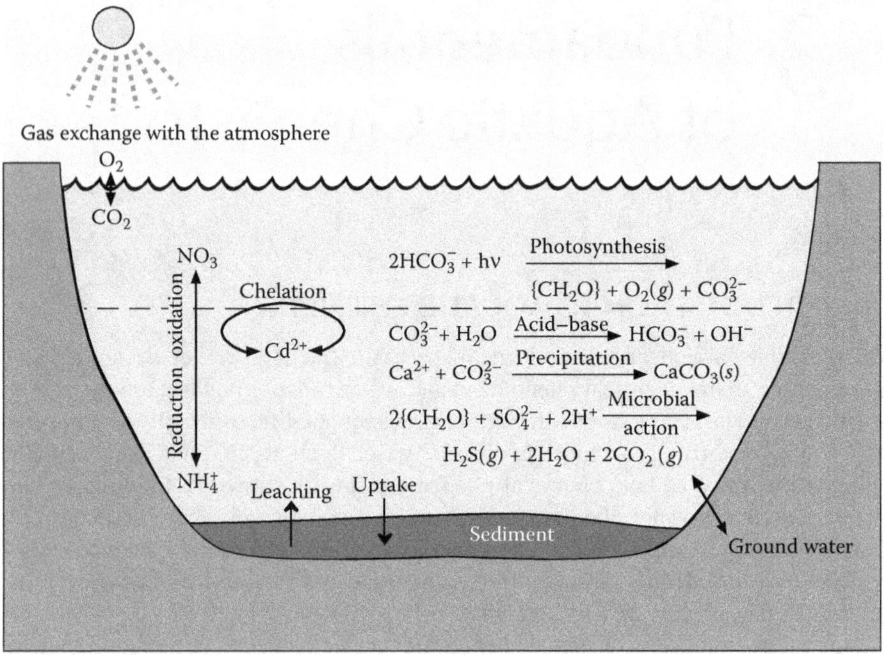

FIGURE 3.1 Major aquatic chemical processes. The body of water shown here is assumed to be thermally stratified with a relatively lower density oxic top layer that is exposed to the atmosphere and that does not mix significantly with an oxygen-deficient lower layer.

measurement of natural water systems. Though greatly simplified, such models are very helpful in visualizing the conditions that determine chemical species and their reactions in natural waters and wastewaters.

3.2 GASES IN WATER

Dissolved gases—O_2 for fish and CO_2 for photosynthetic algae—are crucial to the welfare of living species in water. Some gases in water can also cause problems, such as the death of fish from bubbles of nitrogen formed in the blood caused by exposure to water supersaturated with N_2. As noted in Section 1.3, volcanic carbon dioxide evolved from supersaturated dissolved CO_2 in the waters of Lake Nyos in the African country of Cameroon asphyxiated 1700 people in 1986.

The solubilities of gases in water are calculated with *Henry's Law*, which states that *the solubility of a gas in a liquid is proportional to the partial pressure of that gas in contact with the liquid.* These calculations are discussed in some detail in Chapter 5.

3.2.1 Oxygen in Water

Without an appreciable level of dissolved oxygen, many kinds of aquatic organisms cannot exist in water. Dissolved oxygen is consumed by the degradation of organic matter in water. Many fish kills are caused not from the direct toxicity of pollutants but from a deficiency of oxygen because of its consumption in the biodegradation of pollutants.

Most elemental oxygen comes from the atmosphere, which is 20.95% oxygen by volume of dry air. Therefore, the ability of a body of water to re-oxygenate itself by contact with the atmosphere is an important characteristic. Oxygen is produced by the photosynthetic action of algae, but this process is really not an efficient means of oxygenating water because some of the oxygen formed by photosynthesis during the daylight hours is lost at night when the algae consume oxygen as part of their metabolic processes. When the algae die, the degradation of their biomass also consumes oxygen.

The solubility of oxygen in water depends upon water temperature, the partial pressure of oxygen in the atmosphere, and the salt content of the water. It is important to distinguish between oxygen *solubility*, the maximum dissolved O_2 concentration at equilibrium, and dissolved oxygen *concentration*, which is generally not the equilibrium concentration and is limited by the rate at which oxygen dissolves. The calculation of oxygen solubility as a function of partial pressure is discussed in Section 5.3, where it is shown that the concentration of oxygen in water at 25°C in equilibrium with air at atmospheric pressure is only 8.32 mg/L. Thus, water in equilibrium with air cannot contain a high level of dissolved oxygen compared to many other solute species. If oxygen-consuming processes are occurring in the water, the dissolved oxygen level may rapidly approach zero unless some efficient mechanism for the reaeration of water is operative, such as turbulent flow in a shallow stream or air pumped into the aeration tank of an activated sludge secondary waste treatment facility (see Chapter 8). The problem becomes largely one of kinetics, in which there is a limit to the rate at which oxygen is transferred across the air–water interface. This rate depends upon turbulence, air bubble size, temperature, and other factors.

If organic matter of biological origin is represented by the formula $\{CH_2O\}$, the consumption of oxygen in water by the degradation of organic matter may be expressed by the following biochemical reaction:

$$\{CH_2O\} + O_2 \rightarrow CO_2 + H_2O \tag{3.1}$$

The mass of organic material required to consume the 8.3 mg of O_2 in a liter of water in equilibrium with the atmosphere at 25°C is given by a simple stoichiometric calculation based on Equation 3.1, which yields a value of 7.8 mg of $\{CH_2O\}$. Thus, the microorganism-mediated degradation of only 7 or 8 mg of organic material can completely consume the O_2 in 1 L of water initially saturated with air at 25°C. The depletion of oxygen to levels below those that sustain oxic organisms requires the degradation of even less organic matter at higher temperatures (where the solubility of oxygen is less) or in water not initially saturated with atmospheric oxygen. Furthermore, there are no common aquatic chemical reactions that replenish dissolved oxygen; except for oxygen provided by photosynthesis, it must come from the atmosphere.

The temperature effect on the solubility of gases in water is especially important in the case of oxygen. The solubility of oxygen in water in equilibrium with atmospheric air decreases from 14.74 mg/L at 0°C to 7.03 mg/L at 35°C. At higher temperatures, the decreased solubility of oxygen, combined with the increased respiration rate of aquatic organisms, frequently causes a condition in which a higher demand for oxygen accompanied by lower solubility of the gas in water results in severe oxygen depletion.

3.3 WATER ACIDITY AND CARBON DIOXIDE IN WATER

Acid–base phenomena in water involve loss and acceptance of H^+ ion. Many species act as *acids* in water by releasing H^+ ion, others act as *bases* by accepting H^+, and the water molecule itself does both. An important species in the acid–base chemistry of water is bicarbonate ion, HCO_3^-, which may act as either an acid or a base:

$$HCO_3^- \rightleftarrows CO_3^{2-} + H^+ \quad (3.2)$$

$$HCO_3^- + H^+ \rightleftarrows CO_2(aq) + H_2O \quad (3.3)$$

Acidity as applied to natural water and wastewater is the capacity of the water to neutralize OH^-; it is analogous to alkalinity, the capacity to neutralize H^+, which is discussed in the next section. Although virtually all water has some alkalinity, acidic water is not frequently encountered, except in cases of severe pollution. Acidity generally results from the presence of weak acids, particularly CO_2, but sometimes includes others such as $H_2PO_4^-$, H_2S, proteins, and fatty acids. Acidic metal ions, particularly Fe^{3+}, may also contribute to acidity.

Strong acids are the most important contributors to water pollutant acidity. The term *free mineral acid* is applied to strong acids such as H_2SO_4 and HCl in water. Acid mine water is a common water pollutant that contains an appreciable concentration of free mineral acid. Whereas total acidity is determined by titration with base to the phenolphthalein endpoint (pH 8.2), free mineral acid is determined by titration with base to the methyl orange endpoint (pH 4.3).

As shown by the hydrolysis of hydrated Al^{3+} below, the acidic character of some hydrated metal ions may contribute to acidity. Some industrial wastes, such as spent steel pickling liquor, contain acidic metal ions and often some excess strong acid. The acidity of such wastes must be measured in calculating the amount of lime or other chemicals required to neutralize the acid.

$$Al(H_2O)_6^{3+} \rightleftarrows Al(H_2O)_5OH^{2+} + H^+ \quad (3.4)$$

3.3.1 Carbon Dioxide in Water

The most important weak acid in water is carbon dioxide, CO_2. Because of the presence of carbon dioxide in air and its production from microbial decay of organic matter, dissolved CO_2 is present in virtually all natural waters and wastewaters. Rainfall from even an absolutely unpolluted atmosphere is slightly acidic due to the presence of dissolved CO_2. Dissolution in sea water is an important mechanism for the reduction of atmospheric carbon dioxide, the "greenhouse" gas that makes the greatest contribution to climate warming. Proposals have been made to pump CO_2

from combustion into places in the ocean where seawater is sinking to sequester it from the atmosphere for thousands of years.[2]

Carbon dioxide, and its ionization products, bicarbonate ion (HCO_3^-), and carbonate ion (CO_3^{2-}) have an extremely important influence upon the chemistry of water. Many minerals are deposited as salts of the carbonate ion. Algae in water utilize dissolved CO_2 in the synthesis of biomass. The equilibrium of dissolved CO_2 with gaseous carbon dioxide in the atmosphere,

$$CO_2(\text{water}) \rightleftarrows CO_2(\text{atmosphere}) \tag{3.5}$$

and equilibrium of CO_3^{2-} ion between aquatic solution and solid carbonate minerals,

$$MCO_3(\text{slightly soluble carbonate salt}) \rightleftarrows M^{2+} + CO_3^{2-} \tag{3.6}$$

have a strong buffering effect upon the pH of water.

Carbon dioxide is only about 0.039% by volume of normal dry air. As a consequence of the low level of atmospheric CO_2, water totally lacking in alkalinity (capacity to neutralize H^+, see Section 3.4) in equilibrium with the atmosphere contains only a very low level of carbon dioxide. However, the formation of HCO_3^- and CO_3^{2-} greatly increases the solubility of carbon dioxide. High concentrations of free carbon dioxide in water may adversely affect respiration and gas exchange of aquatic animals. It may even cause death and should not exceed levels of 25 mg/L in water.

A large share of the carbon dioxide found in water is a product of the breakdown of organic matter by bacteria. Even algae, which utilize CO_2 in photosynthesis, produce it through their metabolic processes in the absence of light. As water seeps through layers of decaying organic matter while infiltrating the ground, it may dissolve a great deal of CO_2 produced by the respiration of organisms in the soil. Later, as water goes through limestone formations, it dissolves calcium carbonate because of the presence of the dissolved CO_2:

$$CaCO_3(s) + CO_2(aq) + H_2O \rightleftarrows Ca^{2+} + 2HCO_3^- \tag{3.7}$$

This process is the one by which limestone caves are formed. The implications of the above reaction for aquatic chemistry are discussed in greater detail in Section 3.5.

The concentration of gaseous CO_2 in the atmosphere varies with location and season; it is increasing by about one part per million (ppm) by volume per year. For purposes of calculation here, the concentration of atmospheric CO_2 will be taken as 390 ppm (0.0390%) in dry air. At 25°C, water in equilibrium with unpolluted air containing 390 ppm carbon dioxide has a $CO_2(aq)$ concentration of 1.276×10^{-5} M (see Henry's law calculation of gas solubility in Section 5.3), and this value will be used for subsequent calculations.

Although CO_2 in water is often represented as H_2CO_3, the equilibrium constant for the reaction

$$CO_2(aq) + H_2O \rightleftarrows H_2CO_3 \tag{3.8}$$

is only around 2×10^{-3} at 25°C, so just a small fraction of the dissolved carbon dioxide is actually present as H_2CO_3. In this text, nonionized carbon dioxide in water will be designated simply as CO_2, which in subsequent discussions will stand for the total of dissolved molecular CO_2 and un-dissociated H_2CO_3.

The CO_2–HCO_3^-–CO_3^{2-} system in water may be described by the equations,

$$CO_2(aq) + H_2O \rightleftarrows H^+ + HCO_3^- \tag{3.9}$$

$$K_{a1} = \frac{[H^+][HCO_3^-]}{[CO_2]} = 4.45 \times 10^{-7} \quad pK_{a1} = 6.35 \tag{3.10}$$

$$HCO_3^- \rightleftarrows H^+ + CO_3^{2-} \tag{3.11}$$

$$K_{a2} = \frac{[H^+][CO_3^{2-}]}{[HCO_3^-]} = 4.69 \times 10^{-11} \quad pK_{a2} = 10.33 \tag{3.12}$$

where $pK_a = -\log K_a$. The predominant species formed by CO_2 dissolved in water depends upon pH. This is best shown by a distribution of species diagram with pH as a master variable is illustrated in Figure 3.2. Such a diagram shows the major species present in solution as a function of pH. For CO_2 in aqueous solution, the diagram is a series of plots of the fractions present as CO_2, HCO_3^-, and CO_3^{2-} as a function of pH. These fractions, designated as α_x, are given by the following expressions:

$$\alpha_{CO_2} = \frac{[CO_2]}{[CO_2] + [HCO_3^-] + [CO_3^{2-}]} \tag{3.13}$$

$$\alpha_{HCO_3^-} = \frac{[HCO_3^-]}{[CO_2] + [HCO_3^-] + [CO_3^{2-}]} \tag{3.14}$$

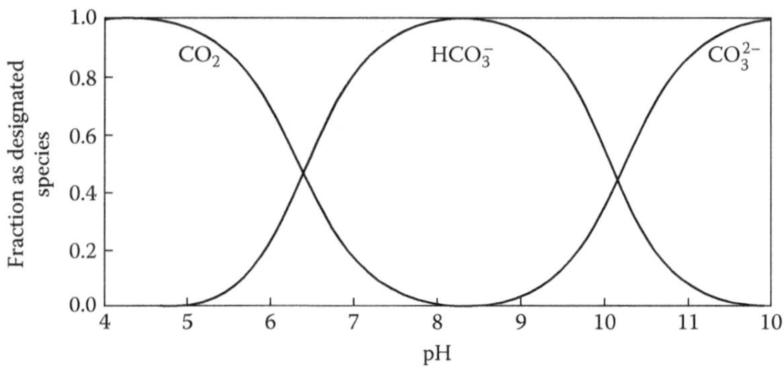

FIGURE 3.2 Distribution of species diagram for the CO_2–HCO_3^-–CO_3^{2-} system in water.

$$\alpha_{CO_3^{2-}} = \frac{[CO_3^{2-}]}{[CO_2] + [HCO_3^-] + [CO_3^{2-}]} \tag{3.15}$$

Substitution of the expressions for K_{a1} and K_{a2} into the α expressions gives the fractions of species as a function of acid dissociation constants and hydrogen ion concentration:

$$\alpha_{CO_2} = \frac{[H^+]^2}{[H^+]^2 + K_{a1}[H^+] + K_{a1}K_{a2}} \tag{3.16}$$

$$\alpha_{HCO_3^-} = \frac{K_{a1}[H^+]}{[H^+]^2 + K_{a1}[H^+] + K_{a1}K_{a2}} \tag{3.17}$$

$$\alpha_{CO_3^{2-}} = \frac{K_{a1}K_{a2}}{[H^+]^2 + K_{a1}[H^+] + K_{a1}K_{a2}} \tag{3.18}$$

Calculations from these expressions show the following:

- For pH significantly below pK_{a1}, α_{CO_2} is essentially 1
- When pH = pK_{a1}, $\alpha_{CO_2} = \alpha_{HCO_3^-}$
- When pH = $1/2(pK_{a1} + pK_{a2})$, $\alpha_{HCO_3^-}$ is at its maximum value of 0.98
- When pH = pK_{a2}, $\alpha_{HCO_3^-} = \alpha_{CO_3^{2-}}$
- For pH significantly above pK_{a2}, $\alpha_{CO_3^{2-}}$ is essentially 1

The distribution of species diagram in Figure 3.2 shows that hydrogen carbonate (bicarbonate) ion (HCO_3^-) is the predominant species in the pH range found in most waters, with CO_2 predominating in more acidic waters.

As mentioned above, the value of $[CO_2(aq)]$ in water at 25°C in equilibrium with air that is 390 ppm CO_2 is 1.276×10^{-5} M. The carbon dioxide dissociates partially in water to produce equal concentrations of H^+ and HCO_3^-:

$$CO_2 + H_2O \rightleftarrows HCO_3^- + H^+ \tag{3.19}$$

The concentrations of H^+ and HCO_3^- are calculated from K_{a1}:

$$K_{a1} = \frac{[H^+][HCO_3^-]}{[CO_2]} = \frac{[H^+]^2}{1.276 \times 10^{-5}} = 4.45 \times 10^{-7} \tag{3.20}$$

$$[H^+] = [HCO_3^-] = (1.276 \times 10^{-5} \times 4.45 \times 10^{-7})^{1/2} = 2.38 \times 10^{-6}$$
$$pH = 5.62$$

This calculation explains why pure water that has equilibrated with the unpolluted atmosphere is slightly acidic with a pH somewhat less than 7.

3.4 ALKALINITY

The capacity of water to accept H^+ ions (protons) is called *alkalinity.* Alkalinity is important in water treatment and in the chemistry and biology of natural waters. Frequently, the alkalinity of water must be known to calculate the quantities of chemicals to be added in treating the water. Highly alkaline water often has a high pH and generally contains elevated levels of dissolved solids. These characteristics may be detrimental for water to be used in boilers, food processing, and municipal water systems. Alkalinity serves as a pH buffer and reservoir for inorganic carbon, thus helping to determine the ability of water to support algal growth and other aquatic life, so alkalinity can be used as a measure of water fertility. Generally, the basic species responsible for alkalinity in water are bicarbonate ion, carbonate ion, and hydroxide ion:

$$HCO_3^- + H^+ \rightleftarrows CO_2 + H_2O \quad (3.21)$$

$$CO_3^{2-} + H^+ \rightleftarrows HCO_3^- \quad (3.22)$$

$$OH^- + H^+ \rightleftarrows H_2O \quad (3.23)$$

Other, usually minor, contributors to alkalinity are ammonia and the conjugate bases of phosphoric, silicic, boric, and organic acids.

At pH values below 7, $[H^+]$ in water detracts significantly from alkalinity, and its concentration must be subtracted in computing the total alkalinity. Therefore, the following equation is the complete equation for alkalinity in a medium where the only contributors to it are HCO_3^-, CO_3^{2-}, and OH^-:

$$[\text{alk}] = [HCO_3^-] + 2[CO_3^{2-}] + [OH^-] - [H^+] \quad (3.24)$$

Alkalinity generally is expressed as *phenolphthalein alkalinity*, corresponding to titration with acid to the pH at which HCO_3^- is the predominant carbonate species (pH 8.3), or *total alkalinity*, corresponding to titration with acid to the methyl orange endpoint (pH 4.3), where both bicarbonate and carbonate species have been converted into CO_2.

It is important to distinguish between high *basicity*, manifested by an elevated pH, and high *alkalinity*, the capacity to accept H^+. Whereas pH is an *intensity* factor, alkalinity is a *capacity* factor. This may be illustrated by comparing a solution of 1.00×10^{-3} M NaOH with a solution of 0.100 M $NaHCO_3$. The sodium hydroxide solution is very basic, pH 11, but a liter of it will neutralize only 1.00×10^{-3} mole of acid. The pH of the $NaHCO_3$ solution is 8.34, much lower than that of the NaOH. However, a liter of the sodium bicarbonate solution will neutralize 0.100 mole of acid; therefore, its alkalinity is 100 times that of the more basic NaOH solution.

In engineering terms, alkalinity frequently is expressed in units of mg/L of $CaCO_3$, based upon the following acid-neutralizing reaction:

$$CaCO_3 + 2H^+ \rightleftarrows Ca^{2+} + CO_2 + H_2O \quad (3.25)$$

The equivalent mass of calcium carbonate is one-half its formula mass. Expressing alkalinity in terms of mg/L of $CaCO_3$ can, however, lead to confusion, and the preferable notation for the chemist is equivalents/L, the number of moles of H^+ neutralized by the alkalinity in a liter of solution.

3.4.1 Contributors to Alkalinity at Different pH Values

Natural water typically has an alkalinity ("[alk]") of 1.00×10^{-3} equivalents per liter (eq/L), meaning that the alkaline solutes in 1 L of the water will neutralize 1.00×10^{-3} moles of acid. The contributions made by different species to alkalinity depend upon pH. This is shown here by calculation of the relative contributions to alkalinity of HCO_3^-, CO_3^{2-}, and OH^- at pH 7.00 and pH 10.00. First, for water at pH 7.00, $[OH^-]$ is too low to make any significant contribution to the alkalinity. Furthermore, as shown in Figure 3.2, at pH 7.00 $[HCO_3^-] \gg [CO_3^{2-}]$. Therefore, the alkalinity is due to HCO_3^- and $[HCO_3^-] = 1.00 \times 10^{-3}$ M. Substitution into the expression for K_{a1} shows that at pH 7.00 and $[HCO_3^-] = 1.00 \times 10^{-3}$ M, the value of $[CO_2(aq)]$ is 2.25×10^{-4} M, higher than the value that arises from water in equilibrium with atmospheric air, but readily reached due to the presence of carbon dioxide from bacterial decay in water and sediments.

Consider next the case of water with the same alkalinity, 1.00×10^{-3} eq/L that has a pH of 10.00. At this higher pH both OH^- and CO_3^{2-} are present at significant concentrations compared to HCO_3^- and the following may be calculated:

$$[\text{alk}] = [HCO_3^-] + 2[CO_3^{2-}] + [OH^-] = 1.00 \times 10^{-3} \quad (3.26)$$

The concentration of CO_3^{2-} is multiplied by 2 because each CO_3^{2-} ion can neutralize $2H^+$ ions. The other two equations that must be solved to get the concentrations of HCO_3, CO_3^{2-}, and OH^- are

$$[OH^-] = \frac{K_W}{[H^+]} = \frac{1.00 \times 10^{-14}}{1.00 \times 10^{-10}} = 1.00 \times 10^{-4} \quad (3.27)$$

and

$$[CO_3^{2-}] = \frac{K_{a2}[HCO_3^-]}{[H^+]} \quad (3.28)$$

Solving these three equations gives $[HCO_3^-] = 4.64 \times 10^{-4}$ M and $[CO_3^{2-}] = 2.18 \times 10^{-4}$ M, so the contributions to the alkalinity of this solution are the following:

$$\begin{aligned} &4.64 \times 10^{-4} \text{ eq/L from } HCO_3^- \\ 2 \times 2.18 \times 10^{-4} = \; &4.36 \times 10^{-4} \text{ eq/L from } CO_3^{2-} \\ &\underline{1.00 \times 10^{-4} \text{ eq/L from}\qquad\quad} \\ \text{alk} = \; &1.00 \times 10^{-3} \text{ eq/L} \end{aligned}$$

3.4.2 Dissolved Inorganic Carbon and Alkalinity

The values given above can be used to show that at the same alkalinity value the concentration of total dissolved inorganic carbon, [C],

$$[C] = [CO_2] + [HCO_3^-] + [CO_3^{2-}] \tag{3.29}$$

varies with pH. At pH 7.00,

$$[C]_{pH\ 7} = 2.25 \times 10^{-4} + 1.00 \times 10^{-3} + 0 = 1.22 \times 10^{-3} \tag{3.30}$$

whereas at pH 10.00,

$$[C]_{pH\ 10} = 0 + 4.64 \times 10^{-4} + 2.18 \times 10^{-4} = 6.82 \times 10^{-4} \tag{3.31}$$

The calculation above shows that the dissolved inorganic carbon concentration at pH 10.00 is only about half that at pH 7.00. This is because at pH 10 major contributions to alkalinity are made by CO_3^{2-} ion, each of which has twice the alkalinity of each HCO_3^- ion, and by OH^-, which does not contain any carbon. The lower inorganic carbon concentration at pH 10 shows that an initially relatively low pH aquatic system (pH 7) can donate dissolved inorganic carbon for use in photosynthesis with a change in pH but none in alkalinity. This pH-dependent difference in dissolved inorganic carbon concentration represents a significant potential source of carbon for algae growing in water which fix carbon by the overall reactions

$$CO_2 + H_2O + h\nu \rightleftarrows \{CH_2O\} + O_2 \tag{3.32}$$

and

$$HCO_3^- + H_2O + h\nu \rightleftarrows \{CH_2O\} + OH^- + O_2 \tag{3.33}$$

As dissolved inorganic carbon is used up to synthesize biomass, $\{CH_2O\}$, the water becomes more basic. The amount of inorganic carbon that can be consumed before the water becomes too basic to allow algal reproduction is proportional to the alkalinity. In going from pH 7.00 to pH 10.00, the amount of inorganic carbon consumed from 1.00 L of water having an alkalinity of 1.00×10^{-3} eq/L is

$$[C]_{pH\ 7} \times 1\ L - [C]_{pH\ 10} \times 1\ L = 1.22 \times 10^{-3}\ mol - 6.82 \times 10^{-4}\ mol = 5.4 \times 10^{-4}\ mol \tag{3.34}$$

This translates to an increase of 5.4×10^{-4} mol/L of biomass. Since the formula mass of $\{CH_2O\}$ is 30, the weight of biomass produced amounts to 16 mg/L. Assuming no input of additional CO_2, at higher alkalinity more biomass is produced for the same change in pH, whereas at lower alkalinity less is produced. Because of this effect, biologists use alkalinity as a measure of water fertility.

3.4.3 Influence of Alkalinity on CO_2 Solubility

The increased solubility of carbon dioxide in water with an elevated alkalinity can be illustrated by comparing its solubility in pure water (alkalinity 0) to its solubility in water initially containing 1.00×10^{-3} M NaOH (alkalinity 1.00×10^{-3} eq/L). The number of moles of CO_2 that will dissolve in a liter of pure water from the atmosphere containing 390 ppm carbon dioxide is

$$\text{Solubility} = [CO_2(aq)] + [HCO_3^-] \tag{3.35}$$

Substituting values calculated in Section 3.3 gives

$$\text{Solubility} = 1.276 \times 10^{-5} + 2.38 \times 10^{-6} = 1.514 \times 10^{-5}\ \text{M}$$

The solubility of CO_2 in water, initially 1.00×10^{-3} M in NaOH, is about 100-fold higher because of uptake of CO_2 by the reaction

$$CO_2(aq) + OH^- \rightarrow HCO_3^- \tag{3.36}$$

so that

$$\text{Solubility} = [CO_2(aq)] + [HCO_3^-] = 1.211 \times 10^{-5} + 1.00 \times 10^{-3} = 1.01 \times 10^{-3}\ \text{M} \tag{3.37}$$

3.5 CALCIUM AND OTHER METALS IN WATER

The kinds and concentrations of metal ions in water are determined largely by the rocks that the water contacts. Metal ions in water, commonly denoted M^{n+}, exist in numerous forms. A bare metal ion, Ca^{2+} for example, cannot exist as a separate entity in water. In order to secure the highest stability of their outer electron shells, metal ions in water are bonded, or *coordinated*, to other species. These may be water molecules or other stronger bases (electron–donor partners) that might be present. Therefore, metal ions in water solution are present in forms such as the *hydrated* metal cation $M(H_2O)_x^{n+}$. Metal ions in aqueous solution seek to reach a state of maximum stability through chemical reactions including acid–base,

$$Fe(H_2O)_6^{3+} \rightleftarrows FeOH(H_2O)_5^{2+} + H^+ \tag{3.38}$$

precipitation,

$$Fe(H_2O)_6^{3+} \rightleftarrows Fe(OH)_3(s) + 3H_2O + 3H^+ \tag{3.39}$$

and oxidation–reduction reactions:

$$Fe(H_2O)_6^{2+} \rightleftarrows Fe(OH)_3(s) + 3H_2O + e^- + 3H^+ \tag{3.40}$$

These provide means through which metal ions in water are transformed to more stable forms. Because of reactions such as these and the formation of dimeric species, such as $Fe_2(OH)_2^{4+}$, the concentration of simple hydrated $Fe(H_2O)_6^{3+}$ ion in water is vanishingly small; the same holds true for many other hydrated metal ions dissolved in water.

3.5.1 Hydrated Metal Ions as Acids

Hydrated metal ions, particularly those with a charge of 3+ or more, tend to lose H^+ ion from the water molecules bound to them in aqueous solution, and fit the definition of Brønsted acids, according to which acids are H^+ donors and bases are H^+ acceptors. The acidity of a metal ion increases with charge and decreases with increasing radius. As shown by the reaction,

$$Fe(H_2O)_6^{3+} \rightleftarrows Fe(H_2O)_5OH^{2+} + H^+ \quad (3.41)$$

hydrated iron(III) ion is an acid, a relatively strong one with a K_{a1} of 8.9×10^{-4}, so that solutions of iron(III) tend to have low pH values. Hydrated trivalent metal ions, such as iron(III), generally are minus at least one hydrogen ion at neutral pH values or above. For tetravalent metal ions, the completely protonated forms, $M(H_2O)_x^{4+}$, are rare even at very low pH values. Commonly, O^{2-} is coordinated to tetravalent metal ions; an example is the vanadium(IV) species, VO^{2+}. Generally, divalent metal ions do not lose a hydrogen ion at pH values below 6, whereas monovalent metal ions such as Na^+ do not act as acids at all, and exist in water solution as simple hydrated ions.

The tendency of hydrated metal ions to behave as acids may have a profound effect upon the aquatic environment. A good example is *acid mine water* (see Chapter 7), which derives part of its acidic character from the acidic nature of hydrated iron(III):

$$Fe(H_2O)_6^{3+} \rightleftarrows Fe(OH)_3(s) + 3H^+ + 3H_2O \quad (3.42)$$

Hydroxide, OH^-, bonded to a metal ion, may function as a bridging group to join two or more metals together as shown by the following dehydration–dimerization process:

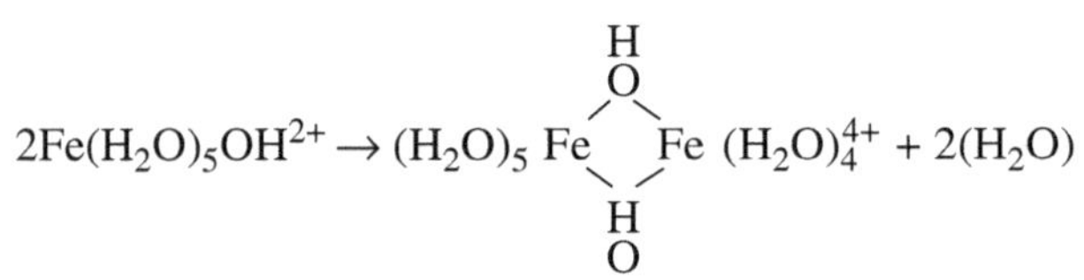
(3.43)

Among the metals other than iron(III)-forming polymeric species with OH^- as a bridging group are Al(III), Be(II), Bi(III), Ce(IV), Co(III), Cu(II), Ga(III), Mo(V), Pb(II), Sc(II), Sn(IV), and U(VI). Additional hydrogen ions may be lost from water molecules bonded to the dimers, furnishing OH^- groups for further bonding and leading to the formation of polymeric hydrolytic species. If the process continues, colloidal hydroxy polymers are formed and, finally, precipitates are produced. This process is thought to be the general one by which hydrated iron(III) oxide, $Fe_2O_3 \cdot x(H_2O)$, [also called iron(III) hydroxide, $Fe(OH)_3$], is precipitated from solutions containing iron(III).

3.5.2 Calcium in Water

Of the cations found in most fresh-water systems, calcium generally has the highest concentration. The chemistry of calcium, although complicated enough, is simpler than that of the transition metal ions found in water. Calcium is a key element in many geochemical processes, and minerals constitute the primary sources of calcium ion in waters. Among the primary contributing minerals are gypsum, $CaSO_4 \cdot 2H_2O$; anhydrite, $CaSO_4$; dolomite, $CaMg(CO_3)_2$; and calcite and aragonite, which are different mineral forms of $CaCO_3$.

Calcium ion, along with magnesium and sometimes iron(II) ion, accounts for *water hardness*. The most common manifestation of water hardness is the curdy precipitate formed by soap in hard water. *Temporary hardness* is due to the presence of calcium and bicarbonate ions in water and may be eliminated by boiling the water:

$$Ca^{2+} + 2HCO_3^- \rightleftarrows CaCO_3(s) + CO_2(g) + H_2O \tag{3.44}$$

Increased temperature may force this reaction to the right by evolving CO_2 gas, and a white precipitate of calcium carbonate may form in boiling water having temporary hardness.

Water containing a high level of carbon dioxide readily dissolves calcium from its carbonate minerals:

$$CaCO_3(s) + CO_2(aq) + H_2O \rightleftarrows Ca^{2+} + 2HCO_3^- \tag{3.45}$$

When this reaction is reversed and CO_2 is lost from the water, calcium carbonate deposits are formed. The concentration of CO_2 in water determines the extent of dissolution of calcium carbonate. The carbon dioxide that water may gain by equilibration with the atmosphere is not sufficient to account for the levels of calcium dissolved in natural waters, especially groundwaters. Rather, the respiration of microorganisms degrading organic matter in water, sediments, and soil,

$$\{CH_2O\} + O_2 \rightleftarrows CO_2 + H_2O \tag{3.46}$$

accounts for the very high levels of CO_2 and HCO_3^- observed in water and is very important in aquatic chemical processes and geochemical transformations.

3.5.3 Dissolved Carbon Dioxide and Calcium Carbonate Minerals

The equilibrium between dissolved carbon dioxide and calcium carbonate minerals is important in determining several natural water chemistry parameters such as alkalinity, pH, and dissolved calcium concentration (Figure 3.3). For fresh water, the typical figures quoted for the concentrations of both HCO_3^- and Ca^{2+} are 1.00×10^{-3} M. It may be shown that these are reasonable values when the water is in equilibrium with limestone, $CaCO_3$, and with atmospheric CO_2. The concentration of CO_2 in water in equilibrium with air has already been calculated as 1.276×10^{-5} M. The other constants needed to calculate $[HCO_3^-]$ and $[Ca^{2+}]$ are the acid dissociation constant for CO_2:

$$K_{a1} = \frac{[H^+][HCO_3^-]}{[CO_2]} = 4.45 \times 10^{-7} \quad pK_{a1} = 6.35 \tag{3.47}$$

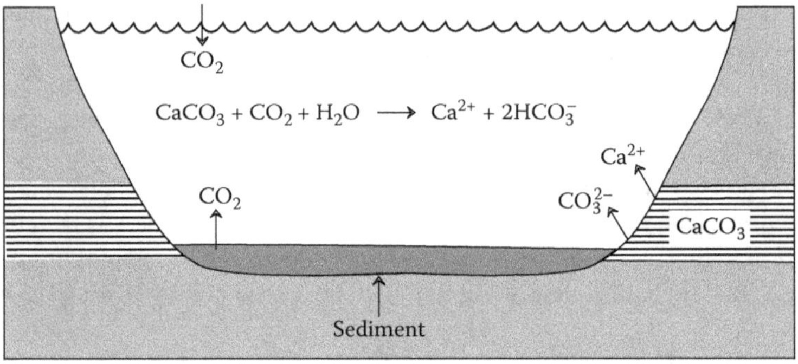

FIGURE 3.3 Carbon dioxide–calcium carbonate equilibria.

the acid dissociation constant of HCO_3^-:

$$K_{a2} = \frac{[H^+][CO_3^{2-}]}{[HCO_3^-]} = 4.69 \times 10^{-11} \quad pK_{a2} = 10.33 \tag{3.48}$$

and the solubility product of calcium carbonate (calcite):

$$K_{sp} = [Ca^{2+}][CO_3^{2-}] = 4.47 \times 10^{-9} \tag{3.49}$$

The reaction between calcium carbonate and dissolved CO_2 is

$$CaCO_3(s) + CO_2(aq) + H_2O \rightleftarrows Ca^{2+} + 2HCO_3^- \tag{3.50}$$

for which the equilibrium expression is as follows:

$$K' = \frac{[Ca^{2+}][HCO_3^-]^2}{[CO_2]} = \frac{K_{sp}K_{a1}}{K_{a2}} = 4.24 \times 10^{-5} \tag{3.51}$$

The stoichiometry of Reaction 3.50 gives a bicarbonate ion concentration that is twice that of calcium. Substitution of the value of CO_2 concentration into the expression for K' yields values of 5.14×10^{-4} M for $[Ca^{2+}]$ and 1.03×10^{-3} for $[HCO_3^-]$. Substitution into the expression for K_{sp} yields 8.70×10^{-6} M for $[CO_3^{2-}]$. When known concentrations are substituted into the product $K_{a1}K_{a2}$,

$$K_{a1}K_{a2} = \frac{[H^+]^2[CO_3^{2-}]}{[CO_2]} = 2.09 \times 10^{-17} \tag{3.52}$$

a value of 5.54×10^{-9} M is obtained for $[H^+]$ (pH 8.26). The alkalinity is essentially equal to $[HCO_3^-]$, which is much higher than $[CO_3^{2-}]$ or $[OH^-]$.

To summarize, for water in equilibrium with solid calcium carbonate and atmospheric CO_2, the following concentrations are calculated:

$$[CO_2] = 1.276 \times 10^{-5}\ M \quad [Ca^{2+}] = 5.14 \times 10^{-4}\ M$$

$$[HCO_3^-] = 1.03 \times 10^{-3}\ M \quad [H^+] = 5.54 \times 10^{-9}\ M$$

$$[CO_3^{2-}] = 8.70 \times 10^{-6}\ M \quad pH = 8.26$$

Factors such as nonequilibrium conditions, high CO_2 concentrations in bottom regions, and increased pH due to algal uptake of CO_2 cause deviations from these values. Nevertheless, they are close to the values found in a large number of natural water bodies.

3.6 COMPLEXATION AND CHELATION

The properties of metals dissolved in water depend largely upon the nature of metal species dissolved in the water. Therefore, *speciation* of metals plays a crucial role in their environmental chemistry in natural waters and wastewaters. In addition to the hydrated metal ions, for example, $Fe(H_2O)_6^{3+}$ and hydroxy species such as $FeOH(H_2O)_5^{2+}$ discussed in the preceding section, metals may exist in water reversibly bound to inorganic anions or to organic compounds as *metal complexes*. For example, a cyanide ion can bond to dissolved iron(II):

$$Fe(H_2O)_6^{2+} + CN^- \rightleftarrows FeCN(H_2O)_5^+ + H_2O \tag{3.53}$$

Additional cyanide ions may bond to the iron to form $Fe(CN)_2$, $Fe(CN)_3^-$, $Fe(CN)_4^{2-}$, $Fe(CN)_5^{3-}$, and $Fe(CN)_6^{4-}$, where the water molecules still bound to the iron(II) are omitted for simplicity. This phenomenon is called *complexation*; the species that binds with the metal ion, CN^- in the example above, is called a *ligand*, and the product in which the ligand is bound with the metal ion is a *complex*, *complex ion*, or *coordination compound*. A special case of complexation in which a ligand bonds in two or more places to a metal ion is called *chelation*. In addition to being present as metal complexes, metals may occur in water as *organometallic* compounds containing carbon-to-metal bonds. The solubilities, transport properties, and biological effects of such species are often vastly different from those of the metal ions themselves. Subsequent sections of this chapter consider metal species with an emphasis upon metal complexation, especially chelation, in which particularly strong metal complexes are formed.

In the example above, the cyanide ion is a *unidentate ligand*, which means that it possesses only one site that bonds to a metal ion. Complexes of unidentate ligands are of relatively little importance in solution in natural waters. Of considerably more importance are complexes with *chelating agents*. A chelating agent has more than one atom that may be bonded to a central metal ion at one time to form a ring structure. Thus, pyrophosphate ion, $P_2O_7^{4-}$, bonds to two sites on a calcium ion to form a chelate:

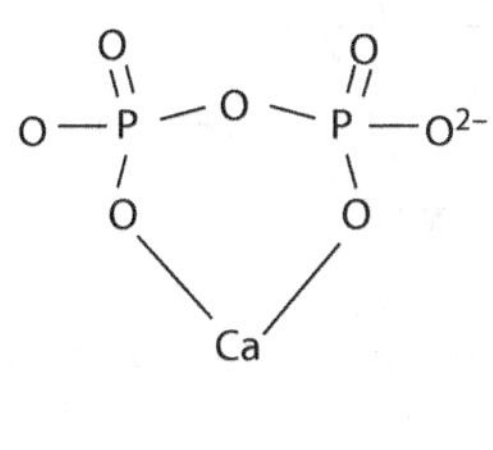

In general, since a chelating agent may bond to a metal ion in more than one place simultaneously (Figure 3.4), chelates are more stable than complexes with unidentate ligands. Stability tends to increase with the number of chelating sites available on the ligand. Structures of metal chelates take a number of different forms, all characterized by rings in various configurations. The structure of a tetrahedrally coordinated chelate of nitrilotriacetate ion is shown in Figure 3.4.

The ligands found in natural waters and wastewaters contain a variety of functional groups which can donate the electrons required to bond the ligand to a metal ion. Among the most common of these groups are:

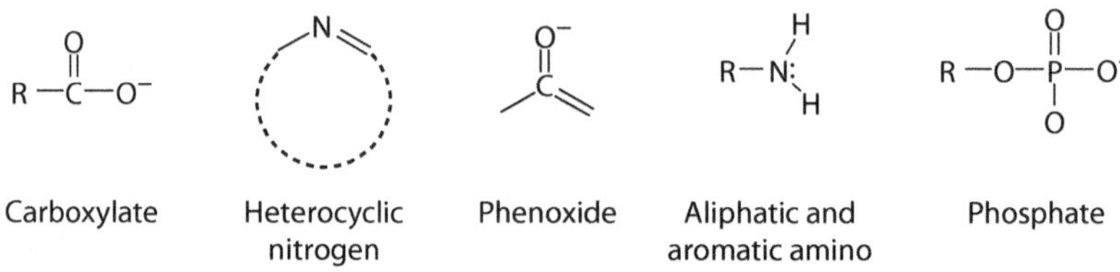

These ligands complex most metal ions found in unpolluted waters and biological systems (Mg^{2+}, Ca^{2+}, Mn^{2+}, Fe^{2+}, Fe^{3+}, Cu^{2+}, Zn^{2+}, VO^{2+}). They also bind to contaminant metal ions such as Co^{2+}, Ni^{2+}, Sr^{2+}, Cd^{2+}, and Ba^{2+}.

Complexation may have a number of effects, including reactions of both ligands and metals. Among the ligand reactions are oxidation–reduction, decarboxylation, hydrolysis, and biodegradation. Complexation may cause changes in oxidation state of the metal and may result in a metal becoming solubilized from an insoluble compound. The formation of insoluble complex compounds removes metal ions from solution. Complexation may strongly influence metals' adsorption, distribution, transport, and fate as well as biochemical effects including bioavailability, toxicity, and plant uptake.[3]

Complex compounds of metals such as iron (in hemoglobin) and magnesium (in chlorophyll) are vital to life processes. Naturally occurring chelating agents, such as

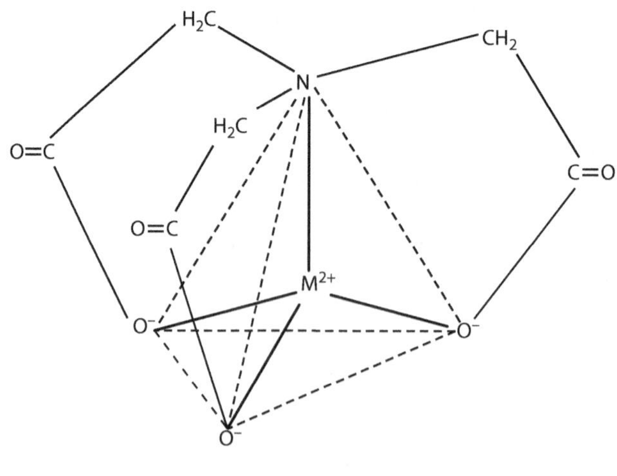

FIGURE 3.4 Nitrilotriacetate chelate of a divalent metal ion in a tetrahedral configuration.

humic substances and amino acids, are found in water and soil. The high concentration of chloride ion in seawater results in the formation of some chloro complexes. Synthetic chelating agents such as sodium tripolyphosphate, sodium ethylenediaminetetraacetate (EDTA), sodium nitrilotriacetate (NTA), and sodium citrate are produced in large quantities for use in metal-plating baths, industrial water treatment, detergent formulations, and food preparation. These compounds may enter aquatic systems through waste discharges.

3.6.1 Occurrence and Importance of Chelating Agents in Water

Chelating agents are common potential water pollutants. These substances can occur in sewage effluent and industrial wastewater such as metal plating wastewater. In addition to pollutant sources, there are natural sources of chelating agents. One such chelating agent is ethylenediaminedisuccinic acid, a metabolite of the soil actinomycete, *Amycolatopsis orientalis.* Unlike a number of common synthetic chelating agents, such as EDTA (see below), ethylenediaminedisuccinic acid is biodegradable and has been used as an extractant for phytoremediation (use of plants to extract pollutants from soil) of sites contaminated with heavy metals.

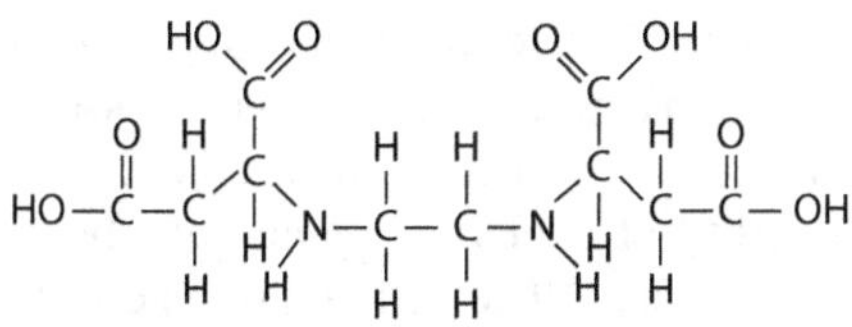

Ethylenediaminedisuccinic acid

The most important pollutant chelating agents are aminopolycarboxylates, of which the most common examples are nitrilotriacetate (NTA, Figure 3.4) and ethylenediaminetetraacetate (EDTA, structure illustrated at the beginning of Section 3.9). Both EDTA and NTA are common water pollutants.[4] Because of their strong bonding to metal ions, the aminopolycarboxylate chelating agents are almost always encountered in the bound form, which has a strong effect on their chemistry and environmental fate and transport. To the extent that EDTA is present in wastewater, it prevents some metals from binding to and settling out with biomass sludge in biological wastewater treatment processes. Therefore, EDTA chelates compose most of the copper, nickel, and zinc in wastewater effluents (most of these metals in excess of strong chelating agents present are incorporated into the sludge).

Chelation by EDTA has been shown to greatly increase the migration rates of radioactive ^{60}Co from pits and trenches used by the Oak Ridge National Laboratory in Oak Ridge, Tennessee, for disposal of intermediate-level radioactive waste. EDTA was used as a cleaning and solubilizing agent for the decontamination of hot cells, equipment, and reactor components. Analysis of water from sample wells in the disposal pits showed EDTA concentrations of 3.4×10^{-7} M. The presence of EDTA 12–15 years after its burial attests to its low rate of biodegradation. In addition to cobalt, EDTA strongly chelates radioactive plutonium and radioisotopes of Am^{3+}, Cm^{3+}, and Th^{4+}. Such chelates with negative charges are much less strongly sorbed by mineral matter and are vastly more mobile than the unchelated metal ions.

Contrary to the above findings, only very low concentrations of chelatable radioactive plutonium were observed in groundwater near the Idaho Chemical Processing Plant's low-level waste disposal well. No plutonium was observed in wells at any significant distance from the disposal well. The waste processing procedure used was designed to destroy any chelating agents in the waste prior to disposal, and no chelating agents were found in the water pumped from the test wells.

The fate of radionuclide metal chelates that have been discarded in soil is obviously important. If some mechanism exists to destroy the chelating agents, the radioactive metals will be much less mobile. Although EDTA is only poorly biodegradable, nitrilotriacetate, NTA, is degraded by the action of *Chlatobacter heintzii* bacteria. In addition to uncomplexed NTA, these bacteria have been shown to degrade NTA that is chelated to metals, including cobalt, iron, zinc, aluminum, copper, and nickel.

Complexing agents in wastewater are of concern primarily because of their ability to solubilize heavy metals from plumbing, and from deposits containing heavy metals. Complexation may increase the leaching of heavy metals from waste disposal sites and reduce the efficiency with which heavy metals are removed with sludge in conventional biological waste treatment. Removal of chelated iron is difficult with conventional municipal water treatment processes. Iron(III) and perhaps several other essential micronutrient metal ions are kept in solution by chelation in algal cultures. The availability of chelating agents may be a factor in determining growth of algae and plants in water.[5] The yellow-brown color of some natural waters is due to naturally occurring chelates of iron.

3.7 BONDING AND STRUCTURE OF METAL COMPLEXES

This section discusses some of the fundamentals of complexation in water. A complex consists of a central metal atom to which neutral or negatively charged ligands possessing electron–donor properties are bonded. The complex may be neutral or may have a positive or negative charge. The ligands are contained within the *coordination sphere* of the central metal atom. Depending upon the type of bonding involved, the ligands within the coordination sphere are held in a definite structural pattern. However, in solution, ligands of many complexes exchange rapidly between solution and the coordination sphere of the central metal ion.

The *coordination number* of a metal atom, or ion, is the number of ligand electron–donor groups that are bonded to it. The most common coordination numbers are 2, 4, and 6. Polynuclear complexes contain two or more metal atoms joined together through bridging ligands, frequently OH, as shown for iron(III) in Reaction 3.43.

3.7.1 Selectivity and Specificity in Chelation

Although chelating agents are never entirely specific for a particular metal ion, some complicated chelating agents of biological origin approach almost complete specificity for certain metal ions. One example of such a chelating agent is ferrichrome, synthesized by and extracted from fungi, which forms extremely stable chelates with iron(III). It has been observed that cyanobacteria of the *Anabaena* species secrete appreciable quantities of iron-selective hydroxamate chelating agents during periods of heavy algal bloom. These photosynthetic organisms readily take up iron chelated by hydroxamate-chelated iron, whereas some competing green algae, such as *Scenedesmus*, do not. Thus, the chelating agent serves a dual function of promoting the growth of certain cyanobacteria while suppressing the growth of competing species, allowing the cyanobacteria to exist as the predominant species. The production of chelating agents selective for iron(III) has been observed in cyanobacterial *Plectonema* and *Spirulina* as well as *Chlorella*, *Scenedesmus*, and *Porphyrium* algae.

3.8 CALCULATIONS OF SPECIES CONCENTRATIONS

The stability of complex ions in solution is expressed in terms of *formation constants*. These can be *stepwise formation constants* (K expressions) representing the bonding of individual ligands to a metal ion, or *overall formation constants* (β expressions) representing the binding of two or more ligands to a metal ion. These concepts are illustrated for complexes of zinc ion with ammonia by the following:

$$Zn^{2+} + NH_3 \rightleftarrows ZnNH_3^{2+} \tag{3.54}$$

$$K_1 = \frac{[ZnNH_3^{2+}]}{[Zn^{2+}][NH_3]} = 3.9 \times 10^2 \text{ (Stepwise formation constant)} \tag{3.55}$$

$$ZnNH_3^{2+} + NH_3 \rightleftarrows Zn(NH_3)_2^{2+} \tag{3.56}$$

$$K_2 = \frac{[Zn(NH_3)_2^{2+}]}{[ZnNH_3^{2+}][NH_3]} = 2.1 \times 10^2 \tag{3.57}$$

$$Zn^{2+} + 2NH_3 \rightleftarrows Zn(NH_3)_2^{2+} \tag{3.58}$$

$$\beta_2 = \frac{[Zn(NH_3)_2^{2+}]}{[Zn^{2+}][NH_3]^2} = K_1K_2 = 8.2 \times 10^4 \text{ (Overall formation constant)} \tag{3.59}$$

[For $Zn(NH_3)_3^{2+}$, $\beta_3 = K_1K_2K_3$ and for $Zn(NH_3)_4^{2+}$, $\beta_4 = K_1K_2K_3K_4$.]

The following sections show some calculations involving chelated metal ions in aquatic systems. Because of their complexity, the details of these calculations may be beyond the needs of some readers, who may choose to simply consider the results. In addition to the complexation itself, consideration must be given to competition of H^+ for ligands, competition among metal ions for ligands, competition among different ligands for metal ions, and precipitation of metal ions by various precipitants. Not the least of the problems involved in such calculations is the lack of accurately known values of equilibrium constants to be used under the conditions being considered, a factor that can yield questionable results from even the most elegant computerized calculations. Furthermore, kinetic factors are often quite important. Such calculations, however, can be quite useful to provide an overall view of aquatic systems in which complexation is important, and as general guidelines to determine areas in which more data should be obtained.

3.9 COMPLEXATION BY DEPROTONATED LIGANDS

Generally, calculation of complex species concentrations are complicated by competition between metal ions and H^+ for ligands. First, consider an example in which the ligand has lost all ionizable hydrogen. At pH values of 11 or above, EDTA is essentially all in the completely ionized tetranegative form, Y^{4-}, illustrated below:

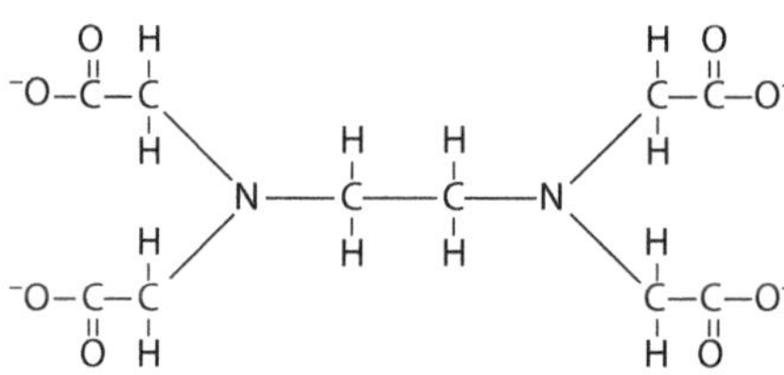

(Y^{4-})

Consider a wastewater with an alkaline pH of 11 containing copper(II) at a total level of 5.0 mg/L and excess uncomplexed EDTA at a level of 200 mg/L (expressed as the disodium salt, $Na_2H_2C_{10}H_{12}O_8N_2 \cdot 2H_2O$, formula mass 372). At this pH, uncomplexed EDTA is present as ionized Y^{4-}. The questions to be asked are: Will most of the copper be present as the EDTA complex? If so, what will be the equilibrium concentration of the hydrated copper(II) ion, Cu^{2+}? To answer the former question it is first necessary to calculate the molar concentration of uncomplexed excess EDTA, Y^{4-}. Since disodium EDTA with a formula mass of 372 is present at 200 mg/L (ppm), the total molar concentration of EDTA as Y^{4-} is 5.4×10^{-4} M. The formation constant K_1 of the copper–EDTA complex CuY^{2-} is

$$K_1 = \frac{[CuY^{2-}]}{[Cu^{2+}][Y^{4-}]} = 6.3 \times 10^{18} \tag{3.60}$$

The ratio of complexed to uncomplexed copper is

$$\frac{[CuY^{2-}]}{[Cu^{2+}]} = [Y^{4-}]K_1 = 5.4 \times 10^{-4} \times 6.3 \times 10^{18} = 3.3 \times 10^{15} \tag{3.61}$$

and therefore, essentially all of the copper is present as the complex ion. The molar concentration of total copper(II) in a solution containing 5.0 mg/L copper(II) is 7.9 × 10^{-5} M, which in this case is essentially all in the form of the EDTA complex. The very low concentration of uncomplexed, hydrated copper(II) ion is given by

$$[Cu^{2+}] = \frac{[CuY^{2-}]}{K_1[Y^{4-}]} = \frac{7.9\times10^{-5}}{6.3\times10^{18}\times5.4\times10^{-4}} = 2.3\times10^{-20}\,M \quad (3.62)$$

In the medium described, the concentration of hydrated Cu^{2+} ion is extremely low compared to total copper(II). Solution phenomena that depend upon the concentration of the hydrated Cu^{2+} ion (such as a physiological effect or an electrode response) would differ greatly in the medium described, compared to the effect observed if all of the copper at a level of 5.0 mg/L were present as Cu^{2+} in a more acidic solution and in the absence of complexing agent. The phenomenon of reducing the concentration of hydrated metal ion to very low values through the action of strong chelating agents is one of the most important effects of complexation in natural aquatic systems.

3.10 COMPLEXATION BY PROTONATED LIGANDS

Generally, complexing agents, particularly chelating compounds, are conjugate bases of Brönsted acids; for example, glycinate anion, $H_2NCH_2CO_2^-$, is the conjugate base of glycine, $^+H_3NCH_2CO_2^-$. Therefore, in many cases hydrogen ion competes with metal ions for a ligand, so that the strength of chelation depends upon pH. In the nearly neutral pH range usually encountered in natural waters, most organic ligands are present in a conjugated acid form.

In order to understand the competition between hydrogen ion and metal ion for a ligand, it is useful to know the distribution of ligand species as a function of pH. Consider nitrilotriacetic acid, commonly designated H_3T, as an example. The trisodium salt of this compound (NTA) is a strong chelating agent that can be used in metal plating, as a detergent phosphate substitute, and in other applications in which strong chelating capability is required. Biological processes are required for NTA degradation, and under some conditions it persists for long times in water. Given the ability of NTA to solubilize and transport heavy metal ions, this material is of potential environmental concern.

Nitrilotriacetic acid, H_3T, loses hydrogen ion in three steps to form the nitrilotriacetate anion, T^{3-}, the structural formula of which is

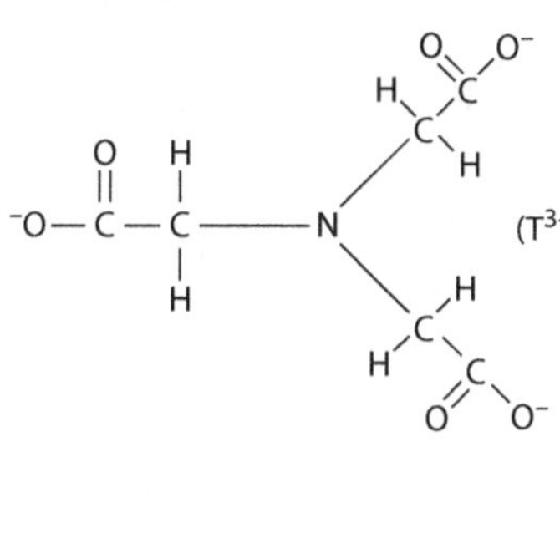

(T^{3-})

The T^{3-} species may coordinate through three $-CO_2^-$ groups and through the nitrogen atom, as shown in Figure 3.4. Note the similarity of the NTA structure to that of EDTA, discussed in Section 3.9. The stepwise ionization of H_3T is described by the following equilibria:

$$H_3T \rightleftarrows H^+ + H_2T^- \tag{3.63}$$

$$K_{a1} = \frac{[H^+][H_2T^-]}{[H_3T]} = 2.18 \times 10^{-2} \quad pK_{a1} = 1.66 \tag{3.64}$$

$$H_2T^- \rightleftarrows H^+ + HT^{2-} \tag{3.65}$$

$$K_{a2} = \frac{[H^+][HT^{2-}]}{[H_2T^-]} = 1.12 \times 10^{-3} \quad pK_{a2} = 2.95 \tag{3.66}$$

$$HT^{2-} \rightleftarrows H^+ + T^{3-} \tag{3.67}$$

$$K_{a3} = \frac{[H^+][T^{3-}]}{[HT^{2-}]} = 5.25 \times 10^{-11} \quad pK_{a3} = 10.28 \tag{3.68}$$

NTA may exist in solution as any one of the four species, H_3T, H_2T^-, HT^{2-}, or T^{3-}, depending upon the pH of the solution. As was shown for the $CO_2/HCO_3^-/CO_3^{2-}$ system in Section 3.3 and Figure 3.2, fractions of NTA species can be illustrated graphically by a diagram of the distribution-of-species with pH as a master (independent) variable. The key points used to plot such a diagram for NTA are given in Table 3.1, and the plot of fractions of species (α values) as a function of pH is shown in Figure 3.5. Examination of the plot shows that the complexing anion T^{3-} is the predominant species only at relatively high pH values, much higher than usually would be encountered in natural waters. The HT^{2-} species has an extremely wide range of predominance, however, spanning the entire normal pH range of ordinary fresh waters.

TABLE 3.1
Fractions of NTA Species at Selected pH Values

pH value	$\alpha_{H_3T^-}$	$\alpha_{H_3T^-}$	$\alpha_{HT^{2-}}$	$\alpha_{T^{3-}}$
pH below 1.00	1.00	0.00	0.00	0.00
$pH = pK_{a1}$	0.49	0.49	0.02	0.00
$pH = 1/2\ (pK_{a1} + pK_{a2})$	0.16	0.68	0.16	0.00
$pH = pK_{a2}$	0.02	0.49	0.49	0.00
$pH = 1/2\ (pK_{a1} + pK_{a3})$	0.00	0.00	1.00	0.00
$pH = pK_{a3}$	0.00	0.00	0.50	0.50
pH above 12	0.00	0.00	0.00	1.00

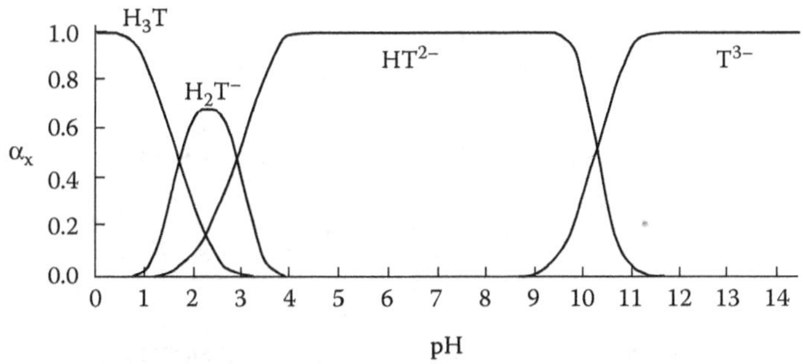

FIGURE 3.5 Plot of fraction of species α_x as a function of pH for NTA species in water.

3.11 SOLUBILIZATION OF LEAD ION FROM SOLIDS BY NTA

Solubilization of toxic heavy metals from solids by strong chelating agents such as NTA is a major concern regarding the introduction of these agents into aquatic ecosystems from sources such as detergents or electroplating wastes. Experimentation is required to determine whether this may be a problem, but calculations are helpful in predicting probable effects. The extent of solubilization of heavy metals depends upon a number of factors, including the stability of the metal chelates, the concentration of the chelating agent in water, pH, and the nature of the insoluble metal deposit. Several example calculations are given here.

Consider first the solubilization of lead from solid $Pb(OH)_2$ by NTA at pH 8.00. As illustrated in Figure 3.5, essentially all uncomplexed NTA is present as HT^{2-} ion at pH 8.00. Therefore, the solubilization reaction is

$$Pb(OH)_2(s) + HT^{2-} \rightleftarrows PbT^- + OH^- + H_2O \tag{3.69}$$

which may be obtained by adding the following reactions:

$$Pb(OH)_2(s) \rightleftarrows Pb^{2+} + 2OH^- \tag{3.70}$$

$$K_{sp} = [Pb^{2+}][OH^-]^2 = 1.61 \times 10^{-20} \tag{3.71}$$

$$HT^{2-} \rightleftarrows H^+ + T^{3-} \tag{3.67}$$

$$K_{a3} = \frac{[H^+][T^{3-}]}{[HT^{2-}]} = 5.25 \times 10^{-11} \tag{3.68}$$

$$Pb^{2+} + T^{3-} \rightleftarrows PbT^- \tag{3.72}$$

$$K_f = \frac{[PbT^-]}{[Pb^{2+}][T^{3-}]} = 2.45 \times 10^{11} \tag{3.73}$$

$$H^+ + OH^- \rightleftarrows H_2O \tag{3.74}$$

$$\frac{1}{K_W} = \frac{1}{[H^+][OH^+]} = \frac{1}{1.00 \times 10^{-14}} \tag{3.75}$$

$$Pb(OH)_2(s) + HT^{2-} \rightleftarrows PbT^- + OH^- + H_2O \tag{3.69}$$

$$K = \frac{[PbT^-][OH^-]}{[HT^{2-}]} = \frac{K_{sp}K_{a3}K_f}{K_W} = 2.07 \times 10^{-5} \tag{3.76}$$

Assume that a sample of water contains 25 mg/L of $N(CH_2CO_2Na)_3$, the trisodium NTA salt, formula mass is 257. The total concentration of both complexed and uncomplexed NTA is 9.7×10^{-5} mmol/mL. Assuming a system in which NTA at pH 8.00 is in equilibrium with solid $Pb(OH)_2$, the NTA may be primarily in the uncomplexed form, HT^{2-}, or in the lead complex, PbT^-. The predominant species may be determined by calculating the $[PbT^-]/[HT^{2-}]$ ratio from the expression for K, noting that at pH 8.00, $[OH^-] = 1.00 \times 10^{-6}$ M.

$$\frac{[PbT^-]}{[HT^{2-}]} = \frac{K}{[OH^-]} = \frac{2.07 \times 10^{-5}}{1.00 \times 10^{-6}} = 20.7 \tag{3.77}$$

Since $[PbT^-]/[HT^{2-}]$ is approximately 20 to 1, most of the NTA in solution is present as the lead chelate. The molar concentration of PbT^- is just slightly less than the 9.7×10^{-5} mmol/mL total NTA present. The atomic mass of lead is 207, so the concentration of lead in solution is approximately 20 mg/L. This reaction is pH-dependent such that the fraction of NTA chelated decreases with increasing pH.

3.11.1 Reaction of NTA with Metal Carbonate

Carbonates are common forms of heavy metal ion solids. Solid lead carbonate, $PbCO_3$, is stable within the pH region and alkalinity conditions often found in natural waters and wastewaters. An example similar to the one in the preceding section may be worked, assuming that equilibrium is established with $PbCO_3$ rather than with solid $Pb(OH)_2$. In this example it is assumed that 25 mg/L of trisodium NTA is in equilibrium with $PbCO_3$ at pH 7.00 and a calculation is made to determine whether the lead will be complexed appreciably by the NTA. The carbonate ion, CO_3^{2-}, reacts with H^+ to form HCO_3^-. As discussed in Section 3.3, the acid–base equilibrium reactions for the $CO_2/HCO_3^-/CO_3^{2-}$ system are

$$CO_2(aq) + H_2O \rightleftarrows H^+ + HCO_3^- \tag{3.9}$$

$$K'_{a1} = \frac{[H^+][HCO_3^-]}{[CO_2]} = 4.45 \times 10^{-7} \tag{3.10}$$

$$HCO_3^- \rightleftarrows H^+ + CO_3^{2-} \tag{3.11}$$

$$K'_{a2} = \frac{[H^+][CO_3^{2-}]}{[HCO_2^-]} = 4.69 \times 10^{-11} \tag{3.12}$$

where the acid dissociation constants of the carbonate species are designated with "′" to distinguish them from the acid dissociation constants of NTA. Figure 3.2 shows that within a pH range of about 7–10 the predominant carbonic species is HCO_3^-; therefore, the CO_3^{2-} released by the reaction of NTA with $PbCO_3$ will go into solution as HCO_3^-:

$$PbCO_3(s) + HT^{2-} \rightleftarrows PbT^- + HCO_3^- \tag{3.78}$$

This reaction and its equilibrium constant are obtained as follows:

$$PbCO_3(s) \rightleftarrows Pb^{2+} + CO_3^{2-} \tag{3.79}$$

$$K_{sp}(s) = [Pb^{2+}][CO_3^{2-}] = 1.48 \times 10^{-13} \tag{3.80}$$

$$Pb^{2+} + T^{3-} \rightleftarrows PbT^- \tag{3.72}$$

$$K_f = \frac{[PbT^-]}{[Pb^{2+}][T^{3-}]} = 2.45 \times 10^{11} \tag{3.73}$$

$$HT^{2-} \rightleftarrows H^+ + T^{3-} \tag{3.67}$$

$$K_{a3} = \frac{[H^+][T^{3-}]}{[HT^{2-}]} = 5.25 \times 10^{-11} \tag{3.68}$$

$$H^+ + CO_3^{2-} \rightleftarrows HCO_3^- \tag{3.81}$$

$$\frac{1}{K'_{a2}} = \frac{[HCO_3^-]}{[H^+][CO_3^{2-}]} = \frac{1}{4.69 \times 10^{-11}} \tag{3.82}$$

$$PbCO_3(s) + HT^{2-} \rightleftarrows PbT^- + HCO_3^- \tag{3.78}$$

$$K = \frac{[PbT^-][HCO_3^-]}{[HT^{2-}]} = \frac{K_{sp}K_{a3}K_f}{K'_{a2}} = 4.06 \times 10^{-2} \tag{3.83}$$

From the expression for K, Equation 3.83, it may be seen that the degree to which $PbCO_3$ is solubilized as PbT^- depends upon the concentration of HCO_3^-. Although this concentration will vary appreciably, the figure commonly used to describe natural waters is a bicarbonate ion concentration of 1.00×10^{-3}, as shown in Section 3.3. Using this value the following may be calculated:

$$\frac{[PbT^-]}{[HT^{2-}]} = \frac{K}{[HCO_3^-]} = \frac{4.06 \times 10^{-2}}{1.00 \times 10^{-3}} = 40.6 \tag{3.84}$$

Thus, under the given conditions, most of the NTA in equilibrium with solid $PbCO_3$ would be present as the lead complex. As in the previous example, at a trisodium NTA level of 25 mg/L, the concentration of soluble lead(II) would be approximately 20 mg/L. At relatively higher concentrations of HCO_3^-, the tendency to solubilize lead would be diminished, whereas at lower concentrations of HCO_3^-, NTA would be more effective in solubilizing lead.

3.11.2 Effect of Calcium Ion upon the Reaction of Chelating Agents with Slightly Soluble Salts

Chelatable calcium ion, Ca^{2+}, which is generally present in natural waters and wastewaters, competes for the chelating agent with a metal in a slightly soluble salt, such as $PbCO_3$. At pH 7.00, the reaction between calcium ion and NTA is

$$Ca^{2+} + HT^{2-} \rightleftarrows CaT^- + H^+ \tag{3.85}$$

described by the following equilibrium expression:

$$K' = \frac{[CaT^-][H^+]}{[Ca^{2+}][HT^{2-}]} = 1.48 \times 10^8 \times 5.25 \times 10^{-11} = 7.75 \times 10^{-3} \tag{3.86}$$

The value of K' is the product of the formation constant of CaT^- (1.48×10^8), and K_{a3} of NTA, 5.25×10^{-11}. The fraction of NTA bound as CaT^- depends upon the concentration of Ca^{2+} and the pH. Typically, $[Ca^{2+}]$ in water is 1.00×10^{-3} M. Assuming this value and pH 7.00, the ratio of NTA present in solution as the calcium complex to that present as HT^{2-} is:

$$\frac{[CaT^-]}{[HT^{2-}]} = \frac{[Ca^{2+}]}{[H^+]} - K' = \frac{1.00 \times 10^{-3}}{1.00 \times 10^{-7}} \times 7.75 \times 10^{-3} = 77.5 \tag{3.87}$$

Therefore, most of the NTA in equilibrium with 1.00×10^{-3} M Ca^{2+} would be present as the calcium complex, CaT^-, which would react with $PbCO_3$ as follows:

$$PbCO_3(s) + CaT^- + H^+ \rightleftarrows Ca^{2+} + HCO_3^- + PbT^- \tag{3.88}$$

$$K'' = \frac{[Ca^{2+}][HCO_3^-][PbT^-]}{[CaT^-][H^+]} \tag{3.89}$$

Reaction 3.88 may be obtained by subtracting Reaction 3.85 from Reaction 3.78, and its equilibrium constant may be obtained by dividing the equilibrium constant of Reaction 3.85 into that of Reaction 3.78:

$$PbCO_3(s) + HT^{2-} \rightleftarrows PbT^- + HCO_3^- \tag{3.78}$$

$$K = \frac{[\mathrm{PbT^-}][\mathrm{HCO_3^-}]}{[\mathrm{HT^{2-}}]} = \frac{K_{\mathrm{sp}}K_{\mathrm{a3}}K_{\mathrm{f}}}{K'_{\mathrm{a2}}} = 4.06\times10^{-2} \tag{3.83}$$

$$-(\mathrm{Ca^{2+} + HT^{2-} \rightleftarrows CaT^- + H^+}) \tag{3.85}$$

$$K' = \frac{[\mathrm{CaT^-}][\mathrm{H^+}]}{[\mathrm{Ca^{2+}}][\mathrm{HT^{2-}}]} = 1.48\times10^{8}\times5.25\times10^{-11} = 7.75\times10^{-3} \tag{3.86}$$

$$\mathrm{PbCO_3}(s) + \mathrm{CaT^-} + \mathrm{H^+} \rightleftarrows \mathrm{Ca^{2+}} + \mathrm{HCO_3^-} + \mathrm{PbT^-} \tag{3.88}$$

$$K'' = \frac{K}{K'} = \frac{4.06\times10^{-2}}{7.75\times10^{-3}} = 5.24 \tag{3.90}$$

Having obtained the value of K'', it is now possible to determine the distribution of NTA between PbT^- and CaT^-. Thus, for water containing NTA chelated to calcium at pH 7.00, a concentration of HCO_3^- of 1.00×10^{-3}, a concentration of Ca^{2+} of 1.00×10^{-3}, and in equilibrium with solid $PbCO_3$, the distribution of NTA between the lead complex and the calcium complex is:

$$\frac{[\mathrm{PbT^-}]}{[\mathrm{CaT^-}]} = \frac{[\mathrm{H^+}]K''}{[\mathrm{Ca^{2+}}][\mathrm{HCO_3^-}]} = 0.524$$

It may be seen that only about 1/3 of the NTA would be present as the lead chelate, whereas under the identical conditions, but in the absence of Ca^{2+}, approximately all of the NTA in equilibrium with solid $PbCO_3$ was chelated to NTA. Since the fraction of NTA present as the lead chelate is directly proportional to the solubilization of $PbCO_3$, differences in calcium concentration will affect the degree to which NTA solubilizes lead from lead carbonate.

3.12 POLYPHOSPHATES AND PHOSPHONATES IN WATER

Phosphorus occurs as many oxoanions, anionic forms in combination with oxygen. Some of these are strong complexing agents. Since about 1930, salts of polymeric phosphorus oxoanions have been used for water treatment, for water softening, and as detergent builders. When used for water treatment, polyphosphates "sequester" calcium ion in a soluble or suspended form. The effect is to reduce the equilibrium concentration of calcium ion and prevent the precipitation of calcium carbonate in installations such as water pipes and boilers. Furthermore, when water is softened properly with polyphosphates, calcium does not form precipitates with soaps or interact detrimentally with detergents.

The simplest form of phosphate is orthophosphate, PO_4^{3-}:

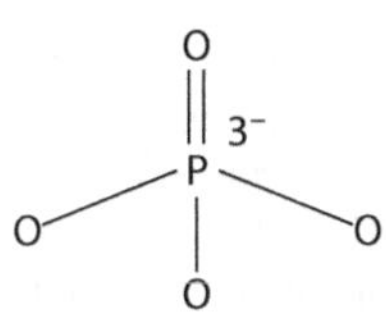

Orthophosphate ion has three sites for attachment of H^+. Orthophosphoric acid, H_3PO_4, has a pK_{a1} of 2.17, a pK_{a2} of 7.31, and a pK_{a3} of 12.36. Because the third hydrogen ion is so difficult to remove from orthophosphate, as evidenced by the very high value of pK_{a3}, very basic conditions are required for PO_4^{3-} to be present at significant levels in water. It is possible for orthophosphate in natural waters to originate from the hydrolysis of polymeric phosphate species.

3.12.1 Polyphosphates

Pyrophosphate ion, $P_2O_7^{4-}$, is the first of a series of unbranched chain polyphosphates produced by the condensation of orthophosphate:

$$2PO_4^{3-} + H_2O \rightleftarrows P_2O_7^{4-} + 2OH^- \tag{3.91}$$

A long series of linear polyphosphates may be formed, the second of which is triphosphate ion, $P_3O_{10}^{5-}$. These species consist of PO_4 tetrahedra with adjacent tetrahedra sharing a common oxygen atom at one corner. The structural formulas of the acidic forms, $H_4P_2O_7$ and $H_5P_3O_{10}$, are given in Figure 3.6.

It is easy to visualize the longer chains composing the higher linear polyphosphates. *Vitreous sodium phosphates* are mixtures consisting of linear phosphate chains from 4 to approximately 18 phosphorus atoms each. Those with intermediate chain lengths comprise the majority of the species present.

3.12.2 Hydrolysis of Polyphosphates

All of the polymeric phosphates hydrolyze to simpler products in water. The rate of hydrolysis depends upon a number of factors, including pH, and the ultimate product is always some form of orthophosphate. The simplest hydrolytic reaction of a polyphosphate is that of pyrophosphoric acid to produce orthophosphoric acid:

$$H_4P_2O_7 + H_2O \rightleftarrows 2H_3PO_4 \tag{3.92}$$

Researchers have found evidence that algae and other microorganisms catalyze the hydrolysis of polyphosphates. Even in the absence of biological activity, polyphos-

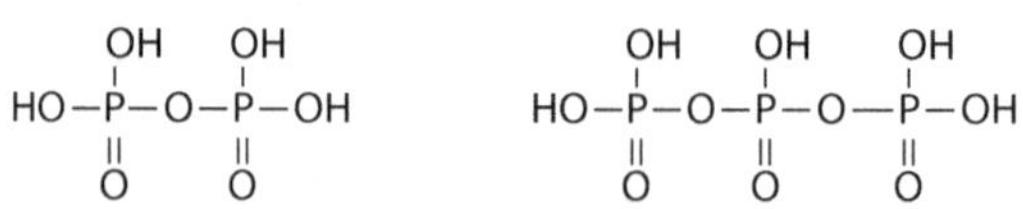

Pyrophosphoric (diphosphoric) acid

Triphosphoric acid

FIGURE 3.6 Structural formulas of diphosphoric acid and triphosphoric acid. For diphosphoric acid, pK_{a1} is quite small (relatively strong acid), whereas pK_{a2} is 2.64, pK_{a3} is 6.76, and pK_{a4} is 9.42. For triphosphoric acid, pK_{a1} and pK_{a2} values are low, pK_{a3} is 2.30, pK_{a4} is 6.50, and pK_{a5} is 9.24. These values reflect the relative ease of removing H^+ from OH groups on P atoms that have no other O^- groups compared to those that already have one O^- group.

phates hydrolyze chemically at a significant rate in water. Therefore, there is much less concern about the possibility of polyphosphates binding to heavy metal ions and transporting them than is the case with organic chelating agents such as NTA or EDTA, which must depend upon microbial degradation for their decomposition.

3.12.3 Complexation by Polyphosphates

In general, chain phosphates are good complexing agents and even form complexes with alkali-metal ions. Ring phosphates form much weaker complexes than do chain species. The different chelating abilities of chain and ring phosphates are due to structural hindrance of bonding by the ring polyphosphates.

3.12.4 Phosphonates

Phosphonate chelating agents are organic compounds structurally analogous to the aminopolycarboxylate chelating agents such as NTA and EDTA discussed earlier in this chapter. The structural formula of the undissociated acid form of a typical phosphonate chelating agent, nitrilotris(methylenephosphonic) acid, NTMP, is the following:

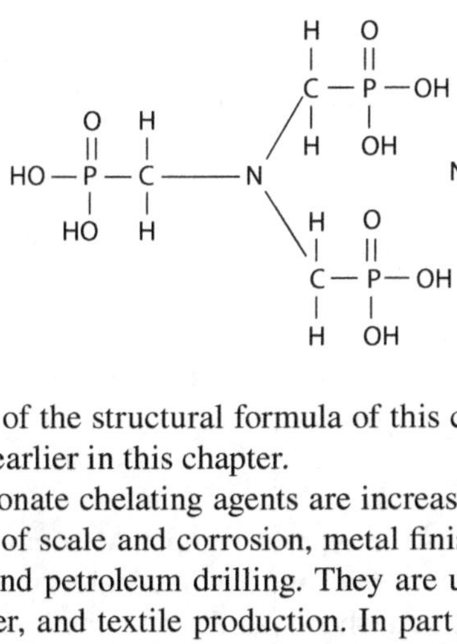

NTMP

Note the similarity of the structural formula of this compound to nitrilotriacetic acid, NTA, discussed earlier in this chapter.

The uses of phosphonate chelating agents are increasing steadily for applications that include inhibition of scale and corrosion, metal finishing, cleaning and laundry agents, ore recovery, and petroleum drilling. They are used in agricultural applications and in pulp, paper, and textile production. In part because of the difficulty in determining low levels of these substances in water, their environmental chemistry is not very well known. Although they are not very biodegradable, they interact strongly with surfaces and are removed with the sludge from biological waste treatment.

3.13 COMPLEXATION BY HUMIC SUBSTANCES

The most important class of complexing agents that occur naturally are the *humic substances*. These are degradation-resistant materials formed during the decomposition of vegetation that occur as deposits in soil, marsh sediments, peat, coal, lignite, or in almost any location where large quantities of vegetation have decayed. Humic substances abound in natural waters and soil.[6] They are commonly classified on the basis of solubility. If a material containing humic substances is extracted with strong base, and the resulting solution is acidified, the products are (1) a nonextractable plant residue called *humin*; (2) a material that precipitates from the acidified extract, called *humic acid*; and (3) an organic material that remains in the acidified solution, called *fulvic acid*. Because of their acid–base, sorptive, and complexing properties, both the soluble and insoluble humic substances have a strong effect upon the properties of water. In general, fulvic acid dissolves in water and exerts its effects as the soluble species. Humin and humic acid remain insoluble and affect water quality through exchange of species, such as cations or organic materials, with water.

Humic substances are high-molecular-mass, polyelectrolytic macromolecules. Molecular masses range from a few hundred for fulvic acid to tens of thousands for the humic acid and humin fractions. These substances contain a carbon skeleton with a high degree of aromatic character and with a large percentage of the molecular mass incorporated in functional groups, most of which contain oxygen. The elementary composition of most humic substances is within the following ranges: C, 45–55%; O, 30–45%, H, 3–6%; N, 1–5%; and S, 0–1%. The terms *humin, humic acid*, and *fulvic acid* do not refer to single compounds but to a wide range of compounds of generally similar origin with many properties in common. Humic substances have been known since before 1800, but their structural and chemical characteristics are still being explained.

Some feeling for the nature of humic substances may be obtained by considering the structure of a hypothetical molecule of fulvic acid shown below:

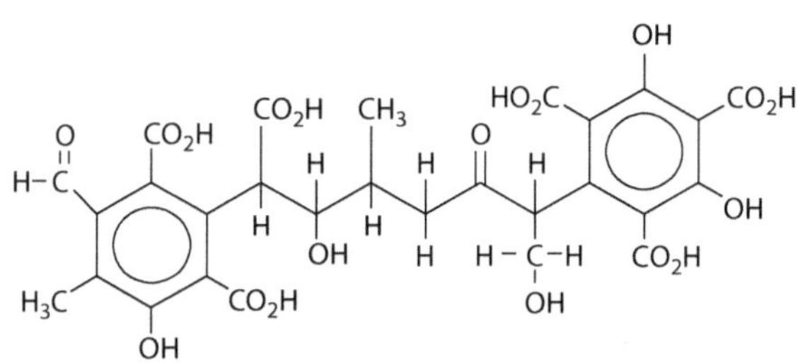

This structure is typical of the type of compound composing fulvic acid. The compound has a formula mass of 666, and its chemical formula may be represented as $C_{20}H_{15}(CO_2H)_6(OH)_5(CO)_2$. As shown in the hypothetical compound, the functional groups that may be present in fulvic acid are carboxyl, phenolic hydroxyl, alcoholic hydroxyl, and carbonyl. The functional groups vary with the particular acid sample. Approximate ranges in units of milliequivalents per gram of acid are: total acidity, 12–14; carboxyl, 8–9; phenolic hydroxyl, 3–6; alcoholic hydroxyl, 3–5; and carbonyl, 1–3. In addition, some methoxyl groups, $-OCH_3$, may be encountered at low levels.

The binding of metal ions by humic substances is one of the most important environmental qualities of humic substances. This binding can occur as chelation between a carboxyl group and a phenolic hydroxyl group, as chelation between two carboxyl groups, or as complexation with a carboxyl group (see Figure 3.7).

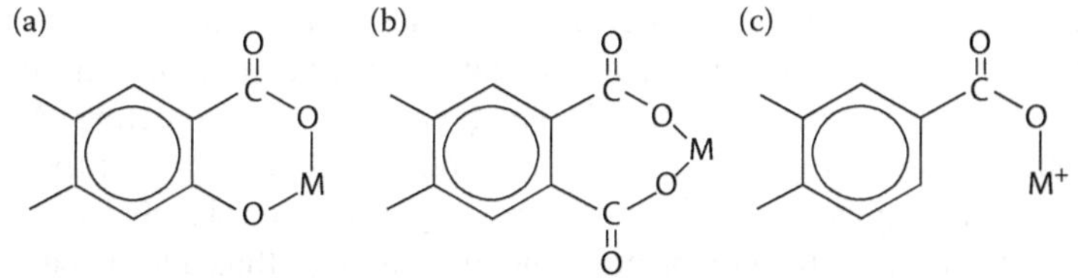

FIGURE 3.7 Binding of a metal ion, M^{2+}, by humic substances (a) by chelation between carboxyl and phenolic hydroxyl, (b) by chelation between two carboxyl groups, and (c) by complexation with a carboxyl group.

One of the more significant characteristics of humic substances is their ability to bind with metal cations. Iron and aluminum are very strongly bound to humic substances, whereas magnesium is rather weakly bound. Other common ions, such as Ni^{2+}, Pb^{2+}, Ca^{2+}, and Zn^{2+}, are intermediate in their binding to humic substances.

The role played by soluble fulvic acid complexes of metals in natural waters is not well known. They probably keep some of the biologically important transition-metal ions in solution, and are particularly involved in iron solubilization and transport. Yellow fulvic acid-type compounds called *Gelbstoffe* and frequently encountered along with soluble iron, are associated with color in water.

Insoluble humins and humic acids, effectively exchange cations with water and may accumulate large quantities of metals. Lignite coal, which is largely a humic acid material, tends to remove some metal ions from water.

Special attention has been given to humic substances since about 1970, following the discovery of *trihalomethanes* (THMs, such as chloroform and dibromochloromethane) in water supplies. It is now generally believed that these suspected carcinogens can be formed in the presence of humic substances during the disinfection of raw municipal drinking water by chlorination (see Chapter 8). The humic substances produce THMs by reaction with chlorine. The formation of THMs can be reduced by removing as much of the humic material as possible prior to chlorination.

3.14 COMPLEXATION AND REDOX PROCESSES

Complexation may have a strong effect upon oxidation–reduction equilibria by shifting reactions, such as that for the oxidation of lead,

$$\mathrm{Pb} \rightleftarrows \mathrm{Pb}^{2+} + 2\mathrm{e}^- \tag{3.93}$$

strongly to the right by binding to the product ion, thus cutting its concentration down to very low levels. Perhaps more important is the fact that upon oxidation,

$$\mathrm{M} + \frac{1}{2}\mathrm{O}_2 \rightleftarrows \mathrm{MO} \tag{3.94}$$

many metals form self-protective coatings of oxides, carbonates, or other insoluble species which prevent further chemical reaction. Copper and aluminum roofing and structural iron are examples of materials which are thus self-protecting. A chelating agent in contact with such metals can result in continual dissolution of the protective coating so that the exposed metal corrodes readily. For example, chelating agents in wastewater may increase the corrosion of metal plumbing, thus adding heavy metals to effluents. Solutions of chelating agents employed to clean metal surfaces in metal plating operations have a similar effect.

LITERATURE CITED

1. Graham, M. C. and J. G. Farmer, Chemistry of freshwaters, *Principles of Environmental Chemistry*, pp. 80–169, Royal Society of Chemistry, Cambridge, UK, 2007.
2. Anon, Carbon conveyer, *Nature Geoscience*, 2, 1, 2009.
3. Nowack, B. and J. M. VanBriesen, Eds., *Biogeochemistry of Chelating Agents*, American Chemical Society, Washington, DC, 2005.
4. Yuan, Z. and J. M. VanBriesen, The formation of intermediates in EDTA and NTA biodegradation, *Environmental Engineering Science*, 23, 533–544, 2006.
5. Bell, P. R. and I. Elmetri, Some chemical factors regulating the growth of *Lyngbya majuscula* in Moreton Bay, Australia: Importance of sewage discharges, *Hydrobiologia*, 592, 359–371, 2007.
6. Van Zomeren, A., A. Costa, J. P. Pinheiro, and R. N. J. Comans, Proton binding properties of humic substances originating from natural and contaminated materials, *Environmental Science & Technology*, 43, 1393–1399, 2009.

SUPPLEMENTARY REFERENCES

Benjamin, M. M., *Water Chemistry*, McGraw-Hill, New York, 2002.

Bianchi, T., *Biogeochemistry of Estuaries*, Oxford University Press, Oxford, UK, 2007.

Brightwell, C., *Marine Chemistry*, T.F.H. Publications, Neptune City, NJ, 2007.

Buffle, J. D., *Complexation Reactions in Aquatic Systems: An Analytical Approach*, Ellis Horwood, Chichester, UK, 1988.

Butler, J. N., *Ionic Equilibrium: Solubility and pH Calculations*, John Wiley & Sons, New York, 1998.

Drever, J. I., *The Geochemistry of Natural Waters: Surface and Groundwater Environments*, 3rd ed., Prentice-Hall, Upper Saddle River, NJ, 1997.

Eby, G. N., *Principles of Environmental Geochemistry*, Thomson-Brooks/Cole, Pacific Grove, GA, 2004.

Essington, M. E., *Soil and Water Chemistry: An Integrative Approach*, Taylor & Francis/CRC Press, 2004.

Faure, G., *Principles and Applications of Geochemistry: A Comprehensive Textbook for Geology Students*, 2nd ed., Prentice-Hall, Upper Saddle River, NJ, 1998.

Ghabbour, E. A. and G. Davies, Eds., *Humic Substances: Nature's Most Versatile Materials*, Garland Publishing, New York, 2003.

Hem, J. D., *Study and Interpretation of the Chemical Characteristics of Natural Water*, 2nd ed., U.S. Geological Survey Paper **1473**, Washington, DC, 1970.

Hessen, D. O. and L. J. Tranvik, Eds., *Aquatic Humic Substances: Ecology and Biogeochemistry*, Springer Verlag, Berlin, 1998.

Holland, H. D. and K. K. Turekian, Eds., *Treatise on Geochemistry*, Elsevier/Pergamon, Amsterdam, 2004.

Howard, A. G., *Aquatic Environmental Chemistry*, Oxford University Press, Oxford, UK, 1998.
Jensen, J. N., *A Problem-Solving Approach to Aquatic Chemistry*, Wiley, New York, 2003.
Jones, J. B. and P. J. Mulholland, Eds., *Streams and Ground Waters*, Academic Press, San Diego, CA, 2000.
Langmuir, D., *Aqueous Environmental Geochemistry*, Prentice-Hall, Upper Saddle River, NJ, 1997.
Morel, F. M. M. and J. G. Hering, *Principles and Applications of Aquatic Chemistry*, Wiley-Interscience, New York, 1993.
Polevoy, S., *Water Science and Engineering*, Krieger Publishing, Melbourne, FL, 2003.
Spellman, F. R., *The Science of Water: Concepts and Applications*, 2nd ed., Taylor & Francis/CRC Press, Boca Raton, FL, 2008.
Steinberg, C., *Ecology of Humic Substances in Freshwaters: Determinants from Geochemistry to Ecological Niches*, Springer, Berlin, 2003.
Stober, I. and K. Bucher, Eds., *Water–Rock Interaction*, Kluwer Academic Publishers, Hingham, MA, 2002.
Stumm, W. and J. J. Morgan, *Aquatic Chemistry: Chemical Equilibria and Rates in Natural Waters*, 3rd ed., John Wiley and Sons, Inc., New York, 1995.
Stumm, W., *Chemistry of the Solid–Water Interface: Processes at the Mineral–Water and Particle–Water Interface in Natural Systems*, John Wiley and Sons, Inc., New York, 1992.
Trimble, S. W., *Encyclopedia of Water Science*, 2nd ed., Taylor & Francis, CRC Press, 2008.
Welch, E. B. and Jean Jacoby, *Pollutant Effects in Fresh Waters: Applied Limnology*, Taylor & Francis, London, 2007.
Weiner, E. R., *Applications of Environmental Aquatic Chemistry: A Practical Guide*, 2nd ed., Taylor & Francis/CRC Press, Boca Raton, FL, 2008.

QUESTIONS AND PROBLEMS

The use of Internet resources is assumed in answering any of the questions. These would include such things as constants and conversion factors as well as additional information needed to complete an answer.

1. Alkalinity is determined by titration with standard acid. The alkalinity is often expressed as mg/L of $CaCO_3$. If V_p mL of acid of normality N are required to titrate V_s mL of sample to the phenolphthalein endpoint, what is the formula for the phenolphthalein alkalinity as mg/L of $CaCO_3$?
2. Exactly 100 pounds of cane sugar (dextrose), $C_{12}H_{22}O_{11}$, were accidentally discharged into a small stream saturated with oxygen from the air at 25°C. How many liters of this water could be contaminated to the extent of removing all the dissolved oxygen by biodegradation?
3. Water with an alkalinity of 2.00×10^{-3} equivalents/L has a pH of 7.00. Calculate $[CO_2]$, $[HCO_3^-]$, $[CO_3^{2-}]$, and $[OH^-]$.
4. Through the photosynthetic activity of algae, the pH of the water in Problem 3 was changed to 10.00. Calculate all the preceding concentrations and the weight of biomass, $\{CH_2O\}$, produced. Assume no input of atmospheric CO_2.
5. Calcium chloride is quite soluble, whereas the solubility product of calcium fluoride, CaF_2, is only 3.9×10^{-11}. A waste stream of 1.00×10^{-3} M HCl

is injected into a formation of limestone, $CaCO_3$, where it comes into equilibrium. Give the chemical reaction that occurs and calculate the hardness and alkalinity of the water at equilibrium. Do the same for a waste stream of 1.00×10^{-3} M HF.

6. For a solution having 1.00×10^{-3} equivalents/L total alkalinity (contributions from HCO_3^-, CO_3^{2-}, and OH^-) at $[H^+] = 4.69 \times 10^{-11}$, what is the percentage contribution to alkalinity from CO_3^{2-}?
7. A wastewater disposal well for carrying various wastes at different times is drilled into a formation of limestone ($CaCO_3$), and the wastewater has time to come to complete equilibrium with the calcium carbonate before leaving the formation through an underground aquifer. Of the following components in the wastewater, the one that would not cause an increase in alkalinity due either to the component itself or to its reaction with limestone, is (a) NaOH, (b) CO_2, (c) HF, (d) HCl, (e) all of the preceding would cause an increase in alkalinity.
8. Calculate the ratio $[PbT^-]/[HT^{2-}]$ for NTA in equilibrium with $PbCO_3$ in a medium having $[HCO_3^-] = 3.00 \times 10^{-3}$ M.
9. If the medium in Problem 8 contained excess calcium such that the concentration of uncomplexed calcium, $[Ca^{2+}]$, were 5.00×10^{-3} M, what would be the ratio $[PbT^-]/[CaT^-]$ at pH 7?
10. A wastewater stream containing 1.00×10^{-3} M disodium NTA, Na_2HT, as the only solute is injected into a limestone ($CaCO_3$) formation through a waste disposal well. After going through this aquifer for some distance and reaching equilibrium, the water is sampled through a sampling well. What is the reaction between NTA species and $CaCO_3$? What is the equilibrium constant for the reaction? What are the equilibrium concentrations of CaT^-, HCO_3^-, and HT^{2-}? (The appropriate constants may be looked up in this chapter.)
11. If the wastewater stream in Problem 10 were 0.100 M in NTA and contained other solutes that exerted a buffering action such that the final pH were 9.00, what would be the equilibrium value of HT^{2-} concentration in mol/L?
12. Exactly 1.00×10^{-3} mol of $CaCl_2$, 0.100 mol of NaOH, and 0.100 mol of Na_3T were mixed and diluted to 1.00 L. What was the concentration of Ca^{2+} in the resulting mixture?
13. How does chelation influence corrosion?
14. The following ligand has more than one site for binding to a metal ion. How many such sites does it have?

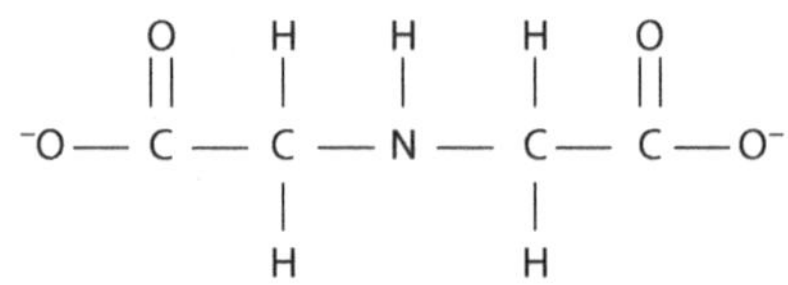

15. If a solution containing initially 25 mg/L trisodium NTA is allowed to come to equilibrium with solid $PbCO_3$ at pH 8.50 in a medium that contains 1.76×10^{-3} M HCO_3^- at equilibrium, what is the value of the ratio of the concentration of NTA bound with lead to the concentration of unbound NTA, $[PbT^-]/[HT^{2-}]$?
16. After a low concentration of NTA has equilibrated with $PbCO_3$ at pH 7.00 in a medium having $[HCO_3^-] = 7.50 \times 10^{-4}$ M, what is the ratio of [PbT]/ $[HT^{2-}]$/?
17. What detrimental effect may dissolved chelating agents have upon conventional biological waste treatment?
18. Why is chelating agent usually added to artificial algal growth media?
19. What common complex compound of magnesium is essential to certain life processes?
20. What is always the ultimate product of polyphosphate hydrolysis?
21. A solution containing initially 1.00×10^{-5} M CaT^- is brought to equilibrium with solid $PbCO_3$. At equilibrium, pH = 7.00, $[Ca^{2+}] = 1.50 \times 10^{-3}$ M, and $[HCO_3^-] = 1.10 \times 10^{-3}$ M. At equilibrium, what is the fraction of total NTA in solution as PbT^-?
22. What is the fraction of NTA present after HT^{2-} has been brought to equilibrium with solid $PbCO_3$ at pH 7.00 in a medium in which $[HCO_3^-] = 1.25 \times 10^{-3}$ M.
23. Assuming levels of atmospheric CO_2 are 390 ppm CO_2, what is the pH of rainwater due to the presence of carbon dioxide? Some estimates are for atmospheric carbon dioxide levels to double in the future. What would be the pH of rainwater if this happens?
24. Assume a sewage treatment plant processing 1 million liters of wastewater per day containing 200 mg/L of degradable biomass, $\{CH_2O\}$. Calculate the volume of dry air at 25°C that must be pumped into the wastewater per day to provide the oxygen required to degrade the biomass (Reaction 3.1).
25. Anaerobic bacteria growing in a lake sediment produced equal molar amounts of carbon dioxide and carbon monoxide according to the biochemical reaction $2\{CH_2O\} \rightarrow CO_2 + CH_4$, so that the water in the lake was saturated with both CO_2 gas and CH_4 gas. In units of $mol \times L^{-1} \times atm^{-1}$ the Henry's law constant for CO_2 is 3.38×10^{-2} and that of CH_4 has a value of 1.34×10^{-3}. At the depth at which the gas was being evolved, the total pressure was 1.10 atm and the temperature was 25°C, so the vapor pressure of water was 0.0313 atm. Calculate the concentrations of dissolved CO_2 and dissolved CH_4.

4 Oxidation–Reduction in Aquatic Chemistry

4.1 THE SIGNIFICANCE OF OXIDATION–REDUCTION

Oxidation–reduction (*redox*) reactions are those involving changes of oxidation states of reactants. Such reactions are easiest to visualize as the transfer of electrons from one species to another. For example, soluble cadmium ion, Cd^{2+}, is removed from wastewater by reaction with metallic iron. The overall reaction is

$$Cd^{2+} + Fe \rightarrow Cd + Fe^{2+} \tag{4.1}$$

This reaction is the sum of two *half-reactions*, a reduction half-reaction in which cadmium ion accepts two electrons and is reduced,

$$Cd^{2+} + 2e^{-} \rightarrow Cd \tag{4.2}$$

and an oxidation half-reaction in which elemental iron is oxidized:

$$Fe \rightarrow Fe^{2+} + 2e^{-} \tag{4.3}$$

When these two half-reactions are added algebraically, the electrons cancel on both sides and the result is the overall reaction given in Equation 4.1.

Oxidation–reduction phenomena are highly significant in the environmental chemistry of natural waters and wastewaters. In a lake, for example, the reduction of oxygen (O_2) by organic matter (represented by $\{CH_2O\}$),

$$\{CH_2O\} + O_2 \rightarrow CO_2 + H_2O \tag{4.4}$$

results in oxygen depletion, which can kill fish. The rate at which sewage is oxidized is crucial to the operation of a waste treatment plant. Reduction of insoluble iron(III) to soluble iron(II),

$$Fe(OH)_3(s) + 3H^{+} + e^{-} \rightarrow Fe^{2+} + 3H_2O \tag{4.5}$$

in a reservoir contaminates the water with dissolved iron, which is hard to remove in the water treatment plant. Oxidation of NH_4^+ to NO_3^- in water

$$NH_4^+ + 2O_2 \rightarrow NO_3^- + 2H^{+} + H_2O \tag{4.6}$$

converts ammonium nitrogen to nitrate, a form more assimilable by algae in the water. Many other examples can be cited of the ways in which the types, rates, and equilibria of redox reactions largely determine the nature of important solute species in water.

This chapter discusses redox processes and equilibria in water. In so doing, it emphasizes the concept of p*E*, analogous to pH and defined as the negative log of electron activity. Low p*E* values are indicative of reducing conditions; high p*E* values reflect oxidizing conditions.

Two important points should be stressed regarding redox reactions in natural waters and wastewaters. First, as is discussed in Chapter 6, "Aquatic Microbial Biochemistry," many of the most important redox reactions are catalyzed by microorganisms. Bacteria are the catalysts by which molecular oxygen reacts with organic matter, iron(III) is reduced to iron(II), and ammonia is oxidized to nitrate ion.

The second important point regarding redox reactions in the hydrosphere is their close relationship to acid–base reactions. Whereas the activity of the H^+ ion is used to express the extent to which water is acidic or basic, the activity of the electron, e^-, is used to express the degree to which an aquatic medium is oxidizing or reducing. Water with a high hydrogen-ion activity, such as runoff from "acid rain," is *acidic*. By analogy, water with a high *electron* activity, such as that in the anoxic digester of a sewage treatment plant, is said to be *reducing*. Water with a low H^+ ion activity (high concentration of OH^-)—such as landfill leachate contaminated with waste sodium hydroxide—is *basic*, whereas water with a low electron activity—highly chlorinated water, for example—is said to be *oxidizing*. Actually, neither free electrons nor free H^+ ions as such are found dissolved in aquatic solution; they are always strongly associated with solvent or solute species. However, the concept of electron activity, like that of hydrogen-ion activity, remains a very useful one to the aquatic chemist.

Many species in water undergo exchange of both electrons and H^+ ions. For example, acid mine water contains the hydrated iron(III) ion, $Fe(H_2O)_6^{3+}$, which readily loses H^+ ion

$$Fe(H_2O)_6^{3+} \rightleftarrows Fe(H_2O)_5OH^{2+} + H^+ \tag{4.7}$$

to contribute acidity to the medium. The same ion accepts an electron

$$Fe(H_2O)_6^{3+} + e^- \rightleftarrows Fe(H_2O)_6^{2+} \tag{4.8}$$

to give iron(II).

Generally, the transfer of electrons in a redox reaction is accompanied by H^+ ion transfer, and there is a close relationship between redox and acid–base processes. For example, if iron(II) loses an electron at pH 7, three hydrogen ions are also lost to form highly insoluble iron(II) hydroxide,

$$Fe(H_2O)_6^{2+} \rightleftarrows e^- + Fe(OH)_3(s) + 3H_2O + 3H^+ \tag{4.9}$$

an insoluble, gelatinous solid.

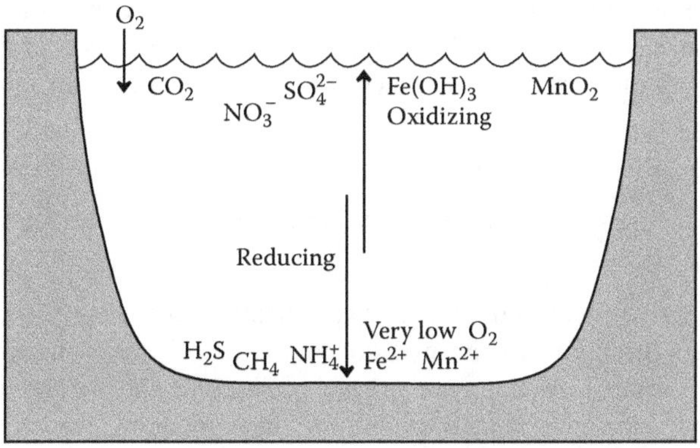

FIGURE 4.1 Predominance of various chemical species in a stratified body of water that has a high oxygen concentration (oxidizing, high p*E*) near the surface and a low oxygen concentration (reducing, low p*E*) near the bottom. (From Manahan, S., *Environmental Chemistry*, 9th ed., Taylor & Francis, Boca Raton, 2010. With permission.)

The stratified body of water shown in Figure 4.1 can be used to illustrate redox phenomena and relationships in an aquatic system. The anoxic sediment layer is so reducing that carbon can be reduced to its lowest possible oxidation state, –4 in CH_4. If the lake becomes anoxic, the hypolimnion may contain elements in their reduced states: NH_4^+ for nitrogen, H_2S for sulfur, and soluble $Fe(H_2O)_6^{2+}$ for iron. Saturation with atmospheric oxygen makes the surface layer a relatively oxidizing medium. If allowed to reach thermodynamic equilibrium, it is characterized by the more oxidized forms of the elements present: CO_2 for carbon, NO_3^- for nitrogen, iron as insoluble $Fe(OH)_3$, and sulfur as SO_4^{2-}. Substantial changes in the distribution of chemical species in water resulting from redox reactions are vitally important to aquatic organisms and have a tremendous influence on water quality.

It should be pointed out that the systems presented in this chapter are assumed to be at equilibrium, a state almost never achieved in any real natural water or wastewater system. Most real aquatic systems are dynamic systems that may approach a steady state, rather than true equilibrium. Nevertheless, the picture of a system at equilibrium is very useful in visualizing trends in natural water and wastewater systems, yet the model is still simple enough to comprehend. It is important to realize the limitations of such a model, however, especially in trying to make measurements of the redox status of water.

4.2 ELECTRON AND REDOX REACTIONS

In order to explain redox processes in natural waters, it is necessary to have an understanding of redox reactions. In a formal sense such reactions can be viewed as the transfer of electrons between species. This section considers such reactions in a simple system. All redox reactions involve changes in the oxidation states of some of the species that take part in the reaction. Consider, for example, a solution containing iron(II) and iron(III) that is sufficiently acidic to prevent precipitation of solid

$Fe(OH)_3$; such a medium might be acid mine water or a steel pickling liquor waste. Suppose that the solution is treated with elemental hydrogen gas over a suitable catalyst to bring about the reduction of iron(III) to iron(II). The overall reaction can be represented as

$$2Fe^{3+} + H_2 \rightleftarrows 2Fe^{2+} + 2H^+ \tag{4.10}$$

The reaction is written with a double arrow, indicating that it is *reversible* and could proceed in either direction; for normal concentrations of reaction participants, this reaction goes to the right. As the reaction goes to the right, the hydrogen is *oxidized* as it changes from an *oxidation state* (number) of 0 in elemental H_2 to a higher oxidation number of +1 in H^+. The oxidation state of iron goes from +3 in Fe^{3+} to +2 in Fe^{2+}; the oxidation number of iron decreases, which means that it is *reduced.*

All redox reactions such as this one can be broken down into a reduction *half-reaction*, in this case

$$2Fe^{3+} + 2e^- \rightleftarrows 2Fe^{2+} \tag{4.11}$$

(for one electron, $Fe^{3+} + e^- \rightleftarrows Fe^{3+}$), and an oxidation half-reaction, in this case

$$H_2 \rightleftarrows 2H^+ + 2e^- \tag{4.12}$$

Note that adding these two half-reactions together gives the overall reaction. *The addition of an oxidation half-reaction and a reduction half-reaction, each expressed for the same number of electrons so that the electrons cancel on both sides of the arrows, gives a whole redox reaction.*

The equilibrium of a redox reaction, that is, the degree to which the reaction as written tends to lie to the right or left, can be deduced from information about its constituent half-reactions. To visualize this, assume that the two half-reactions can be separated into two half-cells of an electrochemical cell as shown for Reaction 4.10 in Figure 4.2.

If the initial activities of H^+, Fe^{2+}, and Fe^{3+} were of the order of 1 (concentrations of 1 M), and if the pressure of H_2 were 1 atm, H_2 would be oxidized to H^+ in the left half-cell, Fe^{3+} would be reduced to Fe^{2+} in the right half-cell, and ions would migrate through the salt bridge to maintain electroneutrality in both half-cells. The net reaction occurring would be Reaction 4.10.

If a voltmeter were inserted in the circuit between the two electrodes, no significant current could flow and the two half-reactions could not take place. However, the voltage registered by the voltmeter would be a measure of the relative tendencies of the two half-reactions to occur. In the left half-cell the oxidation half-reaction

$$H_2 \rightleftarrows 2H^+ + 2e^- \tag{4.12}$$

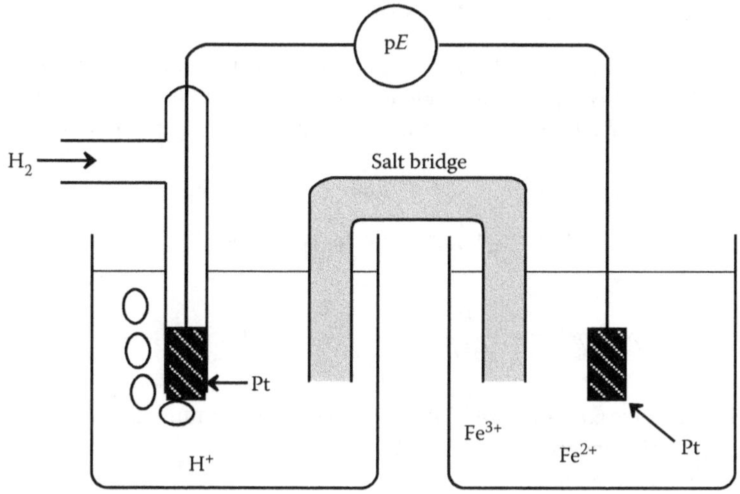

FIGURE 4.2 Electrochemical cell in which the reaction $2Fe^{3+} + 2H^+ \rightleftarrows 2Fe^{2+} + H^+$ can be carried out in two half-cells. (From Manahan, S., *Environmental Chemistry*, 9th ed., Taylor & Francis, Boca Raton, 2010. With permission.)

will tend to go to the right, releasing electrons to the platinum electrode in the half-cell and giving that electrode a relatively negative (–) potential. In the right half-cell, the reduction half-reaction

$$2Fe^{3+} + e^- \rightleftarrows Fe^{2+} \tag{4.11}$$

will tend to go to the right, taking electrons from the platinum electrode in the half-cell and giving that electrode a relatively positive (+) potential. The difference in these potentials is a measure of the "driving force" of the overall reaction. If each of the reaction participants were at unit activity, the potential difference would be 0.77 V.

The left electrode shown in Figure 4.2 is the standard electrode against which all other electrode potentials are compared. It is called the *standard hydrogen electrode* (*SHE*). This electrode has been assigned a potential value of exactly 0 V by convention, and its half-reaction is written as the following:

$$2H^+ + 2e^- \rightleftarrows H_2, \quad E^0 = 0.00\,V \tag{4.13}$$

The measured potential of the right-hand electrode in Figure 4.2 versus the SHE is called the *electrode potential*, *E*. If the Fe^{2+} and Fe^{3+} ions are both at unit activity, the potential is the *standard electrode potential* (according to IUPAC convention, *the standard reduction potential*), E^0. The standard electrode potential for the Fe^{3+}/Fe^{2+} couple is 0.77 V expressed conventionally as follows:

$$Fe^{3+} + e^- \rightleftarrows Fe^{2+}, \quad E^0 = +0.77\,V \tag{4.14}$$

4.3 ELECTRON ACTIVITY AND p*E*

In this book, for the most part, p*E* and pE^0 are used instead of *E* and E^0 to more clearly illustrate redox equilibria in aquatic systems over many orders of magnitude of electron activity in a manner analogous to pH. Numerically, p*E* and pE^0 are simply the following:

$$pE = \frac{E}{2.303RT/F} = \frac{E}{0.0591} \quad \text{(at 25°C)} \tag{4.15}$$

$$pE^0 = \frac{E^0}{2.303RT/F} = \frac{E^0}{0.0591} \quad \text{(at 25°C)} \tag{4.16}$$

where *R* is the molar gas constant, *T* is the absolute temperature, and *F* is the Faraday constant. The "p*E* concept" is explained below.

Just as pH is defined as

$$\text{pH} = -\log(a_{\text{H}}^{+}) \tag{4.17}$$

where a_{H}^{+} is the activity of hydrogen ion in solution, p*E* is defined as

$$pE = -\log(a_{e^-}) \tag{4.18}$$

where a_{e^-} is the activity of the electron in solution. Since hydrogen-ion concentration may vary over many orders of magnitude, pH is a convenient way of expressing a_{H}^{+} in terms of manageable numbers. Similarly, electron activities in water may vary over more than 20 orders of magnitude so that it is convenient to express a_{e^-} as p*E*.

Values of p*E* are defined in terms of the following half-reaction for which pE^0 is defined as exactly zero:*

$$2\text{H}^+(aq) + 2e^- \rightleftarrows \text{H}_2(g), \quad E^0 = +0.00\,\text{V}, \quad pE^0 = 0.00 \tag{4.19}$$

It is relatively easy to visualize the activities of ions in terms of concentration, whereas it is harder to visualize the activity of the electron, and therefore p*E*, in similar terms. For example, at 25°C in pure water, a medium of zero ionic strength, the hydrogen-ion concentration is 1.0×10^{-7} M, the hydrogen-ion *activity* is 1.0×10^{-7}, and the pH is 7.0. The electron activity, however, must be defined in terms of Equation 4.19. When H^+ (*aq*) at unit activity is in equilibrium with hydrogen gas at 1 atm pressure (and likewise at unit activity), the activity of the electron in the medium is exactly 1.00 and the

---

* Thermodynamically, the free-energy change for this reaction is defined as exactly zero when all reaction participants are at unit activity. For ionic solutes, activity—the effective concentration in a sense—approaches concentration at low concentrations and low ionic strengths. The activity of a gas is equal to its partial pressure. Furthermore, the free energy, *G*, decreases for spontaneous processes occurring at constant temperature and pressure. Processes for which the free-energy change, Δ*G*, is zero have no tendency toward spontaneous change and are in a state of equilibrium. Reaction 4.13 is the one upon which free energies of formation of all ions in aqueous solution are based. It also forms the basis for defining free-energy changes for oxidation–reduction processes in water.

pE is 0.0. If the electron activity were increased by a factor of 10 (as would be the case if H^+ (*aq*) at an activity of 0.100 were in equilibrium with H_2 at an activity of 1.00), the electron activity would be 10 and the pE value would be –1.0.

4.4 THE NERNST EQUATION

The *Nernst equation* is used to account for the effect of different activities upon electrode potential. Referring to Figure 4.2, if the Fe^{3+} ion concentration is increased relative to the Fe^{2+} ion concentration, it is readily visualized that the potential and the pE of the right electrode will become more positive because the higher concentration of electron-deficient Fe^{3+} ions clustered around it tends to draw electrons from the electrode. Decreased Fe^{3+} ion or increased Fe^{2+} ion concentration has the opposite effect. Such concentration effects upon E and pE are expressed by the *Nernst equation*. As applied to the half-reaction

$$\mathrm{Fe}^{3+} + e^- \rightleftarrows \mathrm{Fe}^{2+}, \quad E^0 = +0.77\,\mathrm{V}, \quad \mathrm{p}E^0 = 13.2 \tag{4.20}$$

the Nernst equation is the following, where $2.303RT/F = 0.0591$ at 25°C:

$$E = E^0 + \frac{2.303RT}{nF}\log\frac{[\mathrm{Fe}^{3+}]}{[\mathrm{Fe}^{2+}]} = E^0 + \frac{0.0591}{n}\log\frac{[\mathrm{Fe}^{3+}]}{[\mathrm{Fe}^{2+}]} \tag{4.21}$$

where n is the number of electrons involved in the half-reaction (1 in this case), and the activities of Fe^{3+} and Fe^{2+} ions have been taken as their concentrations (a simplification valid for more dilute solutions, which will be made throughout this chapter). Considering that

$$\mathrm{p}E = \frac{E}{2.303RT/F} \quad \text{and} \quad \mathrm{p}E^0 = \frac{E^0}{2.303RT/F}$$

the Nernst equation can be expressed in terms of pE and pE^0

$$\mathrm{p}E = \mathrm{p}E^0 + \frac{1}{n}\log\frac{[\mathrm{Fe}^{3+}]}{[\mathrm{Fe}^{2+}]} \quad (\text{in this case } n = 1) \tag{4.22}$$

The Nernst equation in this form is quite simple and offers some advantages in calculating redox relationships.

If, for example, the value of $[Fe^{3+}]$ is 2.35×10^{-3} M and $[Fe^{2+}] = 7.85 \times 10^{-5}$ M, the value of pE is

$$\mathrm{p}E = 13.2 + \log\frac{2.35\times10^{-3}}{7.85\times10^{-5}} = 14.7 \tag{4.23}$$

As the concentration of Fe^{3+} increases relative to the concentration of Fe^{2+}, the value of pE becomes higher (more positive) and as the concentration of Fe^{2+} increases relative to the concentration of Fe^{3+}, the value of pE becomes lower (more negative).

4.5 REACTION TENDENCY: WHOLE REACTION FROM HALF-REACTIONS

This section discusses how half-reactions can be combined to give whole reactions and how the pE^0 values of the half-reactions can be used to predict the directions in which reactions will go. The half-reactions discussed here are the following:

$$Hg^{2+} + 2e^- \rightleftarrows Hg, \quad pE^0 = 13.35 \tag{4.24}$$

$$Fe^{3+} + e^- \rightleftarrows Fe^{2+}, \quad pE^0 = 13.2 \tag{4.25}$$

$$Cu^{2+} + 2e^- \rightleftarrows Cu, \quad pE^0 = 5.71 \tag{4.26}$$

$$2H^+ + 2e^- \rightleftarrows H_2, \quad pE^0 = 0.00 \tag{4.27}$$

$$Pb^{2+} + 2e^- \rightleftarrows Pb, \quad pE^0 = -2.13 \tag{4.28}$$

Such half-reactions and their pE^0 values can be used to explain observations such as the following: A solution of Cu^{2+} flows through a lead pipe and the lead acquires a layer of copper metal through the reaction

$$Cu^{2+} + Pb \rightarrow Cu + Pb^{2+} \tag{4.29}$$

This reaction occurs because the copper(II) ion has a greater tendency to acquire electrons than the lead ion has to retain them. This reaction can be obtained by subtracting the lead half-reaction, Equation 4.28, from the copper half-reaction, Equation 4.26:

$$Cu^{2+} + 2e^- \rightleftarrows Cu, \qquad pE^0 = 5.71 \tag{4.26}$$

$$-(Pb^{2+} + 2e^- \rightleftarrows Pb), \qquad pE^0 = -2.13 \tag{4.28}$$

$$Cu^{2+} + Pb \rightleftarrows Cu + Pb^{2+}, \qquad pE^0 = 7.84 \tag{4.30}$$

The positive value of pE^0 for this reaction, 7.84, indicates that the reaction tends to go to the right as written. This occurs when lead metal directly contacts a solution of copper(II) ion. Therefore, if a waste solution containing copper(II) ion, a relatively innocuous pollutant, comes into contact with lead in plumbing, toxic lead may go into solution.

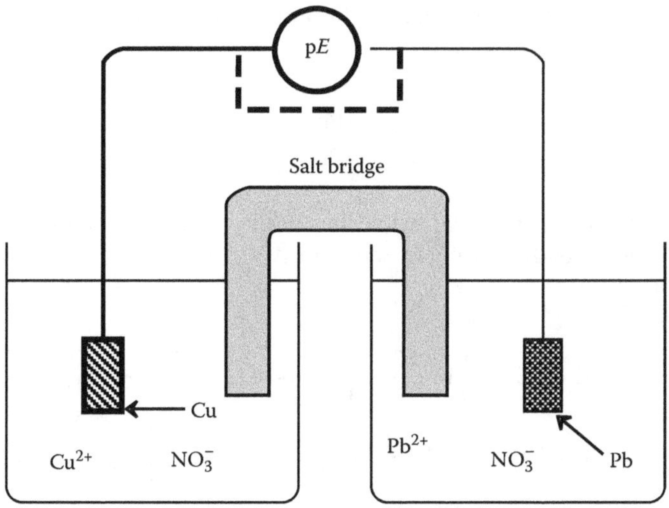

FIGURE 4.3 Cell for the measurement of p*E* between a lead half-cell and a copper half-cell. In this configuration the "p*E*" meter has a very high resistance and current cannot flow. (From Manahan, S., *Environmental Chemistry*, 9th ed., Taylor & Francis, Boca Raton, 2010. With permission.)

In principle, half-reactions may be allowed to occur in separate electrochemical half-cells, as could occur for Reaction 4.30 in the cell shown in Figure 4.3 if the meter (p*E*) were bypassed by an electrical conductor; they are therefore called *cell reactions.*

If the activities of Cu^{2+} and Pb^{2+} are not unity, the direction of the reaction and value of p*E* are deduced from the Nernst equation. For Reaction 4.30 the Nernst equation is

$$\mathrm{p}E = \mathrm{p}E^0 + \frac{1}{n}\log\frac{[\mathrm{Cu}^{2+}]}{[\mathrm{Pb}^{2+}]} = 7.84 + \frac{1}{2}\log\frac{[\mathrm{Cu}^{2+}]}{[\mathrm{Pb}^{2+}]} \tag{4.31}$$

By combining the appropriate half-reactions, it can be shown that copper metal will not cause hydrogen gas to be evolved from solutions of strong acid [hydrogen ion has less attraction for electrons than does copper(II) ion], whereas lead metal, in contrast, will displace hydrogen gas from acidic solutions.

4.6 THE NERNST EQUATION AND CHEMICAL EQUILIBRIUM

Refer again to Figure 4.3. Imagine that instead of the cell being set up to measure the potential between the copper and lead electrodes, the voltmeter used to measure p*E* was removed and the electrodes directly connected with a wire so that the current might flow between them. The reaction

$$\mathrm{Cu}^{2+} + \mathrm{Pb} \rightleftarrows \mathrm{Cu} + \mathrm{Pb}^{2+}, \quad \mathrm{p}E^0 = 7.84 \tag{4.30}$$

will occur until the concentration of lead ion becomes so high, and that of copper ion so low, that the reaction stops. The system is at equilibrium and, since current no longer flows, pE is exactly zero. The equilibrium constant K for the reaction is given by the expression

$$K = \frac{[Pb^{2+}]}{[Cu^{2+}]} \tag{4.32}$$

The equilibrium constant can be calculated from the Nernst equation, noting that under equilibrium conditions pE is zero and $[Cu^{2+}]$ and $[Pb^{2+}]$ are at equilibrium concentrations:

$$\begin{aligned} pE &= pE^0 + \frac{1}{n}\log\frac{[Cu^{2+}]}{[Pb^{2+}]} = 7.84 + \frac{1}{2}\log\frac{[Cu^{2+}]}{[Pb^{2+}]} \\ pE &= 0.00 = 7.84 - \frac{1}{2}\log\frac{[Pb^{2+}]}{[Cu^{2+}]} = 7.84 - \frac{1}{2}\log K \end{aligned} \tag{4.33}$$

Note that the reaction products are placed over reactants in the log term, and a minus sign is placed in front to put the equilibrium constant in the correct form (a purely mathematical operation). The value of $\log K$ obtained from solving the above equation is 15.7.

The equilibrium constant for a redox reaction involving n electrons is given in terms of pE simply by

$$\log K = n(pE^0) \tag{4.34}$$

4.7 THE RELATIONSHIP OF p*E* TO FREE ENERGY

Aquatic systems and the organisms that inhabit them—just like the steam engine or students hoping to pass physical chemistry—must obey the laws of thermodynamics. Bacteria, fungi, and human beings derive their energy from acting as mediators (catalysts) of chemical reactions and extracting a certain percentage of useful energy from them. In predicting or explaining the behavior of an aquatic system, it is helpful to be able to predict the useful energy that can be extracted from chemical reactions in the system, such as microbially mediated oxidation of organic matter to CO_2 and water, or the fermentation of organic matter to methane by anoxic bacteria in the absence of oxygen. Such information may be obtained by knowing the free-energy change, ΔG, for the redox reaction; ΔG, in turn, may be obtained from pE for the reaction. The free-energy change for a redox reaction involving n electrons at an absolute temperature of T is given by

$$\Delta G = -2.303nRT(pE) \tag{4.35}$$

where R is the gas constant. When all reaction participants are in their standard states (pure liquids, pure solids, solutes at an activity of 1.00, gas pressures at 1 atm), ΔG is the standard free-energy change, ΔG^0, given by

$$\Delta G^0 = -2.303nRT(pE^0) \tag{4.36}$$

4.8 REACTIONS IN TERMS OF ONE ELECTRON-MOLE

For comparing free-energy changes between different redox reactions, it is most meaningful to consider the reactions in terms of the transfer of exactly 1 mol of electrons. This concept may be understood by considering two typical and important redox reactions that occur in aquatic systems—nitrification

$$NH_4^+ + 2O_2 \rightleftarrows NO_3^- + 2H^+ + H_2O, \quad pE^0 = 5.85 \tag{4.37}$$

and oxidation of iron(II) to iron(III):

$$4Fe^{2+} + O_2 + 10H_2O \rightleftarrows 4Fe(OH)_3(s) + 8H^+, \quad pE^0 = 7.6 \tag{4.38}$$

What do reactions written in this way really mean? If any thermodynamic calculations are to be made involving the reactions, Reaction 4.37 means that 1 mol of ammonium ion reacts with 2 mol of elemental oxygen to yield 1 mol of nitrate ion, 2 mol of hydrogen ion, and 1 mol of water. Reaction 4.38 is taken to mean that 4 mol of iron(II) ion react with 1 mol of oxygen and 10 mol of water to produce 4 mol of $Fe(OH)_3$ and 8 mol of hydrogen ions. The free-energy changes calculated for these quantities of reaction participants do not enable meaningful comparisons of their free-energy changes. Such comparisons may be made, though, on the common basis of the transfer of 1 mol of electrons, writing each reaction in terms of one electron-mole. The advantage of this approach is illustrated by considering Reaction 4.37, which involves an eight-electron change, and Reaction 4.38, which involves a four-electron change. Rewriting Equation 4.37 for one electron-mole yields

$$\tfrac{1}{8}NH_4^+ + \tfrac{1}{4}O_2 \rightleftarrows \tfrac{1}{8}NO_3^- + \tfrac{1}{4}H^+ + \tfrac{1}{8}H_2O, \quad pE^0 = 5.85 \tag{4.39}$$

whereas Reaction 4.38, when rewritten for one electron-mole rather than four, yields

$$Fe^{2+} + \tfrac{1}{4}O_2 + \tfrac{5}{2}H_2O \rightleftarrows Fe(OH)_3(s) + 2H^+, \quad pE^0 = 7.6 \tag{4.40}$$

From Equation 4.36, the standard free-energy change for a reaction is

$$\Delta G^0 = -2.303nRT(pE^0), \tag{4.36}$$

which, for a one electron-mole reaction is simply

$$\Delta G^0 = -2.303RT(pE^0). \tag{4.36}$$

Therefore, for reactions written in terms of one electron-mole, a comparison of pE^0 values provides a direct comparison of ΔG^0 values.

As shown in Equation 4.34 for a redox reaction involving n electrons, pE^0 is related to the equilibrium constant by

$$\log K = n(\mathrm{p}E^0) \tag{4.41}$$

which for a one electron-mole reaction becomes simply

$$\log K = \mathrm{p}E^0 \tag{4.42}$$

Reaction 4.39, the nitrification reaction written in terms of one electron-mole, has a pE^0 value of +5.85. The equilibrium-constant expression for this reaction is

$$K = \frac{[\mathrm{NO_3^-}]^{1/8}[\mathrm{H^+}]^{1/4}}{[\mathrm{NH_4^+}]^{1/8}\mathrm{PO_2^{1/4}}} \tag{4.43}$$

a computationally cumbersome form for which log K is given very simply as the following:

$$\log K = \mathrm{p}E^0 = 5.85 \quad \text{or} \quad K = 7.08 \times 10^5 \tag{4.44}$$

Table 4.1 is a compilation of pE^0 values for redox reactions that are especially important in aquatic systems. Most of these values are calculated from

TABLE 4.1
pE^0 Values of Redox Reactions Important in Natural Waters (at 25°C)

Reaction	pE^0	pE^0 (W)[a]
1. $\frac{1}{4}O_2(g) + H^+(W) + e^- \rightleftarrows \frac{1}{2}H_2O$	+20.75	+13.75
2. $\frac{1}{5}NO_3^- + \frac{6}{5}H^+ + e^- \rightleftarrows \frac{1}{10}N_2 + \frac{3}{5}H_2O$	+12.65	+21.05
3. $\frac{1}{2}MnO_2 + \frac{1}{2}HCO_3^+(1\times10^{-3}\ M) + \frac{3}{2}H^+(W) + e^- \rightleftarrows \frac{1}{2}MnCO_3(s) + H_2O$	—	+8.5[b]
4. $\frac{1}{2}NO_3^- + H^+(W) + e^- \rightleftarrows \frac{1}{2}NO_2^- + \frac{1}{2}H_2O$	+14.15	+7.15
5. $\frac{1}{8}NO_3^- + \frac{5}{4}H^+(W) + e^- \rightleftarrows \frac{1}{8}NH_4^+ + \frac{3}{8}H_2O$	+14.90	+6.15
6. $\frac{1}{6}NO_2^- + \frac{4}{3}H^+(W) + e^- \rightleftarrows \frac{1}{6}NH_4^+ + \frac{1}{3}H_2O$	+15.14	+5.82
7. $\frac{1}{2}CH_3OH + H^+(W) + e^- \rightleftarrows \frac{1}{2}CH_4(g) + \frac{1}{2}H_2O$	+9.88	+2.88
8. $\frac{1}{4}CH_2O + H^+(W) + e^- \rightleftarrows \frac{1}{4}CH_4(g) + \frac{1}{4}H_2O$	+6.94	−0.06

TABLE 4.1 (continued)
pE^0 Values of Redox Reactions Important in Natural Waters (at 25°C)

Reaction	pE^0	pE^0 (W)[a]
9. $FeOOH(g) + HCO_3^-(1 \times 10^{-3}\ M) + 2H^+(W) + e^- \rightleftarrows FeCO_3(s) + 2H_2O$	—	−1.67[b]
10. $\frac{1}{2}CH_2O + H^+(W) + e^- \rightleftarrows +\frac{1}{2}CH_3OH$	+3.99	−3.01
11. $\frac{1}{6}SO_4^{2-} + \frac{4}{3}H^+(W) + e^- \rightleftarrows \frac{1}{6}S(s) + \frac{2}{3}H_2O$	+6.03	−3.30
12. $\frac{1}{8}SO_4^{2-} + \frac{5}{4}H^+(W) + e^- \rightleftarrows \frac{1}{8}H_2S(g) + \frac{1}{2}H_2O$	+5.75	−3.50
13. $\frac{1}{8}SO_4^{2-} + \frac{9}{8}H^+(W) + e^- \rightleftarrows \frac{1}{8}HS^- + \frac{1}{2}H_2O$	+4.13	−3.75
14. $\frac{1}{2}S(s) + H^+(W) + e^- \rightleftarrows \frac{1}{2}H_2S(g)$	+2.89	−4.11
15. $\frac{1}{8}CO_2 + H^+ + e^- \rightleftarrows \frac{1}{8}CH_4 + \frac{1}{4}H_2O$	+2.87	−4.13
16. $\frac{1}{6}N_2 + \frac{4}{3}H^+(W) + e^- \rightleftarrows \frac{1}{3}NH_4^+$	+4.68	−4.65
17. $H^+(W) + e^- \rightleftarrows \frac{1}{2}H_2(g)$	0.00	−7.00
18. $\frac{1}{4}CO_2(g) + H^+(W) + e^- \rightleftarrows \frac{1}{4}CH_2O + \frac{1}{4}H_2O$	−1.20	−8.20

Source: Reproduced from Stumm, W. and J. J. Morgan, *Aquatic Chemistry*, John Wiley & Sons, New York, 1970, p. 318. With permission.

[a] (W) indicates $a_{H^+} = 1.00 \times 10^{-7}$ M and pE^0(W) is a pE^0 at $a_{H^+} = 1.00 \times 10^{-7}$ M.

[b] These data correspond to $a_{HCO_3^-} = 1.00 \times 10^{-3}$ M rather than unity and, hence, are not exactly pE^0(W); they represent typical aquatic conditions more exactly than pE^0 values do.

thermodynamic data rather than from direct potentiometric measurements in an electrochemical cell, as shown in Figure 4.2. Most electrode systems that might be devised do not give potential responses corresponding to the Nernst equation; that is, they do not behave *reversibly*. It is true that one may place a platinum electrode and a reference electrode in water and measure a potential. This potential, referred to the SHE, is the so-called E_H value. Furthermore, the measured potential will be more positive (more oxidizing) in an oxidizing medium, such as the oxic surface layers of a lake, than in a reducing medium, such as the anaerobic bottom regions of a body of water. However, attaching any quantitative significance to the E_H value measured directly with an electrode is a very dubious practice. Acid mine waters containing relatively high levels of sulfuric acid and dissolved iron give reasonably accurate E_H values by direct measurement, but most aquatic systems do not yield meaningful values of E_H. The most accurate method of assessing the redox status of water is through calculation of *oxidative capacity*, a parameter analogous to buffer capacity for acids and bases and consisting of a single descriptive parameter derived from the sum of the species that can be oxidized or reduced.

4.9 THE LIMITS OF p*E* IN WATER

There are pH-dependent limits to the p*E* values at which water is thermodynamically stable. Water may be either oxidized as

$$2H_2O \rightleftarrows O_2 + 4H^+ + 4e^- \tag{4.45}$$

or reduced as

$$2H_2O + 2e^- \rightleftarrows H_2 + 2OH^- \tag{4.46}$$

These two reactions determine the limits of p*E* in water. On the oxidizing side (relatively more positive p*E* values), the p*E* value is limited by the oxidation of water, half-reaction 4.45. The evolution of hydrogen, Half-reaction 4.46, limits the p*E* value on the reducing side.

The condition under which oxygen from the oxidation of water has a pressure of 1.00 atm can be regarded as the oxidizing limit of water, whereas a hydrogen pressure of 1.00 atm may be regarded as the reducing limit of water. These are *boundary conditions* that enable calculation of the stability boundaries of water. Writing the reverse of Reaction 4.45 for one electron and setting $P_{O_2} = 1.00$ yields:

$$\tfrac{1}{4}O_2 + H^+ + e^- \rightleftarrows \tfrac{1}{2}H_2O, \quad pE^0 = 20.75 \quad \text{(from Table 4.1)} \tag{4.47}$$

Thus, Equation 4.49 defines the pH-dependent oxidizing limit of water. At a specified pH, p*E* values more positive than the one given by Equation 4.49 cannot exist at equilibrium in water in contact with the atmosphere.

$$pE = pE^0 + \log(P_{O_2}^{1/4}[H^+]) \tag{4.48}$$

$$pE = 20.75 - pH \tag{4.49}$$

The p*E*–pH relationship for the reducing limit of water taken at $P_{H_2} = 1$ atm is given by the following derivation:

$$H^+ + e^- \rightleftarrows \tfrac{1}{2}H_2, \quad pE^0 = 0.00 \tag{4.50}$$

$$pE = pE^0 + \log[H^+] \tag{4.51}$$

$$pE = -pH \tag{4.52}$$

For neutral water (pH = 7.00), substitution into Equations 4.52 and 4.49 yields −7.00 to 13.75 for the p*E* range of water. The p*E*–pH boundaries of stability for water are shown by the dashed lines in Figure 4.4 of Section 4.11.

The decomposition of water is very slow in the absence of a suitable catalyst. Therefore, water may have temporary nonequilibrium p*E* values more negative than

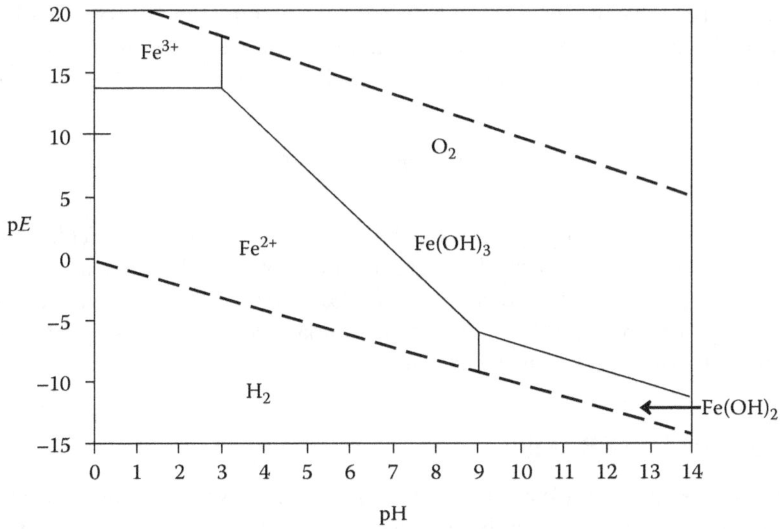

FIGURE 4.4 Simplified p*E*–pH diagram for iron in water. The maximum soluble iron concentration is 1.00×10^{-5} M. (From Manahan, S., *Environmental Chemistry*, 9th ed., Taylor & Francis, Boca Raton, 2010. With permission.)

the reducing limit or more positive than the oxidizing limit. A common example of the latter is a solution of chlorine in water.

4.10 p*E* VALUES IN NATURAL WATER SYSTEMS

Although it is not generally possible to obtain accurate p*E* values by direct potentiometric measurements in natural aquatic systems, in principle, p*E* values may be calculated from the species present in water at equilibrium. An obviously significant p*E* value is that of neutral water in thermodynamic equilibrium with the atmosphere. In water under these conditions, $P_{O_2} = 0.21$ atm and $[H^+] = 1.00 \times 10^{-7}$ M. Substitution into Equation 4.48 yields

$$pE = 20.75 + \log\{(0.21)^{1/4} \times 1.00 \times 10^{-7}\} = 13.8 \tag{4.53}$$

According to this calculation, a p*E* value of around +13 is to be expected for water in equilibrium with the atmosphere, that is, an oxic water. At the other extreme, consider anoxic water in which methane and CO_2 are being produced by microorganisms. Assume $P_{CO_2} = P_{CH_4}$ and that pH = 7.00. The relevant half-reaction is

$$\tfrac{1}{8}CO_2 + H^+ + e^+ \rightleftarrows \tfrac{1}{8}CH_4 + \tfrac{1}{4}H_2O \tag{4.54}$$

for which the Nernst equation is

$$pE = 2.87 + \log\frac{P_{CO_2}^{1/8}[H^+]}{P_{CH_2}^{1/8}} = 2.87 + \log[H^+] = 2.87 - 7.00 = -4.13 \tag{4.55}$$

Note that the pE value of −4.13 does not exceed the reducing limit of water at pH 7.00, which from Equation 4.52 is −7.00. It is of interest to calculate the pressure of oxygen in neutral water at this low pE value of −4.13. Substitution into Equation 4.48 yields

$$-4.13 = 20.75 + \log(P_{O_2}^{1/4} \times 1.00 \times 10^{-7}) \tag{4.56}$$

from which the pressure of oxygen is calculated to be 3.0×10^{-72} atm. This impossibly low figure for the pressure of oxygen means that equilibrium with respect to oxygen partial pressure is not achieved under these conditions. Certainly, under any condition approaching equilibrium between comparable levels of CO_2 and CH_4, the partial pressure of oxygen must be extremely low.

4.11 pE–pH DIAGRAMS

The examples cited so far have shown the close relationships between pE and pH in water. This relationship may be expressed graphically in the form of a pE–pH diagram. Such diagrams show the regions of stability and the boundary lines for various species in water. Because of the numerous species that may be formed, such diagrams may become extremely complicated. For example, if a metal is being considered, several different oxidation states of the metal, hydroxy complexes, and different forms of the solid metal oxide or hydroxide may exist in different regions described by the pE–pH diagram. Most waters contain carbonate, and many contain sulfates and sulfides, so that various metal carbonates, sulfates, and sulfides may predominate in different regions of the diagram. In order to illustrate the principles involved, however, a simplified pE–pH diagram is considered here. The reader is referred to more advanced works on geochemistry and aquatic chemistry for more complicated (and more realistic) pE–pH diagrams.[1,2]

A pE–pH diagram for iron may be constructed assuming a maximum concentration of iron in solution, in this case 1.0×10^{-5} M. The following equilibria will be considered:

$$Fe^{3+} + e^- \rightleftarrows Fe^{2+}, \quad pE^0 = +13.2 \tag{4.57}$$

$$Fe(OH)_2(s) + 2H^+ \rightleftarrows Fe^{2+} + 2H_2O \tag{4.58}$$

$$K_{sp} = \frac{[Fe^{2+}]}{[H^+]^2} = 8.0 \times 10^{12} \tag{4.59}$$

$$Fe(OH)_3(s) + 3H^+ \rightleftarrows Fe^{3+} + 3H_2O \tag{4.60}$$

$$K'_{sp} = \frac{[Fe^{3+}]}{[H^+]^3} = 9.1 \times 10^3 \tag{4.61}$$

[The constants K_{sp} and K'_{sp} are derived from the solubility products of $Fe(OH)_2$ and $Fe(OH)_3$, respectively, and are expressed in terms of $[H^+]$ to facilitate the calculations.]

Note that the formation of species such as $Fe(OH)^{2+}$, $Fe(OH)_2^+$, and solid $FeCO_3$ or FeS, all of which might be of significance in a natural water system, is not considered. Rapid hydrolysis of iron(III) in solution yields solid ferrihydrite, a hydrated iron(III) hydroxide/oxide with a high surface area and strong affinity to coprecipitate metals other than iron.

In constructing the p*E*–pH diagram, several boundaries must be considered. The first two of these are the oxidizing and reducing limits of water (see Section 4.9). At the high p*E* end, the stability limit of water is defined by Equation 4.49 derived previously:

$$pE = 20.75 - \text{pH} \tag{4.49}$$

The low p*E* limit is defined by Equation 4.52:

$$pE = -\text{pH} \tag{4.52}$$

The p*E*–pH diagram constructed for the iron system must fall between the boundaries defined by these two equations.

Below pH 3, Fe^{3+} may exist in equilibrium with Fe^{2+}. The boundary line that separates these two species, where $[Fe^{3+}] = [Fe^{2+}]$, is given by the following calculation:

$$pE = 13.2 + \log\frac{[Fe^{3+}]}{[Fe^{2+}]} \tag{4.62}$$

$$[Fe^{3+}] = [Fe^{2+}] \tag{4.63}$$

$$pE = 13.2 \quad \text{(independent of pH)} \tag{4.64}$$

At p*E* exceeding 13.2, as the pH increases from very low values, $Fe(OH)_3$ precipitates from a solution of Fe^{3+}. The pH at which precipitation occurs depends, of course, upon the concentration of Fe^{3+}. In this example, a maximum soluble iron concentration of 1.00×10^{-5} M has been chosen so that at the $Fe^{3+}/Fe(OH)_3$ boundary, $[Fe^{3+}] = 1.00 \times 10^{-5}$ M. Substitution in Equation 4.61 yields

$$[H^+]^3 = \frac{[Fe^{3+}]}{K'_{sp}} = \frac{1.00\times10^{-5}}{9.1\times10^3} \tag{4.65}$$

$$\text{pH} = 2.99 \tag{4.66}$$

In a similar manner, the boundary between Fe^{2+} and solid $Fe(OH)_2$ may be defined, assuming $[Fe^{2+}] = 1.00 \times 10^{-5}$ M (the maximum soluble iron concentration specified at the beginning of this exercise) at the boundary:

$$[H^+]^2 = \frac{[Fe^{2+}]}{K_{sp}} = \frac{1.00\times10^{-5}}{8.0\times10^{12}} \tag{4.67}$$

$$\text{pH} = 8.95 \tag{4.68}$$

Throughout a wide p*E*–pH range, Fe^{2+} is the predominant soluble iron species in equilibrium with the solid hydrated iron(III) oxide, $Fe(OH)_3$. The boundary between these two species depends upon both p*E* and pH. Substituting Equation 4.61 into Equation 4.62 yields

$$pE = 13.2 + \frac{K'_{sp}[H^+]^3}{[Fe^{2+}]} \tag{4.69}$$

$$pE = 13.2 + \log 9.1 \times 10^3 - \log 1.00 \times 10^{-5} + 3 \times \log[H^+]$$

$$pE = 22.2 - 3\ pH \tag{4.70}$$

The boundary between the solid phases $Fe(OH)_2$ and $Fe(OH)_3$ likewise depends upon both p*E* and pH, but it does not depend upon an assumed value for total soluble iron. The required relationship is derived from substituting both Equations 4.59 and 4.61 into Equation 4.62:

$$pE = 13.2 + \log \frac{K'_{sp}}{K_{sp}[H^+]^2}[H^+]^3$$

$$pE = 13.2 + \log \frac{9.1 \times 10^3}{8.0 \times 10^{12}} + \log[H^+] \tag{4.71}$$

$$pE = 4.3 - pH \tag{4.72}$$

All of the equations needed to prepare the p*E*–pH diagram for iron in water have now been derived. To summarize, the equations are (Equation 4.49), O_2–H_2O boundary; (Equation 4.52), H_2–H_2O boundary; (Equation 4.64), Fe^{3+}–Fe^{2+} boundary; (Equation 4.66), Fe^{3+}–$Fe(OH)_3$ boundary; (Equation 4.68), Fe^{2+}–$Fe(OH)_2$ boundary; (Equation 4.70), Fe^{2+}–$Fe(OH)_3$ boundary; and (Equation 4.72), $Fe(OH)_2$–$Fe(OH)_3$ boundary.

The p*E*–pH diagram for the iron system in water is shown in Figure 4.4. In this system, at a relatively high hydrogen-ion activity and high electron activity (an acidic reducing medium), iron(II) ion, Fe^{2+}, is the predominant iron species; some groundwaters contain appreciable levels of iron(II) under these conditions. (In most natural water systems the solubility range of Fe^{2+} is relatively narrow because of the precipitation of FeS or $FeCO_3$.) At a very high hydrogen-ion activity and low electron activity (an acidic oxidizing medium), Fe^{3+} ion predominates. In an oxidizing medium at lower acidity, solid $Fe(OH)_3$ is the primary iron species present. Finally, in a basic reducing medium, with low hydrogen-ion activity and high electron activity, solid $Fe(OH)_2$ can be stable.

Note that within the pH regions normally encountered in a natural aquatic system (approximately pH 5–9) $Fe(OH)_3$ or Fe^{2+} is the predominant stable iron species. In fact, it is observed that in waters containing dissolved oxygen at any appreciable level (a relatively high p*E*), hydrated iron(III) oxide $Fe(OH)_3$ is essentially the only inorganic iron species found. Such waters contain a high level of suspended iron, but any truly soluble iron must be in the form of a complex (see Chapter 3).

In highly anoxic, low-pE water, appreciable levels of Fe^{2+} may be present. When such water is exposed to atmospheric oxygen, the pE rises and $Fe(OH)_3$ precipitates. The resulting deposits of hydrated iron(III) oxide can stain laundry and bathroom fixtures with a refractory red/brown stain. This phenomenon also explains why red iron oxide deposits are found near pumps and springs that bring deep, anaerobic water to the surface. In shallow wells, where the water may become oxic, solid $Fe(OH)_3$ may precipitate on the well walls, clogging the aquifer outlet. This usually occurs through bacterially mediated reactions, which are discussed in Chapter 6.

One species not yet considered is elemental iron. For the half-reaction

$$Fe^{2+} + 2e^- \rightleftarrows Fe, \quad pE^0 = -7.45 \tag{4.73}$$

the Nernst equation gives pE as a function of $[Fe^{2+}]$

$$pE = -7.45 + \tfrac{1}{2}\log[Fe^{2+}] \tag{4.74}$$

For iron metal in equilibrium with 1.00×10^{-5} M Fe^{2+}, the following pE value is obtained:

$$pE = -7.45 + \tfrac{1}{2}\log 1.00 \times 10^{-5} = -9.95 \tag{4.75}$$

Examination of Figure 4.4 shows that the pE values for elemental iron in contact with Fe^{2+} are below the reducing limit of water. This shows that iron metal in contact with water is thermodynamically unstable with respect to reducing water and going into solution as Fe^{2+}, a factor that contributes to the tendency of iron to undergo corrosion.

4.12 HUMIC SUBSTANCES AS NATURAL REDUCTANTS

Humic substances (see Section 3.13) are oxidation–reduction-active species that may be significant as reducing species in chemical and biochemical processes that occur in natural water and wastewater systems. The capability of humic substances to act as reductants is due primarily to the presence of the quinone/hydroquinone group that acts as an oxidation–reduction couple as follows:

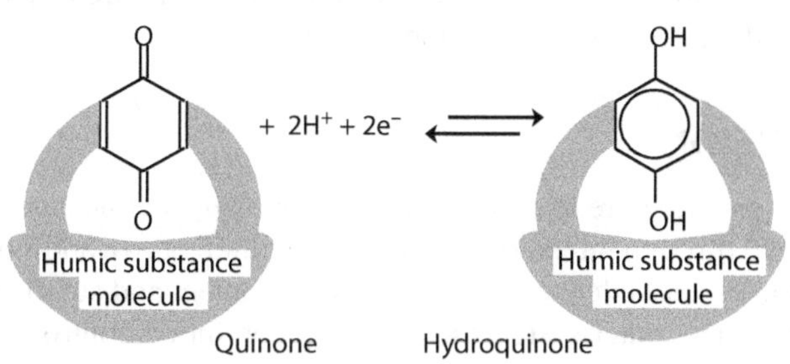

(4.76)

There is some evidence suggesting that soluble humic substances can act as intermediates in the reduction of species in water by solid reductants, such as solid iron(II) species.

It is believed that in addition to being reducing agents, humic substances act as electron shuttles to transfer electrons to iron(III) in the microorganism-mediated bioreduction of iron(III) to iron(II) in water.[3] The process is faster in the presence of high carboxyl ($-CO_2H$) group content in the humic substances, which is indicative of the importance of complexation of iron ions by the humic substance in the reduction process.

4.13 PHOTOCHEMICAL PROCESSES IN OXIDATION–REDUCTION

As discussed in Chapter 1, Section 1.4, absorption of a light photon can introduce a high level of energy into chemical species that can become involved in oxidation–reduction processes in water. An important oxidizing agent in water exposed to sunlight is believed to be *superoxide ion*, $O_2^{\bullet -}$, produced by the action of photochemically excited natural organic matter (humic substance) upon dissolved O_2. This superoxide species is a *free radical*, a generally highly reactive species with an unpaired electron denoted by the dot in the formula $O_2^{\bullet -}$. Superoxide ion is capable of oxidizing inorganically complexed metals such as iron.

An important intermediate involved with photochemically induced oxidation processes in water is hydrogen peroxide, H_2O_2, generated when $O_2^{\bullet -}$ reacts with water. In the presence of iron(II) the Fenton reaction,

$$\text{Fe(II)} + H_2O_2 \rightarrow \text{Fe(III)} + OH^- + HO^\bullet \quad (4.77)$$

generates hydroxyl radical, $HO^\bullet$. Hydroxyl radical is a very reactive species that reacts with even refractory organic species, leading to their oxidation. As a consequence of reactions involving superoxide radical ion and hydroxyl radical, photochemical processes are important in promoting oxidation of oxidizable species in water.

Aqueous suspensions of TiO_2 are photochemically active. When photons of ultraviolet radiation of a wavelength <318 nm ($h\nu$) strike a TiO_2 surface,

$$TiO_2 + h\nu \rightarrow e^-_{cb} + h^+_{vb} \quad (4.78)$$

conduction band electrons (e^-_{cb}) capable of initiating photoreduction and valence band holes (h^+_{vb}) capable of initiating photooxidation are produced. This has led to interest in the use of suspensions of TiO_2 exposed to sunlight to photoreduce and photooxidize pollutant and hazardous waste species.

4.14 CORROSION

One of the most damaging redox phenomena is *corrosion*, defined as the destructive alteration of metal through interactions with its surroundings. In addition to its multibillion dollar annual costs due to destruction of equipment and structures, corrosion introduces metals into water systems and destroys pollution control equipment and waste disposal pipes; it is aggravated by water and air pollutants and corrosive hazardous wastes that exhibit extremes of pH of high salt contents.

Thermodynamically, all commonly used metals are unstable relative to their environments. Elemental metals tend to undergo chemical changes to produce the more stable forms of ions, salts, oxides, and hydroxides. Fortunately, the rates of corrosion are normally slow, so that metals exposed to air and water may endure for long periods of time. However, protective measures are necessary. Sometimes these measures fail; for example, witness the gaping holes in automobile bodies exposed to salt used to control road ice.

Corrosion normally occurs when an electrochemical cell is set up on a metal surface. The area corroded is the anode, where the following oxidation reaction occurs, illustrated for the formation of a divalent metal ion from a metal, M:

$$M \rightarrow M^{2+} + 2e^- \tag{4.79}$$

Several cathodic reactions are possible. One of the most common of these is the reduction of H^+ ion:

$$2H^+ + 2e^- \rightarrow H_2 \tag{4.80}$$

Oxygen may also be involved in cathodic reactions, including reduction to hydroxide, reduction to water, and reduction to hydrogen peroxide:

$$O_2 + 2H_2O + 4e^- \rightarrow 4OH^- \tag{4.81}$$

$$O_2 + 4H^+ + 4e^- \rightarrow 2H_2O \tag{4.82}$$

$$O_2 + 2H_2O + 2e^- \rightarrow 2OH^- + H_2O_2 \tag{4.83}$$

Oxygen may either accelerate corrosion processes by participating in reactions, such as these, or retard them by forming protective oxide films. As discussed in Chapter 6, bacteria are often involved with corrosion.

Corrosion is a very important consideration in domestic water distribution systems. Metals that may be corroded in such systems include iron, copper, lead, and red brass. Corrosion of these materials involves the redox potential of the water as determined by the solutes in it and the solid materials with which it is in contact. Important factors in corrosion include type and concentration of oxidant (usually from added Cl_2), metal release rate, and corrosion scale properties of the corroding metals.

LITERATURE CITED

1. Stumm, W. and J. J. Morgan, *Aquatic Chemistry: Chemical Equilibria and Rates in Natural Waters*, 3rd ed., Wiley, New York, 1995.
2. Garrels, R. M. and C. M. Christ, *Solutions, Minerals, and Equilibria*, Harper and Row, New York, 1965.
3. Rakshit, S., M. Uchimiya, and G. Sposito, Iron(III) bioreduction in soil in the presence of added humic substances, *Soil Science Society of America Journal*, **73**, 65–71, 2009.

4. Stumm, W. and J. J. Morgan, *Aquatic Chemistry*, p. 318, John Wiley & Sons, New York, 1970.
5. Manahan, S., *Environmental Chemistry*, 9th ed., Taylor & Francis, Boca Raton, 2010.

SUPPLEMENTARY REFERENCES

Albarѐ de, F., *Geochemistry*, Cambridge University Press, New York, 2004.

Amatya, S. R. and S. Mika, Influence of Eh/pH—Barriers on releasing/accumulation of manganese and iron at sediment–water interface, *Research Journal of Chemistry and Environment*, **12**, 7–13, 2008.

Auque, L., M. J. Gimeno, J. Gomez, and A. C. Nilsson, Potentiometrically measured Eh in groundwaters from the Scandinavian shield, *Applied Geochemistry*, 1820–1833, 2008.

Baas Becking, L. G. M., I. R. Kaplan, and D. Moore, Limits of the natural environment in terms of pH and oxidation–reduction potentials in natural waters, *Journal of Geology*, **68**, 243–284, 1960.

Bates, R. G., The modern meaning of pH, *Critical Reviews in Analytical Chemistry*, **10**, 247–278, 1981.

Chapelle, F. H., *Ground-Water Microbiology and Geochemistry*, 2nd ed., Wiley, New York, 2000.

Chester, R., *Marine Geochemistry*, Blackwell Science, Oxford, UK, 2000.

Faure, G., *Principles and Applications of Geochemistry: A Comprehensive Textbook for Geology Students*, 2nd ed., Prentice-Hall, Upper Saddle River, NJ, 1998.

Faust, B. C., A review of the photochemical redox reactions of Iron(III) species in atmospheric, oceanic, and surface waters: Influences on geochemical cycles and oxidant formation, in *Aquatic and Surface Photochemistry*, George Helz, R. G. Zepp, and D. G. Crosby, Eds, CRC Press/Lewis Publishers, Boca Raton, FL, pp. 3–37, 1994.

Garrison S., Oxidation reduction reactions, in *The Chemistry of Soils*, 2nd edition, Oxford University Press, New York, pp. 144–173, 2008.

Holland, H. D. and K. K. Turekian, Eds, *Treatise on Geochemistry*, Elsevier/Pergamon, Amsterdam, 2004.

McSween, H. Y., S. M. Richardson, and M. E. Uhle, *Geochemistry: Pathways and Processes*, 2nd ed., Columbia University Press, New York, 2003.

Pankow, J. F., *Aquatic Chemistry Concepts*, Lewis Publishers/CRC Press, Boca Raton, FL, 1991.

Pankow, J. F., *Aquatic Chemistry Problems*, Titan Press, OR, Portland, OR, 1992.

Rowell, D. L., Oxidation and reduction, in *The Chemistry of Soil Processes*, D. J. Greenland and M. H. B. Hayes, Eds, Wiley, Chichester, 1981.

Spellman, F. R., *The Science of Water: Concepts and Applications*, 2nd ed., Taylor & Francis/CRC Press, Boca Raton, FL, 2008.

Stewart, B. A., *Encyclopedia of Water Science*, Marcel Dekker, New York, 2003.

Stumm, W., *Redox Potential as an Environmental Parameter: Conceptual Significance and Operational Limitation*, Third International Conference on Water Pollution Research (Munich, Germany), Water Pollution Control Federation, Washington, DC, 1966.

Sun, L., S. C. Su, L. Chao, and T. Sun, Effects of flooding on changes in Eh, pH and speciation of cadmium and lead in contaminated soil, *Bulletin of Environmental Contamination and Toxicology*, **79**, 2007.

Trimble, S. W., *Encyclopedia of Water Science*, 2nd ed., Taylor & Francis, CRC Press, Boca Raton, FL, 2008.

QUESTIONS AND PROBLEMS

The use of Internet resources is assumed in answering any of the questions. These resources would include such things as constants and conversion factors as well as additional information needed to complete an answer.

1. The acid–base reaction for the dissociation of acetic acid is

$$HOAc + H_2O \rightarrow H_3O^+ + OAc^-$$

with $K_a = 1.75 \times 10^{-5}$. Break this reaction down into two half-reactions involving H^+ ion. Break down the redox reaction

$$Fe^{2+} + H^+ \rightarrow Fe^{3+} + \tfrac{1}{2}H_2$$

into two half-reactions involving the electron. Discuss the analogies between the acid–base and redox processes.

2. Assuming a bicarbonate ion concentration $[HCO_3^-]$ of 1.00×10^{-3} M and a value of 3.5×10^{-11} for the solubility product of $FeCO_3$, what would you expect to be the stable iron species at pH 9.5 and p*E* −8.0, as shown in Figure 4.4?
3. Assuming that the partial pressure of oxygen in water is that of atmospheric O_2, 0.21 atm, rather than the 1.00 atm assumed in deriving Equation 4.49, derive an equation describing the oxidizing p*E* limit of water as a function of pH.
4. Plot log P_{O_2} as a function of p*E* at pH 7.00.
5. Calculate the pressure of oxygen for a system in equilibrium in which $[NH_4^+] = [NO_3^-]$ at pH 7.00.
6. Calculate the values of $[Fe^{3+}]$, p*E*, and pH at the point in Figure 4.4 where Fe^{2+} is at a concentration of 1.00×10^{-5} M, and $Fe(OH)_2$ and $Fe(OH)_3$ are in equilibrium.
7. What is the p*E* value in a solution in equilibrium with air (21% O_2 by volume) at pH 6.00?
8. What is the p*E* value at the point on the Fe^{2+}–$Fe(OH)_3$ boundary line (see Figure 4.4) in a solution with a soluble iron concentration of 1.00×10^{-4} M at pH 6.00?
9. What is the p*E* value in an acid mine water sample having $[Fe^{3+}] = 7.03 \times 10^{-3}$ M and $[Fe^{2+}] = 3.71 \times 10^{-4}$ M?
10. At pH 6.00 and p*E* 2.58, what is the concentration of Fe^{2+} in equilibrium with $Fe(OH)_2$?
11. What is the calculated value of the partial pressure of O_2 in acid mine water of pH 2.00, in which $[Fe^{3+}] = [Fe^{2+}]$?
12. What is the major advantage of expressing redox reactions and half-reactions in terms of exactly one electron-mole?

13. Why are p*E* values that are determined by reading the potential of a platinum electrode versus a reference electrode generally not very meaningful?
14. What determines the oxidizing and reducing limits, respectively, for the thermodynamic stability of water?
15. How would you expect p*E* to vary with depth in a stratified lake?
16. Upon what half-reaction is the rigorous definition of p*E* based?
17. Analysis of water in a sediment sample at equilibrium at pH 7.00 showed $[SO_4^{2-}] = 2.00 \times 10^{-5}$ M and a partial pressure of H_2S of 0.100 atm. Show with appropriate calculations if methane, CH_4, would be expected in the sediment.
18. Choose the correct answer of the following, explain why it is true, and explain why the other choices are untrue: (A) high p*E* is associated with species such as CH_4, NH_4^+, and Fe^{2+}; (B) low p*E* is associated with species such as CO_2, O_2, and NO_3^-; (C) values of p*E* in bodies of water range from about 1×10^{-7} to about 1×10^7; (D) p*E* is a number, but cannot be related to anything real, such as is the case with pH; (E) p*E* uses convenient numbers to express electron activity over many orders of magnitude.
19. Match the following:

A. For 1 electron-mole	1. $Fe(H_2O)_6^{2+} \rightleftarrows e^- + Fe(OH)_3(s) + 3H_2O + 3H^+$
B. Reaction for standard electrode	2. $H_2 \rightleftarrows 2H^+ + 2e^-$
C. At the upper p*E* limit of water	3. $\frac{1}{8}NH_4^+ + \frac{1}{4}O_2 \rightleftarrows \frac{1}{8}NO_3^- + \frac{1}{4}H^+ + H_2O$
D. Formation of a pollutant when anoxic water is brought to the surface	4. $2H_2O \rightleftarrows O_2 + 4H^+ + 4e^-$

20. Of the following, the true statement regarding oxidation–reduction reactions and phenomena in natural water systems is: (A) At a p*E* higher than the oxidizing limit of stability, water decomposes to evolve H_2; (B) the production of CH_4 at a very low p*E* is caused by the action of bacteria; (C) in the p*E*–pH diagram for iron, the region of greatest area is occupied by solid $Fe(OH)_2$; (D) it is easy to accurately measure the p*E* of water with a platinum electrode; (E) there are no p*E*–pH limits for the regions of stability of H_2O.

5 Phase Interactions in Aquatic Chemistry

5.1 CHEMICAL INTERACTIONS INVOLVING SOLIDS, GASES, AND WATER

Homogeneous chemical reactions occurring entirely in aqueous solution are rather rare in natural waters and wastewaters. Instead, most significant chemical and biochemical phenomena in water involve interactions between species in water and another phase. Some of these important interactions are illustrated in Figure 5.1. Several examples of phase interactions in water illustrated by the figure are the following: production of solid biomass through the photosynthetic activity of algae occurs within a suspended algal cell and involves exchange of dissolved solids and gases between the surrounding water and the cell. Similar exchanges occur when bacteria degrade organic matter (often in the form of small particles) in water. Chemical reactions occur that produce solids or gases in water. Iron and many important trace-level elements are transported through aquatic systems as colloidal chemical compounds or are sorbed to solid particles. Pollutant hydrocarbons and some pesticides may be present on the water surface as an immiscible liquid film. Sediment can be washed physically into a body of water.

This chapter discusses the importance of interactions among different phases in aquatic chemical processes. In a general sense, in addition to water, these phases may be divided between *sediments* (bulk solids) and *suspended colloidal material*. The ways in which sediments are formed and the significance of sediments as repositories and sources of aquatic solutes are discussed. Mentioned in earlier chapters, solubilities of solids and gases (Henry's law) are covered here in some detail.

Much of this chapter deals with the behavior of colloidal material that consists of very fine particles of solids, gases, or immiscible liquids suspended in water. Colloidal material is involved with many significant aquatic chemical phenomena. It is very reactive because of its high surface-area-to-volume ratio.

5.2 IMPORTANCE AND FORMATION OF SEDIMENTS

Sediments are the layers of relatively finely divided matter covering the bottoms of rivers, streams, lakes, reservoirs, bays, estuaries, and oceans. Sediments typically consist of mixtures of fine-, medium-, and coarse-grained minerals, including clay, silt, and sand, mixed with organic matter. They may vary in composition from pure mineral matter to predominantly organic matter. Sediments are reservoirs of a variety of biological, chemical, and pollutant detritus in bodies of water and are repositories of pollutants, such as heavy metals and toxic organic compounds. Of particular

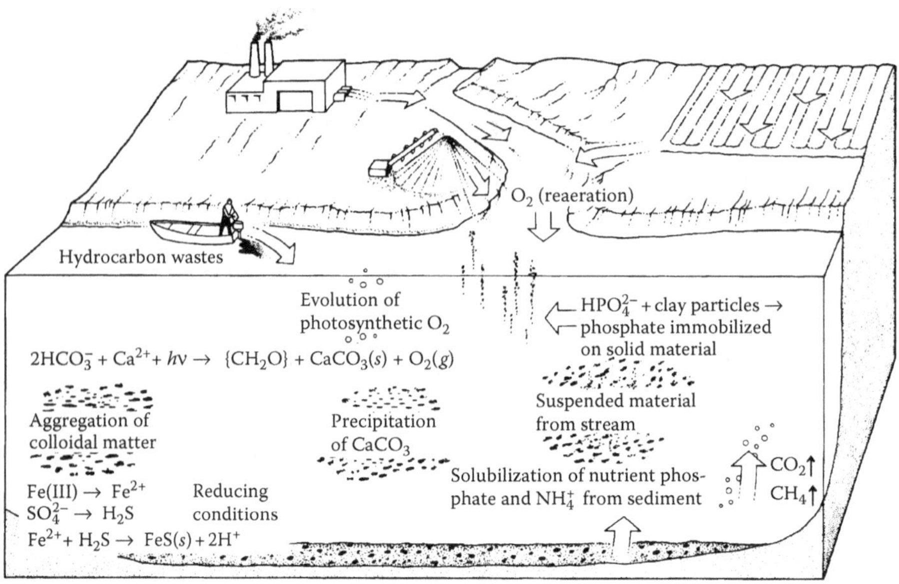

FIGURE 5.1 Most important environmental chemical processes in water involve interactions between water and another phase. (From Manahan, S., *Environmental Chemistry*, 9th ed., Taylor & Francis, Boca Raton, 2010. With permission.)

concern is the transfer of chemical species from sediments into aquatic food chains via organisms that spend significant parts of their life cycles in contact with or living in sediments. Among the sediment-dwelling organisms are various kinds of shellfish (shrimp, crayfish, crab, and clams) and a variety of worms, insects, amphipods, bivalves, and other smaller organisms that are of particular concern because they are located near the base of the food chain.

Pollutant transfer from sediments to organisms may involve an intermediate stage in water solution as well as direct transfer from sediments to organisms. This is probably particularly important for poorly water-soluble organophilic pollutants, such as organohalide pesticides. The portion of substances held in sediments that is most available to organisms is that contained in *pore water*, obtained in microscopic pores within the sediment mass. Pore water is commonly extracted from sediments for measurements of toxicity to aquatic test organisms.

5.2.1 Formation of Sediments

Physical, chemical, and biological processes may all result in the deposition of sediments in the bottom regions of bodies of water. These sediments may eventually be covered and produce sedimentary minerals. Sedimentary material may be simply carried into a body of water by erosion or through sloughing (caving in) of the shore. Thus, clay, sand, organic matter, and other materials may be washed into a lake and settle out as layers of sediment.

Sediments may be formed by simple precipitation reactions, several of which are discussed below. When a phosphate-rich wastewater enters a body of water containing a high concentration of calcium ions, the following reaction occurs to produce solid hydroxyapatite:

$$5Ca^{2+} + H_2O + 3HPO_4^{2-} \rightarrow Ca_5OH(PO_4)_3(s) + 4H^+ \tag{5.1}$$

Calcium carbonate sediment may form when water rich in carbon dioxide and containing a high level of calcium as temporary hardness (see Section 3.5) loses carbon dioxide to the atmosphere:

$$Ca^{2+} + 2HCO_3^- \rightarrow CaCO_3(s) + CO_2(g) + H_2O \tag{5.2}$$

or when the pH is raised by a photosynthetic reaction:

$$Ca^{2+} + 2HCO_3^- + h\nu \rightarrow \{CH_2O\} + CaCO_3(s) + O_2(g) \tag{5.3}$$

Oxidation of reduced forms of an element can result in its transformation to an insoluble species, such as occurs when iron(II) is oxidized to iron(III) to produce a precipitate of insoluble iron(III) hydroxide:

$$4Fe^{2+} + 10H_2O + O_2 \rightarrow 4Fe(OH)_3(s) + 8H^+ \tag{5.4}$$

A decrease in pH can result in the production of an insoluble humic acid sediment from base-soluble organic humic substances in solution (see Section 3.13).

Biological activity is responsible for the formation of some aquatic sediments. Some bacterial species produce large quantities of iron(III) oxide (see Section 6.14) as part of their energy-extracting mediation of the oxidation of iron(II) to iron(III). In anoxic (oxygen-deficient) bottom regions of bodies of water, some bacteria use sulfate ion as an electron receptor:

$$SO_4^{2-} \rightarrow H_2S \tag{5.5}$$

whereas other bacteria reduce iron(III) to iron(II):

$$Fe(OH)_3(s) \rightarrow Fe^{2+} \tag{5.6}$$

These two products can react to give a black layer of iron(II) sulfide sediment:

$$Fe^{2+} + H_2S \rightarrow FeS(s) + 2H^+ \tag{5.7}$$

This frequently occurs during the winter, alternating with the production of calcium carbonate by-product from photosynthesis (Reaction 5.3) during the summer. Under such conditions, a layered bottom sediment is produced composed of alternate layers of black FeS and white $CaCO_3$ as shown in Figure 5.2.

5.2.2 Organic and Carbonaceous Sedimentary Materials

Carbonaceous sediments from organic materials are particularly important because of their affinity for poorly soluble organic water pollutants. (In chemical fate and

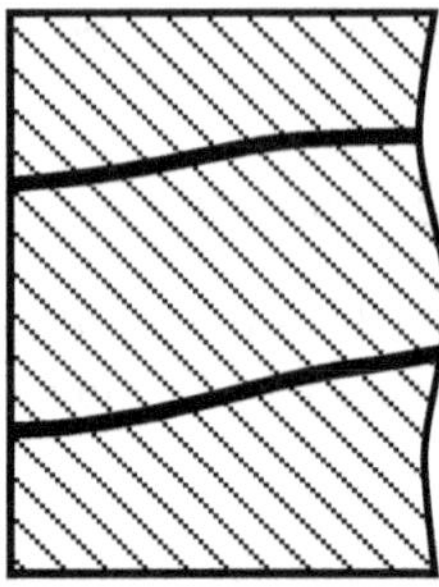

FIGURE 5.2 Alternate layers of FeS and $CaCO_3$ in a lake sediment. This phenomenon has been observed in Lake Zürich in Switzerland. (From Manahan, S., *Environmental Chemistry*, 9th ed., Taylor & Francis, Boca Raton, 2010. With permission.)

transport calculations involving the uptake of organic materials from water by sediments containing organic solids, the sediment–water partition coefficient for the organic substance partitioning between water and soil may be expressed as a function of the fraction of organic matter in the sediment, f_{oc}, and the partition coefficient of the organic contaminant for the pure organic solid, K_{oc}.) Sediment organic carbon comes from biological sources and from fossil fuels. Biological sources may consist of plant, animal, and microbial biomass including materials such as cellulose, lignin, collagen, and cuticle, and the degradation products thereof, especially humic substances. Fossil fuel sources include coal tar, petroleum residues (such as asphalt), soot, coke, charcoal, and coal.

Black carbon is the name given to small particles of carbon left over from the combustion of fossil fuels and biomass. Significant quantities of black carbon are produced by combustion processes and it is found in atmospheric particulate matter, in soil, and in sediments. Elemental carbon has an affinity for organic matter (see the discussion of activated carbon in Chapter 8) and is a significant sink for hydrophobic organic compounds in sediments.[1]

Hydrophobic organic compounds are bound preferentially to sediment organic carbon in sediments. The two most prominent examples of such compounds are polycyclic aromatic hydrocarbons (PAHs) and polychlorinated biphenyls (PCBs). From 60% to 90% of these hydrophobic organic compounds are bound with sediment organic carbon even though typically the sediment organic carbon comprises only 5–7% of the sediments. Sediment organic carbon provides a reservoir for the storage of hydrophobic organic compounds that may persist for many years after the source of pollution is removed. However, compounds held by these solids are relatively less bioavailable and less readily biodegraded than compounds that are in solution or are bound to mineral sedimentary materials.

5.3 SOLUBILITIES

The formation and stabilities of nonaqueous phases in water are strongly dependent upon solubilities. Calculations of the solubilities of solids and gases are addressed in this section.

5.3.1 Solubilities of Solids

Generally, the solubility of a solid in water is of concern when the solid is slightly soluble, often having such a low solubility that it is called "insoluble." In Section 3.11 the solubility of lead carbonate was considered. This salt can introduce toxic lead ion into water by reactions such as

$$PbCO_3(s) \rightleftarrows Pb^{2+} + CO_3^{2-} \quad (3.79)$$

A relatively straightforward calculation of the solubility of an ionic solid can be performed on barium sulfate, which dissolves according to the reaction

$$BaSO_4(s) \rightleftarrows Ba^{2+} + SO_4^{2-} \quad (5.8)$$

for which the equilibrium constant is the following:

$$K_{sp} = [Ba^{2+}][SO_4^{2-}] = 1.23 \times 10^{-10} \quad (5.9)$$

An equilibrium constant in this form that expresses the solubility of a solid that forms ions in water is a *solubility product* and is designated K_{sp}. In the simplest cases, a solubility product can be used alone to calculate the solubility of a slightly soluble salt in water. The solubility (S, mol/L) of barium sulfate is calculated as follows:

$$[Ba^{2+}] = [SO_4^{2-}] = S \quad (5.10)$$

$$[Ba^{2+}][SO_4^{2-}] = S \times S = K_{sp} = 1.23 \times 10^{-10} \quad (5.11)$$

$$S = (K_{sp})^{1/2} = (1.23 \times 10^{-10})^{1/2} = 1.11 \times 10^{-5} \quad (5.12)$$

Even such a simple calculation may be complicated by variations in activity coefficients resulting from differences in ionic strength.

Intrinsic solubilities account for the fact that a significant portion of the solubility of an ionic solid is due to the dissolution of the neutral form of the salt and must be added to the solubility calculated from K_{sp} to obtain the total solubility, as illustrated below for the calculation of the solubility of calcium sulfate. When calcium sulfate dissolves in water that does not have any other sources of calcium or sulfate ions, the two major reactions are

$$CaSO_4(s) \rightleftarrows CaSO_4(aq) \quad (5.13)$$

$$[CaSO_4(aq)] = 5.0 \times 10^{-3}\ M\ (25°C) \quad (5.14)$$

(intrinsic solubility of $CaSO_4$)

$$CaSO_4(s) \rightleftarrows Ca^{2+} + SO_4^{2-} \quad (5.15)$$

$$[Ca^{2+}][SO_4^{2-}] = K_{sp} = 2.6 \times 10^{-5}\ (25°C) \tag{5.16}$$

and the total solubility of $CaSO_4$ is calculated as follows:

$$S = \underset{\substack{\text{Contribution to solubility}\\ \text{from solubility product}}}{[Ca^{2+}]} + \underset{\substack{\text{Contribution to solubility}\\ \text{from intrinsic solubility}}}{[CaSO_4(aq)]}$$

$$\begin{aligned} S &= (K_{sp})^{1/2} + [CaSO_4(aq)] = (2.6 \times 10^{-5})^{1/2} + 5.0 \times 10^{-3} \\ &= 5.1 \times 10^{-3} + 5.0 \times 10^{-3} = 1.01 \times 10^{-2}\ \text{M} \end{aligned} \tag{5.18}$$

It is seen that, in this case, the intrinsic solubility accounts for half of the solubility of the salt.

In Section 3.11, it was seen that solubilities of ionic solids can be very much affected by reactions of cations and anions. It was shown that the solubility of $PbCO_3$ is increased by the chelation of lead ion by NTA,

$$Pb^{2+} + T^{3-} \rightleftarrows PbT^{-} \tag{3.72}$$

increased by reaction of carbonate ion with H^+,

$$H^+ + CO_3^{2-} \rightleftarrows HCO_3^- \tag{5.19}$$

and decreased by the presence of carbonate ion from water alkalinity:

$$CO_3^{2-}(\text{from dissociation of } HCO_3^-) + Pb^{2+} \rightleftarrows PbCO_3(s) \tag{5.20}$$

These examples illustrate that reactions of both cations and anions must often be considered in calculating the solubilities of ionic solids.

5.3.2 Solubilities of Gases

The solubilities of gases in water are described by Henry's law, which states that *at constant temperature the solubility of a gas in a liquid is proportional to the partial pressure of the gas in contact with the liquid.* For a gas, "X," this law applies to equilibria of the type

$$X(g) \rightleftarrows X(aq) \tag{5.21}$$

and does not account for additional reactions of the gas species in water such as

$$NH_3 + H_2O \rightleftarrows NH_4^+ + OH^- \tag{5.22}$$

$$SO_2 + HCO_3^-(\text{from water alkalinity}) \rightleftarrows CO_2 + HSO_3^- \tag{5.23}$$

which may result in much higher solubilities than predicted by Henry's law alone. Mathematically, Henry's law is expressed as

$$[X(aq)] = KP_X \tag{5.24}$$

where $[X(aq)]$ is the aqueous concentration of the gas, P_X is the partial pressure of the gas, and K is the Henry's law constant applicable to a particular gas at a specified temperature. For gas concentrations in units of moles per liter and gas pressures in atmospheres, the units of K are moles per liter per atmosphere. Some values of K for dissolved gases that are significant in water are given in Table 5.1.

In calculating the solubility of a gas in water, a correction must be made for the partial pressure of water by subtracting it from the total pressure of the gas. At 25°C, the partial pressure of water is 0.0313 atm; values at other temperatures are readily obtained from standard handbooks. The concentration of oxygen in water saturated with air at 1.00 atm and 25°C may be calculated as an example of a simple gas solubility calculation. Considering that dry air is 20.95% by volume oxygen, factoring in the partial pressure of water gives the following:

$$P_{O_2} = (1.0000 \text{ atm} - 0.0313 \text{ atm}) \times 0.2095 = 0.2029 \text{ atm} \tag{5.25}$$

$$[O_2(aq)] = K \times P_{O_2} = 1.28 \times 10^{-3} \text{ mol/L/atm} \times 0.2029 \text{ atm} = 2.60 \times 10^{-4} \text{ mol/L} \tag{5.26}$$

The molecular mass of oxygen is 32, so the concentration of dissolved O_2 in water in equilibrium with air under the conditions given above is 8.32 mg/L or 8.32 parts per million (ppm).

The solubilities of gases decrease with increasing temperature. Account is taken of this factor with the *Clausius–Clapeyron* equation,

$$\log \frac{C_2}{C_1} = \frac{\Delta H}{2.303R}\left[\frac{1}{T_1} - \frac{1}{T_2}\right] \tag{5.27}$$

TABLE 5.1
Henry's Law Constants for Some Gases in Water at 25°C

Gas	K (mol/L/atm)
O_2	1.28×10^{-3}
CO_2	3.38×10^{-2}
H_2	7.90×10^{-4}
CH_4	1.34×10^{-3}
N_2	6.48×10^{-4}
NO	2.0×10^{-4}

Source: From Manahan, S., *Environmental Chemistry*, 9th ed., Taylor & Francis, Boca Raton, 2010. With permission.

where C_1 and C_2 denote the gas concentration in water at absolute temperatures of T_1 and T_2, respectively; ΔH is the heat of the solution; and R is the gas constant. The value of R is 1.987 cal/deg/mol, which gives ΔH in units of cal/mol.

5.4 COLLOIDAL PARTICLES IN WATER

Many minerals, some organic pollutants, proteinaceous materials, some algae, and some bacteria are suspended in water as very small particles. Such particles, which have some characteristics of both species in solution and larger particles in suspension, which range in diameter from about 0.001 micrometer (μm) to about 1 μm, and which scatter white light as a light blue hue observed at right angles to the incident light, are classified as *colloidal particles.* The characteristic light-scattering phenomenon of colloids results from their being the same order of size as the wavelength of light and is called the *Tyndall effect.* The unique properties and behavior of colloidal particles are strongly influenced by their physical–chemical characteristics, including high specific area, high interfacial energy, and high surface/charge density ratio. Colloids play a very important role in determining the properties and behavior of natural waters and wastewaters.

5.4.1 Contaminant Transport by Colloids in Water

An important influence of colloids in aquatic chemistry is their ability to transport various kinds of organic and inorganic contaminants. *Colloid-facilitated transport* in which contaminants bound to the surface of colloidal particles can be an important means of moving substances that would otherwise be sorbed to sediments or, in the case of groundwater transport, to aquifer rocks. This mechanism is of concern with respect to circumvention of natural and artificial barriers in the long-term subsurface disposal of some kinds of wastes, such as high-level nuclear wastes including plutonium.[2]

5.4.2 Occurrence of Colloids in Water

Colloids composed of a variety of organic substances (including humic substances), inorganic materials (especially clays), and pollutants occur in natural water and wastewater. These substances have a number of effects, including those on organisms and pollutant transport. The characterization of colloidal materials in water is obviously very important, and a variety of means are used to isolate and characterize those materials. The two most widely used methods are filtration and centrifugation, although other techniques including voltammetry and field-flow fractionation can be used.

5.4.3 Kinds of Colloidal Particles

Colloids may be classified as *hydrophilic colloids*, *hydrophobic colloids*, or *association colloids*. These three classes are briefly summarized below.

Hydrophilic colloids generally consist of macromolecules, such as proteins and synthetic polymers that are characterized by strong interaction with water resulting in spontaneous formation of colloids when they are placed in water. In a sense, hydrophilic colloids are solutions of very large molecules or ions. Suspensions of hydrophilic colloids are less affected by the addition of salts to water than are suspensions of hydrophobic colloids.

Hydrophobic colloids interact to a lesser extent with water and are stable because of their positive or negative electrical charges as shown in Figure 5.3. The charged surface of the colloidal particle and the *counterions* that surround it compose an *electrical double layer*, which causes the particles to repel each other.

Hydrophobic colloids are usually caused to settle from suspension by the addition of salts. Examples of hydrophobic colloids are clay particles, petroleum droplets, and very small gold particles.

Association colloids consist of special aggregates of ions and molecules called *micelles*. To understand how this occurs, consider sodium stearate, a typical soap with the structural formula shown below:

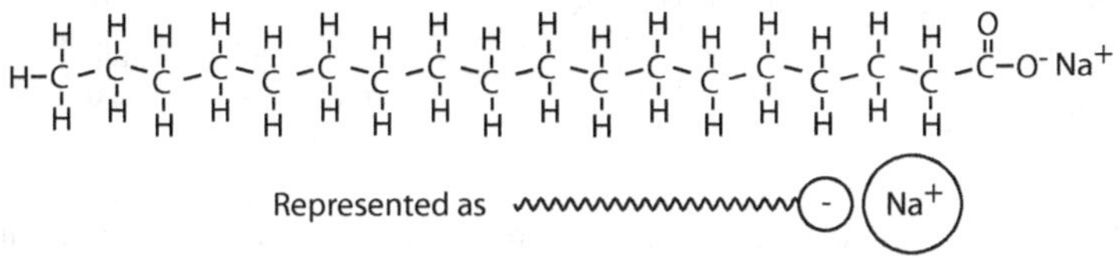

The stearate ion has both a hydrophilic $-CO_2^-$ head and a long organophilic tail, $CH_3(CH_2)_{16}-$. As a result, stearate anions in water tend to form clusters consisting of as many as 100 anions clustered together with their hydrocarbon "tails" on the inside of a spherical colloidal particle and their ionic "heads" on the surface in contact with water and with Na^+ counter-ions. Such *micelles* (Figure 5.4) can be visualized as

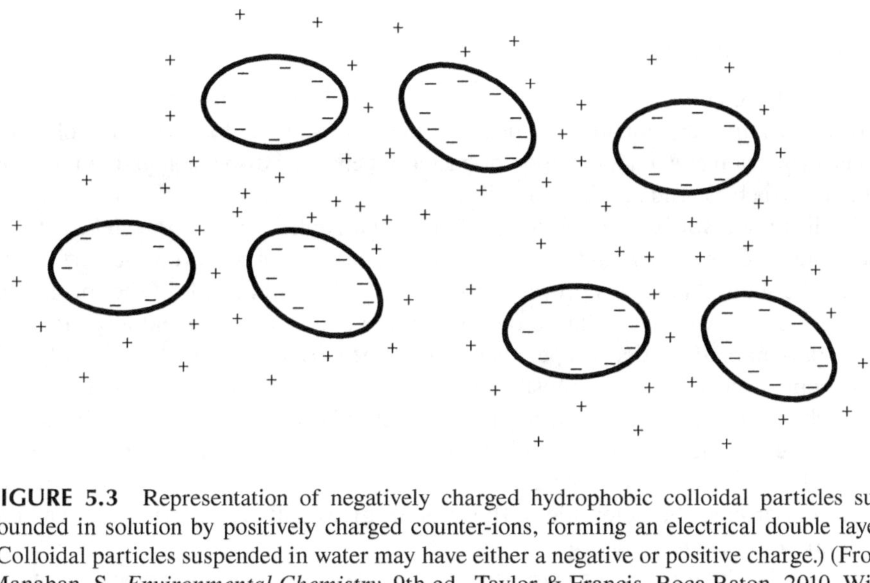

FIGURE 5.3 Representation of negatively charged hydrophobic colloidal particles surrounded in solution by positively charged counter-ions, forming an electrical double layer. (Colloidal particles suspended in water may have either a negative or positive charge.) (From Manahan, S., *Environmental Chemistry*, 9th ed., Taylor & Francis, Boca Raton, 2010. With permission.)

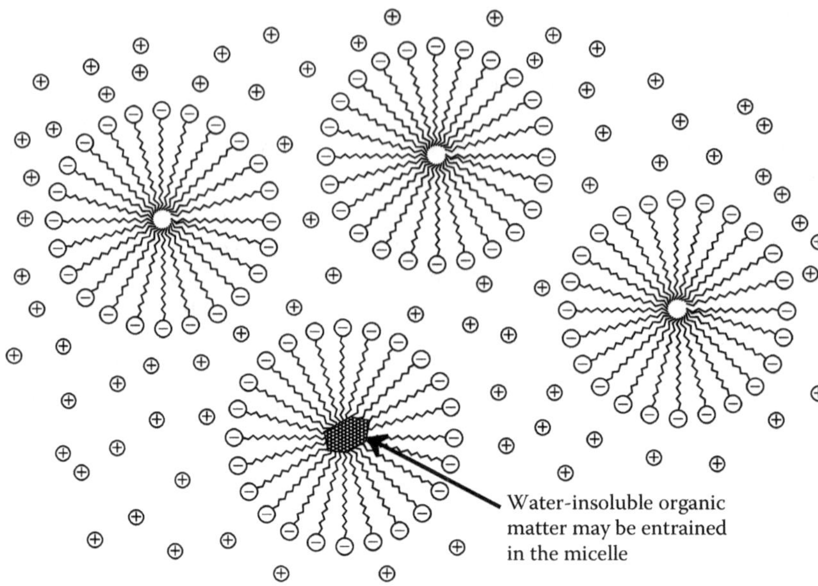

FIGURE 5.4 Representation of colloidal soap micelle particles. (From Manahan, S., *Environmental Chemistry*, 9th ed., Taylor & Francis, Boca Raton, 2010. With permission.)

droplets of oil about 3–4 nanometers (nm) in diameter and covered with ions or polar groups. According to this model, micelles form when a certain concentration of surfactant species, typically around 1×10^{-3} M, is reached. The concentration at which this occurs is called the *critical micelle concentration.*

5.4.4 Colloid Stability

The stability of colloids is a prime consideration in determining their behavior. It is involved in important aquatic chemical phenomena including the formation of sediments, dispersion and agglomeration of bacterial cells, and dispersion and removal of pollutants (such as crude oil from an oil spill).

As discussed above, the two main phenomena contributing to the stabilization of colloids are *hydration* and *surface charge.* The layer of water on the surface of hydrated colloidal particles prevents contact, which would result in the formation of larger units. A surface charge on colloidal particles may prevent aggregation, since like-charged particles repel each other. The surface charge is frequently pH dependent; at around pH 7, most colloidal particles in natural waters are negatively charged. Negatively charged aquatic colloids include algal cells, bacterial cells, proteins, and colloidal petroleum droplets. Natural organic matter in water tends to bind with colloidal particle surfaces and, because of the negatively charged functional groups characteristic of this matter, gives colloidal particles in water a predominantly negative charge.

One of the three major ways in which a particle may acquire a surface charge is by *chemical reaction at the particle surface.* This phenomenon, which frequently involves hydrogen ion and is pH dependent, is typical of hydroxides and oxides and is illustrated for manganese dioxide, MnO_2, in Figure 5.5.

As an illustration of pH-dependent charge on colloidal particle surfaces, consider the effects of pH on the surface charge of hydrated manganese oxide, represented by the chemical formula $MnO_2(H_2O)(s)$. In a relatively acidic medium, the reaction

$$MnO_2(H_2O)(s) + H^+ \rightarrow MnO_2(H_3O)^+(s) \tag{5.28}$$

may occur on the surface giving the particle a net positive charge. In a more basic medium, hydrogen ions may be lost from the hydrated oxide surface to yield negatively charged particles:

$$MnO_2(H_2O)(s) \rightarrow MnO_2(OH)^-(s) + H^+ \tag{5.29}$$

At some intermediate pH value, called the *zero point of charge* (ZPC), colloidal particles of a given hydroxide will have a net charge of zero, which favors aggregation of particles and precipitation of a bulk solid:

$$\text{Number of } MnO_2(H_3O)^+ \text{ sites} = \text{Number of } MnO_2(OH)^- \text{ sites} \tag{5.30}$$

Individual cells of microorganisms that behave as colloidal particles have a charge that is pH dependent. The charge is acquired through the loss and gain of H^+ ions by carboxyl and amino groups on the cell surface:

$$\underset{\text{low pH}}{{}^+H_3N(+\text{cell})CO_2H} \quad \underset{\text{intermediate pH}}{{}^+H_3N(\text{neutral cell})CO_2^-} \quad \underset{\text{high pH}}{H_2N(-\text{cell})CO_2^-}$$

Ion absorption is a second way in which colloidal particles become charged. This phenomenon involves attachment of ions onto the colloidal particle surface by means other than conventional covalent bonding, including hydrogen bonding and London (van der Waal) interactions.

Ion replacement is a third way in which a colloidal particle may gain a net charge; for example, replacement of some of the Si(IV) with Al(III) in the basic SiO_2 chemical unit in the crystalline lattice of some clay minerals as shown in Equation 5.31,

$$[SiO_2] + Al(III) \rightarrow [AlO_2^-] + Si(IV) \tag{5.31}$$

yields sites with a net negative charge. Similarly, replacement of Al(III) by a divalent metal ion such as Mg(II) in the clay crystalline lattice produces a net negative charge.

5.5 COLLOIDAL PROPERTIES OF CLAYS

Clays constitute the most important class of common minerals occurring as colloidal matter in water. The composition and properties of clays are briefly summarized here. *Clays* consist largely of hydrated aluminum and silicon oxides and are

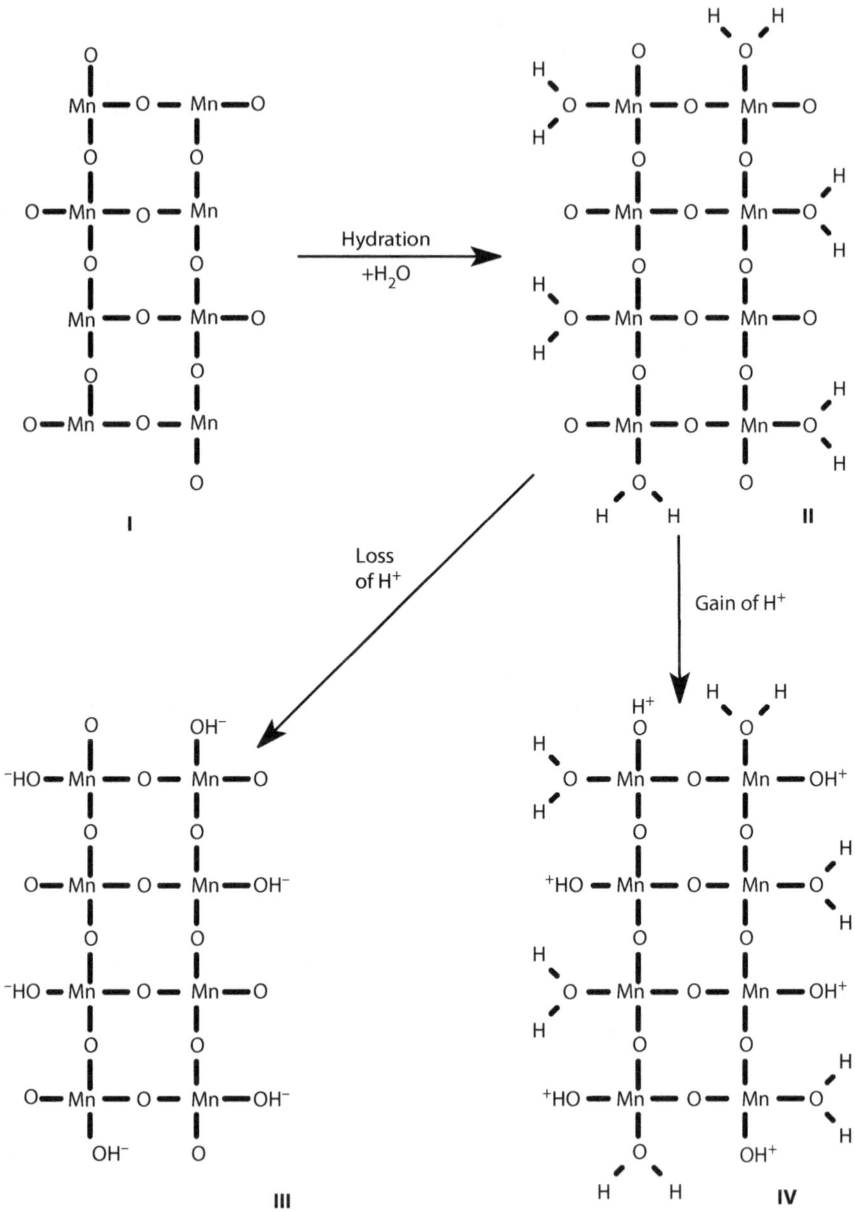

FIGURE 5.5 Acquisition of surface charge by colloidal MnO_2 in water. (I) Anhydrous MnO_2 has two O atoms per Mn atom. (II) Suspended in water as a colloid, it binds to water molecules to form hydrated MnO_2. (III) Loss of H^+ from the bound H_2O yields a negatively charged colloidal particle. (IV) Gain of H^+ by surface O atoms yields a positively charged particle. The former process (loss of H^+ ion) predominates for metal oxides. (From Manahan, S., *Environmental Chemistry*, 9th ed., Taylor & Francis, Boca Raton, 2010. With permission.)

secondary minerals, which are formed by weathering and other processes acting on primary rocks. The general formulas of some common clays are given below:

- Kaolinite: $Al_2(OH)_4Si_2O_5$
- Montmorillonite: $Al_2(OH)_2Si_4O_{10}$
- Nontronite: $Fe_2(OH)_2Si_4O_{10}$
- Hydrous mica: $KAl_2(OH)_2(AlSi_3)O_{10}$

The most common clay minerals are illites, montmorillonites, chlorites, and kaolinites. These clay minerals are distinguished from each other by general chemical formula, structure, and chemical and physical properties. Iron and manganese are commonly associated with clay minerals.

Clays are characterized by layered structures consisting of sheets of silicon oxide alternating with sheets of aluminum oxide. Units of two or three sheets make up *unit layers*. Some clays, particularly the montmorillonites, may absorb large quantities of water between unit layers, a process accompanied by swelling of the clay.

As described in Section 5.4, clay minerals may attain a net negative charge by ion replacement, in which Si(IV) and Al(III) ions are replaced by metal ions of similar size but lesser charge. This negative charge must be compensated for by association of cations with the clay layer surfaces. Since these cations need not fit specific sites in the crystalline lattice of the clay, they may be relatively large ions, such as K^+, Na^+, or NH_4^+. These cations are called *exchangeable cations* and are exchangeable for other cations in water. The amount of exchangeable cations, expressed as milliequivalents (of monovalent cations) per 100 g of dry clay, is called the *cation-exchange capacity* (CEC) of the clay and is a very important characteristic of colloids and sediments that have cation-exchange capabilities.

Because of their structure and high surface area per unit mass, clays have a strong tendency to sorb chemical species from water. Thus, clays play a role in the transport and reactions of biological wastes, organic chemicals, gases, and other pollutant species in water. However, clay minerals may also effectively immobilize dissolved chemicals in water and so exert a purifying action. Some microbial processes occur at clay particle surfaces and, in some cases, sorption of organics by clay inhibits biodegradation. Thus, clay may play a role in the microbial degradation or nondegradation of organic wastes.

5.6 AGGREGATION OF PARTICLES

The processes by which particles aggregate and precipitate from colloidal suspension are quite important in the aquatic environment. For example, the settling of biomass during biological waste treatment depends upon the aggregation of bacterial cells. Other processes involving the aggregation of colloidal particles are the formation of bottom sediments and the clarification of turbid water for domestic or industrial use. Particle aggregation is complicated and may be divided into the two general classes of *coagulation* and *flocculation*. These are discussed below.

Colloidal particles are prevented from aggregating by the electrostatic repulsion of the electrical double layers (adsorbed-ion layer and counter-ion layer). *Coagulation* involves the reduction of this electrostatic repulsion such that colloidal particles of identical materials may aggregate. *Flocculation* uses *bridging compounds*, which form chemically bonded links between colloidal particles and enmesh the particles in relatively large masses called *floc networks*.

Hydrophobic colloids are often readily coagulated by the addition of small quantities of salts that contribute ions to solution. Such colloids are stabilized by electrostatic repulsion. Therefore, the simple explanation of coagulation by ions in solution is that the ions reduce the electrostatic repulsion between particles to such an extent that the particles aggregate. Because of the double layer of electrical charge surrounding a charged particle, this aggregation mechanism is sometimes called *double-layer compression.* It is particularly noticeable in estuaries where sediment-laden fresh water flows into the sea, and is largely responsible for deltas formed where large rivers enter oceans.

The binding of positive ions to the surface of an initially negatively charged colloid can result in precipitation followed by colloid restabilization as shown in Figure 5.6. This kind of behavior is explained by an initial neutralization of the negative surface charge on the particles by sorption of positive ions, allowing coagulation to occur. As more of the source of positive ions is added, their sorption results in the formation of positive colloidal particles.

5.6.1 Flocculation of Colloids by Polyelectrolytes

Polyelectrolytes of both natural and synthetic origin may cause colloids to flocculate. Polyelectrolytes are polymers with a high formula mass that normally contain ionizable functional groups. Typical examples of synthetic polyelectrolytes are shown in Table 5.2.

It can be seen from Table 5.2 that anionic polyelectrolytes have negatively charged functional groups such as $-SO_3^-$ and $-CO_2^-$. Cationic polyelectrolytes have positively charged functional groups, normally H^+ bonded to N. Nonionic polymers that serve as flocculants normally do not have charged functional groups.

Somewhat paradoxically, *anionic* polyelectrolytes may flocculate *negatively charged* colloidal particles, a process that is particularly important in biological systems, for example, in the cohesion of tissue cells, clumping of bacterial cells, and antibody–antigen reactions. The mechanism by which this occurs involves bridging

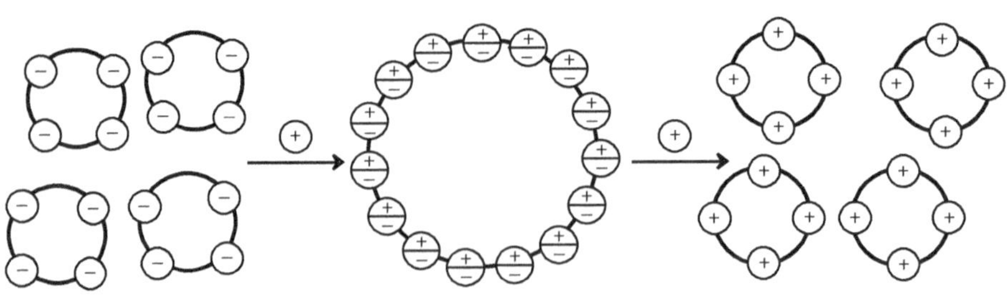

FIGURE 5.6 Aggregation of negatively charged colloidal particles by reaction with positive ions, followed by restabilization as a positively charged colloid. (From Manahan, S., *Environmental Chemistry*, 9th ed., Taylor & Francis, Boca Raton, 2010. With permission.)

TABLE 5.2
Synthetic Polyelectrolytes and Neutral Polymers Used as Flocculants

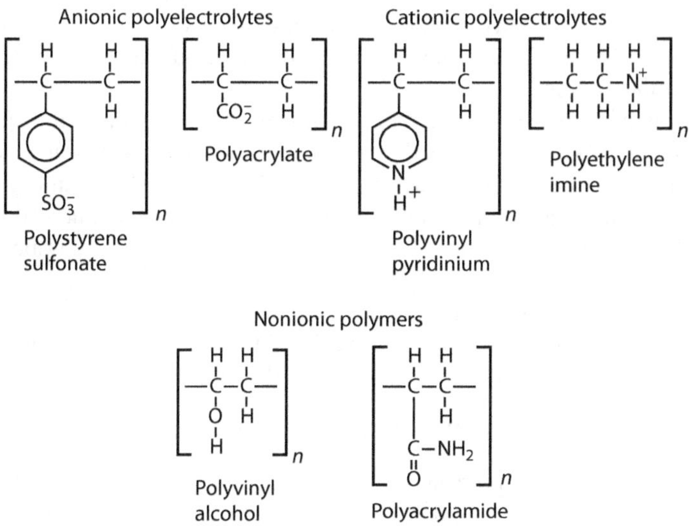

Source: From Manahan, S., *Environmental Chemistry*, 9th ed., Taylor & Francis, Boca Raton, 2010. With permission.

between the colloidal particles by way of the polyelectrolyte anions. It is facilitated by the presence of a low concentration of a metal ion capable of serving as a bridge between the negatively charged anionic polyelectrolytes and negatively charged functional groups on the colloidal particle surface.

5.6.2 Flocculation of Bacteria by Polymeric Materials

The aggregation and settling of microorganism cells is a very important process in aquatic systems and is essential to the function of biological waste treatment systems. In these processes, using bacteria to degrade organic wastes, a significant fraction of the carbon in the waste is removed as *bacterial floc*, consisting of aggregated bacterial cells that have settled from the water. The formation of this floc is obviously an important phenomenon in biological waste treatment. Polymeric substances, including polyelectrolytes, that are formed by the bacteria induce bacterial flocculation.

Within the pH range of normal natural waters (pH 5–9), bacterial cells are negatively charged. The ZPC of most bacteria is within the pH range 2–3. However, even at the ZPC, stable bacterial suspensions may exist. Therefore, surface charge is not necessarily required to maintain bacterial cells in suspension in water, and it is likely that bacterial cells remain in suspension because of the hydrophilic character of their surfaces. As a consequence, some sort of chemical interaction involving bridging species must be involved in bacterial flocculation.

5.7 SURFACE SORPTION BY SOLIDS

Many of the properties and effects of solids in contact with water have to do with the sorption of solutes by solid surfaces. Surfaces in finely divided solids tend to have excess surface energy because of an imbalance of chemical forces among surface atoms, ions, and molecules. Surface energy level may be lowered by a reduction in surface area. Normally this reduction is accomplished by aggregation of particles or by sorption of solute species.

Some kinds of surface interactions can be illustrated with metal oxide surfaces binding with metal ions in water. (Such a surface, its reaction with water, and its subsequent acquisition of a charge by loss or gain of H^+ ion were shown in Figure 5.5 for MnO_2.) Other inorganic solids, such as clays, probably behave much like solid metal oxides. Soluble metal ions, such as Cd^{2+}, Cu^{2+}, Pb^{2+}, or Zn^{2+}, may be bound with metal oxides such as $MnO_2 \cdot xH_2O$ by nonspecific ion-exchange adsorption, complexation with surface –OH groups, coprecipitation in solid solution with the metal oxide, or as a discrete oxide or hydroxide of the sorbed metal. Sorption of metal ions, Mt^{z+}, by complexation to the surface is illustrated by the reaction

$$\text{M–OH} + \text{Mt}^{z+} \rightleftarrows \text{M–OMt}^{z-1} + \text{H}^+ \tag{5.32}$$

and chelation by the following process:

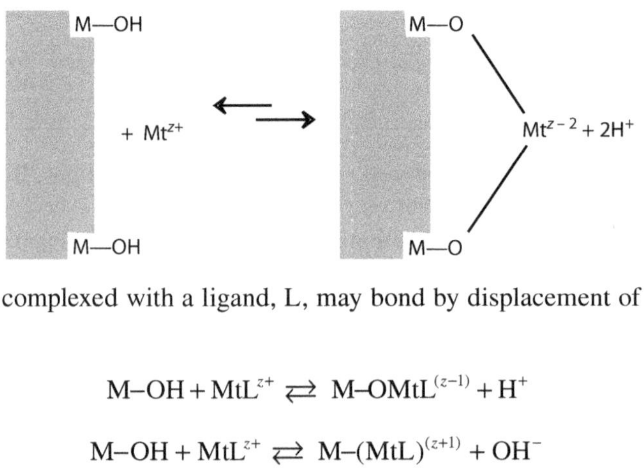

(5.33)

A metal ion complexed with a ligand, L, may bond by displacement of either H^+ or OH^-:

$$\text{M–OH} + \text{MtL}^{z+} \rightleftarrows \text{M–OMtL}^{(z-1)} + \text{H}^+ \tag{5.34}$$

$$\text{M–OH} + \text{MtL}^{z+} \rightleftarrows \text{M–(MtL)}^{(z+1)} + \text{OH}^- \tag{5.35}$$

Furthermore, in the presence of a ligand, dissociation of the complex and sorption of the metal complex and ligand must be considered as shown by the scheme below in which "(sorbed)" represents sorbed species and "(*aq*)" represents dissolved species:

$$\begin{array}{ccc}
\text{Mt}^{z+}(\text{sorbed}) & \rightleftarrows & \text{Mt}^{z+}(aq) \\
\uparrow\downarrow & & \uparrow\downarrow \\
\text{MtL}^{z+}(\text{sorbed}) & \rightleftarrows & \text{MtL}^{z+}(aq) \\
\uparrow\downarrow & & \uparrow\downarrow \\
\text{L}(\text{sorbed}) & \rightleftarrows & \text{L}(aq)
\end{array}$$

Some hydrated metal oxides, such as manganese(IV) oxide and iron(III) oxide, are especially effective in sorbing various species from aquatic solution. The sorption ability is especially pronounced for relatively fresh metal hydroxides or hydrated oxides such as colloidal MnO_2. This oxide usually is produced in natural waters by the oxidation of Mn(II) present in natural waters placed there by the bacterially mediated reduction of manganese oxides in anoxic bottom sediments. Colloidal hydrated manganese(II) oxide can also be produced by the reduction of manganese(VII), which can be deliberately added to water as an oxidant in the form of permanganate salts to diminish taste and odor or to oxidize iron(II).

Freshly precipitated MnO_2 may have a surface area as large as several hundred square meters per gram. The hydrated oxide acquires a charge by loss and gain of H^+ ion and has a ZPC in an acidic pH range between 2.8 and 4.5. Since the pH of most normal natural waters exceeds 4.5, hydrous MnO_2 colloids are usually negatively charged.

The sorption of anions by solid surfaces is harder to explain than the sorption of cations. Phosphates may be sorbed on hydroxylated surfaces by displacement of hydroxides (ion exchange):

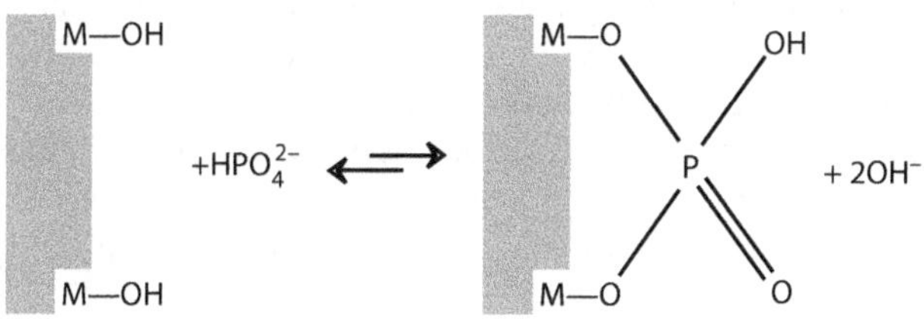

(5.36)

The degree of anion sorption varies. As with phosphate, sulfate may be sorbed by chemical bonding, usually at a pH < 7. Chloride and nitrate are sorbed by electrostatic attraction, such as occurs with positively charged colloidal particles in soil at a low pH. More specific bonding mechanisms may be involved in the sorption of fluoride, molybdate, selenate, selenite, arsenate, and arsenite anions.

5.8 SOLUTE EXCHANGE WITH BOTTOM SEDIMENTS

Bottom sediments are important sources and sinks of inorganic and organic matter in streams, freshwater impoundments, estuaries, and oceans. It is incorrect to consider bottom sediments simply as wet soil. Normal soils are in contact with the atmosphere and are oxic, whereas the environment around bottom sediments is usually anoxic, so sediments are subjected to reducing conditions. Bottom sediments undergo continuous leaching, whereas soils do not. The level of organic matter in sediments is generally higher than that in soils.

One of the most important characteristics of bottom sediments is their ability to exchange cations with the surrounding aquatic medium. CEC measures the capacity of a solid, such as a sediment, to sorb cations. It varies with pH and salt concentration. Another parameter, *exchangeable cation status* (ECS), refers to the amounts of specific ions bonded to a given amount of sediment. Generally, both CEC and ECS are expressed as milliequivalents per 100 g of solid.

Because of the generally anoxic nature of bottom sediments, special care must be exercised in their collection and treatment. Particularly, contact with atmospheric oxygen rapidly oxidizes exchangeable Fe^{2+} and Mn^{2+} to nonexchangeable oxides containing the metals in higher oxidation states as Fe_2O_3 and MnO_2. Therefore, sediment samples must be sealed and frozen as soon as possible after they are collected.

A common method for the determination of CEC consists of (1) treating the sediment with a solution of an ammonium salt so that all exchangeable sites are occupied by NH_4^+ ion; (2) displacing the ammonium ion with a solution of NaCl; and (3) determining the quantity of displaced ammonium ion. The CEC values may then be expressed as the number of milliequivalents of ammonium ion exchanged per 100 g of dried sample. Note that the sample must be dried *after* exchange.

The basic method for the determination of ECS consists of stripping all of the exchangeable metal cations from the sediment sample with ammonium acetate. Metal cations, including Fe^{2+}, Mn^{2+}, Zn^{2+}, Cu^{2+}, Ni^{2+}, Na^+, K^+, Ca^{2+}, and Mg^{2+}, are then determined in the leachate. Exchangeable hydrogen ion is very difficult to determine by direct methods. It is generally assumed that the total CEC minus the sum of all exchangeable cations except hydrogen ion is equal to the exchangeable hydrogen ion.

Freshwater sediments typically have CEC values of 20–30 milliequivalents/100 g. The ECS values for individual cations typically range from <1 to 10–20 milliequivalents/100 g. Sediments are important repositories of metal ions that may be exchanged with surrounding waters. Furthermore, because of their capacity to sorb and release hydrogen ions, sediments have an important pH-buffering effect in some waters.

5.8.1 Trace-Level Metals in Suspended Matter and Sediments

Sediments and suspended particles are important repositories for trace amounts of metals such as chromium, cadmium, copper, molybdenum, nickel, cobalt, and manganese. These metals may be present as discrete compounds, ions held by cation-exchanging clays, bound to hydrated oxides of iron or manganese, or chelated by insoluble humic substances. The form of the metals depends upon p*E*. Examples of specific trace-metal-containing compounds that may be stable in natural waters under oxidizing and reducing conditions are given in Table 5.3. Solubilization of metals from sedimentary or suspended matter is often a function of the complexing agents present. These include amino acids, such as histidine, tyrosine, or cysteine; citrate ion; and, in the presence of seawater, chloride ion. Suspended particles containing trace elements may be in the submicrometer size range. Although less available than metals in true solution, metals held by very small particles are more accessible than those in sediments. Among the factors involved in metal availability are the identity of the metal, its chemical form (type of binding, oxidation state), the nature of the suspended material, the type of organism taking up the metal, and the physical and chemical conditions in the water. The pattern of trace-metal occurrence in suspended matter in relatively unpolluted water tends to correlate well with that of the parent minerals from which the suspended solids

TABLE 5.3
Inorganic Trace Metal Compounds That May Be Stable under Oxidizing and Reducing Conditions

	Discrete Compound That May Be Present	
Metal	**Oxidizing Conditions**	**Reducing Conditions**
Cadmium	$CdCO_3$	CdS
Copper	$Cu_2(OH)_2CO_3$	CuS
Iron	$Fe_2O_3 \cdot x(H_2O)$	FeS, FeS_2
Mercury	HgO	HgS
Manganese	$MnO_2 \cdot x(H_2O)$	MnS, $MnCO_3$
Nickel	$Ni(OH)_2$, $NiCO_3$	NiS
Lead	$2PbCO_3 \cdot Pb(OH)_2$, $PbCO_3$	PbS
Zinc	$ZnCO_3$, $ZnSiO_3$	ZnS

Source: From Manahan, S., *Environmental Chemistry*, 9th ed., Taylor & Francis, Boca Raton, 2010. With permission.

originated; anomalies appear in polluted waters where industrial sources add to the metal content of the stream.

Two inorganic species are particularly important in the sequestration of heavy metals in sediments. One of these is hydrated iron(III) oxide, commonly represented as $Fe_2O_3 \cdot x(H_2O)$, $Fe(OH)_3$, or FeOOH, which coprecipitates heavy metals. This material forms when soluble iron(II) is exposed to oxidizing conditions. Under anoxic conditions the material is reduced to soluble iron(II) (Reaction 5.6), which releases bound heavy metals. The second important solid that binds heavy metals consists of *acid-volatile sulfides*, especially FeS. Most heavy metals have a stronger affinity for sulfide than does iron(II), so heavy metals tend to displace Fe(II) from this compound. The conditions conducive to reduction of hydrated iron(III) oxide leading to release of sequestered heavy metal ions are also conducive to formation of FeS, which may release sulfide to bind such ions in an insoluble form.

The toxicities of heavy metals in sediments and their availability to organisms are very important in determining the environmental effects of heavy metals in aquatic systems. Many sediments are anoxic, so that microbial reduction of sulfate to sulfide leads to a preponderance of metal sulfides in sediments. The very low solubilities of sulfides tend to limit bioavailability of metals in anoxic sediments. However, exposure of such sediments to air, and subsequent oxidation of sulfide to sulfate, can release significant amounts of heavy metals. Dredging operations can expose anoxic sediments to air, leading to oxidation of sulfides and release of metals such as lead, mercury, cadmium, zinc, and copper.

5.8.2 Phosphorus Exchange with Bottom Sediments

Phosphorus is one of the key elements in aquatic chemistry and is the limiting nutrient in the growth of algae under many conditions. Exchange with sediments plays a role in making phosphorus available for algae and contributes, therefore, to eutrophication. Sedimentary phosphorus may be classified into the following types:

- *Phosphate minerals*, particularly hydroxyapatite, $Ca_5OH(PO_4)_3$.
- *Nonoccluded phosphorus*, such as orthophosphate ion bound to the surface of SiO_2 or $CaCO_3$. Such phosphorus is generally more soluble and more available than occluded phosphorus (below).
- *Occluded phosphorus* consisting of orthophosphate ions contained within the matrix structures of amorphous hydrated oxides of iron and aluminum and amorphous aluminosilicates. Such phosphorus is not as readily available as nonoccluded phosphorus.
- *Organic phosphorus* incorporated within aquatic biomass, usually of algal or bacterial origin.

In some waters receiving heavy loads of domestic or industrial wastes, inorganic polyphosphates (e.g., from detergents) may be present in sediments. Runoff from fields where liquid polyphosphate fertilizers have been used can result in polyphosphates sorbed on sediments. The action of microorganisms on phosphate minerals, especially iron phosphates, can release phosphate to water.[3]

5.8.3 Organic Compounds on Sediments and Suspended Matter

Many organic compounds interact with suspended material and sediments in bodies of water. Colloids can play a significant role in the transport of organic pollutants in surface waters, through treatment processes, and even, to a limited extent, in groundwater. Settling of suspended material containing sorbed organic matter carries organic compounds into the sediment of a stream or lake. For example, this phenomenon is largely responsible for the presence of herbicides in sediments containing contaminated soil particles eroded from cropland. Some organics are carried into sediments by the remains of organisms or by fecal pellets from zooplankton that have accumulated organic contaminants.

Suspended particulate matter affects the mobility of organic compounds sorbed to particles. Furthermore, sorbed organic matter undergoes chemical degradation and biodegradation at different rates and by different pathways compared to organic matter in solution. There is, of course, a vast variety of organic compounds that get into water and they react with sediments in different ways, the type and strength of binding varying with the type of compound.

The most common types of sediments considered for their organic binding abilities are clays, organic (humic) substances, and complexes between clay and humic substances. Both clays and humic substances act as cation exchangers. Therefore, these materials sorb cationic organic compounds through ion exchange. This is a relatively strong sorption mechanism, greatly reducing the mobility and biological activity of the organic compound. When sorbed by clays, cationic organic compounds are generally held between the layers of the clay mineral structure where their biological activity is essentially zero.

Since most sediments lack strong anion-exchange sites, negatively charged organics are not held strongly at all. Thus, these compounds are relatively mobile and biodegradable in water despite the presence of solids.

The degree of sorption of organic compounds is generally inversely proportional to their water solubility. The more water-insoluble compounds tend to be taken up strongly by lipophilic ("fat-loving") solid materials, such as humic substances (see Section 3.13). Compounds having a relatively high vapor pressure can be lost from water or solids by evaporation. When this happens, photochemical processes (see Chapter 1, Section 1.4) can play an important role in their degradation.

The herbicide 2,4-dichlorophenoxyacetic acid (2,4-D) has been studied extensively in regard to sorption reactions. Most of these studies have dealt with pure clay minerals, however, whereas soils and sediments are likely to have a strong clay–fulvic acid complex component. The sorption of 2,4-D by such a complex can be described using an equation of the Freundlich isotherm type

$$X = KC^n \tag{5.37}$$

where X is the amount sorbed per unit weight of solid, C is the concentration of 2,4-D in water solution at equilibrium, and n and K are constants. These values are determined by plotting log X versus log C. If a Freundlich-type equation is obeyed, the plot will be linear with a slope of n and an intercept of log K.

Sorption of comparatively nonvolatile hydrocarbons by sediments not only removes these materials from contact with aquatic organisms but also greatly retards their biodegradation. Aquatic plants produce some of the hydrocarbons that are found in sediments. Photosynthetic organisms, for example, produce quantities of *n*-heptadecane. The distribution of hydrocarbon types in sediments can be used to distinguish natural from pollutant sources.

The sorption of neutral species like petroleum obviously cannot be explained by ion-exchange processes. It probably involves phenomena such as van der Waal's forces (a term sometimes invoked when the true nature of an attractive force is not understood, but generally regarded as consisting of induced dipole–dipole interaction involving a neutral molecule), hydrogen bonding, charge-transfer complexation, and hydrophobic interactions.

In some cases, pollutant compounds become covalently bounded as *bound residues* to humic substances in sediments and soil. Among the kinds of organic compounds for which there is evidence of bound residues in sediments are pesticides and their metabolites, plasticizers, halogenated aromatic compounds, and nitro compounds.[4] It is very difficult to remove such residues from humic substances thermally, biochemically, or by exposure to acid or base (hydrolysis). Therefore, the formation of bound residues by persistent organic pollutants during humification of organic matter is an immobilization and detoxification process that is significant compared with biodegradation and mineralization in effectively removing such pollutants from the environment. The binding is thought to occur through the action of enzymes from some organisms. These enzymes are extracellular enzymes (those acting outside the cell) that act as oxidoreductases, which catalyze oxidation–reduction reactions. Microbial phenoloxidase enzymes act upon aromatic substrates from lignin decomposition, which are present in humic substances produced from lignin precursors, and also act upon aromatic xenobiotic aromatic compounds, such as anilines and phenols. This can result in the binding of xenobiotic aromatic compounds with humic substances as illustrated below for the coupling of the pollutant 2,4-dichlorophenol to an aryl ring on a humic substance molecule:

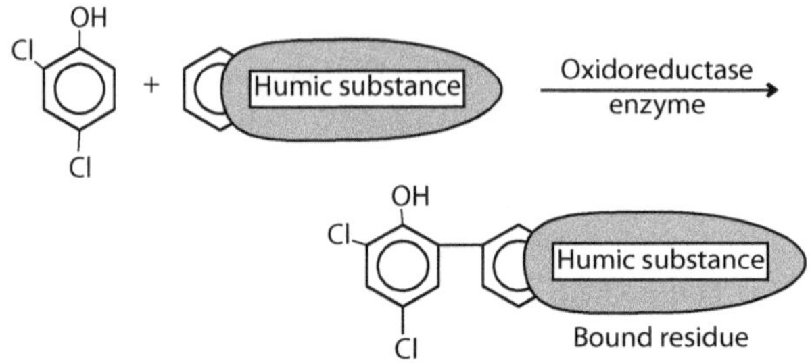

(5.38)

Obviously, uptake of organic matter by suspended and sedimentary material in water is an important phenomenon. This phenomenon significantly reduces the overall toxicity of pesticides in water. Biodegradation is generally slowed down appreciably, however, by binding of a substance to a solid. In certain intensively farmed areas, there is a very high accumulation of pesticides in the sediments of streams, lakes, and reservoirs. The sorption of pesticides by solids and the resulting influence on their biodegradation is an important consideration in the licensing of new pesticides.

The transfer of surface water to groundwater often results in sorption of some water contaminants by soil and mineral material. To take advantage of this purification effect, some municipal water supplies are drawn from beneath the surface of natural or artificial river banks as a first step in water treatment. The movement of water from waste landfills to aquifers is also an important process in which pollutants in the landfill leachate may be sorbed by the solid material through which the water passes.

Surfactant alkylphenolpolyethoxylates and their biodegradation product alkylphenols (see Section 7.10), which have many industrial and domestic uses, are discharged in significant quantities in wastewater treatment plant effluents. These compounds have the potential to disrupt the endocrine systems of aquatic organisms and even humans, tend to accumulate in sediments, and may be poorly biodegradable under anoxic conditions. Levels of these compounds up to as high as 14 mg/kg have been found in sediments in the United States and some fish from the United Kingdom have shown tissue levels approaching 1 μg/kg.

5.8.4 Bioavailability of Sediment Contaminants

An important consideration with respect to contaminants in sediments and soils is *bioavailability*, defined generally as the degree to which a substance can be absorbed into the system of an organism. Bioavailability is an especially important consideration in assessing the hazards posed to organisms by pollutants in sediments and soils. In some cases, it may be best to not attempt to remediate these media if the

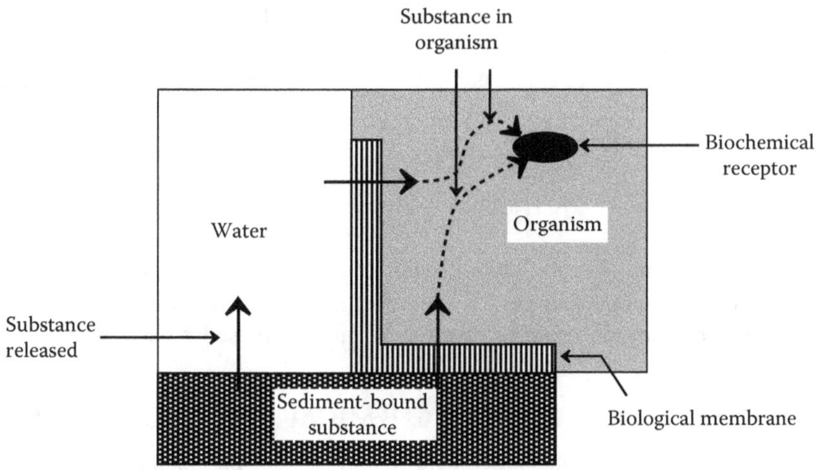

FIGURE 5.7 Transfer of contaminants in sediment to organisms where they may have an effect by acting on a biochemical receptor. (From Manahan, S., *Environmental Chemistry*, 9th ed., Taylor & Francis, Boca Raton, 2010. With permission.)

substances of concern have a low bioavailability. In other cases, very high bioavailability of toxic substances may make them particularly dangerous.

Figure 5.7 shows the overall process by which a contaminant may get into a microorganism from a sediment. A substance bound in a sediment may be released into water, then passed across a biological membrane into an organism, or it may transfer directly across a membrane from sediment to an organism. Once inside the organism, a potentially toxic substance may migrate to a biochemical receptor in the organism such as an enzyme in the organism to exert a toxic effect. These processes may actually occur within an organism when contaminated soil or sediment material is ingested and traverses the gut wall.

5.9 INTERSTITIAL WATER

Interstitial water or *pore water* consists of water held by sediments. The solutes in interstitial water reflect the chemical and biochemical conditions in the sediments. These solutes include species capable of undergoing oxidation/reduction, reduced metal ions, nutrients such as NH_4^+, and soluble organic compounds. These species include all of the products expected from the decomposition and mineralization of planktonic biomass, largely through the activity of anoxic bacteria in the sediments. Circulation is strongly curtailed in sediments so that there is a pronounced vertical gradient in the species present. At sediment surfaces that may contact water with some oxygen dissolved in it, some more oxidized species are encountered, whereas deeper in the sediments reduced species predominate.

Interstitial water is an important reservoir for gases in natural water systems. Generally, the gas concentrations in interstitial waters are different from those in the overlying water. Typically, interstitial water at the surface of the sediment contains significant amounts of N_2 and relatively little CH_4, which requires anoxic conditions for its production and biodegrades in contact with oxygen. At depths of around 1 m in sediments, CH_4 levels are high and N_2 concentrations are low because nitrogen is stripped from the interstitial water by microbially produced methane and carbon dioxide, which are produced by the anoxic fermentation of organic matter:

$$2\{CH_2O\} \rightarrow CH_4(g) + CO_2(g) \tag{5.39}$$

The concentrations of argon and nitrogen are much lower at a depth of 1 m than they are at the sediment surface. This finding may be explained by the stripping action of the fermentation-produced methane rising to the sediment surface.

5.10 PHASE INTERACTIONS IN CHEMICAL FATE AND TRANSPORT

Environmental chemical fate and transport within and among the spheres of the environment is a very important aspect of environmental chemistry. The hydrosphere is arguably the most important environmental sphere with respect to chemical fate and transport processes, many of which involve distribution of species between phases. It is appropriate at this point to consider chemical fate and transport in the hydrosphere. This is an environmentally important topic that increasingly utilizes sophisticated mathematical models and calculations that are beyond the scope of this book. For more details and example calculations, the reader is referred to a comprehensive reference work on the subject.[5]

5.10.1 RIVERS

The physical movement of chemical species in a river is largely by advection due to the gravitational movement of masses of water downstream. This results in relatively rapid mixing and dilution because of *velocity shear* in which parcels of water move at different rates in a river where water in contact with the banks and bottom moves more slowly than water in the middle of the stream. Also contributing to this mixing are turbulent mixing and diffusive transport of dissolved species and colloidal particles. The net result of these processes is that a pollutant introduced into the river in a relatively concentrated "plug" becomes spread out as it flows downstream.

5.10.2 LAKES AND RESERVOIRS

Lakes and reservoirs are relatively quiet compared with rivers, but there is still movement within such bodies of water. Water enters by stream inflow, inflow from underground springs and aquifers, and directly by rainfall. It exits by outflow, such as over a spillway in a reservoir, infiltration into ground, and evaporation. The balance between input and loss of water in a body of water means that the water has a *hydraulic residence time.* A major factor in mixing processes in lakes is the influence of wind. This occurs because windblown surface water moves at a rate that is 2–3% of the wind velocity, a phenomenon called *wind drift.* Water moved toward one side of a lake by wind drift must return some distance below the water surface as a *return current.* In a

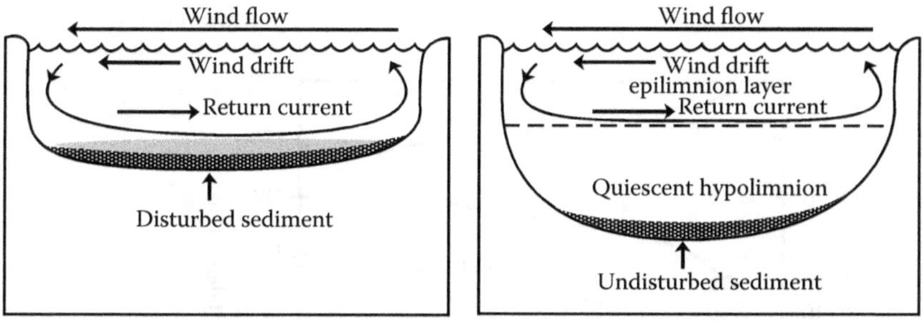

FIGURE 5.8 Mixing due to wind in a shallow unstratified lake (left) and in a stratified lake (right). In the former case in which the return current contacts the sediment, the sediment may be stirred, enabling release of substances to the water. (From Manahan, S., *Environmental Chemistry*, 9th ed., Taylor & Francis, Boca Raton, 2010. With permission.)

shallow, unstratified body of water, the return current flows along the bottom where it contacts and may agitate bottom sediments. In stratified lakes (see Figure 2.4) the circulation of water occurs in the upper epilimnion layer as shown in Figure 5.8.

5.10.3 Exchange with the Atmosphere

Exchange of chemical species between water and overlying air is an important process. It is the means by which atmospheric oxygen enters water to provide oxygen needed by fish. Carbon dioxide required for algal growth may come from the air. Air pollutants such as acid gases may enter water from the atmosphere. Under circumstances of high algal photosynthetic activity, oxygen produced by algae is released to the air. Decay of organic matter can oversaturate water with carbon dioxide, requiring its release to the air. Anoxic microbial processes in sediments can produce hydrogen sulfide and methane that are released to air. Volatile organic water pollutants may move from water to the atmosphere.

The air–water interface is the boundary across which species move and is crucial in determining the rate of exchange of materials. Current models of these processes assume that there is a thin, stationary liquid film on the surface of the water in direct contact with a thin, stationary layer of air. Since both of these layers may be only a few micrometers thick, molecular diffusion is the only mechanism for movement of solute species in them. Immediately below the thin surface film of water, turbulent diffusion mixes water solutes, and immediately above the thin air film, turbulent diffusion mixes the species in air.

5.10.4 Exchange with Sediments

Sediments are very important in chemical fate and transport in the hydrosphere. This is because substances, including pollutants such as heavy metals or hydrophobic organics, bind with particles as they settle through the water column and are incorporated into sediments. The settling flux density, J, is equal to the mass of a substance transported across an area through which the settling occurs per unit time

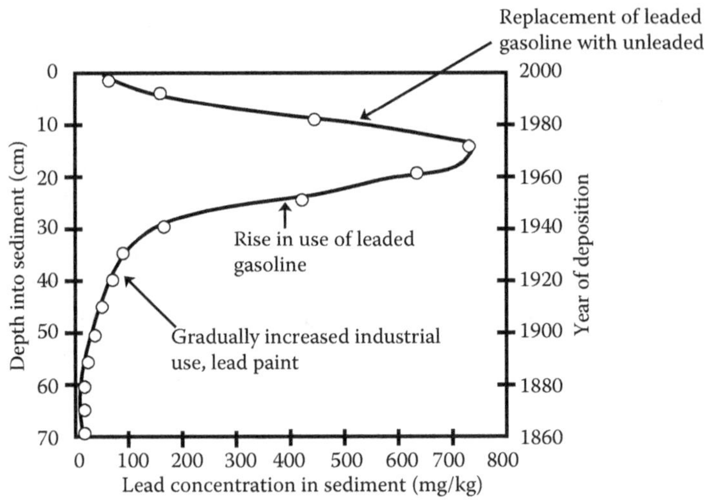

FIGURE 5.9 Typical sediment record for lead deposition reflecting the sharp increase in lead from gasoline and a drop of lead after phaseout of leaded gasoline. (From Manahan, S., *Environmental Chemistry*, 9th ed., Taylor & Francis, Boca Raton, 2010. With permission.)

and is given by the product of the rate at which sedimentary materials settle and the concentration of the substance in question in the settling materials:

$$J = (\text{sediment deposition rate}) \times (\text{substance concentration in particles}) \quad (5.40)$$

Incorporation of pollutants by settling from the water column into sediments is a particularly important mode of chemical fate and transport under quiescent conditions during which very small (colloidal) particles are aggregating together (flocculation). If undisturbed, the layers of sediments can provide a record of pollution. A hypothetical, though typical, such plot is shown for lead in Figure 5.9. Although sediments are normally repositories of pollutants and reduce their environmental harm, they can also provide sources of pollutants that can be mobilized by physical, chemical, or biochemical processes. For example, mercury precipitated in sediments can be mobilized as soluble methylmercury species by the action of anoxic bacteria in the oxygen-deficient sediments (see Section 7.3).

LITERATURE CITED

1. Ghosh, U., The role of black carbon in influencing availability of PAHs in sediments, *Human and Ecological Risk Assessment*, 13, 276–285, 2007.
2. Massoudieh, A. and T. R. Ginn, Modeling colloid-facilitated transport of multi-species contaminants in unsaturated porous media, *Journal of Contaminant Hydrology*, 92, 162–183, 2007.
3. Huang, T.-L., X.-C. Ma, C. Hai-Bing, and B.-B. Chai, Microbial effects on phosphorus release in aquatic sediments, *Water Science and Technology*, 58, 1285–1289, 2008.
4. Schwarzbauer, J., M. Ricking, B. Gieren, and R. Keller, Anthropogenic organic contaminants incorporated into the non-extractable particulate matter of riverine sediments from

the Teltow Canal (Berlin). In E. Lichtfouse, J. Schwarzbauer, and R. Didier, Eds, *Environmental Chemistry*, Springer, Berlin, 329–352, 2005.
5. Gulliver, J. S., *Introduction to Chemical Transport in the Environment*, Cambridge University Press, New York, 2007.
6. Manahan, S., *Environmental Chemistry*, 9th ed., Taylor & Francis, Boca Raton, 2010.

SUPPLEMENTARY REFERENCES

Allen, H. E., Ed., *Metal Contaminated Aquatic Sediments*, Ann Arbor Press, Chelsea, MI, 1995.

Barnes, G. and I. Gentle, *Interfacial Science: An Introduction*, Oxford University Press, New York, 2005.

Beckett, R., Ed., *Surface and Colloid Chemistry in Natural Waters and Water Treatment*, Plenum, New York, 1990.

Berkowitz, B., *Contaminant Geochemistry*, Springer, New York, 2007.

Bianchi, T. S., *Biogeochemistry of Estuaries*, Oxford University Press, New York, 2007.

Birdi, K. S., Ed., *Handbook of Surface and Colloid Chemistry*, 2nd ed., CRC Press, Boca Raton, FL, 2003.

Burdige, D. J., *Geochemistry of Marine Sediments*, Princeton University Press, Princeton, NJ, 2006.

Calabrese, E. J., P. T. Kostecki, and J. Dragun, Eds, *Contaminated Soils, Sediments, and Water, Volume 10, Successes and Challenges*, Springer, New York, 2006.

Eby, G. N., *Principles of Environmental Geochemistry*, Thomson-Brooks/Cole, Pacific Grove, CA, 2004.

Evans, R. D., J. Wisniewski, and J. R. Wisniewski, Eds, *The Interactions between Sediments and Water*, Kluwer, Dordrecht, The Netherlands, 1997.

Golterman, H. L., Ed., *Sediment–Water Interaction 6*, Kluwer, Dordrecht, The Netherlands, 1996.

Gustafsson, O. and P. M. Gschwend, Aquatic colloids: Concepts, definitions, and current challenges, *Limnology and Oceanography*, 42, 519–528, 1997.

Holland, H. D., and K K. Turekian, Eds, *Treatise on Geochemistry*, Elsevier/Pergamon, Amsterdam, 2004.

Holmberg, K., D. O. Shah, and M. J. Schwuger, Eds, *Handbook of Applied Surface and Colloid Chemistry*, Wiley, New York, 2002.

Hunter, R. J., *Foundations of Colloid Science*, 2nd ed., Oxford University Press, New York, 2001.

John V. W., *Essentials of Geochemistry*, 2nd ed., Jones and Bartlett Publishers, Sudbury, MA, 2009.

Jones, M. N. and N. D. Bryan, Colloidal properties of humic substances, *Advances in Colloid and Interfacial Science*, 78, 1–48, 1998.

Jones, S. J. and L. E. Frostick, Eds, *Sediment Flux to Basins: Causes, Controls and Consequences*, Geological Society, London, 2002.

McSween, H. Y., S. M. Richardson, and M. E. Uhle, *Geochemistry: Pathways and Processes*, 2nd ed., Columbia University Press, New York, 2003.

Mudroch, A., J. M. Mudroch, and P. Mudroch, Eds, *Manual of Physico-Chemical Analysis of Aquatic Sediments*, CRC Press, Boca Raton, FL, 1997.

Myers, D., *Surfaces, Interfaces, and Colloids: Principles and Applications*, 2nd ed., Wiley, New York, 1999.

Shchukin, E. D., *Colloid and Surface Chemistry*, Elsevier, Amsterdam, 2001.

Stumm, W., L. Sigg, and B. Sulzberger, *Chemistry of the Solid–Water Interface*, Wiley, New York, 1992.

Stumm, W. and J. J. Morgan, *Aquatic Chemistry: Chemical Equilibria and Rates in Natural Waters*, 3rd ed., Wiley, New York, 1995.

U.S. Environmental Protection Agency, Contaminated sediment in water, 2003, available at http://www.epa.gov/waterscience/cs/

Wilkinson, K. J. and J. R. Lead, Eds, *Environmental Colloids and Particles: Behaviour, Separation and Characterisation*, Wiley, Hoboken, NJ, 2007.

QUESTIONS AND PROBLEMS

The use of Internet resources is assumed in answering any of the questions. These resources would include such things as constants and conversion factors as well as additional information needed to complete an answer.

1. A sediment sample was taken from a lignite strip-mine pit containing highly alkaline (pH 10) water. Cations were displaced from the sediment by treatment with HCl. A total analysis of cations in the leachate yielded, on the basis of millimoles per 100 g of dry sediment, 150 mmol of Na^+, 5 mmol of K^+, 20 mmol of Mg^{2+}, and 75 mmol of Ca^{2+}. What is the CEC of the sediment in milliequivalents per 100 g of dry sediment? Why does H^+ not have to be considered in this case?
2. What is the value of $[O_2(aq)]$ for water saturated with a mixture of 50% O_2, and 50% N_2 by volume at 25°C and a total pressure of 1.00 atm?
3. Of the following, the least likely mode of transport of iron(III) in a normal stream is: (a) bound to suspended humic material, (b) bound to clay particles by cation-exchange processes, (c) as suspended Fe_2O_3, (d) as soluble Fe^{3+} ion, and (e) bound to colloidal clay–humic substance complexes.
4. How does freshly precipitated colloidal iron(III) hydroxide interact with many divalent metal ions in solution?
5. What stabilizes colloids composed of bacterial cells in water?
6. The solubility of oxygen in water is 14.74 mg/L at 0°C and 7.03 mg/L at 35°C. Estimate the solubility at 50°C.
7. What is thought to be the mechanism by which bacterial cells aggregate?
8. What is a good method for the production of freshly precipitated MnO_2?
9. A sediment sample was equilibrated with a solution of NH_4^+ ion, and the NH_4^+ was later displaced by Na^+ for analysis. A total of 33.8 milliequivalents of NH_4^+ were bound to the sediment and later displaced by Na^+. After drying, the sediment weighed 87.2 g. What was its CEC in milliequivalents per 100 g?
10. A sediment sample with a CEC of 67.4 milliequivalents per 100 g was found to contain the following exchangeable cations in milliequivalents per 100 g: Ca^{2+}, 21.3; Mg^{2+}, 5.2; Na^+, 4.4; K^+, 0.7. The quantity of hydrogen ion, H^+, was not measured directly. What was the ECS of H^+ in milliequivalents per 100 g?
11. What is the meaning of ZPC as applied to colloids? Is the surface of a colloidal particle totally without charged groups at the ZPC?

12. The concentration of methane in an interstitial water sample was found to be 150 mL/L at standard temperature and pressure of 0°C and 1 atm (STP). Assuming that the methane was produced by the fermentation of organic matter, $\{CH_2O\}$, what mass of organic matter was required to produce the methane in a liter of the interstitial water?
13. What is the difference between CEC and ECS?
14. Match the sedimentary mineral on the left with its conditions of formation on the right:

a. $FeS(s)$	1. May be formed when anoxic water is exposed to O_2
b. $Ca_5OH(PO_4)_3$	2. May be formed when oxic water becomes anoxic.
c. $Fe(OH)_3$	3. Photosynthesis by-product
d. $CaCO_3$	4. May be formed when wastewater containing a particular kind of contaminant flows into a body of very hard water

15. In terms of their potential for reactions with species in solution, how might metal atoms, M, on the surface of a metal oxide, MO, be described?
16. Air is 20.95% oxygen by volume. If air at 1.0000 atm pressure is bubbled through water at 25°C, what is the partial pressure of O_2 in the water?
17. The volume percentage of CO_2 in a mixture of that gas with N_2 was determined by bubbling the mixture at 1.00 atm and 25°C through a solution of 0.0100 M $NaHCO_3$ and measuring the pH. If the equilibrium pH was 6.50, what was the volume percentage of CO_2?
18. For what purpose is a polymer with the following general formula used?

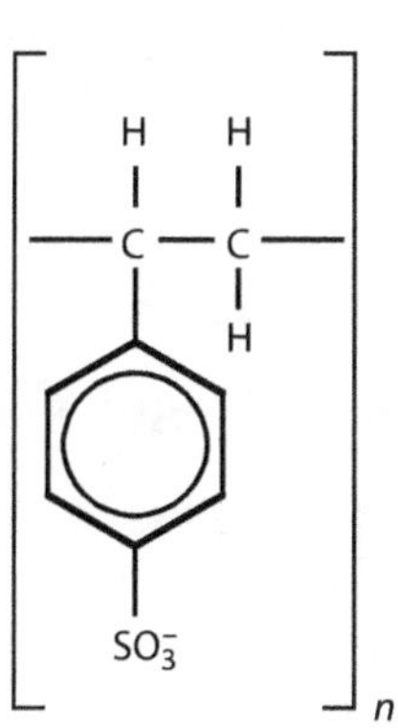

19. Of the following statements, the one that is true regarding colloids is: (A) Hydrophilic colloids consist of aggregates of relatively small molecules; (B) hydrophobic colloids do not have electrical charges; (C) hydrophilic colloids are those formed by clusters of species, such as $H_3C(CH_2)_{16}CO_2^-$; (D) association colloids form micelles; (E) the electrical charges of hydrophobic colloids are insignificant.

20. For a slightly soluble divalent metal sulfate, MSO_4, $K_{sp} = 9.00 \times 10^{-14}$. An excess of pure solid MSO_4 was equilibrated with pure water to give a solution that contained 6.45×10^{-7} mol/L of dissolved M. Considering these observations the true statement is: (A) MSO_4 has a significant degree of intrinsic solubility; (B) the solubility product, alone, accurately predicts solubility; (C) the value of the solubility product is in error; (D) the concentration of "M" in water must have been in error; (E) the only explanation for the observations is formation of HSO_4^{2-}.
21. Of the following, the *incorrect* statement regarding sediments and their formation is: (A) Physical, chemical, and, biological processes may all result in the deposition of sediments in the bottom regions of bodies of water; (B) indirectly, photosynthesis can result in formation of $CaCO_3$ sediment; (C) oxidation of Fe^{2+} ion can result in formation of an insoluble species that can be incorporated into sediment; (D) sediments typically consist of mixtures of clay, silt, sand, organic matter, and various minerals, and may vary in composition from pure mineral matter to predominantly organic matter; (E) FeS that gets into sediment tends to form at the surface of water in contact with O_2.
22. Given that at 25°C, the Henry's law constant for oxygen is 1.28×10^{-3} mol/L/atm and the partial pressure of water vapor is 0.0313 atm, what is the value of $[O_2(aq)]$ for water saturated with a mixture of 33.3% O_2 and 66.7% N_2 by volume at 25°C and a total pressure of 1.00 atm in units of mol/L?
23. Match the following regarding colloids:

A. Hydrophilic colloids	1. $CH_3CO_2^-Na^+$
B. Association colloids	2. Macromolecular proteins
C. Hydrophobic colloids	3. Often removed by addition of salt
D. Noncolloidal	4. $CH_3(CH_2)_{16}CO_2^-Na^+$

6 Aquatic Microbial Biochemistry

6.1 AQUATIC BIOCHEMICAL PROCESSES

Microorganisms—*bacteria, fungi, protozoa,* and *algae*—are living catalysts that enable a vast number of chemical processes to occur in water and soil. A majority of the important chemical reactions that take place in water, particularly those involving organic matter and oxidation–reduction processes, occur through bacterial intermediaries. Algae are the primary producers of biological organic matter (biomass) in water. Microorganisms are responsible for the formation of many sediment and mineral deposits; they also play the dominant role in secondary waste treatment. Some of the effects of microorganisms on the chemistry of water in nature are illustrated in Figure 6.1.

Pathogenic microorganisms must be eliminated from water purified for domestic use. In the past, major epidemics of typhoid, cholera, and other water-borne diseases resulted from pathogenic microorganisms in water supplies. Even today, constant vigilance is required to ensure that water for domestic use is free of pathogens.

Most of this chapter is concerned with aquatic chemical transformations mediated by microorganisms. Although they are not involved in such transformations, special mention should be made of viruses in water. Viruses cannot grow by themselves, but reproduce in the cells of host organisms. They are only about 1/30–1/20 the size of bacterial cells, and they cause a number of diseases, such as polio, viral hepatitis, and perhaps cancer. It is thought that many of these diseases are waterborne.

Because of their small size (0.025–0.100 μm) and biological characteristics, viruses are difficult to isolate and culture. They often survive municipal water treatment, including chlorination. Thus, although viruses have no effect on the overall environmental chemistry of water, they are an important consideration in the treatment and use of water.

Microorganisms are divided into the two broad categories of *prokaryotes* and *eukaryotes*; the latter have well-defined cell nuclei enclosed by a nuclear membrane, whereas the former lack a nuclear membrane and the nuclear genetic material is more diffuse in the cell. Other differences between these two classes of organisms include location of cell respiration, means of photosynthesis, means of motility, and reproductive processes. All classes of microorganisms produce *spores*, metabolically inactive bodies that form and survive under adverse conditions in a "resting" state until conditions favorable for growth occur.

Fungi, protozoa, and bacteria (with the exception of photosynthetic bacteria and protozoa) are classified as *reducers*, which break down chemical compounds to more simple species and thereby extract the energy needed for their growth and

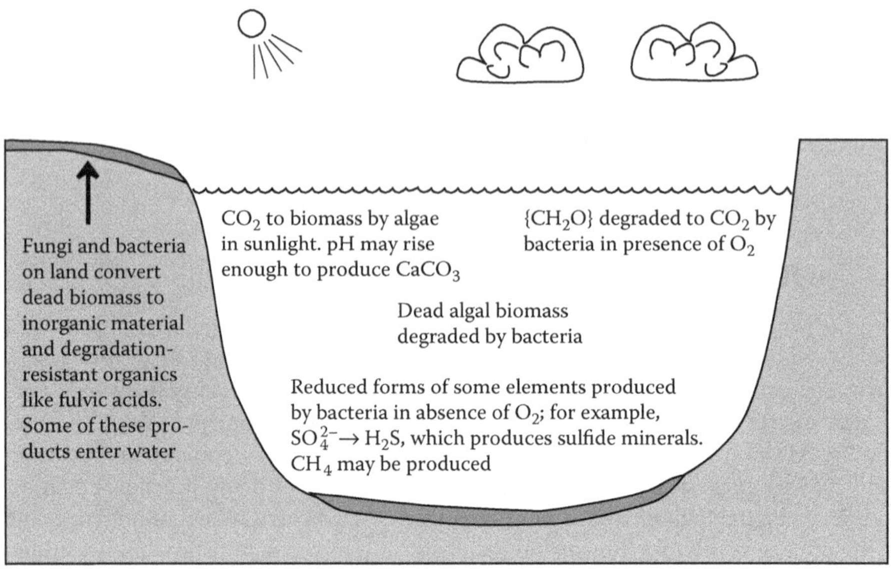

FIGURE 6.1 Effects of microorganisms on the chemistry of water in nature. (From Manahan, S., *Environmental Chemistry*, 9th ed., Taylor & Francis, Boca Raton, 2010. With permission.)

metabolism. Algae are classified as *producers* because they utilize light energy and store it as chemical energy. In the absence of sunlight, however, algae utilize chemical energy for their metabolic needs. In a sense, therefore, bacteria, protozoa, and fungi may be looked upon as environmental catalysts, whereas algae function as aquatic solar fuel cells.

All microorganisms can be put into one of the four following classifications based on the sources of energy and carbon that they utilize: chemoheterotrophs, chemoautotrophs, photoheterotrophs, and photoautotrophs. These classifications are based on (1) the energy source and (2) the carbon source utilized by the organism. *Chemotrophs* use chemical energy derived from oxidation–reduction reactions of simple inorganic chemical species for their energy needs. *Phototrophs* utilize light energy from photosynthesis. *Heterotrophs* obtain their carbon from other organisms; *autotrophs* use carbon dioxide and ionic carbonates for the C that they require. Figure 6.2 summarizes the classifications into which microorganisms may be placed with these definitions.

6.1.1 Microorganisms at Interfaces

Aquatic microorganisms tend to grow at interfaces. Many such microorganisms grow on solids that are suspended in water or are present in sediments. Large populations of aquatic bacteria typically reside on the surface of water at the air–water interface. In addition to being in contact with air that aerobic microorganisms need for their metabolic processes, this interface also accumulates food in the form of

Energy source → / Carbon source ↓	Chemical	Photochemical (light)
Organic matter	**Chemoheterotrophs** All fungi and protozoans, most bacteria. Chemoheterotrophs use organic sources for both energy and carbon	**Photoheterotrophs** A few specialized bacteria that use photoenergy, but are dependent on organic matter for a carbon source
Inorganic carbon (CO_2, HCO_3^-)	**Chemoautotrophs** Use CO_2 for biomass and oxidize substances such as H_2 (*Pseudomonas*), NH_4^+ (*Nitrosomonas*), S (*Thiobacillus*) for energy	**Photoautotrophs** Algae, and photosynthetic bacteria including cyanobacteria that use light energy to convert CO_2 (HCO_3^-) to biomass by photosynthesis

FIGURE 6.2 Classification of microorganisms among chemoheterotrophs, chemoautotrophs, photoheterotrophs, and photoautotrophs. (From Manahan, S., *Environmental Chemistry*, 9th ed., Taylor & Francis, Boca Raton, 2010. With permission.)

lipids (oils, fats), polysaccharides, and proteins. Bacteria at this interface are generally different from those in the body of water and may have a hydrophobic cell character. When surface bubbles burst, bacteria at the air–water interface can be incorporated into aerosol water droplets and carried by wind. This is a matter of some concern with respect to sewage treatment plants as a possible vector for spreading disease-causing microorganisms.

6.2 ALGAE

For the purposes of discussion here, *algae* may be considered as generally microscopic organisms that subsist on inorganic nutrients and produce organic matter from carbon dioxide by photosynthesis. In addition to single cells, algae grow as filaments, sheets, and colonies. Some algae, particularly the marine kelps, are huge multicellular organisms. The study of algae is called *phycology.*

The four main classes of unicellular algae of importance in environmental chemistry are the following.

- *Chrysophyta*, which contain pigments that give these organisms a yellow-green or golden-brown color. Chrysophyta are found in both freshwater and marine systems. They store food as carbohydrate or oil. The most well known of these algae are *diatoms*, characterized by silica-containing cell walls.
- *Chlorophyta*, commonly known as green algae, are responsible for most of the primary productivity in freshwaters.
- *Pyrrophyta*, commonly known as dinoflagellates, are motile with structures that enable them to move about in water, thus giving them characteristics of

protozoa (Section 6.4). Pyrrophyta occur in both marine and freshwater environments. "Blooms" of *Gymnodinium* and *Gonyaulax* species release toxins that cause harmful "red tides."

- *Euglenophyta* likewise exhibit characteristics of both plants and animals. Although capable of photosynthesis, these algae are not exclusively photoautotrophic (see Figure 6.2), and they utilize biomass from other sources for at least part of their carbon needs.

The general nutrient requirements of algae are carbon (obtained from CO_2 or HCO_3^-), nitrogen (generally as NO_3^-), phosphorus (as some form of orthophosphate), sulfur (as SO_4^{2-}), and trace elements including sodium, potassium, calcium, magnesium, iron, cobalt, and molybdenum.

In a highly simplified form, the production of organic matter by algal photosynthesis is described by the reaction

$$CO_2 + H_2O + h\nu \rightarrow \{CH_2O\} + O_2(g) \tag{6.1}$$

where $\{CH_2O\}$ represents a unit of carbohydrate and $h\nu$ stands for the energy of a quantum of light. A reasonably accurate representation of the overall formula of the algae *Chlorella* is $C_{5.7}H_{9.8}O_{2.3}NP_{0.06}$. Using this formula for algal biomass exclusive of the phosphorus, the overall reaction for photosynthesis is

$$5.7CO_2 + 3.4H_2O + NH_3 + h\nu \rightarrow C_{5.7}H_{9.8}O_{2.3}N + 6.25O_2(g) \tag{6.2}$$

In the absence of light, algae metabolize organic matter in the same manner as do nonphotosynthetic organisms. Thus, algae may satisfy their metabolic demands by utilizing chemical energy from the degradation of stored starches or oils, or from the consumption of algal protoplasm itself. In the absence of photosynthesis, the metabolic process consumes oxygen, so during the hours of darkness an aquatic system with a heavy growth of algae may become depleted in oxygen.

Symbiotic relationships of algae with other organisms are common. There are even reports of unicellular green algae growing inside hairs on polar bears, which are hollow for purposes of insulation; the sight of a green polar bear is alleged to have driven more than one Arctic explorer to the brink of madness. The most common symbiotic relationship involving algae is that of *lichen* in which algae coexist with fungi; both kinds of organisms are woven into the same thallus (tubular vegetative unit). The fungus provides moisture and nutrients required by the algae, which generates food photosynthetically. Lichen are involved in weathering processes of rocks (see Chapter 1, Section 1.3).

The main role of algae in aquatic systems is the production of biomass. This occurs through photosynthesis, which fixes carbon dioxide and inorganic carbon from dissolved carbonate species as organic matter, thus providing the basis of the food chain for other organisms in the system. The production of moderate amounts of biomass is beneficial to other organisms in the aquatic system. Under some conditions, the growth of algae can produce metabolites that are responsible for odor and even toxicity in water.

An interesting aspect of the growth of oceanic plankton including algae is their uptake of micronutrient metals resulting in extremely low concentrations of essential metals in surface seawater. In some cases the resultant low availability of nutrient metals limits photosynthetic rates in seawater. Some microorganisms in seawater increase bioavailability of nutrient metals and accelerate their cycling in water by releasing strong chelating agents and catalyzing oxidation/reduction reactions that put nutrient metals in a more soluble and available form.

6.3 FUNGI

Fungi are nonphotosynthetic, often filamentous, organisms exhibiting a wide range of morphology (structure). The study of fungi is called *mycology*. Some fungi are as simple as the microscopic unicellular yeasts, whereas other fungi form large, intricate toadstools. The microscopic filamentous structures of fungi generally are much larger than bacteria, and usually are 5–10 μm in width. Fungi are aerobic or oxic (oxygen-requiring) organisms and generally can thrive in more acidic media than can bacteria. They are also more tolerant of higher concentrations of heavy metal ions than are most bacteria.

Perhaps the most important function of fungi in the environment is the breakdown of cellulose in wood and other plant materials. To accomplish this, fungal cells secrete an extracellular enzyme (exoenzyme), *cellulase*, that hydrolyzes insoluble cellulose to soluble carbohydrates that can be absorbed by the fungal cell.

Fungi do not grow well in water. However, they play an important role in determining the composition of natural waters and wastewaters because of the large amount of their decomposition products that enter water. An example of such a product is humic material, which interacts with hydrogen ions and metals (see Section 3.13).

6.4 PROTOZOA

Protozoa are microscopic animals consisting of single eukaryotic cells. The numerous kinds of protozoa are classified on the bases of morphology (physical structure), means of locomotion (flagella, cilia, pseudopodia), presence or absence of chloroplasts, presence or absence of shells, ability to form cysts (consisting of a reduced-size cell encapsulated in a relatively thick skin that can be carried in the air or by animals in the absence of water), and ability to form spores. Protozoa occur in a wide variety of shapes and their movement in the field of a microscope is especially fascinating to watch. Some protozoa contain chloroplasts and are photosynthetic.

Protozoa play a relatively small role in environmental biochemical processes, but are nevertheless significant in the aquatic and soil environment for the following reasons.

- Several devastating human diseases, including malaria, sleeping sickness, and some kinds of dysentery, are caused by protozoa that are parasitic to the human body
- Parasitic protozoa can cause debilitating, even fatal, diseases in livestock and wildlife

- Vast limestone ($CaCO_3$) deposits have been formed by the deposition of shells from the *foramifera* group of protozoa
- Protozoa are active in the oxidation of degradable biomass, particularly in sewage treatment
- Protozoa may affect bacteria active in degrading biodegradable substances by "grazing" on bacterial cells

Although they are single celled, protozoa have a fascinating variety of structures that enable them to function. The protozoal cell membrane is protected and supported by a relatively thick pellicle, or by a mineral shell that may act as an exoskeleton. Food is ingested through a structure called a cytosome from which it is concentrated in a cytopharynx or oral groove, and then digested by enzymatic action in a food vacuole. Residue from food digestion is expelled through a cytopyge and soluble metabolic products, such as urea or ammonia, are eliminated by a contractile vacuole, which also expels water from the cell interior.

6.5 BACTERIA

Bacteria are single-celled prokaryotic microorganisms that may be shaped as rods (*bacillus*), spheres (*coccus*), or spirals (*vibrios, spirilla, spirochetes*). Bacteria cells may occur individually or grow as groups ranging from two to millions of individual cells. Most bacteria fall into the size range of 0.5–3.0 μm. However, considering all species, a size range of 0.3–50 μm is observed. Characteristics of most bacteria include a semirigid cell wall, motility with flagella for those capable of movement, unicellular nature (although clusters of cloned bacterial cells are common), and multiplication by binary fission in which each of two daughter cells is genetically identical to the parent cell. Like other microorganisms, bacteria produce spores.

The metabolic activity of bacteria is greatly influenced by their small size. Their surface-to-volume ratio is extremely large, so that the inside of a bacterial cell is highly accessible to a chemical substance in the surrounding medium. Thus, for the same reason that a finely divided catalyst is more efficient than a more coarse one, bacteria may bring about very rapid chemical reactions compared to those mediated by larger organisms. Bacteria excrete exoenzymes that break down solid food material to soluble components which can penetrate bacterial cell walls, where the digestion process is completed.

Although individual bacteria cells cannot be seen by the naked eye, bacterial colonies arising from individual cells are readily visible. A common method of counting individual bacterial cells in water consists of spreading a measured volume of an appropriately diluted water sample on a plate of agar gel containing bacterial nutrients. Wherever a viable bacterial cell adheres to the plate, a bacterial colony consisting of many cells will grow. These visible colonies are counted and related to the number of cells present initially. Because bacterial cells may already be present in groups, and because individual cells may not live to form colonies or even have the ability to form colonies on a plate, plate counts tend to underestimate the number of viable bacteria.

6.5.1 Autotrophic and Heterotrophic Bacteria

Bacteria may be divided into two main categories, autotrophic and heterotrophic. Autotrophic bacteria are not dependent on organic matter for growth and thrive in a completely inorganic medium; they use carbon dioxide or other carbonate species as a carbon source. A number of sources of energy may be used, depending on the species of bacteria; however, a biologically mediated chemical reaction always supplies the energy.

An example of autotrophic bacteria is *Gallionella*. In the presence of oxygen, these bacteria are grown in a medium consisting of NH_4Cl, phosphates, mineral salts, CO_2 (as a carbon source), and solid FeS (as an energy source). It is believed that the following is the energy-yielding reaction for this species:

$$4FeS(s) + 9O_2 + 10H_2O \rightarrow 4Fe(OH)_3(s) + 4SO_4^{2-} + 8H^+ \tag{6.3}$$

Starting with the simplest inorganic materials, autotrophic bacteria must synthesize all of the complicated proteins, enzymes, and other materials needed for their life processes. It follows, therefore, that the biochemistry of autotrophic bacteria is quite sophisticated. Because of their consumption and production of a wide range of minerals, autotrophic bacteria are involved in many geochemical transformations.

Heterotrophic bacteria depend on organic compounds, both for energy and for carbon to build their biomass. They are much more common in occurrence than autotrophic bacteria. Heterotrophic bacteria are the microorganisms primarily responsible for the breakdown of pollutant organic matter in water, and of organic wastes in biological waste-treatment processes.

Some bacteria are capable of photosynthesis to supply their energy and carbon. The most common such bacteria in water are the cyanobacteria. These organisms were once thought to be algae and were called blue-green algae. They can grow prolifically under some conditions and may impart such disagreeable tastes and odors to water that they may make it unfit for domestic consumption during cyanobacteria blooms.

A type of cyanobacteria of unique importance consists of the marine cyanobacteria from the *Prochlorococcus* genus.[1] Only discovered in 1988, these bacteria are the smallest photosynthetic organisms known, only about 0.5 μm in size. They are the most abundant photosynthetic organisms in the oceans (therefore, on the entire planet), constituting 40–50% of phytoplankton biomass in ocean waters in the region between 40° north and 40° south. One of the two major strains of *Prochlorococcus* lives near the surface where light is abundant, whereas a second strain conducts photosynthesis to depths of 200 m reached by only a fraction of 1% of incident light. Despite their small cell size, *Prochlorococcus* produce a significant fraction of photosynthetically generated biomass and are very important in marine food chains. Because they fix vast quantities of carbon dioxide, they have a potentially significant role in the alleviation of greenhouse warming effects. They are noted for their rapid genetic adaptation abilities, which hopefully will enable them to continue functioning efficiently under altered conditions, such as reduced pH in ocean waters, caused by elevated carbon dioxide levels in the atmosphere.

6.5.2 Oxic and Anoxic Bacteria

Another classification system for bacteria depends on their requirement for molecular oxygen. Oxic (aerobic bacteria) require oxygen as an electron receptor:

$$O_2 + 4H^+ + 4e^- \rightarrow 2H_2O \tag{6.4}$$

Anaerobic bacteria, or anoxic bacteria, function only in the complete absence of molecular oxygen. Frequently, molecular oxygen is quite toxic to anoxic bacteria. Anoxic bacteria have gained increasing attention for their ability to degrade organic wastes.

A third class of bacteria, facultative bacteria, utilize free oxygen when it is available and use other substances as electron receptors (oxidants) when molecular oxygen is not available. Common oxygen substitutes in water are nitrate ion (see Section 6.11) and sulfate ion (see Section 6.12).

6.5.3 Marine Bacteria

Most of the attention that has been given to bacteria in water has been focused on bacteria found in freshwater. Recently, more attention has been given to marine bacteria including those in ocean sediments. An example of such bacteria is *Salinospora*, a type of actinomycete bacteria that thrive in ocean sediments under dark, cold, high pressure, saline conditions. Known freshwater and terrestrial actinomycetes have been the major source of antibiotics such as streptomycin and vancomycin and the possibility of finding new antibiotics and even anticancer drugs in marine actinomycetes has generated intense interest in these organisms.

6.6 THE PROKARYOTIC BACTERIAL CELL

Figure 6.3 illustrates a generic prokaryotic bacterial cell. Bacterial cells are enclosed in a *cell wall*, which holds the contents of the bacterial cell and determines the shape of the cell. The cell wall in many bacteria is frequently surrounded by a *slime layer* (capsule). This layer protects the bacteria and helps the bacterial cells to adhere to surfaces.

The *cell membrane* or *cytoplasmic membrane* composed of protein and phospholipid occurs as a thin layer only about 7 nm in thickness on the inner surface of the cell wall enclosing the cellular cytoplasm. The cytoplasmic membrane is of crucial importance to cell function in that it controls the nature and quantity of materials transported into and out of the cell. It is also very susceptible to damage from some toxic substances.

Folds in the cytoplasmic membrane called *mesosomes* serve several functions. One of these is to increase the surface area of the membrane to enhance transport of materials through it. Another function is to act as a site for division of the cell during reproduction. Bacterial DNA is separated at the mesosome during cell division.

Hairlike *pili* on the surface of a bacterial cell enable the cell to stick to surfaces. Specialized *sex pili* enable nucleic acid transfer between bacterial cells during an

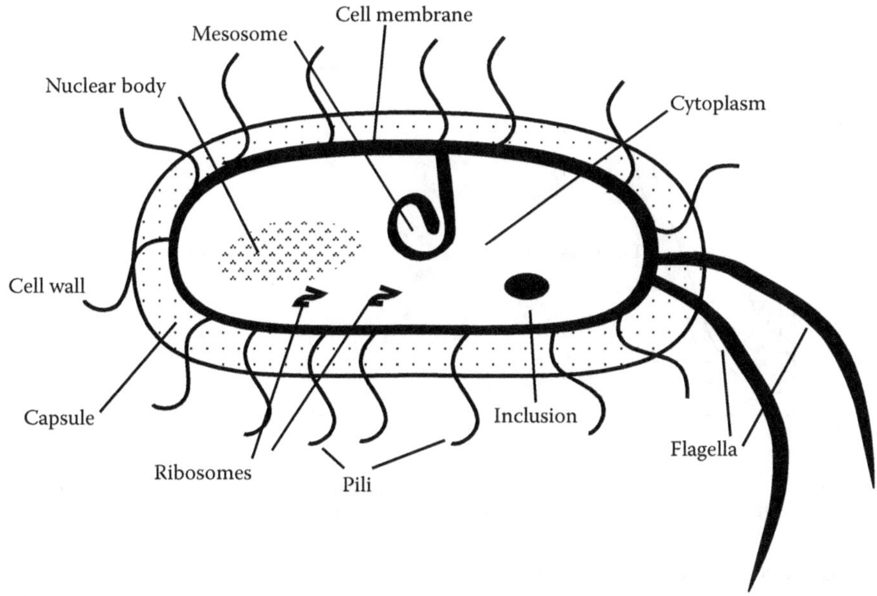

FIGURE 6.3 Generic prokaryotic bacterial cell illustrating major cell features. (From Manahan, S., *Environmental Chemistry*, 9th ed., Taylor & Francis, Boca Raton, 2010. With permission.)

exchange of genetic material. Somewhat similar to pili—but larger, more complex, and fewer in number—are *flagella*, moveable appendages that cause bacterial cells to move by their whipping action. Bacteria with flagella are termed *motile.*

Bacterial cells are filled with an aqueous solution and suspension containing proteins, lipids, carbohydrates, nucleic acids, ions, and other materials. Collectively, these materials are referred to as *cytoplasm*, the medium in which the cell's metabolic processes are carried out. The major cell bodies suspended in cytoplasm are the following.

- *Nuclear body* consisting of a single DNA macromolecule that controls metabolic processes and reproduction
- *Inclusions* of reserve food material consisting of fats, carbohydrates, and even elemental sulfur
- *Ribosomes*, which are sites of protein synthesis and which contain protein and RNA

6.7 KINETICS OF BACTERIAL GROWTH

The population size of bacteria and unicellular algae as a function of time in a growth culture is illustrated by Figure 6.4, which shows a *population curve* for a bacterial culture. Such a culture is started by inoculating a rich nutrient medium with a small number of bacterial cells. The population curve consists of four regions. The first region is characterized by little bacterial reproduction and is called the *lag phase*. The

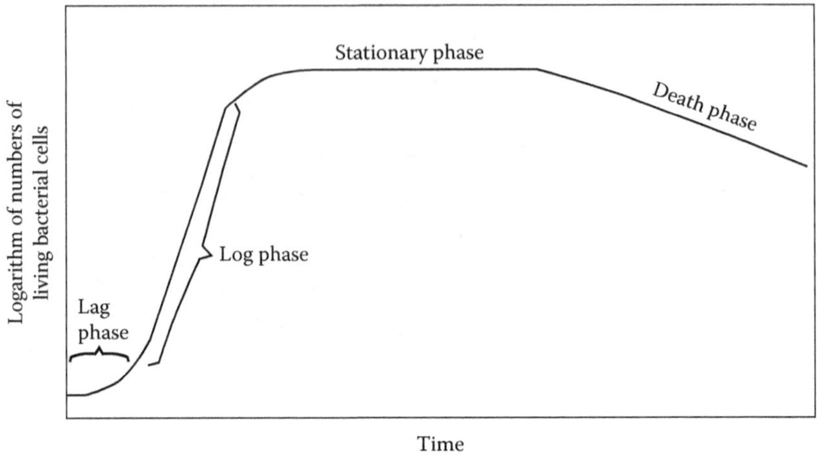

FIGURE 6.4 Population curve for a bacterial culture. (From Manahan, S., *Environmental Chemistry*, 9th ed., Taylor & Francis, Boca Raton, 2010. With permission.)

lag phase occurs because the bacteria must become acclimated to the new medium. Following the lag phase comes a period of very rapid bacterial growth. This is the *log phase*, or exponential phase, during which the population doubles over a regular time interval called the *generation time*. This behavior can be described by a mathematical expression that applies when the growth rate is proportional to the number of individuals present and there are no limiting factors such as death or lack of food:

$$\frac{dN}{dt} = kN \tag{6.5}$$

$$\ln\left(\frac{N}{N_0}\right) = kt \quad \text{or} \quad N = N_0 e^{kt} \tag{6.6}$$

where N is the population at time t and N_0 is the population at time $t = 0$. Thus, another way of describing population growth during the log phase is to say that the logarithm of bacterial population increases linearly with time. The generation time, or doubling time, is (ln 2)/k, analogous to the half-life of radioactive decay. Fast growth during the log phase can cause very rapid microbial transformations of chemical species in water.

The log phase terminates and the *stationary phase* begins when a limiting factor is encountered. Typical factors limiting growth are depletion of an essential nutrient, buildup of toxic material, and exhaustion of oxygen. During the stationary phase, the number of viable cells remains virtually constant. After the stationary phase, the bacteria begin to die faster than they reproduce, and the population enters the *death phase*.

6.8 BACTERIAL METABOLISM

Bacteria obtain the energy and raw materials needed for their metabolic processes and reproduction by mediating chemical reactions. Nature provides a large number of such reactions, and bacterial species have evolved that utilize many of these. As a consequence of their participation in such reactions, bacteria are involved in many biogeochemical processes in water and soil. Bacteria are essential participants in many important elemental cycles in nature, including those of nitrogen, carbon, and sulfur. They are responsible for the formation of many mineral deposits, including some of iron and manganese. On a smaller scale, some of these deposits form through bacterial action in natural water systems and even in pipes used to transport water.

Bacterial metabolism addresses the biochemical processes by which chemical species are modified in bacterial cells. It is basically a means of deriving energy and cellular material from nutrient substances. Figure 6.5 summarizes the essential features of bacterial metabolism. The two major divisions of bacterial metabolism are catabolism, energy-yielding degradative metabolism that breaks macromolecules down to their small monomeric constituents, and anabolism, synthetic metabolism in which small molecules are assembled into large ones. A key distinction among bacteria has to do with the terminal electron acceptor in the electron transport chain involved in the process by which bacteria gain energy by oxidizing food materials. If the terminal electron acceptor is molecular O_2, the process is *oxic respiration*. If it is another reducible species, commonly including SO_4^{2-}, NO_3^-, HCO_3^-, or iron(III), the process is called *anoxic respiration*. As examples, *Desulfovibrio* bacteria convert SO_4^{2-} to H_2S, *Methanobacterium* reduce HCO_3^- to CH_4, and assorted bacteria reduce NO_3^- to NO_2^-, N_2O, N_2, or NH_4^+.

6.8.1 Factors Affecting Bacterial Metabolism

Bacterial metabolic reactions are mediated by enzymes, biochemical catalysts endogenous to living organisms that enable metabolic processes to occur. Enzymatic processes in bacteria are essentially the same as those in other organisms. At this point, however, it is useful to review several factors that influence bacterial enzyme activity and, therefore, bacterial growth.

Figure 6.6 illustrates the effect of *substrate concentration* on enzyme activity, where a substrate is a substance on which an enzyme acts. It is seen that enzyme

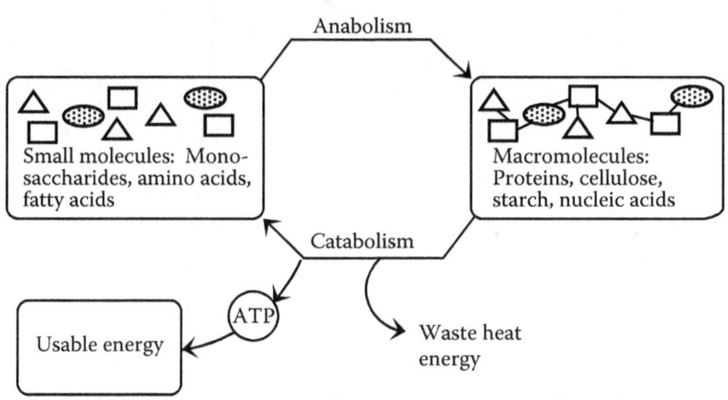

FIGURE 6.5 Bacterial metabolism and energy production. (From Manahan, S., *Environmental Chemistry*, 9th ed., Taylor & Francis, Boca Raton, 2010. With permission.)

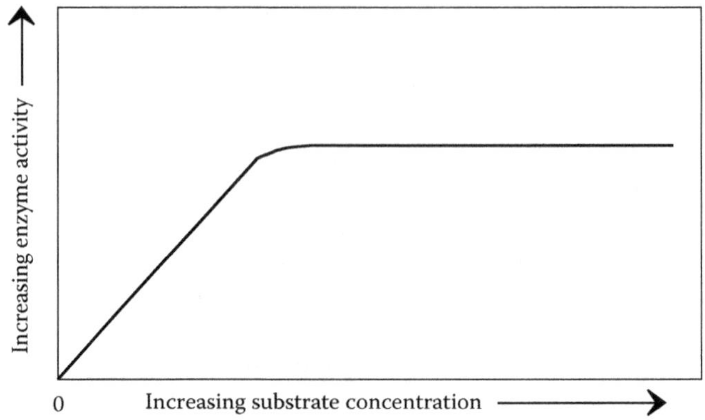

FIGURE 6.6 Effect of increasing substrate concentration on enzyme activity. Bacterial metabolism parallels such a plot. (From Manahan, S., *Environmental Chemistry*, 9th ed., Taylor & Francis, Boca Raton, 2010. With permission.)

activity increases in a linear fashion up to a value that represents saturation of the enzyme activity. Beyond this concentration, increasing substrate levels do not result in increased enzyme activity. This kind of behavior is reflected in bacterial activity, which increases with available nutrients up to a saturation value. Superimposed on this plot in a bacterial system is increased bacterial population which, in effect, increases the amount of available enzyme.

Figure 6.7 shows the effect of *temperature* on enzyme activity and on bacterial growth and metabolism. Over a relatively short range of temperature in which

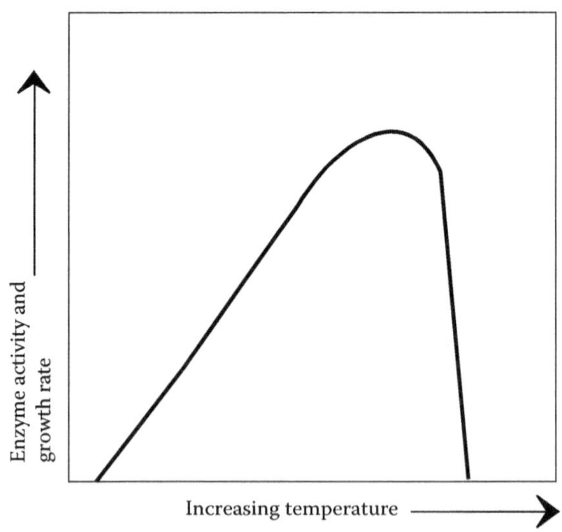

FIGURE 6.7 Enzyme activity as a function of temperature. A plot of bacterial growth versus temperature has the same shape. (From Manahan, S., *Environmental Chemistry*, 9th ed., Taylor & Francis, Boca Raton, 2010. With permission.)

the bacteria thrive, enzyme activity increases with temperature. The curve shows a maximum growth rate with an optimum temperature that is skewed toward the high-temperature end of the curve, and exhibits an abrupt drop-off beyond the temperature maximum. This occurs because enzymes are destroyed by being denatured at temperatures not far above the optimum. Bacteria show different temperature optima. *Psychrophilic bacteria* are bacteria having temperature optima below approximately 20°C. The temperature optima of *mesophilic bacteria* lie between 20°C and 45°C. Bacteria having temperature optima above 45°C are called *thermophilic bacteria.* The temperature range for optimum growth of different kinds of bacteria is remarkably wide, with some bacteria being able to grow at 0°C, and some thermophilic bacteria existing in boiling hot water.

There has been much interest in the adaptation of thermophilic bacteria and their enzymes to industrial applications because, as shown by Figure 6.7, enzyme activity generally increases with increasing temperature.[2] An example of such an application is the potential use of a heat-resistant catalase enzyme from a thermophilic bacterium isolated from a hot spring pool in Yellowstone National Park to catalyze the breakdown of hydrogen peroxide in industrial bleaching wastewaters. The application of hydrogen peroxide as a bleaching agent has increased greatly in recent years because it is generally regarded as more of a "green chemical" than chlorine and hypochlorite that it has been replacing. It is important to break down spent hydrogen peroxide to oxygen and water before it is discharged, but catalase enzymes from conventional sources are inhibited or destroyed at the elevated temperatures and pH values typical of the bleaching solutions. However, the catalase enzyme from thermophilic bacteria can function well at 70°C and a pH up to 10.

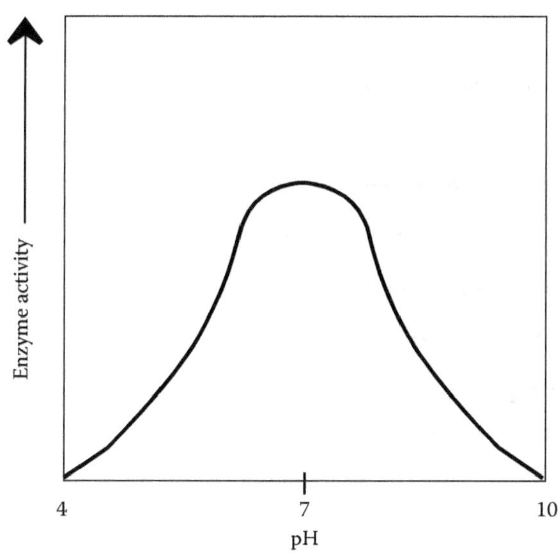

FIGURE 6.8 Enzyme activity and bacterial growth rate as a function of pH. (From Manahan, S., *Environmental Chemistry*, 9th ed., Taylor & Francis, Boca Raton, 2010. With permission.)

Figure 6.8 is a plot of pH versus bacterial enzyme activity. Although the optimum pH will vary somewhat, enzymes typically have a pH optimum around neutrality. Enzymes tend to become denatured at pH extremes. For some bacteria, such as those that generate sulfuric acid by the oxidation of sulfide or that produce organic acids by fermentation of organic matter, the pH optimum may be quite acidic.

6.8.2 Microbial Oxidation and Reduction

The metabolic processes by which bacteria obtain their energy involve mediation of oxidation–reduction reactions. The most environmentally important oxidation–reduction reactions occurring in water and soil through the action of bacteria are summarized in Table 6.1. Much of the remainder of this chapter is devoted to a discussion of important redox reactions mediated by bacteria, particularly those summarized in Table 6.1.

6.9 MICROBIAL TRANSFORMATIONS OF CARBON

Carbon is an essential life element and composes a high percentage of the dry mass of microorganisms. For most microorganisms, the bulk of net energy-yielding or energy-consuming metabolic processes involve changes in the oxidation state of carbon. These chemical transformations of carbon have important environmental

TABLE 6.1
Principal Microbially Mediated Oxidation and Reduction Reactions

Oxidation	$pE^0(w)$[a]
(1) $\frac{1}{4}\{CH_2O\}+\frac{1}{4}H_2O \rightleftarrows \frac{1}{4}CO_2+H^+(w)+e^-$	−8.20
(1a) $\frac{1}{2}HCOO^- \rightleftarrows \frac{1}{2}CO_2(g)+\frac{1}{2}H^+(w)+e^-$	−8.73
(1b) $\frac{1}{2}\{CH_2O\}+\frac{1}{2}H_2O \rightleftarrows \frac{1}{2}HCOO^-+\frac{3}{2}H^+(w)+e^-$	−7.68
(1c) $\frac{1}{2}CH_3OH \rightleftarrows \frac{1}{2}\{CH_2O\}+H^+(w)+e^-$	−3.01
(1d) $\frac{1}{2}CH_4(g)+\frac{1}{2}H_2O \rightleftarrows \frac{1}{2}CH_3OH+H^+(w)+e^-$	−2.88
(2) $\frac{1}{8}HS^-+\frac{1}{2}H_2O \rightleftarrows \frac{1}{8}SO_4^{2-}+\frac{9}{8}H^+(w)+e^-$	−3.75
(3) $\frac{1}{8}NH_4^++\frac{3}{8}H_2O \rightleftarrows \frac{1}{8}NO_3^-+\frac{5}{4}H^+(w)+e^-$	+6.16
(4)[a] $FeCO_3(s)+2H_2O \rightleftarrows FeOOH(s)+HCO_3^-(10^{-3})+2H^+(w)+e^-$	−1.67
(5)[a] $\frac{1}{2}MnCO_3(s)+H_2O \rightleftarrows \frac{1}{2}MnO_2+\frac{1}{2}HCO_3^-(10^{-3})+\frac{3}{2}H^+(w)+e^-$	−8.5
Reduction	
(A) $\frac{1}{4}O_2(g)+H^+(w)+e^- \rightleftarrows \frac{1}{2}H_2O$	+13.75
(B) $\frac{1}{5}NO_3^-+\frac{6}{5}H^+(w)+e^- \rightleftarrows \frac{1}{10}N_2+\frac{3}{5}H_2O$	+12.65

TABLE 6.1 (continued)
Principal Microbially Mediated Oxidation and Reduction Reactions

Reduction	$pE^0(w)$[a]
(C) $\frac{1}{8}NO_3^- + \frac{5}{4}H^+(w) + e^- \rightleftarrows \frac{1}{8}NH_4^+ + \frac{3}{8}H_2O$	+6.15
(D) $\frac{1}{2}\{CH_2O\} + H^+(w) + e^- \rightleftarrows \frac{1}{2}CH_3OH$	−3.01
(E) $\frac{1}{8}SO_4^{2-} + \frac{9}{8}H^+(w) + e^- \rightleftarrows \frac{1}{8}HS^- + \frac{1}{2}H_2O$	−3.75
(F) $\frac{1}{8}CO_2(g) + H^+(w) + e^- \rightleftarrows \frac{1}{8}CH_4(g) + \frac{1}{4}H_2O$	−4.13
(G) $\frac{1}{6}N_2 + \frac{4}{3}H^+(w) + e^- \rightleftarrows \frac{1}{3}NH_4^+$	−4.68

Sequence of Microbial Mediation

Model 1: Excess of organic material (water initially contains O_2, NO_3^-, SO_4^{2-}, and HCO_3^-)
Examples: Hypolimnion of a eutrophic lake, sediments, sewage treatment plant digester.

	Combination	$pE^0(w)$[b]	$\Delta G^0(w)$ (kcal)
Oxic respiration	(1) + (A)	21.95	−29.9
Denitrification	(1) + (B)	20.85	−28.4
Nitrate reduction	(1) + (C)	14.36	−19.6
Fermentation[c]	(1b) + (D)	4.67	−6.4
Sulfate reduction	(1) + (E)	4.45	−5.9
Methane fermentation	(1) + (F)	4.07	−5.6
N-fixation	(1) + (G)	3.52	−4.8

Model 2: Excess O_2 [water initially contains organic matter SH^-, NH_4^+, and possibly Fe(II) and Mn(II)]
Examples: Oxic waste treatment, self-purification in streams, epilimnion of lake.

	Combination	$pE^0(w)$[b]	$\Delta G^0(w)$ (kcal)
Oxic respiration	(A) + (1)	21.95	−29.9
Sulfide oxidation	(A) + (2)	17.50	−23.8
Nitrification	(A) + (3)	7.59	−10.3
Iron(II) oxidation[d]	(A) + (4)	15.42	21.0
Manganese(II) oxidation[d]	(A) + (5)	5.75	−7.2

Source: Reproduced from Stumm, W. and J. J. Morgan, *Aquatic Chemistry*, pp. 336–337, Wiley-Interscience, New York, 1970. With permission.

[a] These pE^0 values are at H^+ ion activity of 1.00×10^{-7}; $H^+(w)$ designates water in which $[H^+] = 1.00 \times 10^{-7}$. pE^0 values for half-reactions (1) through (5) are given for reduction, although the reaction is written as an oxidation.

[b] pE^0 values = log K(w) for a reaction written for a one-electron transfer. The term K(w) is the equilibrium constant for the reaction in which the activity of the hydrogen ion has been set at 1.00×10^{-7} and incorporated into the equilibrium constant.

[c] Fermentation is interpreted as an organic redox reaction where one organic substance is reduced by oxidizing another organic substance (e.g., alcoholic fermentation; the products are metastable thermodynamically with respect to CO_2 and CH_4).

[d] The data for $pE^0(w)$ or $\Delta G^0(w)$ of these reactions correspond to an activity of HCO_3^- ion of 1.00×10^{-3} rather than unity.

implications. For example, when algae and other plants fix CO_2 as carbohydrate, represented as $\{CH_2O\}$,

$$CO_2 + H_2O + h\nu \rightarrow \{CH_2O\} + O_2(g) \tag{6.7}$$

carbon changes from the +4 to the 0 oxidation state. Energy from sunlight is stored as chemical energy in organic compounds. However, when the algae die, bacterial decomposition occurs through oxic respiration in the reverse of the biochemical process represented by the above reaction for photosynthesis, energy is released, and oxygen is consumed.

In the presence of oxygen, the principal energy-yielding reaction of bacteria is the oxidation of organic matter. Since it is generally more meaningful to compare reactions on the basis of the reaction of one electron-mole, the oxic degradation of organic matter is conveniently written as

$$\tfrac{1}{4}\{CH_2O\} + \tfrac{1}{4}O_2(g) \rightarrow \tfrac{1}{4}CO_2 + \tfrac{1}{4}H_2O \tag{6.8}$$

for which the free-energy change is –29.9 kcal (see oxic respiration, Table 6.1). From this general type of reaction, bacteria and other microorganisms extract the energy needed to carry out their metabolic processes, to synthesize new cell material, for reproduction, and for locomotion.

Partial microbial decomposition of organic matter is a major step in the production of peat, lignite, coal, oil shale, and petroleum. Under reducing conditions, particularly below water, the oxygen content of the original plant material (approximate empirical formula, $\{CH_2O\}$) is lowered, leaving materials with relatively higher carbon contents.

6.9.1 Methane-Forming Bacteria

The production of methane in anoxic (oxygenless) sediments is favored by high organic levels and low nitrate and sulfate levels. Methane production plays a key role in local and global carbon cycles as the final step in the anoxic decomposition of organic matter. This process is the source of about 80% of the methane entering the atmosphere.

The carbon from microbially produced methane can come from either the reduction of CO_2 or the fermentation of organic matter, particularly acetate. The anoxic production of methane can be represented in the following simplified manner. When carbon dioxide acts as an electron receptor in the absence of oxygen, methane gas is produced:

$$\tfrac{1}{8}CO_2 + H^+ + e^- \rightarrow \tfrac{1}{8}CH_4 + \tfrac{1}{4}H_2O \tag{6.9}$$

This reaction is mediated by methane-forming bacteria. When organic matter is degraded microbially, the half-reaction for one electron-mole of $\{CH_2O\}$ is

$$\tfrac{1}{4}\{CH_2O\} + \tfrac{1}{4}H_2O \rightarrow \tfrac{1}{4}CO_2 + H^+ + e^- \tag{6.10}$$

Adding half-reactions 6.9 and 6.10 yields the overall reaction for the anoxic degradation of organic matter by methane-forming bacteria, which involves a free-energy change of –5.55 kcal per electron-mole:

$$\tfrac{1}{4}\{CH_2O\} \rightarrow \tfrac{1}{8}CH_4 + \tfrac{1}{8}CO_2 \qquad (6.11)$$

This reaction, in reality a series of complicated processes, is a *fermentation reaction*, defined as a redox process in which both the oxidizing agent and reducing agent are organic substances. It may be seen that only about one-fifth as much free energy is obtained from one electron-mole of methane formation as from a one electron-mole reaction involving complete oxidation of the organic matter, Reaction 6.8.

Methane formation is a valuable process responsible for the degradation of large quantities of organic wastes, both in biological waste-treatment processes (see Chapter 8) and in nature. Methane production is used in biological waste treatment plants to further degrade excess sludge from the activated sludge process. In the bottom regions of natural waters, methane-forming bacteria degrade organic matter in the absence of oxygen. This eliminates organic matter, which would otherwise require oxygen for its biodegradation. If this organic matter were transported to oxic water containing dissolved O_2, it would exert a BOD. Methane production is a very efficient means for the removal of BOD. The reaction

$$CH_4 + 2O_2 \rightarrow CO_2 + 2H_2O \qquad (6.12)$$

shows that 1 mol of methane requires 2 mols of oxygen for its oxidation to CO_2. Therefore, the production of 1 mol of methane and its subsequent evolution from water are equivalent to the removal of 2 mols of oxygen demand. In a sense, therefore, the removal of 16 g (1 mol) of methane is equivalent to the addition of 64 g (2 mols) of available oxygen to the water.

In favorable cases, methane fuel can be produced cost-effectively as a renewable resource from anoxic digestion of organic wastes. Some installations use cattle feedlot wastes. Methane is routinely generated by the action of anoxic bacteria and is used for heat and engine fuel at sewage treatment plants (see Chapter 8). Methane produced underground in municipal landfills is being tapped by some municipalities; however, methane seeping into basements of buildings constructed on landfill containing garbage has caused serious explosions and fires.

6.9.2 Bacterial Utilization of Hydrocarbons

Methane is oxidized under oxic conditions by a number of strains of bacteria. One of these, *Methanomonas*, is a highly specialized organism that cannot use any material other than methane as an energy source. Methanol, formaldehyde, and formic acid are intermediates in the microbial oxidation of methane to carbon dioxide. As discussed in Section 6.10, several types of bacteria can degrade higher hydrocarbons and use them as energy and carbon sources.

6.9.3 Microbial Utilization of Carbon Monoxide

Carbon monoxide is removed from the atmosphere by contact with soil. Since neither sterilized soil nor green plants grown under sterile conditions show any capacity to remove carbon monoxide from air, this ability must be because of microorganisms in the soil. Fungi capable of CO metabolism include some commonly occurring strains of the ubiquitous *Penicillium* and *Aspergillus*. It is also possible that some bacteria are involved in CO removal. However some microorganisms metabolize CO, other aquatic and terrestrial organisms produce this gas.

6.10 BIODEGRADATION OF ORGANIC MATTER

The biodegradation of organic matter in the aquatic and terrestrial environments is a crucial environmental process. Some organic pollutants are biocidal; for example, effective fungicides must be antimicrobial in action. Therefore, in addition to killing harmful fungi, fungicides frequently harm beneficial saprophytic fungi (fungi that decompose dead organic matter) and bacteria. Herbicides, which are designed for plant control, and insecticides, which are used to control insects, generally do not have any detrimental effect on microorganisms.

The biodegradation of organic matter by microorganisms occurs by way of a number of stepwise, microbially catalyzed reactions. These reactions will be discussed individually with examples.

6.10.1 Oxidation

Oxidation occurs by the action of oxygenase enzymes (those that enable oxidation reactions to occur in organisms). *Epoxidation* is a major step in many oxidation mechanisms and consists of adding an oxygen atom between two C atoms in an unsaturated system as shown for benzene below:

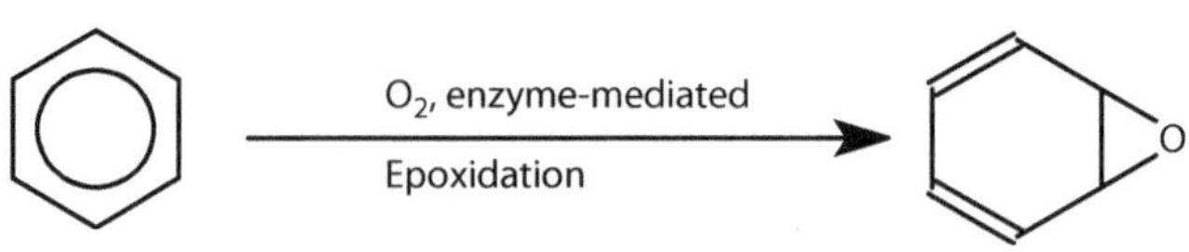

(6.13)

Epoxidation is a particularly important means of metabolic attack upon aromatic rings that abound in many xenobiotic compounds and is part of the process by which potentially carcinogenic PAH compounds such as benzo(a)pyrene are biodegraded.[3] Epoxidation of an aromatic ring may be followed by *ring cleavage*, an important step in the biodegradation of aromatic compounds:

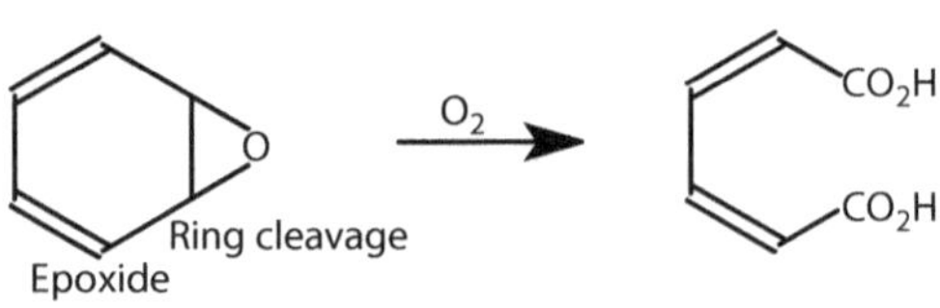

(6.14)

6.10.1.1 Microbial Oxidation of Hydrocarbons

The degradation of hydrocarbons by microbial oxidation is an important environmental process because it is the primary means by which petroleum wastes are eliminated from water and soil. Bacteria capable of degrading hydrocarbons include *Micrococcus*, *Pseudomonas*, *Mycobacterium*, and *Nocardia*.

The most common initial step in the microbial oxidation of alkanes involves conversion of a terminal $-CH_3$ group to a $-CO_2H$ group. After formation of a carboxylic acid from the alkane, further oxidation normally occurs by β-oxidation:

$$CH_3CH_2CH_2CH_2CO_2H + 3O_2 \rightarrow CH_3CH_2CO_2H + 2CO_2 + 2H_2O \quad (6.15)$$

A complicated cycle with a number of steps is involved in β-oxidation. The residue at the end of each cycle is an organic acid with two fewer carbon atoms than its precursor at the beginning of the cycle.

Hydrocarbons vary significantly in their biodegradability, and microorganisms show a strong preference for straight-chain hydrocarbons. A major reason for this preference is that branching inhibits β-oxidation at the site of the branch. The presence of a quaternary carbon (below) particularly inhibits alkane degradation.

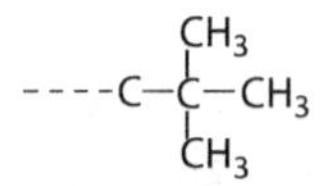

Source: From Manahan, S., *Environmental Chemistry*, 9th ed., Taylor & Francis, Boca Raton, 2010. With permission.

The biodegradation of petroleum is essential for the elimination of oil spills (of the order of a million metric tons per year). This oil is degraded by both marine bacteria and filamentous fungi. In some cases, the rate of degradation is limited by available nitrate and phosphate and it has been observed that the microbial degradation of hydrocarbons from oil spills is enhanced by providing nitrogen, phosphorus, and potassium nutrients. The physical form of crude oil makes a large difference in its degradability. Degradation in water occurs at the water–oil interface and in layers of oil floating on top of water and exposed to the atmosphere. Therefore, thick layers of crude oil prevent contact with bacterial enzymes and O_2.

Hydroxylation often accompanies microbial oxidation. It is the attachment of –OH groups to hydrocarbon chains or rings. In the biodegradation of foreign compounds, hydroxylation often follows epoxidation as shown by the following rearrangement reaction for benzene epoxide:

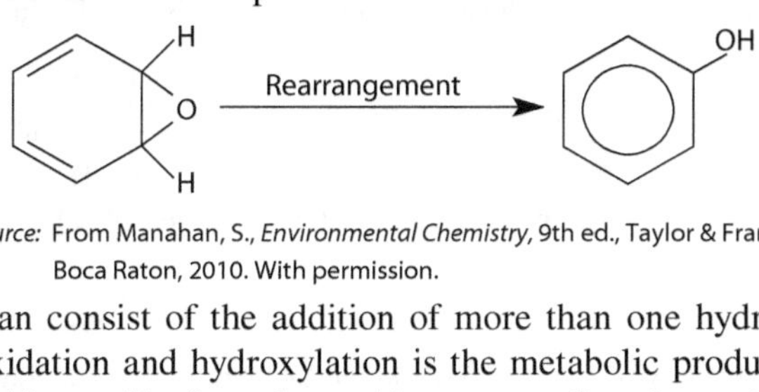

Source: From Manahan, S., *Environmental Chemistry*, 9th ed., Taylor & Francis, Boca Raton, 2010. With permission.

Hydroxylation can consist of the addition of more than one hydroxide group. An example of epoxidation and hydroxylation is the metabolic production of carcinogenic 7,8-diol-9,10-epoxide from benzo(a)pyrene, a five-ring polycyclic aromatic hydrocarbon shown in Chapter 12, Figure 12.8.

6.10.2 Other Biochemical Processes in Biodegradation of Organics

Hydrolysis, which involves the addition of H_2O to a molecule accompanied by cleavage of the molecule into two products, is a major step in microbial degradation of many pollutant compounds, especially pesticidal esters, amides, and organophosphate esters. The types of enzymes that bring about hydrolysis are *hydrolase enzymes*; those that enable the hydrolysis of esters are called *esterases*, whereas those that hydrolyze amides are *amidases*. At least one species of *Pseudomonas* hydrolyzes malathion in a type of hydrolysis reaction typical of those by which pesticides are degraded:

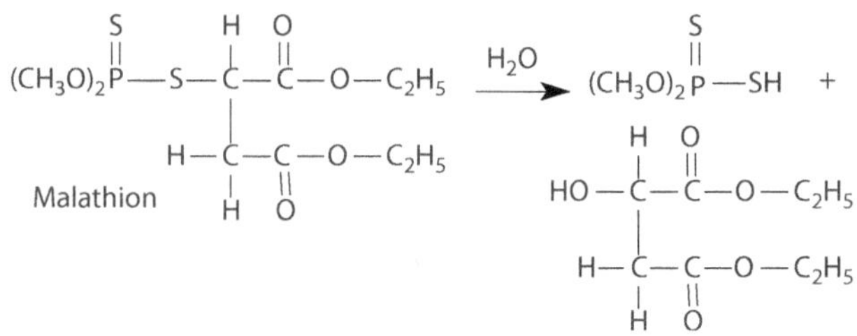

(6.16)

Reductions are carried out by *reductase enzymes*; for example, nitroreductase enzyme catalyzes the reduction of the nitro group. Table 6.2 gives the major kinds of functional groups reduced by microorganisms.

TABLE 6.2
Functional Groups That Undergo Microbial Reduction

Reactant	Process	Product
R—C(=O)—H	Aldehyde reduction	R—CH(H)—OH
R—C(=O)—R′	Ketone reduction	R—C(OH)(H)—R′
R—S(=O)—R′	Sulfoxide reduction	R—S—R′
R—SS—R′	Disulfide reduction	R—SH, R′—SH
R(H)C=C(H)R′	Alkene reduction	R—CH(H)—CH(H)—R′
R—NO_2	Nitro group reduction	R—NO, R—NH_2, R—N(H)(OH)

Source: From Manahan, S., *Environmental Chemistry*, 9th ed., Taylor & Francis, Boca Raton, 2010. With permission.

Dehalogenation reactions of organohalide compounds involve the bacterially mediated replacement of a covalently bound halogen atom (F, Cl, Br, I) with –OH, and are discussed in more detail in Section 6.13.

Many environmentally significant organic compounds contain alkyl groups, such as the methyl ($-CH_3$) group, attached to atoms of O, N, and S. An important step in the microbial metabolism of many of these compounds is *dealkylation*, replacement of alkyl groups by H as shown in Figure 6.9. Examples of these kinds of reactions include O-dealkylation of methoxychlor insecticides, N-dealkylation of carbaryl insecticide, and S-dealkylation of dimethyl sulfide. Alkyl groups removed by dealkylation usually are attached to oxygen, sulfur, or nitrogen atoms; those attached to carbon are normally not removed directly by microbial processes.

6.11 MICROBIAL TRANSFORMATIONS OF NITROGEN

Some of the most important microorganism-mediated chemical reactions in aquatic and soil environments are those involving nitrogen compounds.[4] They are summarized in the *nitrogen cycle* shown in Figure 6.10. This cycle describes the dynamic processes through which nitrogen is interchanged among the atmosphere, organic matter, and inorganic compounds. It is one of nature's most vital dynamic processes.

Among the biochemical transformations in the nitrogen cycle are nitrogen fixation, whereby molecular nitrogen is fixed as organic nitrogen; nitrification, the process of oxidizing ammonia to nitrate; nitrate reduction, the process by which nitrogen in nitrate ion is reduced to form compounds having nitrogen in a lower oxidation state; and denitrification, the reduction of nitrate and nitrite to N_2, with a resultant net loss of nitrogen gas to the atmosphere. Each of these important chemical processes will be discussed separately.

6.11.1 Nitrogen Fixation

The overall microbial process for *nitrogen fixation*, the binding of atmospheric nitrogen in a chemically combined form

$$3\{CH_2O\} + 2N_2 + 3H_2O + 4H^+ \rightarrow 3CO_2 + 4NH_4^+ \tag{6.17}$$

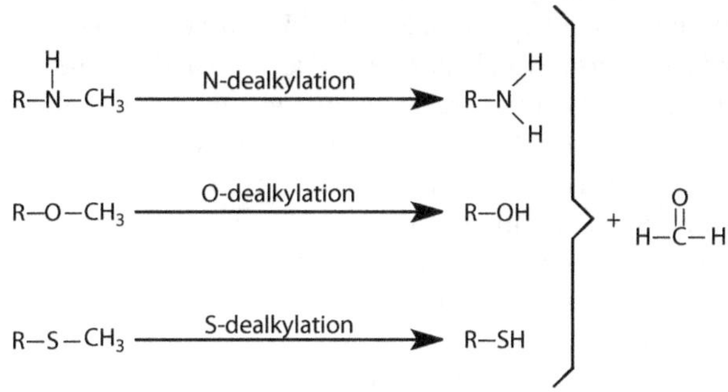

FIGURE 6.9 Metabolic dealkylation reactions shown for the removal of CH_3 from N, O, and S atoms in organic compounds. (From Manahan, S., *Environmental Chemistry*, 9th ed., Taylor & Francis, Boca Raton, 2010. With permission.)

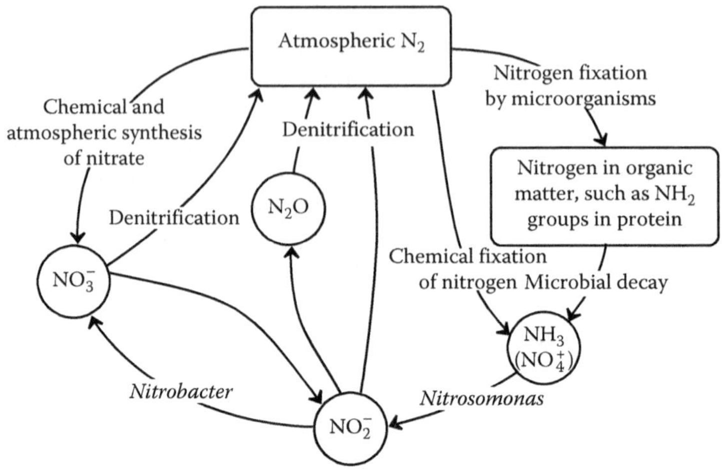

FIGURE 6.10 The nitrogen cycle. (From Manahan, S., *Environmental Chemistry*, 9th ed., Taylor & Francis, Boca Raton, 2010. With permission.)

is actually quite a complicated process that is the subject of extensive investigation. Biological nitrogen fixation is a key biochemical process in the environment and is essential for plant growth in the absence of synthetic fertilizers.

Only a few species of aquatic microorganisms have the ability to fix atmospheric nitrogen. Among the aquatic bacteria having this capability are photosynthetic bacteria, *Azotobacter*, several species of *Clostridium*, and cyanobacteria. In most natural freshwater systems, however, the fraction of nitrogen fixed by organisms in the water relative to that originating from the decay of organic material, fertilizer runoff, and other external sources is relatively low.

The best-known and most important form of nitrogen-fixing bacteria is *Rhizobium*, which enjoys a symbiotic (mutually advantageous) relationship with leguminous plants such as clover or alfalfa. The *Rhizobium* bacteria are found in root nodules, special structures bound to the roots of legumes and connected directly to the vascular (circulatory) system of the plant, enabling the bacteria to derive photosynthetically produced energy directly from the plant. Thus, the plant provides the energy required to break the strong triple bonds in the dinitrogen molecule, converting the nitrogen to a reduced form which is directly assimilated by the plant. When the legumes die and decay, NH_4^+ ion is released and is converted by microorganisms to nitrate ion, which is assimilable by other plants. Some of the ammonium ion and nitrate released may be carried into natural water systems.

Some nonlegume angiosperms fix nitrogen through the action of actinomycetes bacteria contained in root nodules. Shrubs and trees in the nitrogen-fixing category are abundant in fields, forests, and wetlands throughout the world. Their rate of nitrogen fixation is comparable to that of legumes.

Free-living bacteria associated with some grasses are stimulated by the grasses to fix nitrogen. One such bacterium is *Spirillumlipoferum*. In tropical surroundings, the amount of reduced nitrogen fixed by such bacteria can amount to the order of 100 kg per hectare per year.

Because of the cost of energy required to fix nitrogen synthetically, efforts are underway to increase the efficiency of natural means of nitrogen fixation. One approach uses recombinant DNA methodologies in attempts to transfer the nitrogen-fixing capabilities of nitrogen-fixing bacteria directly to plant cells. Although a fascinating possibility, this transfer has not yet been achieved on a practical basis. The other approach uses more conventional plant breeding and biological techniques in attempts to increase the range and effectiveness of the symbiotic relationship existing between some plants and nitrogen-fixing bacteria.

One matter of concern is that successful efforts to increase nitrogen fixation may upset the global nitrogen balance. Total annual global fixation of nitrogen is now more than 50% higher than the preindustrial level of 150 million metric tons estimated for 1850. Potential accumulation of excess fixed nitrogen is the subject of some concern because of aquatic nitrate pollution and microbial production of N_2O gas. Some atmospheric scientists fear that excess N_2O gas may be involved in depletion of the protective atmospheric ozone layer (see Chapter 1, Section 1.4).

6.11.2 Nitrification

Nitrification, the conversion of N(–III) to N(V), is a very common and extremely important process in water and in soil. Aquatic nitrogen in thermodynamic equilibrium with air is in the +5 oxidation state as NO_3^-, whereas in most biological compounds, nitrogen is present as N(–III), such as $-NH_2$ in amino acids. The equilibrium constant of the overall nitrification reaction, written for one electron-mole

$$\tfrac{1}{4}O_2 + \tfrac{1}{8}NH_4^+ \rightarrow \tfrac{1}{8}NO_3^- + \tfrac{1}{4}H^+ + \tfrac{1}{8}H_2O \tag{6.18}$$

is $10^{7.59}$ (Table 6.1), showing that the reaction is highly favored from a thermodynamic viewpoint.

Nitrification is especially important in nature because nitrogen is absorbed by plants primarily as nitrate. When fertilizers are applied in the form of ammonium salts or anhydrous ammonia, a microbial transformation to nitrate enables maximum assimilation of nitrogen by the plants.

In nature, nitrification is catalyzed by two groups of bacteria, *Nitrosomonas* and *Nitrobacter*. *Nitrosomonas* bacteria bring about the transition of ammonia to nitrite

$$NH_3 + \tfrac{3}{2}O_2 \rightarrow H^+ + NO_2^- + H_2O \tag{6.19}$$

whereas *Nitrobacter* mediates the oxidation of nitrite to nitrate:

$$NO_2^- + \tfrac{1}{2}O_2 \rightarrow NO_3^- \tag{6.20}$$

Both of these highly specialized types of bacteria are *obligate aerobes*; that is, they function only in the presence of molecular O_2. These bacteria are also *chemolithotrophic*, meaning that they can utilize oxidizable inorganic materials as electron donors in oxidation reactions to yield needed energy for metabolic processes.

For the oxic conversion of one electron-mole of ammoniacal nitrogen to nitrite ion at pH 7.00,

$$\tfrac{1}{4}O_2 + \tfrac{1}{6}NH_4^+ \rightarrow \tfrac{1}{6}NO_2^- + \tfrac{1}{3}H^+ + \tfrac{1}{6}H_2O \tag{6.21}$$

the free-energy change is −10.8 kcal. The free-energy change for the oxic oxidation of one electron-mole of nitrite ion to nitrate ion:

$$\tfrac{1}{4}O_2 + \tfrac{1}{2}NO_2^- \rightarrow \tfrac{1}{2}NO_3^- \tag{6.22}$$

is −9.0 kcal. Both steps of the nitrification process involve an appreciable yield of free energy.

6.11.3 Nitrate Reduction

As a general term, *nitrate reduction* refers to microbial processes by which nitrogen in chemical compounds is reduced to lower oxidation states. In the absence of free oxygen, nitrate may be used by some bacteria as an alternate electron receptor. The most complete possible reduction of nitrogen in nitrate ion involves the acceptance of eight electrons by the nitrogen atom, with the consequent conversion of nitrate to ammonia (+V to −III oxidation state). Nitrogen is an essential component of protein, and any organism that utilizes nitrogen from nitrate for the synthesis of protein must first reduce the nitrogen to the −III oxidation state (ammoniacal form). However, incorporation of nitrogen into protein generally is a relatively minor use of the nitrate undergoing microbially mediated reactions and is more properly termed nitrate assimilation.

Nitrate ion functioning as an electron receptor usually produces NO_2^-:

$$\tfrac{1}{2}NO_3^- + \tfrac{1}{4}\{CH_2O\} \rightarrow \tfrac{1}{2}NO_2^- + \tfrac{1}{4}H_2O + \tfrac{1}{4}CO_2 \tag{6.23}$$

The free-energy yield per electron-mole is only about two-third of the yield when oxygen is the oxidant; however, nitrate ion is a good electron receptor in the absence of O_2. One of the factors limiting the use of nitrate ion in this function is its generally low concentration in most waters. Furthermore, nitrite, NO_2^-, is relatively toxic and tends to inhibit the growth of many bacteria after building up to a certain level. Sodium nitrate has been used as a "first-aid" treatment in sewage lagoons that have become oxygen deficient. It provides an emergency source of oxygen to reestablish normal bacterial growth.

Nitrate ion can be an effective oxidizing agent for a number of species in water that are oxidized by the action of microorganisms. Therefore, nitrate salts are sometimes added as an alternate source of oxygen in the biological treatment of oxidizable wastes.

6.11.4 Denitrification

An important special case of nitrate reduction is *denitrification*, in which the reduced nitrogen product is a nitrogen-containing gas, usually N_2. At pH 7.00, the free-energy change per electron-mole of reaction

$$\tfrac{1}{5}NO_3^- + \tfrac{1}{4}\{CH_2O\} + \tfrac{1}{5}H^+ \rightarrow \tfrac{1}{10}N_2 + \tfrac{1}{4}CO_2 + \tfrac{7}{20}H_2O \tag{6.24}$$

is –2.84 kcal. The free-energy yield per mole of nitrate reduced to N_2 (five electron-moles) is lower than that for the reduction of the same quantity of nitrate to nitrite. More important, however, the reduction of a nitrate ion to N_2 gas consumes five electrons, compared to only two electrons for the reduction of NO_3^- to NO_2^-.

Denitrification is an important process in nature. It is the mechanism by which fixed nitrogen is returned to the atmosphere. Denitrification is also used in advanced water treatment for the removal of nutrient nitrogen (see Chapter 8). Because nitrogen gas is a nontoxic volatile substance that does not inhibit microbial growth, and since nitrate ion is a very efficient electron acceptor, denitrification allows the extensive growth of bacteria under anoxic conditions.

Loss of nitrogen to the atmosphere may also occur through the formation of N_2O and NO by bacterial action on nitrate and nitrite catalyzed by the action of several types of bacteria. Production of N_2O relative to N_2 is enhanced during denitrification in soils by increased concentrations of NO_3^-, NO_2^-, and O_2.

6.11.5 Competitive Oxidation of Organic Matter by Nitrate Ion and Other Oxidizing Agents

The successive oxidation of organic matter by dissolved O_2, NO_3^-, and SO_4^{2-} brings about an interesting sequence of nitrate-ion levels in sediments and hypolimnion waters initially containing O_2 but lacking a mechanism for reaeration. This is shown in Figure 6.11, where concentrations of dissolved O_2, NO_3^-, and SO_4^{2-} are plotted as a function of total organic matter metabolized. This behavior can be explained by the following sequence of biochemical processes:

$$O_2 + \text{organic matter} \rightarrow \text{products} \tag{6.25}$$

$$NO_3^- + \text{organic matter} \rightarrow \text{products} \tag{6.26}$$

$$SO_4^{2-} + \text{organic matter} \rightarrow \text{products} \tag{6.27}$$

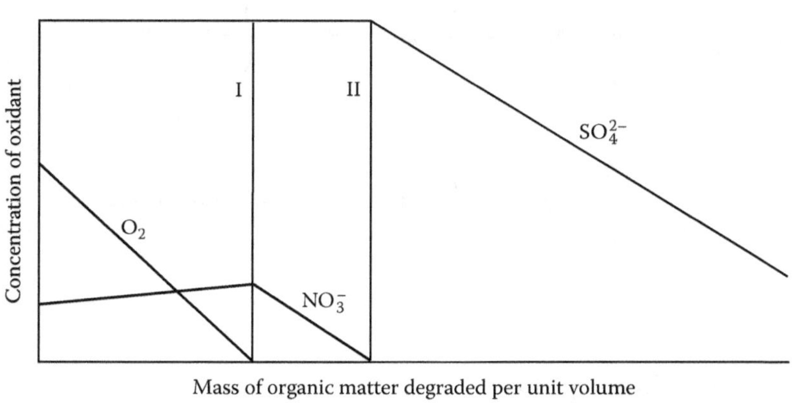

FIGURE 6.11 Oxidation of organic matter by O_2, NO_3^-, and SO_4^{2-}. (From Manahan, S., *Environmental Chemistry*, 9th ed., Taylor & Francis, Boca Raton, 2010. With permission.)

So long as some O_2 is present, some nitrate may be produced from nitrogenous organic matter. After exhaustion of molecular oxygen, nitrate is the favored oxidizing agent, and its concentration falls from a maximum value (I) to zero (II). Sulfate, which is usually present in a large excess over the other two oxidants, then becomes the favored electron receptor, enabling biodegradation of organic matter to continue.

6.12 MICROBIAL TRANSFORMATIONS OF PHOSPHORUS AND SULFUR

6.12.1 Phosphorus Compounds

Biodegradation of phosphorus compounds is important in the environment for two reasons. The first of these is that it provides a source of algal nutrient orthophosphate from the hydrolysis of polyphosphates (see Section 3.12). Secondly, biodegradation deactivates toxic organophosphate compounds, such as the organophosphate insecticides.

The organophosphorus compounds of greatest environmental concern tend to be sulfur-containing *phosphorothionate* and *phosphordithioate* ester insecticides. These are discussed in more detail in Chapter 7, Section 7.11, and structural formulas of several representative examples are shown in Figure 7.7. The biodegradation of these compounds is an important environmental chemical process. Fortunately, unlike the organohalide insecticides that they largely displaced until less toxic classes of compounds came into general use, the organophosphates readily undergo biodegradation and do not bioaccumulate.

Hydrolysis is an important step in the biodegradation of phosphorothionate, phosphorodithioate, and phosphate ester insecticides leading to their detoxification and loss of insecticidal activity. Hydrolysis of these insecticides is shown by the following general reactions where R is an alkyl group, Ar is a substituent group that is frequently aromatic, and X is either S or O:

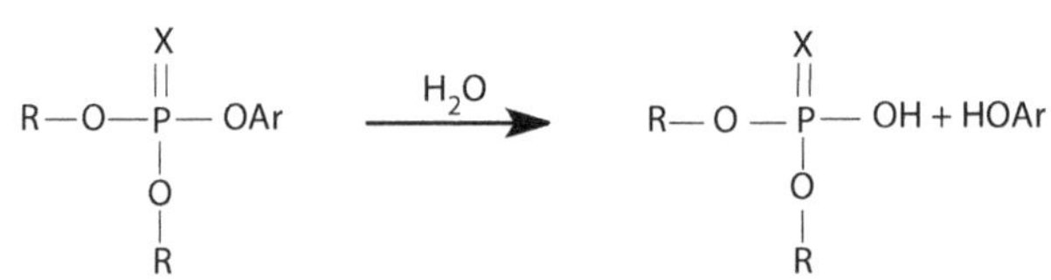

(6.28)

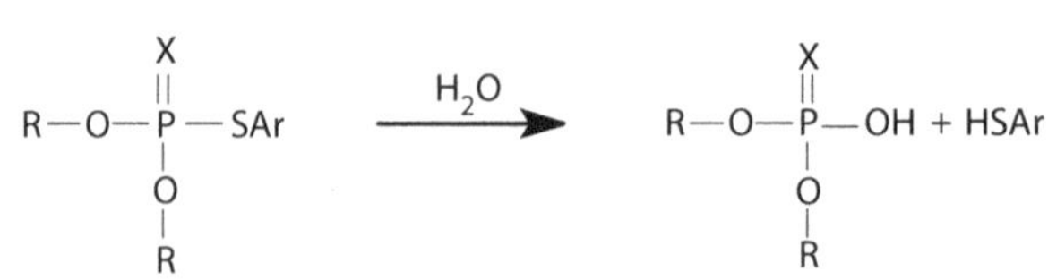

(6.29)

6.12.2 Sulfur Compounds

Sulfur compounds are very common in water. Sulfate ion, SO_4^{2-}, is found in varying concentrations in practically all natural waters. Organic sulfur compounds, both those of natural origin and pollutant species, are very common in natural aquatic systems, and the degradation of these compounds is an important microbial process. Sometimes the degradation products, such as odiferous and toxic H_2S, cause serious problems with water quality.

Sulfur in living material is present primarily in its most reduced state, for example, as the hydrosulfide group, –SH. When organic sulfur compounds are decomposed by bacteria, the initial sulfur product is generally the reduced form, H_2S. Some bacteria produce and store elemental sulfur from sulfur compounds. In the presence of oxygen, some bacteria convert reduced forms of sulfur to the oxidized form in SO_4^{2-} ion.

6.12.2.1 Oxidation of H_2S and Reduction of Sulfate by Bacteria

In the presence of oxygen some bacteria can oxidize sulfide (H_2S) to sulfate and in the absence of oxygen other bacteria can reduce sulfate to sulfide.[5] The bacteria *Desulfovibrio* can reduce inorganic sulfate ion to H_2S. In so doing, they utilize sulfate as an electron acceptor in the oxidation of organic matter. The overall reaction for the microbially mediated oxidation of biomass with sulfate is

$$SO_4^{2-} + 2\{CH_2O\} + 2H^+ \rightarrow H_2S + 2CO_2 + 2H_2O \tag{6.30}$$

and it requires other bacteria besides *Desulfovibrio* to oxidize organic matter completely to CO_2. Because of the high concentration of sulfate ion in seawater, bacterially mediated formation of H_2S causes pollution problems in some coastal areas and is a major source of atmospheric sulfur. In waters where sulfide formation occurs, the sediment is often black in color due to the formation of FeS.

Some bacteria such as purple sulfur bacteria and green sulfur bacteria can oxidize sulfur in hydrogen sulfide to higher oxidation states. The oxic colorless sulfur bacteria may use molecular oxygen to oxidize H_2S to elemental sulfur and can oxidize elemental S and thiosulfate ($S_2O_3^{2-}$) to sulfate.

Oxidation of sulfur in a low oxidation state to sulfate ion produces sulfuric acid, a strong acid. One of the colorless sulfur bacteria, *Thiobacillus thiooxidans*, is tolerant of one normal acid solutions; a remarkable acid tolerance. When elemental sulfur is added to excessively alkaline soils, the acidity is increased because of a microorganism-mediated reaction, which produces sulfuric acid. Elemental sulfur may be deposited as granules in the cells of purple sulfur bacteria and colorless sulfur bacteria. Such processes are important sources of elemental sulfur deposits.

6.12.3 Microorganism-Mediated Degradation of Organic Sulfur Compounds

Sulfur occurs in many types of biological compounds. As a consequence, organic sulfur compounds of natural and pollutant origin are very common in water. The degradation of these compounds is an important microbial process having a strong effect upon water quality.

Among some of the common sulfur-containing functional groups found in aquatic organic compounds are hydrosulfide (–SH), disulfide (–SS–), sulfide (–S–), sulfoxide

($-\overset{O}{\overset{\|}{S}}-$), sulfonic acid ($-SO_2OH$), thioketone ($-\overset{S}{\overset{\|}{C}}-$), and thiazole (a heterocyclic sulfur group). Protein contains some amino acids with sulfur functional groups—cysteine,

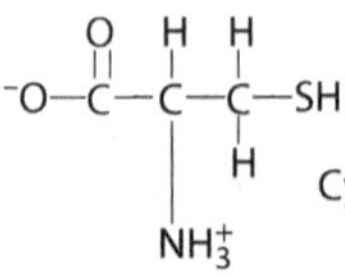

Cysteine

cystine, and methionine—whose breakdown is important in natural waters. The amino acids are readily degraded by bacteria and fungi. The biodegradation of sulfur-containing amino acids can result in production of volatile organic sulfur compounds such as methane thiol, CH_3SH, and dimethyl disulfide, CH_3SSCH_3. These compounds have strong, unpleasant odors. Their formation, in addition to that of H_2S, accounts for much of the odor associated with the biodegradation of sulfur-containing organic compounds.

Hydrogen sulfide is formed from a large variety of organic compounds through the action of a number of different kinds of microorganisms. A typical sulfur-cleavage reaction producing H_2S is the conversion of cysteine to pyruvic acid through the action of cysteine desulfhydrase enzyme in bacteria:

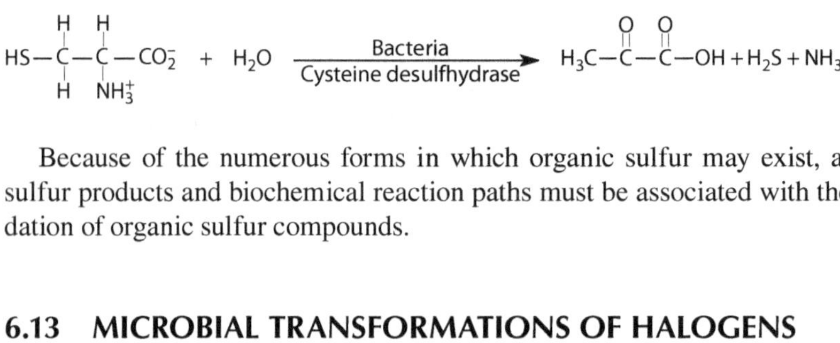

(6.31)

Because of the numerous forms in which organic sulfur may exist, a variety of sulfur products and biochemical reaction paths must be associated with the biodegradation of organic sulfur compounds.

6.13 MICROBIAL TRANSFORMATIONS OF HALOGENS AND ORGANOHALIDES

Microbially mediated reactions are important in the environmental degradation of pollutant organohalide compounds.[6] *Dehalogenation* reactions involving the replacement of a halogen atom, for example,

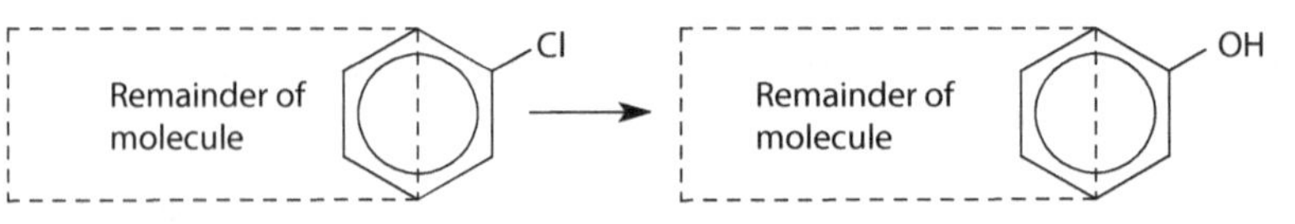

represent a major pathway for the biodegradation of organohalide hydrocarbons. In some cases, organohalide compounds serve as sole carbon sources, sole energy sources, or electron acceptors for anoxic bacteria. Microorganisms need not utilize a particular organohalide compound as a sole carbon source in order to cause its degradation. This is due to the phenomenon of *cometabolism*, which results from a lack of specificity in the microbial degradation processes. Thus, bacterial degradation of small amounts of an organohalide compound may occur while the microorganism involved is metabolizing much larger quantities of another substance.

Some anoxic bacteria can reductively dechlorinate chlorinated aliphatic and aromatic compounds using relatively highly chlorinated compounds as electron acceptors as illustrated by the reaction

$$\{CH_2O\} + H_2O + 2Cl\text{–}R \rightarrow CO_2 + 2H^+ + 2Cl^- + 2H\text{–}R \tag{6.32}$$

where Cl–R represents a site of chlorine substitution on a chlorinated hydrocarbon molecule and H–R represents a site of hydrogen substitution. The net result of this process called *dehalorespiration* is the replacement of Cl by H on chlorinated hydrocarbons.

Microbially mediated dichloroelimination from 1,1,2,2-tetrachloroethane in the anoxic biodegradation of this compound can produce one of three possible isomers of dichloroethylene; the process followed by successive hydrogenolysis reactions can produce vinyl chloride and ethene (ethylene):

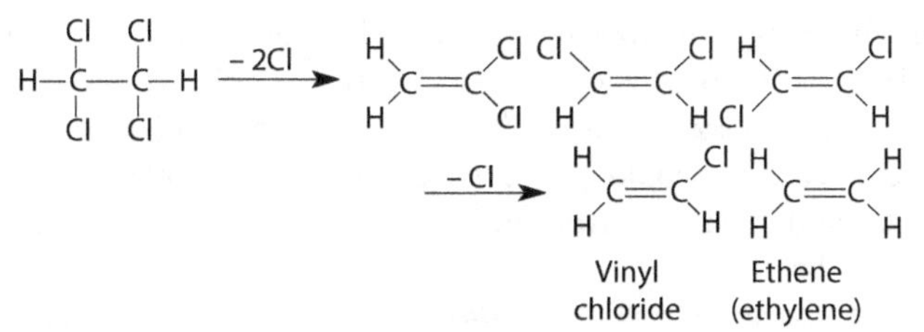

(6.33)

Successive hydrogenolysis reactions of 1,1,2,2-tetrachloroethane can produce ethane derivatives with 3, 2, 1, and 0 chlorine atoms

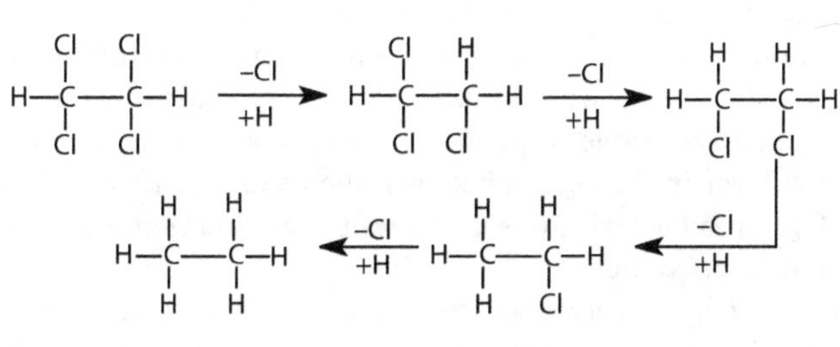

(6.34)

Bioconversion of DDT to replace Cl with H yields DDD:

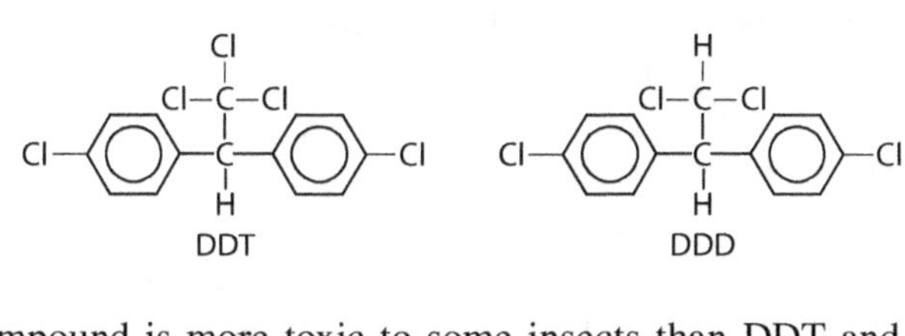

The latter compound is more toxic to some insects than DDT and has even been manufactured as a pesticide.

6.14 MICROBIAL TRANSFORMATIONS OF METALS AND METALLOIDS

Some bacteria, including *Ferrobacillus*, *Gallionella*, and some forms of *Sphaerotilus*, utilize iron compounds to obtain energy for their metabolic needs. These bacteria catalyze the oxidation of iron(II) to iron(III) by molecular oxygen:

$$4Fe(II) + 4H^+ + O_2 \rightarrow 4Fe(III) + 2H_2O \tag{6.35}$$

The carbon source for some of these bacteria is CO_2. Since they do not require organic matter for carbon, and they derive energy from the oxidation of inorganic matter, these bacteria may thrive in environments where organic matter is absent.

The microorganism-mediated oxidation of iron(II) is not a particularly efficient means of obtaining energy for metabolic processes. For the reaction

$$FeCO_3(s) + \tfrac{1}{4}O_2 + \tfrac{3}{2}H_2O \rightarrow Fe(OH)_3(s) + CO_2 \tag{6.36}$$

the change in free energy is approximately 10 kcal/electron-mole. Approximately 220 g of iron(II) must be oxidized to produce 1.0 g of cell carbon. The calculation assumes CO_2 as a carbon source and a biological efficiency of 5%. The production of only 1.0 g of cell carbon would produce approximately 430 g of solid $Fe(OH)_3$. It follows that large deposits of hydrated iron(III) oxide form in areas where iron-oxidizing bacteria thrive.

Some of the iron bacteria, notably *Gallionella*, secrete large quantities of hydrated iron(III) oxide in intricately branched structures. The bacterial cell grows at the end of a twisted stalk of the iron oxide. Individual cells of *Gallionella*, photographed through an electron microscope, have shown that the stalks consist of a number of strands of iron oxide secreted from one side of the cell (Figure 6.12).

At nearly neutral pH values, bacteria deriving energy by mediating the air oxidation of iron(II) must compete with direct chemical oxidation of iron(II) by O_2. The latter process is relatively rapid at pH 7. As a consequence, these bacteria tend to grow in a narrow layer in the region between the oxygen source and the source of iron(II). Therefore, iron bacteria are sometimes called *gradient organisms*, and they grow at intermediate pE values.

Bacteria are strongly involved in the oceanic manganese cycle. Manganese nodules, a potentially important source of manganese, copper, nickel, and cobalt that occur on ocean floors, yield different species of bacteria which enzymatically mediate both the oxidation and reduction of manganese.

6.14.1 Acid Mine Waters

One consequence of bacterial action on metal compounds is acid mine drainage, one of the most common and damaging problems in the aquatic environment. Many waters flowing from coal mines and draining from the "gob piles" left over from coal processing and washing are practically sterile due to high acidity.

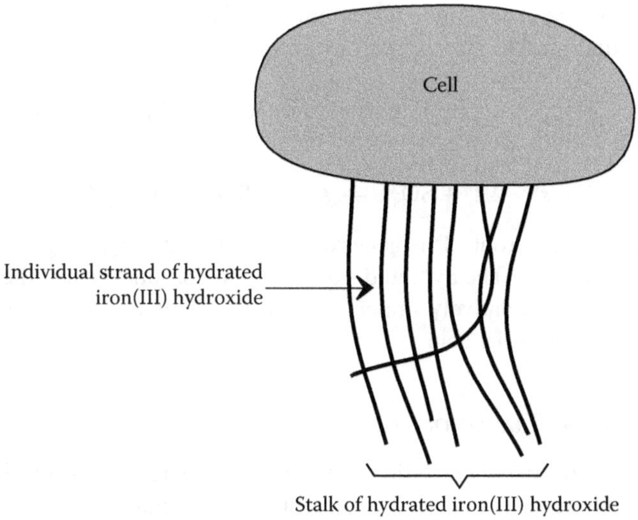

FIGURE 6.12 Sketch of a cell of *Gallionella* showing iron(III) oxide secretion. (From Manahan, S., *Environmental Chemistry*, 9th ed., Taylor & Francis, Boca Raton, 2010. With permission.)

Acid mine water results from the presence of sulfuric acid produced by the oxidation of pyrite, FeS_2. Microorganisms are closely involved in the overall process, which consists of several reactions and involves a number of microbial species that can exist under acidic conditions. The first of these reactions is the oxidation of pyrite, which is followed by the oxidation of iron(II) ion to iron(III) ion:

$$2FeS_2(s) + 2H_2O + 7O_2 \rightarrow 4H^+ + 4SO_4^{2-} + 2Fe^{2+} \qquad (6.37)$$

$$4Fe^{2+} + O_2 + 4H^+ \rightarrow 4Fe^{3+} + 2H_2O \qquad (6.38)$$

a process that occurs very slowly at low pH values found in acid mine waters. Below pH 3.5, the iron oxidation is catalyzed by the iron bacterium *Thiobacillus ferrooxidans*, and in the pH range 3.5–4.5 it may be catalyzed by a variety of *Metallogenium*, a filamentous iron bacterium. Other bacteria that may be involved in acid mine water formation are *Thiobacillus thiooxidans* and *Ferrobacillus ferrooxidans*. The Fe^{3+} ion further dissolves pyrite

$$FeS_2(s) + 14Fe^{3+} + 8H_2O \rightarrow 15Fe^{2+} + 2SO_4^{2-} + 16H^+ \qquad (6.39)$$

which in conjunction with Reaction 6.38 constitutes a cycle for the dissolution of pyrite. $Fe(H_2O)_6^{3+}$ is an acidic ion and at pH values much above 3, the iron(III) precipitates as the hydrated iron(III) oxide:

$$Fe^{3+} + 3H_2O \rightarrow Fe(OH)_3(s) + 3H^+ \qquad (6.40)$$

The beds of streams afflicted with acid mine drainage often are covered with "yellowboy," an unsightly deposit of amorphous, semigelatinous $Fe(OH)_3$. The most damaging component of acid mine water, however, is sulfuric acid. It is directly toxic and has other undesirable effects such as excessive weathering of minerals that it contacts.

Limestone, $CaCO_3$, is commonly used to treat acid mine water. When acid mine water is treated with limestone, the following reaction occurs:

$$CaCO_3(s) + 2H^+ + SO_4^{2-} \rightarrow Ca^{2+} + SO_4^{2-} + H_2O + CO_2(g) \quad (6.41)$$

Unfortunately, because iron(III) is generally present, $Fe(OH)_3$ precipitates as the pH is raised (Reaction 6.40). The hydrated iron(III) oxide product covers the particles of carbonate rock with a relatively impermeable layer. This armoring effect prevents further neutralization of the acid.

6.14.2 Microbial Transitions of Selenium

Directly below sulfur in the periodic table, selenium is subject to bacterial oxidation and reduction. These transitions are important because selenium is a crucial element in nutrition, particularly of livestock. Diseases related to either selenium excesses or deficiency have been reported in at least half of the states of the United States and in 20 other countries, including the major livestock producing countries. Livestock in New Zealand, in particular, suffer from selenium deficiency.

Microorganisms are closely involved with the selenium cycle, and microbial reduction of oxidized forms of selenium has been known for some time. Reductive processes under anoxic conditions can reduce both SeO_3^{2-} and SeO_4^{2-} ions to elemental selenium, which can accumulate as a sink for selenium in anoxic sediments. Some bacteria such as selected strains of *Thiobacillus* and *Leptothrix* can oxidize elemental selenium to selenite, SeO_3^{2-}, thus remobilizing this element from deposits of Se(0).

The major volatile selenium species emitted to the atmosphere by microbial processes in water and soil is dimethylselenide, $(CH_3)_2Se$. (The single largest source of natural sulfur discharged to the atmosphere is the analogous sulfur compound, biogenic dimethyl sulfide, $(CH_3)_2S$, produced by marine microorganisms). The major biological selenium compound that is the precursor to dimethylselenide formation is selenomethionine:

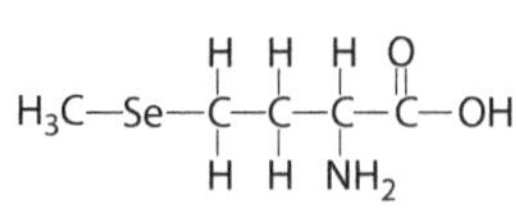

Selenomethionine

6.14.3 Microbial Corrosion

Corrosion involving the deterioration of iron as well as other metals is a redox phenomenon and was discussed in Section 4.14. Much corrosion is bacterial in nature.[7] Bacteria involved with corrosion set up their own electrochemical cells in which a portion of the surface of the metal being corroded forms the anode of the cell and is oxidized. Structures called *tubercles* form in which bacteria pit and corrode metals as shown in Figure 6.13.

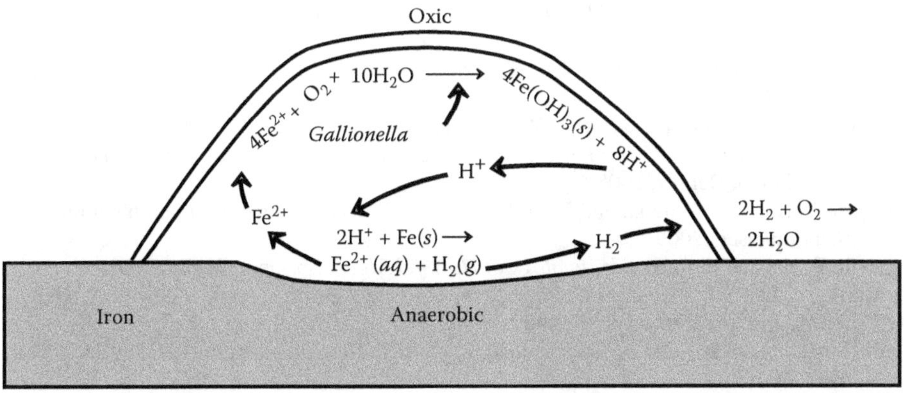

FIGURE 6.13 Tubercle in which the bacterially mediated corrosion of iron occurs through the action of *Gallionella*. (From Manahan, S., *Environmental Chemistry*, 9th ed., Taylor & Francis, Boca Raton, 2010. With permission.)

LITERATURE CITED

1. Vaulot, D., W. Eikrem, M. Viprey, and H. Moreau, The diversity of small eukaryotic phytoplankton (≤3 μm) in marine ecosystems, *FEMS Microbiology Reviews*, 32, 795–820, 2008.
2. Koskinen, P. E. P., S. R. Beck, J. Orlygsson, and J. A. Puhakka, Ethanol and hydrogen production by two thermophilic, anaerobic bacteria isolated from icelandic geothermal areas, *Biotechnology and Bioengineering*, 101(4), 679–690, 2008.
3. Diab, E. A., Phytoremediation of polycyclic aromatic hydrocarbons (PAHs) in a polluted desert soil, with special reference to the biodegradation of the carcinogenic PAHs, *Australian Journal of Basic and Applied Sciences*, 2, 757–762, 2008.
4. Jetten, M. S. M., The microbial nitrogen cycle, *Environmental Microbiology*, 10, 2903–2909, 2008.
5. Pereyra, L. P., S. R. Hiibel, A. Pruden, and K. F. Reardon, Comparison of microbial community composition and activity in sulfate-reducing batch systems remediating mine drainage, *Biotechnology and Bioengineering*, 101, 702–713, 2008.
6. Scheutz, C., N. D. Durant, P. Dennis, M. H. Hansen, T. Joergensen, R. Jakobsen, E. E. Cox, and P. L. Bjerg, Concurrent ethene generation and growth of dehalococcoides containing vinyl chloride reductive dehalogenase genes during an enhanced reductive dechlorination field demonstration, *Environmental Science and Technology*, 42, 9302–9309, 2008.
7. Nizhegorodov, S. Yu., S. A. Voloskov, V. A. Trusov, L. M. Kaputkina, and T. A. Syur, Corrosion of steels due to the action of microorganisms, *Metal Science and Heat Treatment*, 50, 191–195, 2008.

SUPPLEMENTARY REFERENCES

Amsler, C. D., Ed., *Algal Chemical Ecology*, Springer, Berlin, 2008.

Bitton, G., *Wastewater Microbiology*, 3rd ed., Wiley-Liss, New York, 2005.

Bitton, G., *Encyclopedia of Environmental Microbiology*, Wiley, New York, 2002.

Chapelle, F. H., *Ground-Water Microbiology and Geochemistry*, 2nd ed., Wiley, New York, 2001.

Crawford, R. L. and C. J. Hurst, Eds, *Manual of Environmental Microbiology*, ASM Press, Washington, DC, 2002.

Deacon, J. W., *Introduction to Modern Mycology*, Blackwell Science Inc., Cambridge, MA, 1997.
Gerardi, M. H., *Microscopic Examination of the Activated Sludge Process*, Wiley, Hoboken, NJ, 2008.
Gerardi, M. H., *Wastewater Bacteria*, Wiley-Interscience, Hoboken, NJ, 2006.
Glymph, T., *Wastewater Microbiology: A Handbook for Operators*, American Water Works Association, Denver, 2005.
Graham, L. E., L. W. Wilcox, and J. Graham, *Algae*, 2nd ed., Pearson/Benjamin Cummings, San Francisco, 2009.
Granéli, E. and J. T. Turner, Eds, *Ecology of Harmful Algae*, Springer, Berlin, 2008.
Howard, A. D., Ed., *Algal Modelling: Processes and Management*, Kluwer Academic Publishing, the Netherlands, 1999.
Kim, M.-B., Ed., *Progress in Environmental Microbiology*, Nova Biomedical Books, New York, 2008.
Leadbetter, J. R., Ed., *Environmental Microbiology*, Elsevier Academic Press, Amsterdam, 2005.
Lee, R. E., *Phycology*, 4th ed., Cambridge University Press, New York, 2008.
León, R., A. Galván, and E. Fernández, *Transgenic Microalgae as Green Cell Factories*, Springer, New York, 2007.
Madigan, M. T. and J. M. Martinko, *Brock Biology of Microorganisms*, 11th ed., Prentice Hall, Upper Saddle River, NJ, 2006.
Manahan, S., *Environmental Chemistry*, 9th ed., Taylor & Francis, Boca Raton, 2010.
Mara, D., Ed., *Handbook of Water and Wastewater Microbiology*, Academic Press, San Diego, CA, 2003.
McKinney, R. E., *Microbiology for Sanitary Engineers*, McGraw-Hill, New York, 1962.
Mitchell, R., *Water Pollution Microbiology*, Vol. 1, Wiley-Interscience, New York, 1970.
Mitchell, R., *Water Pollution Microbiology*, Vol. 2, Wiley-Interscience, New York, 1978.
Mohandas, A. and I. S. Bright Singh, Eds, *Frontiers in Applied Environmental Microbiology*, A.P.H. Publishing Corporation, New Delhi, 2002.
Nester, E. W., *Microbiology: A Human Perspective*, 4th ed., McGraw-Hill, Boston, 2004.
Sigee, D. C., *Freshwater Microbiology: Biodiversity and Dynamic Interactions of Microorganisms in the Aquatic Environment*, Wiley, Hoboken, NJ, 2005.
Stumm, W., and J. J. Morgan, *Aquatic Chemistry*, pp. 336–337, Wiley-Interscience, New York, 1970.
Sutton, B., Ed., *A Century of Mycology*, Cambridge University Press, New York, 1996.
Talaro, K. P., *Foundations in Microbiology: Basic Principles*, 7th ed., McGraw-Hill, Dubuque, IA, 2009.
Varnam, A. H. and M. G. Evans, *Environmental Microbiology*, Manson, London, 2000.
Wheelis, M. L., *Principles of Modern Microbiology*, Jones and Bartlett Publishers, Sudbury, MA, 2008.

QUESTIONS AND PROBLEMS

The use of Internet resources is assumed in answering any of the questions. These would include such things as constants and conversion factors as well as additional information needed to complete an answer.

1. As $CH_3CH_2CH_2CH_2CO_2H$ biodegrades in several steps to carbon dioxide and water, various chemical species are observed. What stable chemical species would be observed as a result of the first step of this degradation process?

2. Which of the following statements is true regarding the production of methane in water: (a) it occurs in the presence of oxygen, (b) it consumes oxygen, (c) it removes biological oxygen demand from the water, (d) it is accomplished by oxic bacteria, (e) it produces more energy per electron-mole than does oxic respiration.
3. At the time zero, the cell count of a bacterial species mediating oxic respiration of wastes was 1×10^6 cells/L. At 30 min it was 2×10^6; at 60 min it was 4×10^6; at 90 min, 7×10^6; at 120 min, 10×10^6; and at 150 min, 13×10^6. From these data, which of the following logical conclusions would you draw? (a) The culture was entering the log phase at the end of the 150-min period, (b) the culture was in the log phase throughout the 150-min period, (c) the culture was leaving the log phase at the end of the 150-min period, (d) the culture was in the lag phase throughout the 150-min period, (e) the culture was in the death phase throughout the 150-min period.
4. What may be said about the biodegradability of a hydrocarbon containing the following structure?

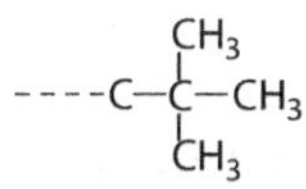

Source: From Manahan, S., *Environmental Chemistry*, 9th ed., Taylor & Francis, Boca Raton, 2010. With permission.

5. Suppose that the anoxic fermentation of organic matter, $\{CH_2O\}$, in water yields 15.0 L of CH_4 (at standard temperature and pressure). How many grams of oxygen would be consumed by the oxic respiration of the same quantity of $\{CH_2O\}$? (Recall the significance of 22.4 L in chemical reaction of gases.)
6. What mass of $FeCO_3(s)$, using Reaction (A) + (4) in Table 6.1, gives the same free energy yield as 1.00 g of organic matter, using Reaction (A) + (1), when oxidized by oxygen at pH 7.00?
7. How many bacteria would be produced after 10 h by one bacterial cell, assuming exponential growth with a generation time of 20 min?
8. Referring to Reaction 6.18, calculate the concentration of ammonium ion in equilibrium with oxygen in the atmosphere and 1.00×10^{-5} M NO_3^- at pH 7.00.
9. When a bacterial nutrient medium is inoculated with bacteria grown in a markedly different medium, the lag phase (Figure 6.4) often is quite long, even if the bacteria eventually grow well in the new medium. Can you explain this behavior?
10. Most plants assimilate nitrogen as nitrate ion. However, ammonia (NH_3) is a popular and economical fertilizer. What essential role do bacteria play when ammonia is used as a fertilizer? Do you think any problems might occur when using ammonia in a waterlogged soil lacking oxygen?
11. Why is the growth rate of bacteria as a function of temperature (Figure 6.7) not a symmetrical curve?

12. Discuss the analogies between bacteria and a finely divided chemical catalyst.
13. Would you expect autotrophic bacteria to be more complex physiologically and biochemically than heterotrophic bacteria? Why?
14. Wastewater containing 8 mg/L O_2 (atomic mass O = 16), 1.00×10^{-3} M NO_3^-, and 1.00×10^{-2} M soluble organic matter, $\{CH_2O\}$, is stored isolated from the atmosphere in a container richly seeded with a variety of bacteria. Assume that denitrification is one of the processes that will occur during storage. After the bacteria have had a chance to do their work, which of the following statements will be true? (a) No $\{CH_2O\}$ will remain, (b) some O_2 will remain, (c) some NO_3^- will remain, (d) denitrification will have consumed more of the organic matter than oxic respiration, (e) the composition of the water will remain unchanged.
15. Of the four classes of microorganisms—algae, fungi, bacteria, and virus—which has the least influence on water chemistry?
16. Figure 6.3 shows the main structural features of a bacterial cell. Which of these do you think might cause the most trouble in water-treatment processes such as filtration or ion exchange, where the maintenance of a clean, unfouled surface is critical? Explain.
17. A bacterium capable of degrading 2,4-D herbicide was found to have its maximum growth rate at 32°C. Its growth rate at 12°C was only 10% of the maximum. Do you think there is another temperature at which the growth rate would also be 10% of the maximum? If you believe this to be the case, of the following temperatures, choose the one at which it is most plausible for the bacterium to also have a growth rate of 10% of the maximum: 52°C, 37°C, 8°C, and 20°C.
18. The day after a heavy rain washed a great deal of cattle feedlot waste into a farm pond, the following counts of bacteria were obtained:

Time	Thousands of Viable Cells/mL	Time	Thousands of Viable Cells/mL
6:00 a.m.	0.10	11:00 a.m.	0.40
7:00 a.m.	0.11	12:00 Noon	0.80
8:00 a.m.	0.13	1:00 p.m.	1.60
9:00 a.m.	0.16	2:00 p.m.	3.20
10:00 a.m.	0.20		

To which portion of the bacterial growth curve, Figure 6.3, does this time span correspond?

19. Addition of which two half-reactions in Table 6.1 is responsible for (a) elimination of an algal nutrient in secondary sewage effluent using methanol as a carbon source, (b) a process responsible for a bad-smelling pollutant when bacteria grow in the absence of oxygen, (c) a process that converts a common form of commercial fertilizer to a form that most crop plants can absorb, (d) a process responsible for the elimination of organic matter from wastewater in the aeration tank of an activated sludge sewage-treatment

plant, (e) a characteristic process that occurs in the anoxic digester of a sewage treatment plant.

20. What is the surface area in square meters of 1.00 g of spherical bacterial cells, 1.00 μm in diameter, having a density of 1.00 g/cm^3?
21. What is the purpose of exoenzymes in bacteria?
22. Match each species of bacteria listed in the left column with its function on the right.

a. *Spirillumlipoferum*	1. Reduces sulfate to H_2S
b. *Rhizobium*	2. Catalyzes oxidation of Fe^{2+} to Fe^{3+}
c. *Thiobacillusferrooxidans*	3. Fixes nitrogen in grasses
d. *Desulfovibrio*	4. On legume roots

23. What factors favor the production of methane in anoxic surroundings?
24. Below are listed three kinds of microorganisms, and below that are listed with numbers several chemical species or energy sources. In parentheses to the *left* of each kind of microorganism, place the numbers corresponding to at least *two* things that the microorganism might *need* or *use*. In parentheses to the *right* of each kind of microorganism, place the numbers corresponding to at least *two* things that the microorganism might produce.
( ) Algae ( )
( ) Aerobic, nonphotosynthetic, autotrophic *Gallionella* bacteria ( )
( ) Anaerobic, heterotrophic bacteria ( )

1. CO_2, 2. $h\nu$, 3. O_2, 4. $\{CH_2O\}$, 5. CH_4, 6. $Fe(OH)_3$, 7. electron acceptor other than O_2.

25. Bacteria growing in the log phase on a waste in water were assayed at a particular time, t_0, and found to number 3.01×10^5 cells/mL. At 90 min after t_0 the count was 2.41×10^6 cells/mL. The best estimate of the following of the bacterial population at 60 min after t_0 is (A) 4.20×10^5, (B) 8.25×10^5, (C) 6.48×10^5, (D) 1.20×10^6, and (E) 3.21×10^6.
26. Consider a type of bacteria that derive energy by mediating the oxidation of sulfides, such as H_2S, FeS, or FeS_2, with molecular oxygen, O_2. Such bacteria (A) are most likely heterotrophic, (B) cannot be autotrophic, (C) should be thermophilic, (D) should be acid-tolerant, (E) cannot exist.
27. The growth rate of bacteria, "*G*," was plotted versus an unidentified parameter, "*X*," giving the following values in which both *G* and *X* are in arbitrary units:

X	*G*	*X*	*G*
5	100	30	600
10	200	35	700
15	300	40	700
20	400	45	350
25	500	50	25

Based on the above data X is most likely (A) time, showing the bacteria with a generation time of 5 units of X, (B) nutrient concentration, (C) waste product, (D) pH, (E) temperature.

28. Of the following, the *untrue* statement pertaining to biodegradation of organic matter is (A) epoxidation consists of adding an oxygen atom between two carbon atoms, (B) oxidation of hydrocarbon chains tends to occur two carbon atoms at a time, (C) esterases are a specific category of hydrolase ions, (D) carbon atoms bonded to three or four other carbon atoms are especially susceptible to microbial epoxidation, (E) exoenzymes are involved in the biodegradation of cellulose.
29. Of the following, the *untrue* statement is (A) all fungi and protozoans are chemoheterotrophs, (B) photoheterotrophs that use photoenergy, but are dependent on organic matter as a carbon source, are especially abundant and widespread, (C) chemoautotrophs use CO_2 for biomass and oxidize substances such as NH_4^+ for energy, (D) algae are photoautotrophs, (E) some bacteria, such as cyanobacteria, perform photosynthesis.
30. The reaction, $4FeS(s) + 9O_2 + 10H_2O \rightarrow 4Fe(OH)_3(s) + 4SO_4^{2-} + 8H^+$, (A) illustrates sulfate reduction, (B) is a means for autotrophic *Gallionella* bacteria to obtain energy, (C) illustrates the action of heterotrophic bacteria, (D) is not mediated by bacteria, (E) is written for one electron-mole.

7 Water Pollution

7.1 NATURE AND TYPES OF WATER POLLUTANTS

Throughout history, the quality of drinking water has been a factor in determining human welfare. Fecal pollution of drinking water has frequently caused waterborne diseases that have decimated the populations of whole cities. Unwholesome water polluted by sewage has caused great hardship for people forced to drink it or use it for irrigation. Although waterborne diseases are now well controlled in technologically advanced countries, the shortage of safe drinking water is still a major problem in regions afflicted by strife and poverty.

An ongoing concern with water safety now is the potential presence of chemical pollutants. These may include organic chemicals, inorganics, and heavy metals from industrial, urban runoff, and agricultural sources. Water pollutants can be divided into some general categories, as summarized in Table 7.1. Most of these categories of pollutants, and several subcategories, are discussed in this chapter.

7.1.1 Markers of Water Pollution

Markers of water pollution are substances that show the presence of pollution sources. These include herbicides indicative of agricultural runoff, fecal coliform bacteria that are characteristic of pollution from sewage, and pharmaceuticals, pharmaceutical metabolites, and even caffeine that show contamination by domestic wastewater.

Biomarkers of water pollution are organisms that live in or are closely associated with bodies of water and provide evidence of pollution either from accumulation of water pollutants or their metabolites or from effects on the organism due to pollutant exposure. Fish are the most common bioindicators of water pollution, and fish lipid (fat) tissue is commonly analyzed for persistent organic water pollutants.

An organism that has been described as "a worldwide sentinel species to assess and monitor environmental pollution in rivers, lakes, reservoirs, and estuaries"[1] is the osprey (*Pandionhaliaetus*), a large raptor bird with a wingspan that can exceed 1.5 m and a mass up to 2 kg. Found in all continents throughout the world except for Antarctica, the osprey feeds almost exclusively on fish. In addition to these characteristics, other aspects of the osprey that make it a good indicator species are that it is well adapted to human landscapes where pollution is most likely to occur, it is at the top of the aquatic food web through which persistent pollutants undergo bioaccumulation and biomagnification, it is sensitive to many pollutants, and it has a relatively long life span. This bird builds highly visible nests that are evenly dispersed over wide areas, usually remains with a single nest, and tolerates disturbance of nests for brief periods. The osprey is very sensitive to some pollutants, and although populations are now comfortably high, they were almost wiped out by the effects of DDT before that insecticide was banned. Chemical and biochemical analyses of osprey feathers, eggs, blood, and organs

TABLE 7.1
General Types of Water Pollutants

Class of Pollutant	Significance
Trace elements	Health, aquatic biota, toxicity
Heavy metals	Health, aquatic biota, toxicity
Organically-bound metals	Metal transport
Radionuclides	Toxicity
Inorganic pollutants	Toxicity, aquatic biota
Asbestos	Human health
Algal nutrients	Eutrophication
Acidity, alkalinity, salinity (in excess)	Water quality, aquatic life
Trace organic pollutants	Toxicity
Polychlorinated biphenyls	Possible biological effects
Pesticides	Toxicity, aquatic biota, wildlife
Petroleum wastes	Effect on wildlife, esthetics
Sewage, human, and animal wastes	Water quality, oxygen levels
Biochemical oxygen demand	Water quality, oxygen levels
Pathogens	Health effects
Detergents	Eutrophication, wildlife, esthetics
Chemical carcinogens	Incidence of cancer
Sediments	Water quality, aquatic biota, wildlife
Taste, odor, and color	Esthetics

Source: From Manahan, S., *Environmental Chemistry*, 9th ed., Taylor & Francis, Boca Raton, 2010. With permission.

along with observations of behavior, nesting habits, and populations have been used to assess water pollution.

7.2 ELEMENTAL POLLUTANTS

Table 7.2 lists the more important *trace elements* (those encountered at levels of a few ppm or less) found in natural waters. Typical of the behavior of many substances in the aquatic environment, some of these are essential plant and animal nutrients at low levels but toxic at higher levels. Several of them, such as lead or mercury, have such toxicological and environmental significance that they are discussed in detail in separate sections.

A few of the *heavy metals* are among the most harmful of the elemental pollutants and are of particular concern because of their toxicities to humans. These elements are in general the transition metals, and some of the representative elements, such as lead and tin, are placed in the lower right-hand corner of the periodic table. Heavy metals include essential elements like iron as well as toxic metals like cadmium and mercury. Most of them have a tremendous affinity for sulfur and disrupt enzyme function by forming bonds with sulfur groups in enzymes. Protein carboxylic acid ($-CO_2H$) and amino ($-NH_2$) groups are also chemically bound by heavy metals. Cadmium, copper, lead, and mercury ions bind to cell membranes, hindering transport processes through

TABLE 7.2
Important Trace Elements in Natural Waters

Element	Sources	Effects and Significance
Arsenic	Mining by-product, chemical waste	Toxic, possibly carcinogenic
Beryllium	Coal, industrial wastes	Toxic
Boron	Coal, detergents, wastes	Toxic
Chromium	Metal plating	Essential as Cr(III), toxic as Cr(VI)
Copper	Metal plating, mining, industrial waste	Essential trace element, toxic to plants and algae at higher levels
Fluorine (F^-)	Natural geological sources, wastes, water additive	Prevents tooth decay at around 1 mg/L, toxic at higher levels
Iodine (I^-)	Industrial wastes, natural brines, seawater intrusion	Prevents goiter
Iron	Industrial wastes, corrosion, acid mine water, microbial action	Essential nutrient, damages fixtures by staining
Lead	Industrial waste, mining, fuels	Toxic, harmful to wildlife
Manganese	Industrial wastes, acid mine water, microbial action	Toxic to plants, damages fixtures, by staining
Mercury	Industrial waste, mining, coal	Toxic, mobilized as methyl mercury compounds by anaerobic bacteria
Molybdenum	Industrial wastes, natural sources	Essential to plants, toxic to animals
Selenium	Natural sources, coal	Essential at lower levels, toxic at higher levels
Zinc	Industrial waste, metal plating, plumbing	Essential element, toxic to plants at higher levels

Source: From Manahan, S., *Environmental Chemistry*, 9th ed., Taylor & Francis, Boca Raton, 2010. With permission.

the cell wall. Heavy metals may also precipitate phosphate biocompounds or catalyze their decomposition. Some heavy metals are quite toxic.

Some of the *metalloids*, elements on the borderline between metals and nonmetals, are significant water pollutants. Arsenic, selenium, and antimony are of particular interest.

Inorganic chemicals' manufacture has the potential to contaminate water with trace elements. Among the industries regulated for potential trace element pollution of water are those producing chlor-alkali, hydrofluoric acid, sodium dichromate (sulfate process and chloride ilmenite process), aluminum fluoride, chrome pigments, copper sulfate, nickel sulfate, sodium bisulfate, sodium hydrosulfate, sodium bisulfite, titanium dioxide, and hydrogen cyanide.

7.3 HEAVY METALS

7.3.1 Cadmium

Pollutant *cadmium* in water may arise from mining wastes and industrial discharges, especially from metal plating. Chemically, cadmium is very similar to zinc, and these two metals frequently undergo geochemical processes together. Both metals are found in water in the +2 oxidation state.

The effects of acute cadmium poisoning in humans include high blood pressure, kidney damage, damage to testicular tissue, and destruction of red blood cells. Much of the physiological action of cadmium is due to its chemical similarity to zinc. Cadmium may replace zinc in some enzymes, thereby altering the stereostructure of the enzyme and impairing its catalytic activity, causing disease symptoms.

Cadmium and zinc are common water and sediment pollutants in harbors surrounded by industrial installations. Concentrations of more than 100 ppm dry weight sediment have been found in harbor sediments. Typically, during periods of calm in the summer when the water stagnates, the anaerobic bottom layer of harbor water has a low soluble Cd concentration because microbial reduction of sulfate by organic matter, $\{CH_2O\}$, produces sulfide, which precipitates cadmium as insoluble cadmium sulfide:

$$2\{CH_2O\} + SO_4^{2-} + H^+ \rightarrow 2CO_2 + HS^- + 2H_2O \tag{7.1}$$

$$CdCl^+(\text{chloro complex in seawater}) + HS^- \rightarrow CdS(s) + H^+ + Cl^- \tag{7.2}$$

Mixing of bay water from outside the harbor and harbor water by high winds during the winter results in desorption of cadmium from harbor sediments by oxic bay water. This dissolved cadmium is carried out into the bay where it is absorbed by suspended solid materials, which then becomes incorporated with the bay sediments. This is an example of the sort of complicated interaction of hydraulic, chemical solution–solid, and microbiological factors involved in the transport and distribution of a pollutant in an aquatic system.

7.3.2 Lead

Inorganic *lead* arising from a number of industrial and mining sources and formerly from leaded gasoline occurs in water in the +2 oxidation state. In addition to pollutant sources, lead-bearing limestone and galena (PbS) contribute lead to natural waters in some locations. Evidence from hair samples and other sources indicates that body burdens of this toxic metal have decreased during recent decades, largely the result of less lead used in plumbing and other products that come in contact with food or drink.

Acute lead poisoning in humans causes severe dysfunction in the kidneys, reproductive system, liver, and the brain and central nervous system and results in sickness or death. Lead poisoning from environmental exposure is thought to have caused mental retardation in many children. Mild lead poisoning causes anemia. The victim may have headaches and sore muscles, and may feel generally fatigued and irritable.

Except in isolated cases, lead is probably not a major problem in drinking water, although the potential exists in cases where old lead pipes are still in use. Lead used to be a constituent of solder and some pipe-joint formulations, such that household water does have some contact with lead. Water that has stood in household plumbing for some time may have elevated levels of lead (along with zinc, cadmium, and copper) and should be drained for a while before use.

7.3.3 Mercury

Because of its toxicity, mobilization as methylated forms by anoxic bacteria, and other pollution factors, mercury generates a great deal of concern as a heavy-metal pollutant. Mercury is found as a trace component of many minerals, with continental rocks containing an average of around 80 parts per billion, or slightly less, of this element. Fossil fuel coal and lignite contain mercury, often at levels of 100 parts per billion or even higher, and emissions from the combustion of these fuels are a major source of environmental mercury.

Metallic mercury once was commonly used as an electrode in the electrolytic generation of chlorine gas, in laboratory vacuum apparatus, and in other applications, and significant quantities of inorganic mercury(I) and mercury(II) compounds were used annually. Organic mercury compounds used to be widely applied as pesticides, particularly fungicides. These mercury compounds included aryl mercurials such as phenyl mercuric dimethyldithiocarbamate,

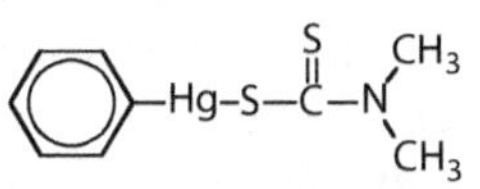

(formerly used in paper mills as a slimicide and as a mold retardant for paper) and alkyl-mercurials such as ethylmercuric chloride, C_2H_5HgCl, that was used as a seed fungicide. Because of concern over the health and environmental effects of mercury, these uses have been severely curtailed in recent years. As of 2008, approximately 3800 metric tons of mercury were traded worldwide each year.

Globally, one of the greatest sources of mercury pollution has been mercury used to extract gold from gold-bearing ores. It is estimated that each year 15 million miners in approximately 40 developing countries use 650–1000 metric tons of mercury to extract gold. This use subjects many local areas to mercury contamination and contributes significantly to the global environmental burden of mercury as well as exposing the miners, many of whom are children, to toxic mercury. The problem has been made worse in recent years as industrialized countries have reduced mercury use with the result that surplus mercury has been made available on the world market for gold extraction. In recognition of this problem, in 2008 the European Union adopted a mercury export ban to take effect in 2011 and the United States adopted a similar ban to be effective in 2013.

The toxicity of mercury was tragically illustrated in the Minamata Bay area of Japan during the period 1953–1960. A total of 111 cases of mercury poisoning and 43 deaths were reported among people who had consumed seafood from the bay that had been contaminated with mercury waste from a chemical plant that drained into Minamata Bay. Congenital defects were observed in 19 babies whose mothers had consumed seafood contaminated with mercury. The level of metal in the contaminated seafood was 5–20 ppm.

Among the toxicological effects of mercury are neurological damage, including irritability, paralysis, blindness, or insanity; chromosome breakage; and birth defects. The milder symptoms of mercury poisoning such as depression and irritability have a psychopathological character. Because of the resemblance of these symptoms to common human behavioral problems, mild mercury poisoning may escape detection. Some forms of mercury are relatively nontoxic and were formerly used as medicines, for example, in the treatment of syphilis. Other forms of mercury, particularly organic compounds, are highly toxic.

Because there are few major natural sources of mercury, and since most inorganic compounds of this element are relatively insoluble, it was assumed for some time that mercury was not a serious water pollutant. However, in 1970, alarming mercury levels were discovered in fish in Lake Saint Clair located between Michigan and Ontario, Canada. A subsequent survey by the U.S. Federal Water Quality Administration revealed a number of other waters contaminated with mercury. It was found that several chemical plants, particularly caustic chemical manufacturing operations, were each releasing up to 14 or more kilograms of mercury in wastewaters each day.

The unexpectedly high concentrations of mercury found in water and in fish tissues result from the formation of soluble monomethylmercury ion, CH_3Hg^+, and volatile dimethylmercury, $(CH_3)_2Hg$, by anoxic bacteria in sediments. Mercury from these compounds becomes concentrated in fish lipid (fat) tissue and the concentration factor from water to fish may exceed 10^3. The methylating agent by which inorganic mercury is converted to methylmercury compounds is methylcobalamin, a vitamin B_{12} analog:

$$HgCl_2 \xrightarrow{\text{Methylcobalamin}} CH_3HgCl + Cl^- \tag{7.3}$$

It is believed that the bacteria that synthesize methane produce methylcobalamin as an intermediate in the synthesis. Thus, waters and sediments in which anoxic decay is occurring provide the conditions under which methylmercury production occurs. In neutral or alkaline waters, volatile dimethylmercury, $(CH_3)_2Hg$, may be formed.

7.4 METALLOIDS

The most significant water pollutant metalloid element is arsenic, a toxic element that has been the chemical villain of more than a few murder plots. Acute arsenic poisoning can result from the ingestion of more than about 100 mg of the element. Chronic poisoning occurs with the ingestion of small amounts of arsenic over a long period of time. There is some evidence that this element is also carcinogenic.

Arsenic occurs in the Earth's crust at an average level of 2–5 ppm. The combustion of fossil fuels, particularly coal, introduces large quantities of arsenic into the environment, much of it reaching natural waters. Arsenic occurs with phosphate minerals and enters into the environment along with some phosphorus compounds. Some formerly used pesticides, particularly those from before World War II, contain highly toxic arsenic compounds. The most common of these are lead arsenate, $Pb_3(AsO_4)_2$; sodium arsenite, Na_3AsO_3; and Paris green, $Cu_3(AsO_3)_2$. Another major source of arsenic is mine tailings. Arsenic produced as a by-product of copper, gold, and lead refining exceeds the commercial demand for arsenic, and it accumulates as waste material.

Like mercury, arsenic may be converted to more mobile and toxic methyl derivatives by bacteria, first by reduction of H_3AsO_4 to H_3AsO_3 then by methylation to produce $CH_3AsO(OH)_2$ (methylarsinic acid), $(CH_3)_2AsO(OH)$ (dimethyarsinic acid), and $(CH_3)_2AsH$ (dimethylarsine).

In what may have been the largest mass poisoning of a human population in history, between 35 million and 77 million people of the 125 million inhabitants of Bangladesh have been exposed to potentially toxic levels of arsenic in drinking water. This catastrophic public health problem has resulted from well-intentioned programs funded initially by the United Nations Children's Fund to install shallow tube wells that provided a source of drinking water free from disease-causing pathogens. By 1987, numerous cases of arsenic-induced skin lesions characterized by pigmentation changes, predominantly on the upper chest, arms, and legs, and keratoses of the palms of the hands and soles of the feet were being observed. These effects were characteristic of arsenic poisoning, leading to the discovery that arsenic-contaminated drinking water from the tube wells was responsible for this. Since the initial discovery of arsenic poisoning, huge numbers of new cases have been revealed and it is possible that tens of thousands of people in Bangladesh will die prematurely because of exposure to arsenic in drinking water. Other countries with acute problems from arsenic in drinking water include Vietnam (where large numbers of tube wells were drilled more recently than those in Bangladesh), Argentina, Chile, China, Mexico, Taiwan, and Thailand.

The geochemical conditions that result in arsenic contamination of water are often associated with the presence of iron, sulfur, and organic matter in (alluvial) deposits produced by water. Iron released from rocks eroded by river water forms iron oxide deposits on rock particle surfaces. The iron oxide accumulates arsenic and concentrates it from the river water. These particles then get buried along with degradable organic matter in sediments, and the insoluble iron(III) in the iron oxides is converted to soluble iron(II) by the anaerobic reducing conditions under which the organic matter biodegrades. This releases the bound arsenic that can get into well water.

7.5 ORGANICALLY BOUND METALS AND METALLOIDS

An appreciation of the strong influence of complexation and chelation on heavy metals' behavior in natural waters and wastewaters may be gained by reading Sections 3.6 through 3.13, which deal with that subject. Methylmercury formation is discussed in Section 7.3. Both topics involve the combination of metals and organic entities in water. It must be stressed that the interaction of metals with organic compounds is of utmost importance in determining the role played by the metal in an aquatic system.

There are two major types of metal–organic interactions to be considered in an aquatic system. The first of these is complexation, usually chelation when organic ligands are involved. As discussed in Chapter 3, complexation and chelation involve reversible binding of a metal ion with a ligand species.

Organometallic compounds, on the other hand, contain metals bound to organic entities by way of a carbon atom and do not dissociate reversibly at lower pH or greater dilution. Furthermore, the organic component, and sometimes the particular oxidation state of the metal involved, may not be stable apart from the organometallic compound. The major categories of organometallic compounds encountered in the environment are (1) those in which the organic group is an alkyl group such as ethyl in tetraethyl lead, $Pb(C_2H_5)_4$; (2) carbonyls, some of which are quite volatile and toxic, having carbon monoxide bonded to metals; and (3) those in which the organic group is a π electron donor, such as ethylene (C_2H_4) or benzene (C_6H_6).

Combinations of the three general types of compounds outlined above exist, the most prominent of which are arene carbonyl species, in which a metal atom is bonded to both an aryl entity such as benzene and to several carbon monoxide molecules.

A large number of compounds exist that have at least one bond between the metal and a C atom on an organic group, as well as other covalent or ionic bonds between the metal and atoms other than carbon. An example is monomethylmercury chloride, CH_3HgCl, in which the organometallic CH_3Hg^+ ion is ionically bonded to the chloride anion. Another class of compounds with organometallic character are those that have organic groups bonded to a metal atom through atoms other than carbon. An example of such a compound is isopropyl titanate, $Ti(i\text{-}OC_3H_7)_4$, in which the hydrocarbon groups are bonded to the metal by way of oxygen atoms.

The interaction of trace metals with organic compounds in natural waters is too vast an area to cover in detail in this chapter; however, it may be noted that metal–organic interactions may involve organic species of both pollutant (such as EDTA) and natural (such as fulvic acids) origin. These interactions are influenced by, and sometimes play a role in, redox equilibria, formation and dissolution of precipitates, colloid formation and stability, acid–base reactions, and microorganism-mediated reactions in water. Metal–organic interactions may increase or decrease the toxicity of metals in aquatic ecosystems, and they have a strong influence on the growth of algae in water.

7.5.1 Organotin Compounds

Of all the metals, tin has had the greatest number of organometallic compounds in commercial use, with global production that reached 40,000 metric tons per year before uses were restricted because of concerns over water pollution. In addition to synthetic organotin compounds, methylated tin species can be produced biologically in the environment. Figure 7.1 gives some examples of the many known organotin compounds.

Major industrial uses of organotin compounds in the past have included applications of tin compounds in fungicides, acaricides, disinfectants, antifouling paints, stabilizers to lessen the effects of heat and light in polyvinylchloride (PVC) plastics, catalysts, and precursors for the formation of films of SnO_2 on glass. Tributyl tin (TBT) chloride and related TBT compounds have bactericidal, fungicidal, and insecticidal properties and formerly were of particular environmental significance because of their use as industrial biocides. In addition to TBT chloride, other TBT compounds used as biocides include the hydroxide, the naphthenate, bis(tributyltin)

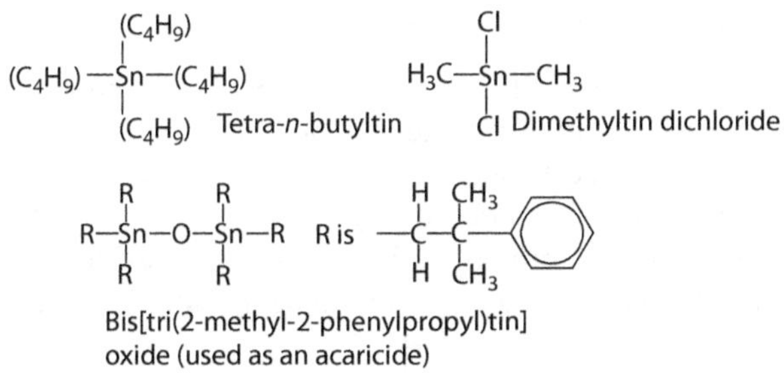

FIGURE 7.1 Examples of organotin compounds. (From Manahan, S., *Environmental Chemistry*, 9th ed., Taylor & Francis, Boca Raton, 2010. With permission.)

oxide, and tris(tributylstannyl) phosphate. TBT was once widely used in boat and ship hull coatings to prevent the growth of fouling organisms. Other applications have included preservation of wood, leather, paper, and textiles. Antifungal TBT compounds have been used as slimicides in cooling tower water.

Obviously the many applications of organotin compounds for a variety of uses have posed a significant potential for environmental pollution. Because of their applications near or in contact with bodies of water, organotin compounds are potentially significant water pollutants and have been linked to endocrine disruption in shellfish, oysters, and snails. Because of such concerns, several countries, including the United States, England, and France, prohibited TBT application on vessels smaller than 25 m in length during the 1980s. In response to concerns over water pollution, in 2001 the International Maritime Organization agreed to ban organotin antifouling paints on ships and yachts. Provisions of this agreement came into effect in September 2008.

7.6 INORGANIC SPECIES

Some important inorganic water pollutants were mentioned in Sections 7.2 through 7.4 as part of the discussion of pollutant trace elements. Inorganic pollutants that contribute acidity, alkalinity, or salinity to water are considered separately in this chapter. Still another class is that of algal nutrients. This leaves unclassified, however, some important inorganic pollutant species, of which cyanide ion, CN^-, is probably the most important. Others include ammonia, carbon dioxide, hydrogen sulfide, nitrite, and sulfite.

7.6.1 Cyanide

Cyanide, a deadly poisonous substance, exists in water as HCN, a weak acid, K_a of 6×10^{-10}. The cyanide ion has a strong affinity for many metal ions, forming relatively less toxic ferrocyanide, $Fe(CN)_6^{4-}$, with iron(II), for example. Volatile HCN is very toxic and has been used in gas chamber executions in the United States.

Cyanide is widely used in industry, especially for metal cleaning and electroplating. It is also one of the main gas and coke scrubber effluent pollutants from gas works and coke ovens. Cyanide is also used in certain mineral-processing operations. Numerous fish kills have resulted from the discharge of cyanide from mineral processing operations into waterways.

7.6.2 Ammonia and Other Inorganic Pollutants

Excessive levels of ammoniacal nitrogen cause water-quality problems. *Ammonia* is the initial product of the decay of nitrogenous organic wastes, and its presence frequently indicates the presence of such wastes. It is a normal constituent of low-pE groundwaters and is sometimes added to drinking water as an aid to disinfection, where it reacts with chlorine to provide residual chlorine (see Section 8.11). Since the pK_a of ammonium ion, NH_4^+, is 9.26, most ammonia in water is present as NH_4^+ rather than as NH_3.

Hydrogen sulfide, H_2S, is a product of the anoxic decay of organic matter containing sulfur. It is also produced in the anoxic reduction of sulfate by microorganisms (see Chapter 6) and is evolved as a gaseous pollutant from geothermal waters. Wastes from chemical plants, paper mills, textile mills, and tanneries may also contain H_2S. Its presence is easily detected by its characteristic rotten-egg odor. In water, H_2S is a weak diprotic acid with pK_{a1} of 6.99 and pK_{a2} of 12.92; S^{2-} is not present in normal natural waters. The sulfide ion has tremendous affinity for many heavy metals, and precipitation of metallic sulfides often accompanies production of H_2S.

Free *carbon dioxide*, CO_2, is frequently present in water at high levels due to decay of organic matter. It is also added to softened water during water treatment as part of a recarbonation process (see Chapter 8). Excessive carbon dioxide levels may make water more corrosive and may be harmful to aquatic life.

Nitrite ion, NO_2^-, occurs in water as an intermediate oxidation state of nitrogen over a relatively narrow pE range. Nitrite is added to some industrial process water as a corrosion inhibitor. However, it rarely occurs in drinking water at levels over 0.1 mg/L.

Sulfite ion, SO_3^{2-}, is found in some industrial wastewaters. Sodium sulfite is commonly added to boiler feedwaters as an oxygen scavenger:

$$2SO_3^{2-} + O_2 \rightarrow 2SO_4^{2-} \tag{7.4}$$

Since pK_{a1} of sulfurous acid is 1.76 and pK_{a2} is 7.20, sulfite exists as either HSO_3^- or SO_3^{2-} in natural waters, depending on the pH. It may be noted that hydrazine, N_2H_4, also functions as an oxygen scavenger:

$$N_2H_4 + O_2 \rightarrow 2H_2O + N_2(g) \tag{7.5}$$

Perchlorate ion, ClO_4^-, emerged as a water pollution problem in some areas in the 1990s when advances in ion chromatography enabled its detection in the low parts per billion range of concentrations. Ammonium perchlorate, NH_4ClO_4, has been widely manufactured as an oxidizer in solid rocket propellants and contamination from ammonium perchlorate manufacturing facilities has been regarded as the major source of contamination. Perchlorate in water is very unreactive and all common perchlorate salts other than $KClO_4$ are soluble, so it is difficult to remove. Physiologically, it competes with iodide ion and diminishes essential iodide uptake by the thyroid. The U.S. Environmental Protection Agency has recommended a drinking water standard for perchlorate of around 1 part per billion.

7.6.3 Asbestos in Water

The toxicity of inhaled asbestos is well established. The fibers scar lung tissue and cancer eventually develops, often 20 or 30 years after exposure. It is not known for sure whether asbestos is toxic in drinking water. This has been a matter of considerable concern because of the dumping of taconite (iron ore tailings) containing asbestos-like fibers into Lake Superior. The fibers have been found in drinking waters of cities around the lake. After having dumped the tailings into Lake Superior since 1952, the Reserve Mining Company at Silver Bay on Lake Superior solved the problem in 1980 by constructing a 6-square-mile containment basin inland from the lake. This $370-million facility keeps the taconite tailings covered with a 3-m layer of water to prevent the escape of fiber dust.

7.7 ALGAL NUTRIENTS AND EUTROPHICATION

The term *eutrophication*, derived from the Greek word meaning "well-nourished," describes a condition of lakes or reservoirs involving excess algal growth. Although some algal productivity is necessary to support the food chain in an aquatic ecosystem, excess growth under eutrophic conditions may eventually lead to severe deterioration of the body of water. The first step in eutrophication of a body of water is an input of plant nutrients (Table 7.3) from watershed runoff or sewage. The nutrient-rich body of water then produces a great deal of plant biomass by photosynthesis, along with a smaller amount of animal biomass. Dead biomass accumulates in the bottom

TABLE 7.3
Essential Plant Nutrients: Sources and Functions

Nutrient	Source	Function
	Macronutrients	
Carbon (CO_2)	Atmosphere, decay	Biomass constituent
Hydrogen	Water	Biomass constituent
Oxygen	Water	Biomass constituent
Nitrogen (NO_3^-)	Decay, pollutants, atmosphere	Protein constituent (from nitrogen- fixing organisms)
Phosphorus	Decay, minerals (phosphate)	DNA/RNA constituent pollutants
Potassium	Minerals, pollutants	Metabolic function
Sulfur (sulfate)	Minerals	Proteins, enzymes
Magnesium	Minerals	Metabolic function
Calcium	Minerals	Metabolic function
	Micronutrients	
B, Cl, Co, Cu, Fe, Mo, Mn, Na, Si, V, Zn	Minerals, pollutants	Metabolic function and/or constituent of enzymes

Source: From Manahan, S., *Environmental Chemistry*, 9th ed., Taylor & Francis, Boca Raton, 2010. With permission.

of the lake, where it partially decays, recycling nutrient carbon dioxide, phosphorus, nitrogen, and potassium. If the lake is not too deep, bottom-rooted plants begin to grow, accelerating the accumulation of solid material in the basin. Eventually a marsh is formed, which finally fills in to produce a meadow or forest.

Eutrophication is often a natural phenomenon; for instance, it is basically responsible for the formation of huge deposits of coal and peat. However, human activity can greatly accelerate the process. To understand why this is so, consider that most of the nutrients required for plant and algae growth shown in Table 7.3 are available in adequate amounts from natural sources. The nutrients most likely to be limiting are the "fertilizer" elements: nitrogen, phosphorus, and potassium. All these are present in sewage and are, of course, found in runoff from heavily fertilized fields. They are also constituents of various kinds of industrial wastes. Each of these elements can also come from natural sources—phosphorus and potassium from mineral formations, and nitrogen fixed by bacteria, cyanobacteria, or discharge of lightning in the atmosphere.

In some cases, nitrogen or even carbon may be the limiting nutrients, the presence of which determines the rate of algal growth. This is particularly true of nitrogen in seawater. In seawater, micronutrients, particularly iron, may be limiting.

In most cases in freshwaters, the single plant nutrient most likely to be limiting is phosphorus, and it is generally regarded as the culprit in excessive eutrophication. Household detergents were once a common source of phosphate in wastewater, and eutrophication control has concentrated on eliminating phosphates from detergents, removing phosphate at the sewage treatment plant, and preventing phosphate-laden sewage effluents from entering bodies of water enabling the excessive growth of algae that can cause eutrophication.

7.8 ACIDITY, ALKALINITY, AND SALINITY

Aquatic biota are sensitive to extremes of pH. Largely because of osmotic effects, they cannot live in a medium having a salinity to which they are not adapted. Thus, a freshwater fish soon succumbs in the ocean, and sea fish normally cannot live in freshwater. Excess salinity soon kills plants not adapted to it. There are, of course, ranges in salinity and pH in which organisms live. As shown in Figure 7.2, these ranges frequently may be represented by a reasonably symmetrical curve, along the fringes of which an organism may live without really thriving. These curves do not generally exhibit a sharp cutoff at one end or the other, as does the high-temperature end of the curve representing the growth of bacteria as a function of temperature (Figure 6.7).

The most common source of *pollutant acid* in water is acid mine drainage. The sulfuric acid in such drainage arises from the microbial oxidation of pyrite or other sulfide minerals as described in Chapter 6. The values of pH encountered in acid-polluted water may fall below 3, a condition deadly to most forms of aquatic life except the culprit bacteria mediating the pyrite and iron(II) oxidation, which thrive under very low pH conditions. Industrial wastes frequently have the potential to contribute strong acid to water. Sulfuric acid produced by the air oxidation of pollutant sulfur dioxide enters natural waters as acidic rainfall (see Chapter 1, Figure 1.8). In cases where the water does not have contact with a basic mineral, such as limestone, the water pH may become dangerously low. This condition occurs in some Canadian lakes, for example.

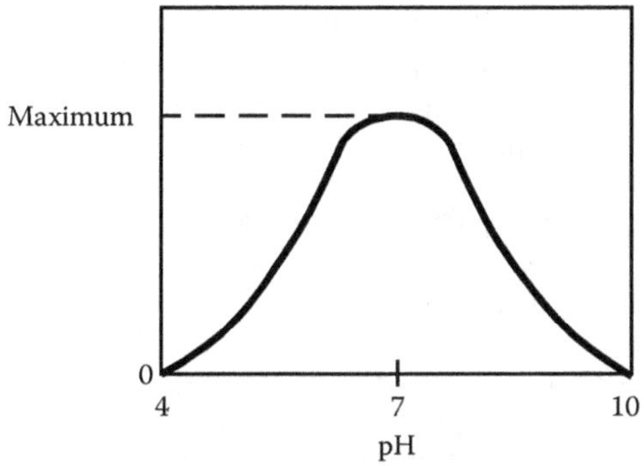

FIGURE 7.2 A generalized plot of the growth of an aquatic organism as a function of pH. (From Manahan, S., *Environmental Chemistry*, 9th ed., Taylor & Francis, Boca Raton, 2010. With permission.)

Excess *alkalinity*, and frequently accompanying high pH, generally are not introduced directly into water from anthropogenic sources. However, in many geographic areas, the soil and mineral strata are alkaline and impart a high alkalinity to water. Human activity can aggravate the situation, for example, by exposure of alkaline overburden from strip mining to surface water or groundwater. Excess alkalinity in water is manifested by a characteristic fringe of white salts at the edges of a body of water or on the banks of a stream.

Water *salinity* may be increased by a number of human activities. Water passing through a municipal water system inevitably picks up salt from sources such as recharging water softeners with sodium chloride. Salts can leach from spoil piles. One of the major environmental constraints on the production of shale oil, for example, is the high percentage of leachable sodium sulfate in piles of spent shale. Careful control of these wastes is necessary to prevent further saline pollution of water in areas where salinity is already a problem. Irrigation adds a great deal of salt to water, a phenomenon responsible for the Salton Sea in California, and is a source of conflict between the United States and Mexico over saline contamination of the Rio Grande and Colorado rivers. Irrigation and intensive agricultural production have caused saline seeps in some of the Western states. These occur when water seeps into a slight depression in tilled, sometimes irrigated, fertilized land, carrying salts (particularly sodium, magnesium, and calcium sulfates) along with it. The water evaporates in the dry summer heat, leaving behind a salt-laden area that no longer supports much plant growth. With time, these areas spread, destroying the productivity of crop land.

7.9 OXYGEN, OXIDANTS, AND REDUCTANTS

Oxygen is a vitally important species in water (see Chapter 2, Section 2.10). In water, oxygen is consumed rapidly by the oxidation of organic matter, $\{CH_2O\}$:

$$\{CH_2O\} + O_2 \xrightarrow{\text{Microorganisms}} CO_2 + H_2O \tag{7.6}$$

Unless the water is reaerated efficiently, as by turbulent flow in a shallow stream, it rapidly loses oxygen and will not support higher forms of aquatic life.

In addition to the microorganism-mediated oxidation of organic matter, oxygen in water may be consumed by the biooxidation of nitrogenous material:

$$NH_4^+ + 2O_2 \rightarrow 2H^+ + NO_3^- + H_2O \tag{7.7}$$

and by the chemical or biochemical oxidation of chemical reducing agents:

$$4Fe^{2+} + O_2 + 10H_2O \rightarrow 4Fe(OH)_3(s) + 8H^+ \tag{7.8}$$

$$2SO_3^{2-} + O_2 \rightarrow 2SO_4^{2-} \tag{7.9}$$

All these processes contribute to the deoxygenation of water.

The degree of oxygen consumption by microbially mediated oxidation of contaminants in water is called the BOD (biochemical or biological oxygen demand). This parameter is commonly measured by determining the quantity of oxygen utilized by suitable aquatic microorganisms during a 5-day period.

The addition of oxidizable pollutants to streams produces a typical oxygen sag curve as shown in Figure 7.3. Initially, a well-aerated, unpolluted stream is relatively free of oxidizable material; the oxygen level is high; and the bacterial population is relatively low. With the addition of oxidizable pollutants, the oxygen level drops because re-aeration cannot keep up with oxygen consumption. In the decomposition zone, the bacterial population rises. The septic zone is characterized by a high bacterial population and very low oxygen levels. The septic zone terminates when the oxidizable pollutant is exhausted, and then the recovery zone begins. In the recovery zone, the bacterial population decreases and the dissolved oxygen level increases until the water regains its original condition.

Although BOD is a reasonably realistic measure of water quality insofar as oxygen is concerned, the test for determining it is time consuming and cumbersome to

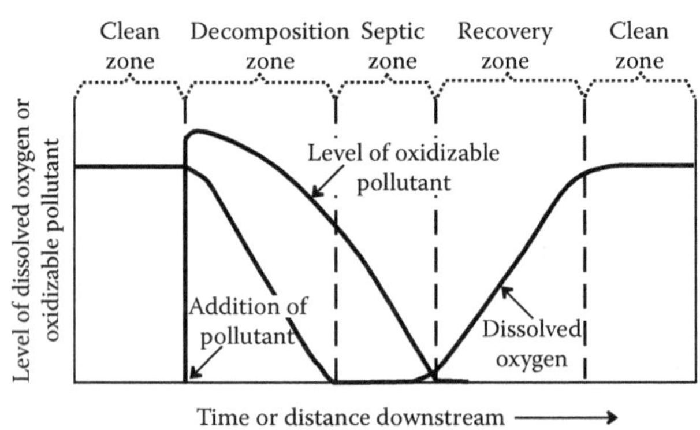

FIGURE 7.3 Oxygen sag curve resulting from the addition of oxidizable pollutant material to a stream. (From Manahan, S., *Environmental Chemistry*, 9th ed., Taylor & Francis, Boca Raton, 2010. With permission.)

perform. Total organic carbon (TOC) is frequently measured by catalytically oxidizing carbon in the water and measuring the CO_2 that is evolved. It has become popular because TOC is readily determined instrumentally.

7.10 ORGANIC POLLUTANTS

7.10.1 Bioaccumulation of Organic Pollutants

An important characteristic of organic water pollutants, especially those that have an affinity for lipid (fat) tissue and that resist biodegradation, is the *bioconcentration factor* (BCF), which is defined as the ratio of a substance's concentration in the tissue of an aquatic organism to the concentration of the substance in the water where the organism lives. It assumes that exposure is through water only and that the concentration in water is stable over a long period of time. A related parameter, the *bioaccumulation factor* (BAF) is defined in the same way, except that it assumes that both the organism and its food are exposed similarly to a pollutant over a long period of time. Although literally thousands of such factors involving hundreds of species and substances that are taken up by organisms from water have been measured or estimated, there is considerable uncertainty in these numbers; however, they remain indicative of the pollution potential of persistent organic compounds.[2]

7.10.2 Sewage

As shown in Table 7.4, sewage from domestic, commercial, food-processing, and industrial sources contains a wide variety of pollutants, including organic pollutants. Some of these pollutants, particularly oxygen-demanding substances (see Section 7.9)—oil, grease, and solids—are removed by primary and secondary sewage-treatment processes. Others, such as salts, heavy metals, and refractory (degradation-resistant) organics, are not efficiently removed.

Disposal of inadequately treated sewage can cause severe problems. For example, offshore disposal of sewage, once commonly practiced by coastal cities, results in the formation of beds of sewage residues. Municipal sewage typically contains about 0.1% solids, even after treatment, and these settle out in the ocean in a typical pattern, illustrated in Figure 7.4. The warm sewage water rises in the cold hypolimnion and is carried in one direction or another by tides or currents. It does not rise above the thermocline (metalimnion); instead, it spreads out as a cloud from which the solids rain down on the ocean floor. Aggregation of sewage colloids is aided by dissolved salts in seawater (see Chapter 5), thus promoting the formation of sludge-containing sediments.

Another major disposal problem with sewage is the sludge produced as a product of the sewage treatment process (see Chapter 8). This sludge contains organic material, which continues to degrade slowly; refractory organics; and heavy metals. The amounts of sludge produced are truly staggering. For example, the city of Chicago produces about 3 million tons of sludge each year. A major consideration in the safe disposal of such amounts of sludge is the presence of potentially dangerous components such as heavy metals.

TABLE 7.4
Some of the Primary Constituents of Sewage from a City Sewage System

Constituent	Potential Sources	Effects in Water
Oxygen-demanding substances	Mostly organic materials, particularly human feces and urine	Consume dissolved oxygen
Refractory organics	Industrial wastes, household products	Toxic to aquatic life
Viruses	Human wastes	Cause disease (possibly cancer); major deterrent to sewage recycle through water systems
Detergents	Household detergents	Esthetics, prevent grease and oil removal, toxic to aquatic life
Phosphates	Detergents	Algal nutrients
Grease and oil	Cooking, food processing, industrial wastes	Esthetics, harmful to some aquatic life
Salts	Human wastes, water softeners, industrial wastes	Increase water salinity
Heavy metals	Industrial wastes, chemical laboratories	Toxicity
Chelating agents	Some detergents, industrial wastes	Heavy metal ion solubilization and transport
Solids	All sources	Esthetics, harmful to aquatic life

Source: From Manahan, S., *Environmental Chemistry*, 9th ed., Taylor & Francis, Boca Raton, 2010. With permission.

Careful control of sewage sources is needed to minimize sewage pollution problems. Particularly, heavy metals and refractory organic compounds need to be controlled at the source to enable use of sewage, or treated sewage effluents, for irrigation, recycling to the water system, or groundwater recharge.

Soaps, detergents, and associated chemicals are potential sources of organic pollutants. These pollutants are discussed briefly here.

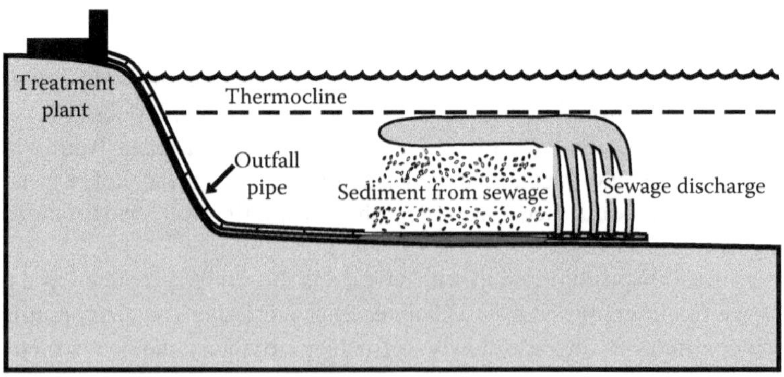

FIGURE 7.4 Settling of solids from an ocean-floor sewage effluent discharge. (From Manahan, S., *Environmental Chemistry*, 9th ed., Taylor & Francis, Boca Raton, 2010. With permission.)

7.10.3 Soaps, Detergents, and Detergent Builders

7.10.3.1 Soaps

Soaps are salts of higher fatty acids, such as sodium stearate, $C_{17}H_{35}COO^-Na^+$. The cleaning action of soap results largely from its emulsifying power and its ability to lower the surface tension of water. This concept may be understood by considering the dual nature of the soap anion. An examination of its structure shows that the stearate ion consists of an ionic carboxyl "head" and a long hydrocarbon "tail":

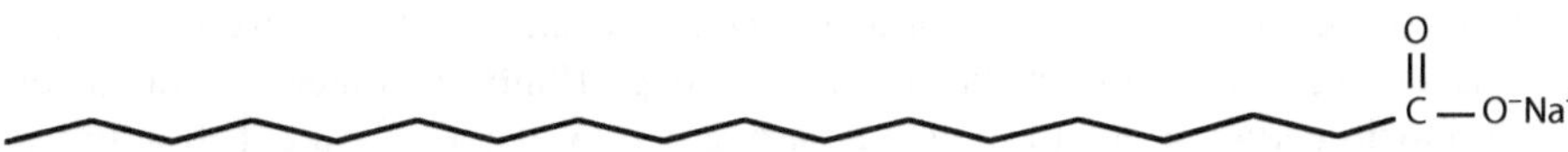

In the presence of oils, fats, and other water-insoluble organic materials, the tendency is for the "tail" of the anion to dissolve in the organic matter, whereas the "head" remains in aquatic solution. Thus, the soap emulsifies, or suspends, organic material in water. In the process, the anions form colloidal soap micelles, as shown in Figure 5.4.

The primary disadvantage of soap as a cleaning agent comes from its reaction with divalent cations to form insoluble salts of fatty acids:

$$2C_{17}H_{35}COO^-Na^+ + Ca^{2+} \rightarrow Ca(C_{17}H_{35}CO_2)_2(s) + 2Na^+ \tag{7.10}$$

These insoluble solids, usually salts of magnesium or calcium, are not at all effective as cleaning agents. In addition, the insoluble "curds" form unsightly deposits on clothing and in washing machines. If sufficient soap is used, all of the divalent cations may be removed by their reaction with soap, and the water containing excess soap will have good cleaning qualities. This is the approach commonly used when soap is employed with unsoftened water in the bathtub or wash basin, where the insoluble calcium and magnesium salts can be tolerated. However, in applications such as washing clothing, the water must be softened by the removal of calcium and magnesium or their complexation by substances such as polyphosphates (see Section 3.12).

Although the formation of insoluble calcium and magnesium salts has resulted in the essential elimination of soap as a cleaning agent for clothing, dishes, and most other materials, it has distinct advantages from the environmental standpoint. As soon as soap gets into sewage or an aquatic system, it generally precipitates as calcium and magnesium salts. Hence, any effects that soap might have in solution are eliminated. With eventual biodegradation, the soap is completely eliminated from the environment. Therefore, aside from the occasional formation of unsightly scum, soap does not cause any substantial pollution problems.

7.10.3.2 Detergents

Synthetic *detergents* have good cleaning properties and do not form insoluble salts with "hardness ions" such as calcium and magnesium. Such synthetic detergents have the additional advantage of being the salts of relatively strong acids and, therefore, they do not precipitate out of acidic waters as insoluble acids, an undesirable characteristic of soaps. The potential of detergents to contaminate water is high because of their heavy use throughout the consumer, institutional, and industrial markets. Over 1 billion pounds of detergent surfactants are consumed annually in the U.S. household market alone, with slightly more consumed in Europe. Most of this material, along with the other ingredients associated with detergent formulations, is discarded with wastewater.

The key ingredient of detergents is the *surfactant* or surface-active agent, which acts in effect to make water "wetter" and a better cleaning agent. Surfactants concentrate at interfaces of water with gases (air), solids (dirt), and immiscible liquids (oil). They do so because of their *amphiphilic structure*, meaning that one part of the molecule is a polar or ionic group (head) with a strong affinity for water, and the other part is a hydrocarbon group (tail) with an aversion to water. This kind of structure is illustrated below for the molecule of alkyl benzene sulfonate (ABS) surfactant:

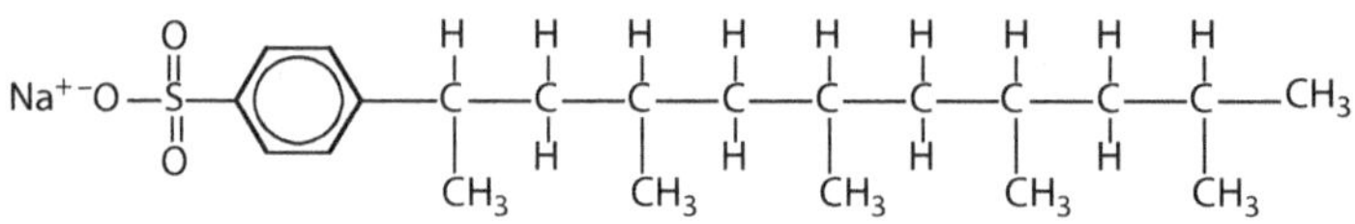

Until the early 1960s, ABS was the most common surfactant used in detergent formulations. However, it suffered the distinct disadvantage of being only very slowly biodegradable because of its branched-chain structure (see Section 6.10). An objectionable manifestation of the nonbiodegradable detergents was the "head" of foam that began to appear in glasses of drinking water in areas where sewage was recycled through the domestic water supply. Spectacular beds of foam appeared near sewage outflows and in sewage treatment plants and at least one fatality was reported when an individual fell in a bed of foam at a sewage treatment plant and was asphyxiated by the gases in the foam. Occasionally, the entire aeration tank of an activated sludge plant would be smothered by a blanket of foam. Among the other undesirable effects of persistent detergents upon waste-treatment processes were lowered surface tension of water, deflocculation of colloids, flotation of solids, emulsification of grease and oil, and destruction of useful bacteria. Consequently, ABS was replaced by a biodegradable surfactant known as linear alkyl sulfonate (LAS).

LAS, α-benzenesulfonate, has the general structural formula illustrated below where the benzene ring may be attached at any point on the alkyl chain except at the ends:

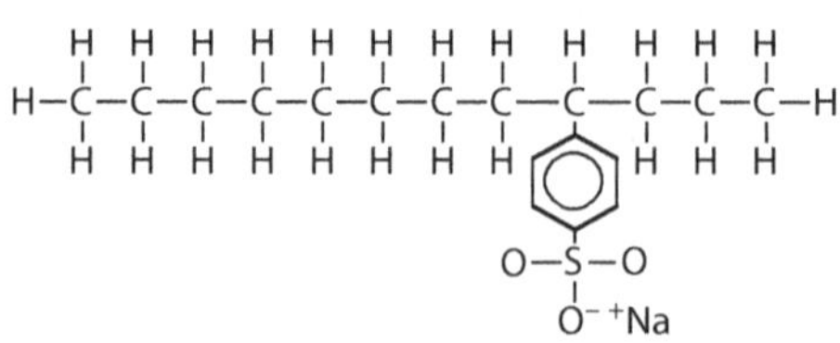

LAS

LAS is more biodegradable than ABS because the alkyl portion of LAS is not branched and does not contain the tertiary carbon which is so detrimental to biodegradability. Since LAS has replaced ABS in detergents, the problems arising from the surface-active agent in the detergents (such as toxicity to fish fingerlings) have greatly diminished and the levels of surface-active agents found in water have decreased markedly.

Some detergent surfactants are nonionic. One type that has proven troublesome consists of the alkyl phenol polyethoxylates:

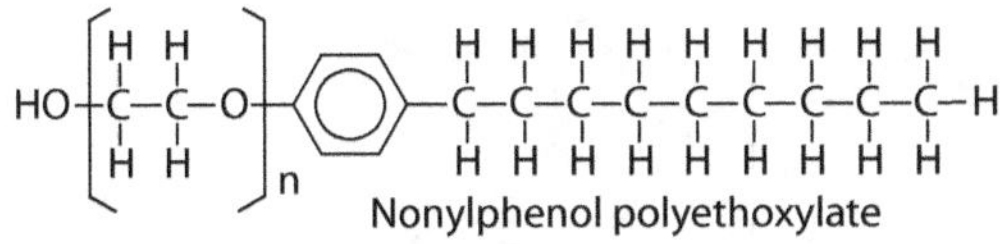

Nonylphenol polyethoxylate

The alkyl phenol polyethoxylates are very useful as detergents, dispersing agents, emulsifiers, solubilizers, and wetting agents, leading to annual use of millions of kilograms in the United States. These substances and their degradation products in which the polyethoxylate chains have been shortened by microbially mediated hydrolysis tend to survive biological sewage treatment and to be discharged with the sewage effluent. They also accumulate in sewage sludge, much of which is disposed on agricultural lands. These products are thought to be xenoestrogens (estrogen hormone mimickers) and their potential entry into the food chain from sludge-treated soil is of concern. Because of these concerns, the uses of these compounds have been severely restricted in some European countries.

Detergent *builders* added to detergents to bind to hardness ions, making the detergent solution alkaline and greatly improving the action of the detergent surfactant, can cause environmental problems. A commercial solid detergent contains only 10–30% surfactant. Detergent formulations also contain complexing agents added to complex calcium and to function as builders. Other ingredients include ion exchangers, alkalies (sodium carbonate), anticorrosive sodium silicates, amide foam stabilizers, soil-suspending carboxymethylcellulose, bleaches, fabric softeners, enzymes, optical brighteners, fragrances, dyes, and diluent sodium sulfate. The polyphosphates formerly used in builders have caused the most concern as environmental pollutants, although these problems have largely been resolved as polyphosphates have been phased out.

Increasing demands on the performance of detergents have led to the use of enzymes in detergent formulations destined for both domestic and commercial applications. To a degree, enzymes can take the place of chlorine and phosphates, both of which can have detrimental environmental consequences. Lipases and cellulases are the most useful enzymes for detergent applications.

7.10.4 Naturally Occurring Chlorinated and Brominated Compounds

Although halogenated organic compounds in water, such as those discussed as pesticides in Section 7.11, are normally considered to be from anthropogenic sources, more than 2000 such compounds have been identified from natural sources.[3] These are produced largely by marine species, especially some kinds of red algae, probably as chemical defense agents. Some marine microorganisms, worms, sponges, and tunicates are also known to produce organochlorine and organobromine compounds.

Various ones of these compounds have been detected in samples in Arctic regions including air, fish, seabird eggs, marine mammals, and Eskimo women's milk. An example of such a compound commonly encountered in the marine environment is 1,2′-bi-1H-pyrrole,2,3,3′,4,4′,5,5′-heptachloro-1′-methyl-($C_9H_3N_2Cl_7$), and its analogous bromine compound:

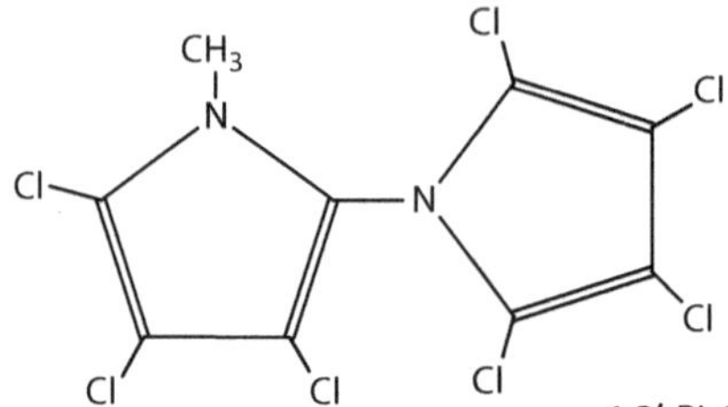

1,2′-Bi-1H-pyrrole,2,3,3′,4,4′,5,5′-heptachloro-1′-methyl

7.10.5 Microbial Toxins

Bacteria and protozoa in water can produce toxins that can cause illness or even death. Toxins produced in rivers, lakes, and reservoirs by cyanobacteria including *Anabaena*, *Microcystis*, and *Nodularia* have caused adverse health effects in Australia, Brazil, England, and elsewhere in the world. There are about 40 species of cyanobacteria that produce toxins from six chemical groups. *Cylindrospermopsin* toxin (below) produced by cyanobacteria has poisoned people who have drunk water contaminated by the toxin.

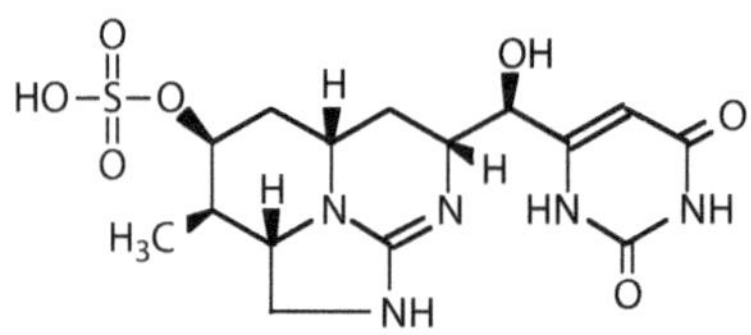

Cylindrospermopsin

Most of the protozoans that produce toxins belong to the order *Dinoflagellata*, which are predominantly marine species. The cells of these organisms are enclosed in cellulose envelopes, which often have beautiful patterns on them. Among the effects caused by toxins from these organisms are gastrointestinal, respiratory, and skin disorders in humans; mass kills of various marine animals; and paralytic conditions caused by eating infested shellfish.

The marine growth of dinoflagellates is characterized by occasional incidents in which they multiply at such an explosive rate that they color the water yellow, olive-green, or red by their vast numbers. In 1946, some sections of the Florida coast became so afflicted by "red tide" that the water became viscous and for many miles the beaches were littered with the remains of dead fish, shellfish, turtles, and other marine organisms. The sea spray in these areas became so irritating that coastal schools and resorts were closed.

The greatest danger to humans from dinoflagellata toxins comes from the ingestion of shellfish, such as mussels and clams, that have accumulated the protozoa from sea water. In this form, the toxic material is called paralytic shellfish poison. As little as 4 mg of this toxin, the amount found in several severely infested mussels or clams, can be fatal to a human. The toxin depresses respiration and affects the heart, resulting in complete cardiac arrest in extreme cases.

7.11 PESTICIDES IN WATER

The introduction of DDT during World War II marked the beginning of a period of very rapid growth in pesticide use. Pesticides are employed for many different purposes. Chemicals used in the control of invertebrates include *insecticides*, *molluscicides* for the control of snails and slugs, and *nematicides* for the control of microscopic roundworms. Vertebrates are controlled by *rodenticides* that kill rodents, *avicides* are used to repel birds, and *piscicides* are used in fish control. *Herbicides* are used to kill plants, particularly weeds in agricultural crops. Plant *growth regulators*, *defoliants*, and *plant desiccants* are used for various purposes in the cultivation of plants. *Fungicides* are used against fungi, *bactericides* against bacteria, *slimicides* against slime-causing organisms in water, and *algicides* against algae. As of the mid-1990s, U.S. agriculture used about 365 million kg of pesticides per year, whereas about 900 million kg of insecticides were used in nonagricultural applications including forestry, landscaping, gardening, food distribution, and home pest control. Insecticide production has remained about level during the last three or four decades. However, insecticides and fungicides are the most important pesticides with respect to human exposure in food because they are applied shortly before or even after harvesting. Herbicide production has increased as chemicals have increasingly replaced cultivation of land in the control of weeds and now accounts for the majority of agricultural pesticides. The potential exists for large quantities of pesticides to enter water either directly, in applications such as mosquito control or indirectly, primarily from drainage of agricultural lands.

Several classes of pesticides and other chemicals are of particular concern as water pollutants because of their potential effects. These are (1) highly biodegradation-resistant compounds, (2) known or probable carcinogens, (3) toxicants with adverse reproductive or developmental effects, (4) neurotoxins including cholinesterase inhibitors, (5) substances with high acute toxicities, and (6) known groundwater contaminants. Table 7.5 lists some of the most commonly used pesticides that may be of concern as water pollutants.

Degradation (often hydrolysis) products of some pesticides are encountered in water at levels similar to, or even higher than, the parent pesticides. In some cases, the degradation products are more toxic than the parent pesticides. An example of a pesticide degradation product commonly found in water is aminomethyl phosphonic acid produced from herbicidal glyphosate (Figure 7.13) which, under the brand name of Roundup, is the most widely produced pesticide in the world.

7.11.1 NATURAL PRODUCT INSECTICIDES, PYRETHRINS, AND PYRETHROIDS

Plants provide several significant classes of insecticides including *nicotine* from tobacco, *rotenone* from certain legume roots, and *pyrethrins* (see structural formulas

TABLE 7.5
Pesticides That May Be Found as Water Pollutants[a]

Pesticide	Use	Compound Type
Alachlor	Herbicide	Chloroacetanilide
Aldicarb	Insecticide	Carbamate
Allethrin	Insecticide	Pyrethroid
Atrazine	Herbicide	Triazine
Azadirachtin	Insecticide, nematicide	Complex botanical compound
Azinphos-methyl	Insecticide	Organophosphate
Azoxystrobin	Fungicide	Strobin
Captan	Fungicide	Thiophthalimide
Carbaryl	Insecticide, plant growth regulator, nematicide	Carbamate
Carbofuran	Insecticide, nematicide	Carbamate
Chlorothalonil	Fungicide	Substituted benzene
Chlorpyrifos	Insecticide, nematicide	Organophosphate
Cypermethrin	Insecticide	Pyrethroid
Deltamethrin	Insecticide	Pyrethroid
Diazinon	Insecticide	Organophosphorus
Diclofop-methyl	Herbicide	Chlorophenoxy acid/ester
Diuron	Herbicide	Urea
EPTC	Herbicide	Thiocarbamate
Ethephon	Plant growth regulator	Organophosphate
Fenvalerate	Insecticide	Pyrethroid
Fluquinconazole	Fungicide	Azole
Glyphosate	Herbicide	Phosphonoglycine
Iprodione	Fungicide	Dicarboximide
Kresoxim-methyl	Fungicide	Strobin
Linuron	Herbicide	Urea
Malathion	Insecticide	Organophosphate
Mecoprop	Herbicide	Chlorphenoxy compound
Metam-sodium	Fumigant, herbicide, microbiocide, algacide	Dithiocarbamate
Methiocarb	Insecticide, molluscicide	Carbamate
Metolachlor	Herbicide	Chloroacetanilide
Metribuzin	Herbicide	Triazinone
Phosmet	Insecticide	Organophosphate
Pirimicarb	Insecticide	Carbamate
Prometon	Herbicide	Triazine
Propachlor	Herbicide	Chloroacetanilide
Propanil	Herbicide	Anilide
Propiconazole	Fungicide	Azole
Simazine	Herbicide	Triazine
Spinosad	Insecticide	Compound from bacteria
Tebuconazole	Fungicide	Azole
Tebuthiuron	Herbicide	Urea
Terbacil	Herbicide	Uracil

TABLE 7.5 (continued)
Pesticides That May Be Found as Water Pollutants[a]

Pesticide	Use	Compound Type
Terbuthylazine	Algicide, herbicide, microbiocide	Triazine
Thifensulfuron-methyl	Herbicide	Sulfonylurea
Thiophanate-methyl	Fungicide	Benzimidazole
Tri-allate	Herbicide	Thiocarbamate
Triclopyr	Herbicide	Chloropyridinyl
Trifloxystrobin	Fungicide	Strobin
Trifluralin	Herbicide	2,6-Dinitroaniline

Source: From Manahan, S., *Environmental Chemistry*, 9th ed., Taylor & Francis, Boca Raton, 2010. With permission.

[a] For additional information see Pesticide Action Network: http://www.pesticideinfo.org/

in Figure 7.5). Because of the ways in which they are applied and their biodegradabilities, these substances are unlikely to be significant water pollutants.

Pyrethrins and their synthetic analogs represent both the oldest and newest of insecticides. Extracts of dried chrysanthemum or pyrethrum flowers, which contain pyrethrin I and related compounds, have been known for their insecticidal properties for a long time, and may have even been used as botanical insecticides in China almost 2000 years ago. The most important commercial sources of insecticidal pyrethrins are chrysanthemum varieties grown in Kenya. Pyrethrins have several advantages as insecticides, including facile enzymatic degradation, which makes them relatively safe for mammals; ability to rapidly paralyze ("knock down") flying insects; and good biodegradability characteristics.

Synthetic analogs of the pyrethrins, *pyrethroids*, have been widely produced as insecticides during recent years. The first of these was allethrin, and another common example is fenvalerate (see structural formulas in Figure 7.5). Other examples of insecticidal pyrethroids that have water pollution potential include cypermethrin and deltamethrin.

7.11.2 DDT and Organochlorine Insecticides

Chlorinated hydrocarbon or organochlorine insecticides are hydrocarbon compounds in which various numbers of hydrogen atoms have been replaced by Cl atoms (Figure 7.6). Of the organochlorine insecticides, the most notable has been DDT, which was used in massive quantities following World War II. It has a low acute toxicity to mammals, although there is some evidence that it might be carcinogenic. It is a very persistent insecticide and accumulates in food chains. It has been banned in the United States since 1972.

Many organochlorine insecticides were widely used in past decades but are now banned because of their toxicities, and particularly their accumulation and persistence

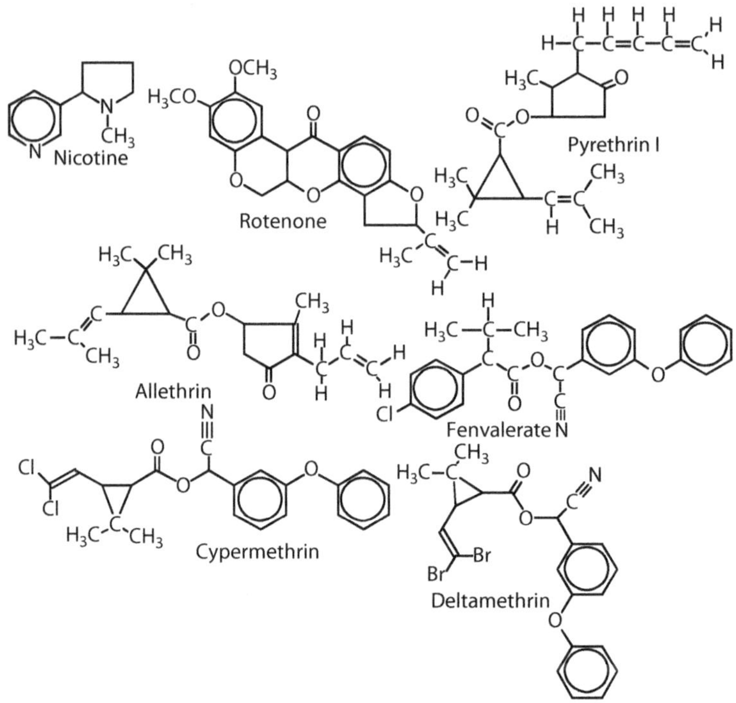

FIGURE 7.5 Common botanical insecticides and synthetic analogs of the pyrethrins. (From Manahan, S., *Environmental Chemistry*, 9th ed., Taylor & Francis, Boca Raton, 2010. With permission.)

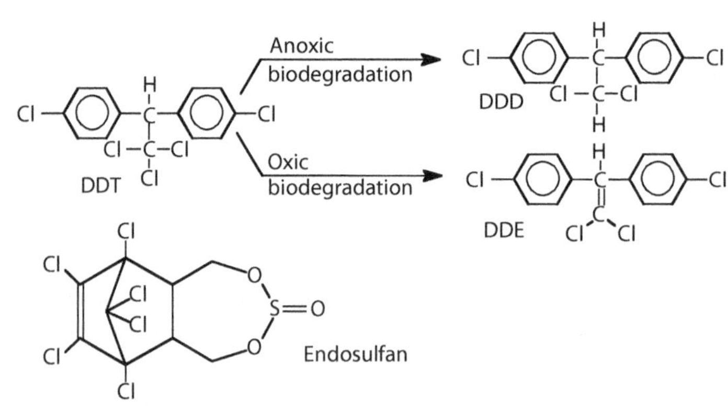

FIGURE 7.6 Examples of organochlorine insectides. DDT was the most notable of these because of its history of environmental effects. Endosulfan has been one of the last of these to be phased out of use. (From Manahan, S., *Environmental Chemistry*, 9th ed., Taylor & Francis, Boca Raton, 2010. With permission.)

in food chains. Now largely of historical interest, these include methoxychlor (once a popular replacement for DDT), dieldrin, endrin, chlordane, aldrin, dieldrin/endrin, heptachlor, toxaphene, lindane, and endosulfan (one of the last to be phased out of general use).

7.11.3 Organophosphate Insecticides

Organophosphate insecticides are insecticidal organic compounds that contain phosphorus, some of which are organic esters of orthophosphoric acid, such as paraoxon:

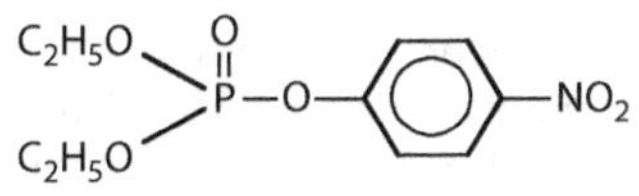

More commonly, insecticidal phosphorus compounds are phosphorothionate and phosphorodithioate compounds, such as those shown in Figure 7.7, which have an =S group rather than an =O group bonded to P.

The toxicities of organophosphate insecticides vary a great deal. Their major toxic effect is the inhibition of acetylcholinesterase, an enzyme essential for nerve function. For example, as little as 120 mg of parathion has been known to kill an adult human, and a dose of 2 mg has killed a child. Most accidental poisonings have occurred by absorption through the skin. Since its use began, several hundred people have been killed by parathion. In contrast, *malathion* shows how differences in structural formula can cause pronounced differences in the properties of organophosphate pesticides. Malathion has two carboxyester linkages, which are hydrolyzable by carboxylase enzymes to relatively nontoxic products as shown by the following reaction:

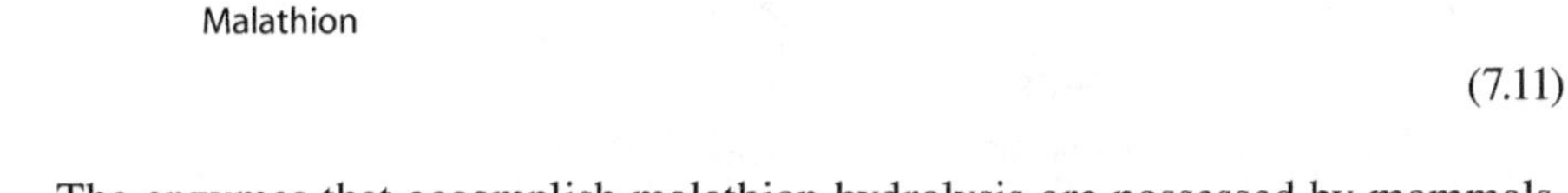

(7.11)

The enzymes that accomplish malathion hydrolysis are possessed by mammals, but not by insects, so mammals can detoxify malathion, whereas insects cannot. The result is that malathion has selective insecticidal activity. For example, although malathion is a very effective insecticide, its LD_{50} (dose required to kill 50% of test subjects) for adult male rats is about 100 times that of parathion, reflecting the much lower mammalian toxicity of malathion compared to some of the more toxic organophosphate insecticides, such as parathion.

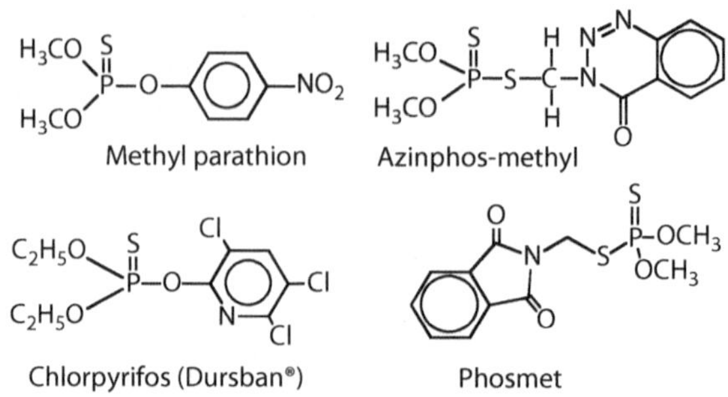

FIGURE 7.7 Examples of phosphorothionate (methyl parathion and chlorpyrifos) and phosphorodithioate (azinphos-methylphosmet) insecticides. (From Manahan, S., *Environmental Chemistry*, 9th ed., Taylor & Francis, Boca Raton, 2010. With permission.)

Unlike the organohalide compounds they largely displaced, the organophosphates readily undergo biodegradation and do not bioaccumulate. Because of their high biodegradability and restricted use, organophosphates are of comparatively little significance as water pollutants.

7.11.4 CARBAMATES

Pesticidal organic derivatives of carbamic acid, for which the formula is shown in Figure 7.8, are known collectively as *carbamates*. Carbamate pesticides have been widely used because some are more biodegradable than the formerly popular organochlorine insecticides, and have lower dermal toxicities than most common organophosphate pesticides.

Carbaryl has been widely used as an insecticide on lawns or gardens. It has a low toxicity to mammals. *Carbofuran* has a high water solubility and acts as a plant systemic insecticide. As a plant systemic insecticide, it is taken up by the roots and leaves of plants so that insects are poisoned by the plant material on which they feed. *Pirimicarb* has been widely used in agriculture as a systemic aphicide. Unlike many carbamates, it is rather persistent, with a strong tendency to bind to soil.

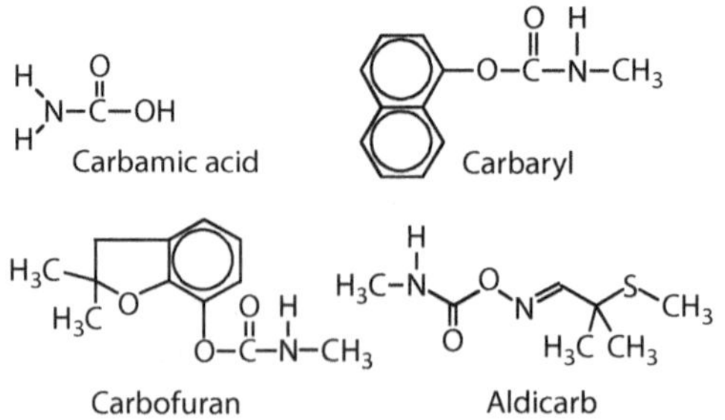

FIGURE 7.8 Carbamic acid and three insecticidal carbamates. (From Manahan, S., *Environmental Chemistry*, 9th ed., Taylor & Francis, Boca Raton, 2010. With permission.)

Carbamates are toxic to animals because they inhibit acetylcholinesterase. Unlike some of the organophosphate insecticides, they do so without the need for undergoing a prior biotransformation and are therefore classified as direct inhibitors. Their inhibition of acetylcholinesterase is relatively reversible. Loss of acetylcholinesterase inhibition activity may result from hydrolysis of the carbamate ester, which can occur metabolically.

7.11.5 Fungicides

Fungicides are widely applied to cereal and food crops to prevent fungal infections of these crops. Because of this, fungicides have the potential to contaminate water. Structural formulas of three commonly used fungicides are shown in Figure 7.9. Of the ones shown, chlorothalonil has been used for more than 30 years with annual applications in the United States of more than 5 million kg. It is typically applied at a rate of 1 kg per hectare per application with 4–9 applications per year. The strobilurin fungicides exemplified by azoxystrobin and the triazole fungicides exemplified by propiconazole came into use during the 1990s and have been effective, although some problems with resistance developed by target organisms have been encountered.

7.11.6 Herbicides

Herbicides are applied over millions of acres of farmland worldwide and are widespread water pollutants as a result of this intensive use. Herbicides are commonly found in surface water and groundwater. Among those commonly encountered in surface water and groundwater are atrazine, simazine, and cyanazine widely used to control weeds on corn and soybeans in the "Corn Belt" states of Kansas, Nebraska, Iowa, Illinois, and Missouri as well as agricultural regions throughout the world. Other herbicides found in water include prometon, metolachlor, metribuzin,

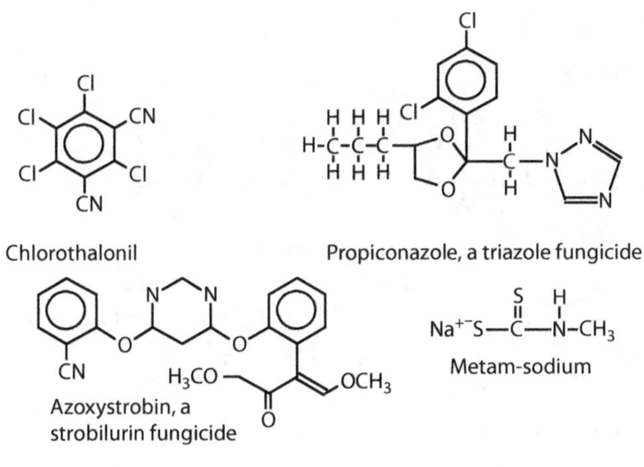

FIGURE 7.9 Four examples of widely used fungicides that may be of concern as potential water pollutants. (From Manahan, S., *Environmental Chemistry*, 9th ed., Taylor & Francis, Boca Raton, 2010. With permission.)

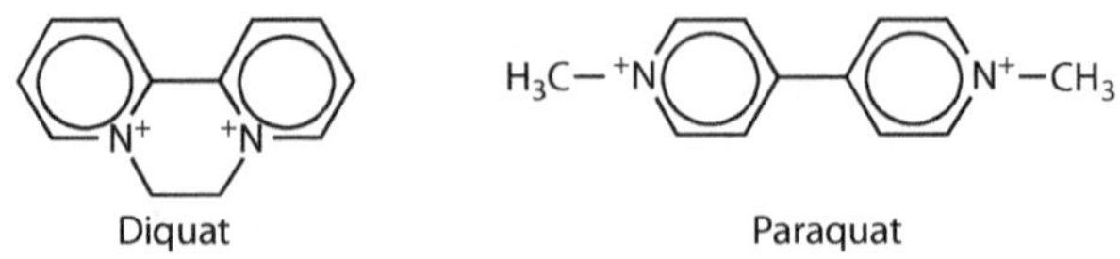

FIGURE 7.10 The two major bipyridilium herbicides (cation forms). (From Manahan, S., *Environmental Chemistry*, 9th ed., Taylor & Francis, Boca Raton, 2010. With permission.)

tebuthiuron, trifluran, alachlor, and the atrazine metabolites deisopropylatrazine and deethylatrazine. Although glyphosate is a very widely used herbicide to control weeds on crops genetically engineered to resist its effects, it is rarely found at levels of concern in water because of its very strong affinity for soil solids.

7.11.6.1 Bipyridilium Compounds

As shown by the structures in Figure 7.10, a bipyridilium compound contains two pyridine rings per molecule. The two important pesticidal compounds of this type are the herbicides *diquat* and *paraquat*, the structural formulas of which are illustrated in Figure 7.10.

Other members of this class of herbicides include chlormequat, morfamquat, and difenzoquat. Applied directly to the plant tissue, these compounds rapidly destroy plant cells and give the plant a frostbitten appearance. However, they bind tenaciously to soil, especially the clay mineral fraction, which results in rapid loss of herbicidal activity so that sprayed fields can be planted within a day or two of herbicide application.

Paraquat, which was registered for use in 1965, has been one of the most used of the bipyridilium herbicides. Highly toxic, it is reputed to have "been responsible for

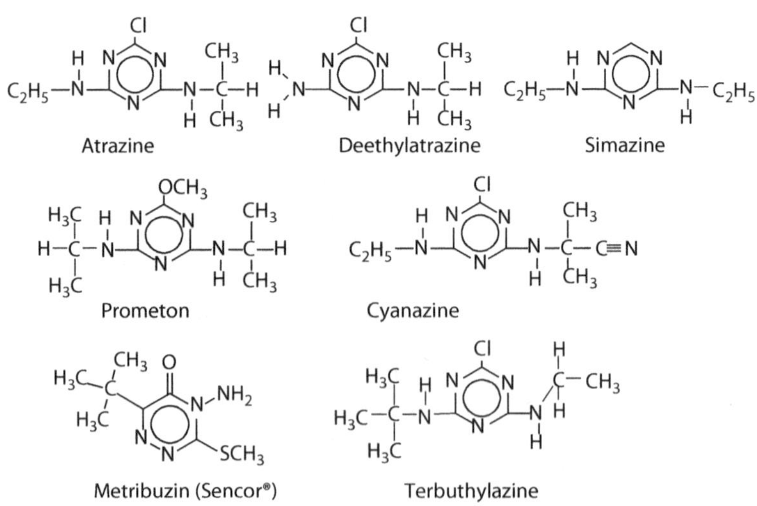

FIGURE 7.11 Triazine herbicides. These compounds are commonly encountered as water pollutants in agricultural areas where they are widely used. (From Manahan, S., *Environmental Chemistry*, 9th ed., Taylor & Francis, Boca Raton, 2010. With permission.)

hundreds of human deaths."[4] Exposure to fatal or dangerous levels of paraquat can occur by all pathways, including inhalation of spray, skin contact, ingestion, and even suicidal hypodermic injections. Despite these possibilities and its widespread application, paraquat is used safely without ill effects when proper procedures are followed.

Because of its widespread use as a herbicide, the possibility exists of substantial paraquat contamination of food. Drinking water contamination by paraquat has also been observed.

7.11.6.2 Herbicidal Heterocyclic Nitrogen Compounds

A number of important herbicides contain three heterocyclic nitrogen atoms in ring structures and are therefore called *triazines* (Figure 7.11). Triazine herbicides inhibit photosynthesis. Selectivity is gained by the inability of target plants to metabolize and detoxify the herbicide. The most long established and common example of this class is atrazine, commonly applied to kill weeds in corn, and a widespread water pollutant in corn-growing regions. Another member of this class is metribuzin, which is widely used on soybeans, sugarcane, and wheat.

7.11.6.3 Chlorophenoxy Herbicides

The chlorophenoxy herbicides, including 2,4-D and 2,4,5-trichlorophenoxyacetic acid (2,4,5-T) shown in Figure 7.12, were manufactured on a large scale for weed and brush control and as military defoliants. At one time the latter was of particular concern because of contaminant 2,3,7,8-tetrachlorodibenzo-*p*-dioxin (TCDD) (see figures) present as a manufacturing by-product.

7.11.6.4 Miscellaneous Herbicides

Many herbicides that are environmentally significant do not fall into the classifications described above. The most commonly used of these that are likely to be encountered as water pollutants are shown in Figure 7.13.

Nitroaniline herbicides are characterized by the presence of NO_2 and a substituted $-NH_2$ group on a benzene ring as shown for trifluralin. This class of herbicides is widely represented in agricultural applications and includes benefin (Balan®), oryzalin (Surflan®), pendimethalin (Prowl®), and fluchoralin (Basalin®).

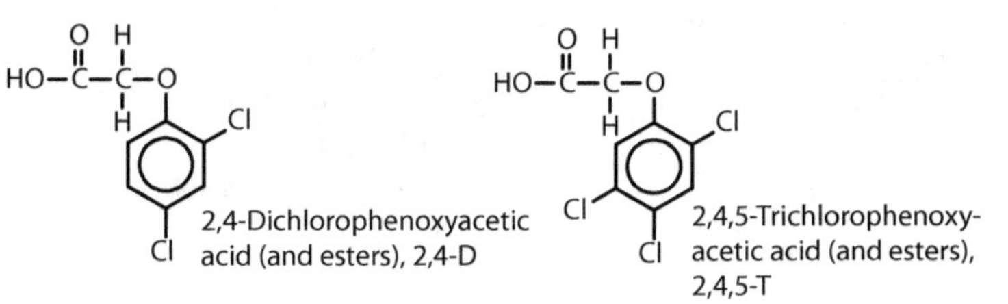

FIGURE 7.12 Chlorophenoxy herbicides. (From Manahan, S., *Environmental Chemistry*, 9th ed., Taylor & Francis, Boca Raton, 2010. With permission.)

A wide variety of chemicals have been used as herbicides, and have been potential water pollutants. One such compound is R-mecoprop (Figure 7.13). Other types of herbicides include substituted ureas, carbamates, and thiocarbamates.

Until about 1960, arsenic trioxide and other inorganic arsenic compounds (see Section 7.4) were employed to kill weeds. Because of the incredibly high rates of application of up to several hundred kilograms per acre, and because arsenic is nonbiodegradable, the potential still exists for arsenic pollution of surface water and groundwater from fields formerly dosed with inorganic arsenic. Organic arsenicals, such as cacodylic acid, have also been widely applied to kill weeds:

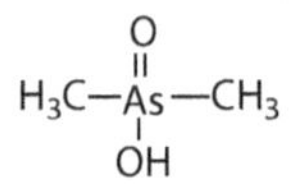

Hydroxydimethylarsine oxide, cacodylic acid

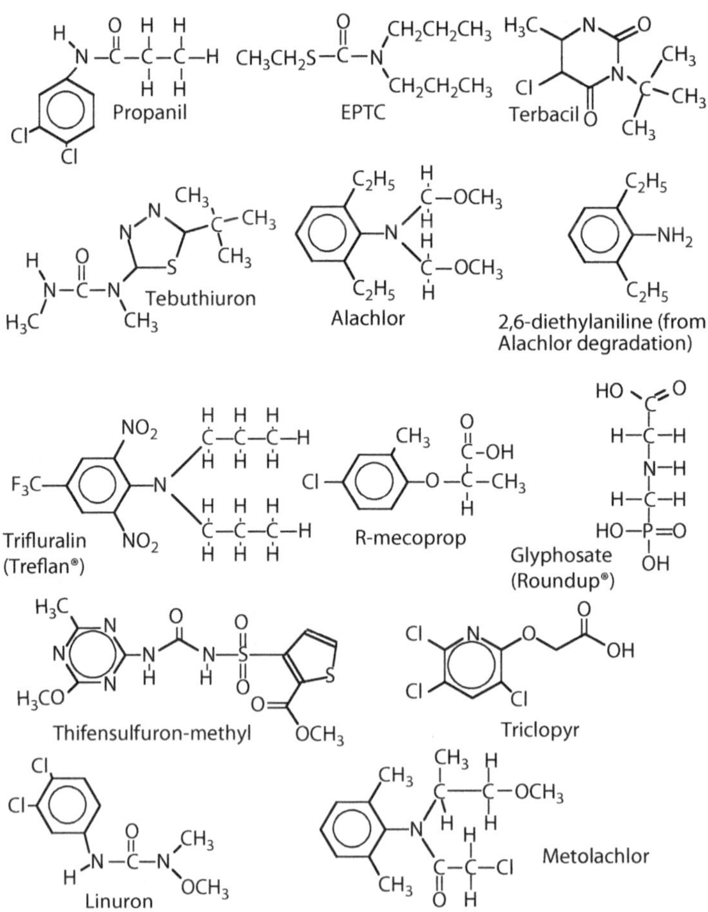

FIGURE 7.13 Miscellaneous herbicides, some of which are commonly found as water pollutants. (From Manahan, S., *Environmental Chemistry*, 9th ed., Taylor & Francis, Boca Raton, 2010. With permission.)

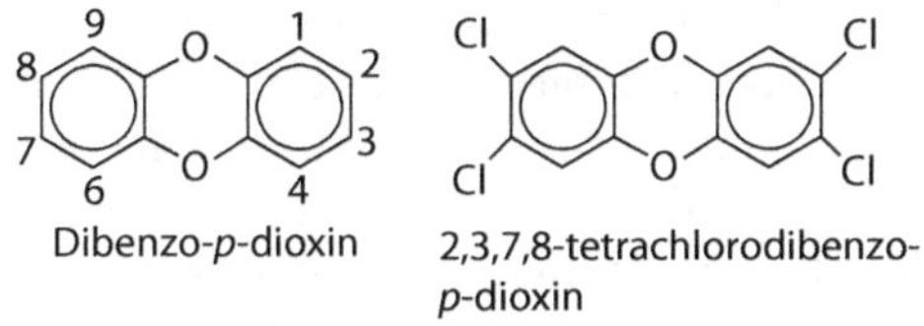

FIGURE 7.14 Dibenzo-*p*-dioxin and TCDD, often called simply "dioxin." In the structure of dibenzo-*p*-dioxin, each number refers to a numbered carbon atom to which an H atom is bound, and the names of derivatives are based upon the carbon atoms where another group has been substituted for the H atoms, as is seen by the structure and name of 2,3,7,8-tetrachlorodibenzo-*p*-dioxin. (From Manahan, S., *Environmental Chemistry*, 9th ed., Taylor & Francis, Boca Raton, 2010. With permission.)

7.11.7 By-Products of Pesticide Manufacture

A number of water pollution and health problems have been associated with the manufacture of organochlorine pesticides. The most notorious by-products of pesticide manufacture are *polychlorinated dibenzodioxins*. From 1 to 8 Cl atoms may be substituted for H atoms on dibenzo-*p*-dioxin (Figure 7.14), giving a total of 75 possible chlorinated derivatives. Commonly referred to as "dioxins," these species have a high environmental and toxicological significance. Of the dioxins, the most notable pollutant and hazardous waste compound is TCDD, often referred to simply as *dioxin*. This compound was produced as a low-level contaminant in the manufacture of some aryl, oxygen-containing organohalide compounds such as chlorophenoxy herbicides (mentioned previously in this section) synthesized by processes used until the 1960s.

TCDD has a very low vapor pressure, a high melting point of 305°C, and a water solubility of only 0.2 μg/L. It is chemically unreactive, stable thermally up to about 700°C, and is poorly biodegradable. It is very toxic to some animals, with an LD_{50} of only about 0.6 μg/kg body mass in male guinea pigs. (The type and degree of its toxicity to humans are largely unknown; it is known to cause a severe skin condition called chloracne.) Because of its properties, TCDD is a stable, persistent environmental pollutant and a hazardous waste constituent of considerable concern. It has been identified in some municipal incineration emissions, in which it is believed to form when chlorine from the combustion of organochlorine compounds reacts with carbon in the incinerator.

TCDD contamination has resulted from improper waste disposal, the most notable case of which resulted from the spraying of waste oil mixed with TCDD on roads and horse arenas in Missouri in the early 1970s. Contamination of the soil in Times Beach, Missouri, resulted in the whole town being bought out and its topsoil dug up and incinerated at a cost exceeding $100 million.

One of the greater environmental disasters ever to result from pesticide manufacture involved the production of Kepone, structural formula

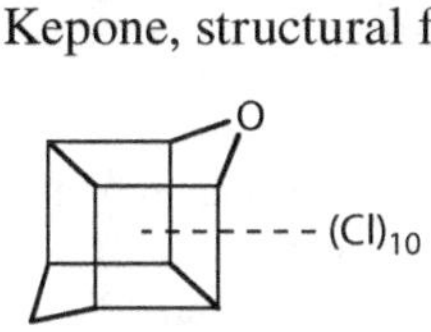

This pesticide has been used for the control of banana-root borer, tobacco wireworm, ants, and cockroaches. Kepone exhibits acute, delayed, and cumulative toxicity in birds, rodents, and humans, and it causes cancer in rodents. It was manufactured in Hopewell, Virginia, during the mid-1970s. During this time, workers were exposed to Kepone and are alleged to have suffered health problems as a result. As much as 53,000 kg of kepone may have been dumped through the sewage system of Hopewell and into the James River. The cost of dredging the river to remove this waste was estimated at a prohibitive cost of several billion dollars.

7.12 POLYCHLORINATED BIPHENYLS

First discovered as environmental pollutants in 1966, PCB compounds have been found throughout the world in water, sediments, bird tissue, and fish tissue. These compounds constitute an important class of special wastes. They are made by substituting from 1 to 10 Cl atoms onto the biphenyl aryl structure as shown on the left in Figure 7.15. This substitution can produce 209 different compounds (congeners), of which one example is shown on the right in Figure 7.15.

Polychlorinated biphenyls have very high chemical, thermal, and biological stability; low vapor pressure; and high dielectric constants. These properties have led to the use of PCBs as coolant-insulation fluids in transformers and capacitors; for the impregnation of cotton and asbestos; as plasticizers; and as additives to some epoxy paints. The same properties that made extraordinarily stable PCBs so useful also contributed to the widespread dispersion and accumulation of these substances in the environment. By regulations issued in the United States under the authority of the Toxic Substances Control Act passed in 1976, the manufacture of PCBs was discontinued in the United States, and their uses and disposal were strictly controlled. Some degree of biodegradation of PCBs in the environment does occur.[5]

Substitutes for PCBs for electrical applications have been developed. Disposal of PCBs from discarded electrical equipment and other sources have caused problems, particularly since PCBs can survive ordinary incineration by escaping as vapors through the smokestack. However, they can be destroyed by special incineration processes.

PCBs are especially prominent pollutants in the sediments of the Hudson River as a result of waste discharges from two capacitor manufacturing plants that operated about 60 km upstream from the southernmost dam on the river from 1950 to 1976. The river sediments downstream from the plants exhibit PCB levels of about 10 ppm, 1–2 orders of magnitude higher than levels commonly encountered in river and

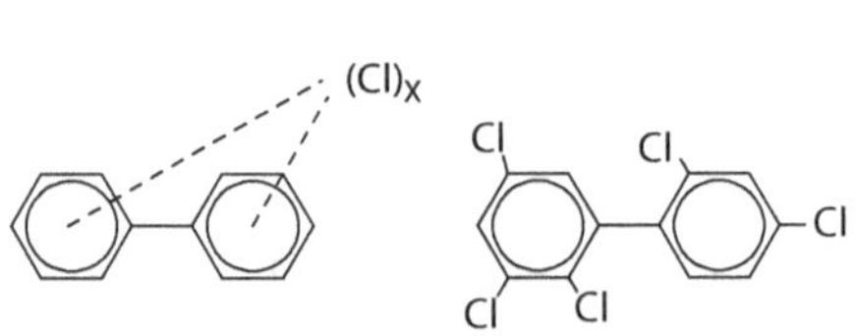

FIGURE 7.15 General formula of polychlorinated biphenyls (left, where X may range from 1 to 10) and a specific 5-chlorine congener (right). (From Manahan, S., *Environmental Chemistry*, 9th ed., Taylor & Francis, Boca Raton, 2010. With permission.)

estuary sediments. In 2002, General Electric Co. was ordered to dredge and decontaminate sections of the Hudson River polluted with PCBs at a cost exceeding $100 million. As of 2009, the actual cleanup was barely underway.

7.13 EMERGING WATER POLLUTANTS, PHARMACEUTICALS, AND HOUSEHOLD WASTES

The continued development of new products used for various purposes has led to interest in *emerging pollutants* of various kinds which may be of concern in water. Prominent among these are *nanomaterials* consisting of very small entities in the 1–100 nm size range. Nanomaterials of various kinds have unique properties of high thermal stability, low permeability, high strength, and high conductivity. These and other properties are leading to uses in electronics, automobiles, apparel, sunscreens, cosmetics, water purification, and other products. It is anticipated that the use of nanomaterials in drug delivery will increase rapidly in the future. The commercial uses of nanomaterials are in their infancy, but very rapid growth is underway. Little is known about the potential pollution effects and toxicities of nanomaterials, so their potential effects as water pollutants are of significant concern.

Another class of emerging pollutants consists of *siloxanes* (commonly called silicones) including octamethylcyclotetrasiloxane, decamethylcyclopentasiloxane, and dodecamethylcyclohexasiloxane. Siloxanes are thermally and chemically very stable leading to their uses as coolants in transformers, protective encapsulating materials in semiconductors, lubricants, coatings, and sealants. Siloxanes are widely used in personal care products including deodorants, cosmetics, soaps, hair conditioners, and hair dyes and other products such as water repellent windshield coatings, detergent antifoaming agents, and even food additives. Siloxanes are resistant to biodegradation and as a result are encountered in water that has received wastewater.

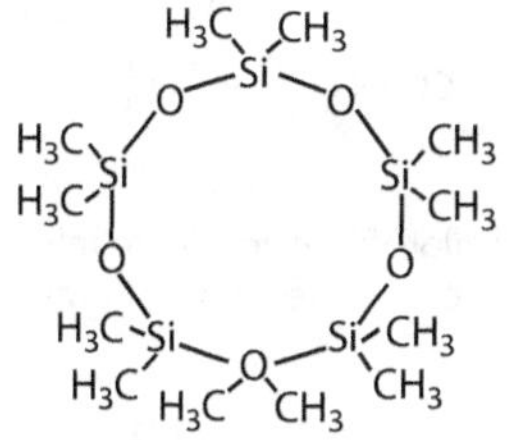

Decamethylcyclopentasiloxane, a cyclic siloxane

Disinfection by-products are of some concern as water pollutants. These are compounds containing halogens and nitrogen that result from reactions of water disinfectants including chlorine, hypochlorite, and chlorine dioxide. In addition to exposure in drinking water, humans may be exposed to these substances through skin contact in bathing or swimming and as vapors emitted from water during showers. Brominated and iodated compounds are formed in chlorinated water by reactions of bromide or iodide in the water, usually present in areas of seawater or saline groundwater intrusion.

The most common disinfection by-products are the *trihalomethanes*—chloroform ($CHCl_3$), dibromochloromethane ($CHClBr_2$), bromodichloromethane ($CHCl_2Br$), and tribromomethane ($CHBr_3$). These are all by-products of water chlorination and are Group B carcinogens (shown to cause cancer in laboratory animals). By far the most abundant of these in water systems is trichloromethane (chloroform). Dibromochloromethane is regarded as posing about 10 times the risk of cancer as chloroform, and tribromomethane is thought to pose just a slightly greater cancer risk than chloroform. As of 2009, the maximum allowable limit for total trihalomethanes in drinking water was 100 μg/L.

Various substances associated with household wastes are found in water, especially in treated sewage discharges. These materials include steroids, surfactants, flame retardants, fragrances, plasticizers, and pharmaceuticals and their metabolites. It should be noted that, although significant numbers of such compounds are found, they are generally at subpart-per-billion levels and are detectable only by the remarkable capability of modern analytical instrumentation. A study of "exotic" organics found in groundwater and water supplies[6] detected a variety of substances including cholesterol, nicotine metabolite cotinine, β-sitosterol (a natural plant sterol), 1,7-dimethylxanthine (caffeine metabolite), bisphenol-A plasticizer, and fire retardant (2-chloroethyl) phosphate.

Pharmaceutical compounds and their partial degradation products are discharged with sewage as wastes from human ingestion and from being discarded with wastewater.[7] The quantities of these substances in sewage in developed countries can reach of the order of 100 metric tons per year. Levels of common pharmaceuticals of around 1 μg/L have been observed in river water. Although these are relatively low values, they are of some concern because of the biological activity inherent to pharmaceutical products, which are increasingly designed for higher potency, bioavailability, and resistance to degradation. Figure 7.16 shows some of the most

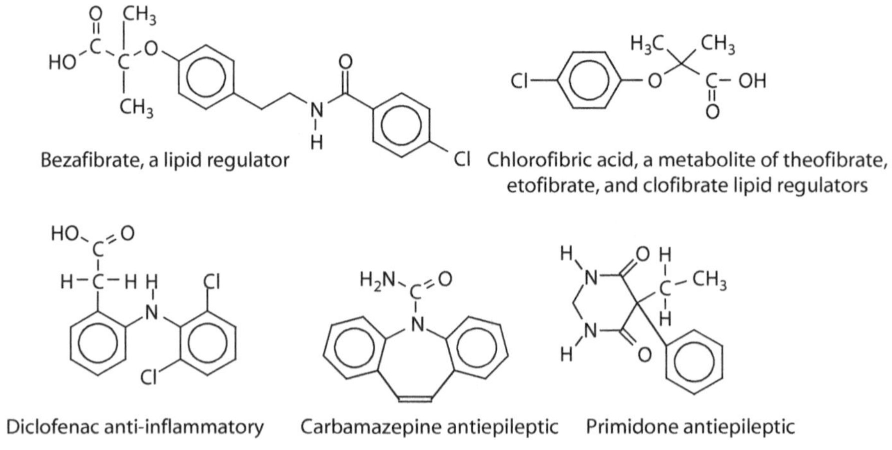

FIGURE 7.16 Pharmaceuticals and metabolites of concern as water pollutants. (From Manahan, S., *Environmental Chemistry*, 9th ed., Taylor & Francis, Boca Raton, 2010. With permission.)

common pharmaceuticals and their degradation products that have been observed in water.

The most obvious effect of pharmaceuticals and their metabolic products in water has been the feminization of male fish observed downstream from treated sewage discharges resulting from estrogenic compounds in wastewater. First noted in England and later in the United States and Europe, these male fish have been observed to produce proteins associated with egg production by female fish and to produce early-stage eggs in their testes. These effects are largely attributed to residues of synthetic 17α-ethinylestradiol and the natural hormone 17β-estradiol, used in oral contraceptives. The metabolically-formed glucuronide and sulfate conjugates of 17α-ethinylestradiol are excreted with urine and cleaved by bacteria in water to regenerate the original compound.

7.13.1 Bactericides

Bactericides used in cleaning and consumer products may be encountered in water. One of the most common of these is triclosan,

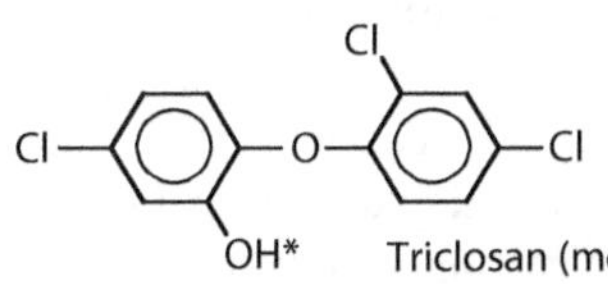

Triclosan (methyl triclosan has a $-CH_3$ group substituted for the H designated H*)

widely used in antibacterial soaps as well as other consumer items such as shampoos, deodorants, lotions, toothpastes, sportswear, shoes, carpets, and even refuse containers. This compound and its methyl derivative have been found in natural waters in Switzerland.[8]

7.13.2 Estrogenic Substances in Wastewater Effluents

A class of water pollutants of particular concern commonly found in sewage and even treated sewage effluent are *estrogenic substances* that can disrupt the crucial endocrine gland activities that regulate the metabolism and reproductive functions of organisms. Aquatic organisms including fish, frogs, and reptiles such as alligators exposed to such substances may exhibit reproductive dysfunction, alterations in secondary sex characteristics, and abnormal serum steroid levels. Such substances include exogenous estrogenic substances among which are 17α-ethynyl estradiol, diethylstilbestrol, mestranol, levonorgestrel, and norethindrone used in oral contraceptives, treatment of hormonal disorders, and cancer treatment. Some synthetic substances also act as estrogen disruptors. Of prime concern as water pollutants are nonionic surfactant polyethoxylates, mentioned in the discussion of detergents above, and their major degradation product, persistent nonylphenol. Although these substances are orders of magnitude less potent than hormonal substances, annual usage of millions of kilograms of nonionic surfactants make them a significant factor as water pollutants.

7.13.3 Biorefractory Organic Pollutants

Millions of tons of organic compounds are manufactured globally each year. Significant quantities of several thousand such compounds appear as water pollutants. Most of these compounds, particularly the less biodegradable ones, are substances to which living organisms have not been exposed until recent years. Often, their effects on organisms are not known, particularly for long-term exposures at very low levels. The potential exists for synthetic organics to cause genetic damage, cancer, or other ill effects. On the positive side, organic pesticides enable a level of agricultural productivity without which millions would starve. Synthetic organic chemicals are increasingly taking the place of natural products in short supply. Thus it is seen that organic chemicals are essential for the operation of a modern society. Because of their potential danger, however, acquisition of knowledge about their environmental chemistry must have a high priority.

Biorefractory organics are the organic compounds of most concern in wastewater, particularly when they are found in sources of drinking water. These are poorly biodegradable substances and are sometimes referred to as *persistent organic pollutants* (POP), prominent among which are aryl or chlorinated hydrocarbons. Biorefractory compounds that may be found in water include benzene, chloroform, methyl chloride, styrene, tetrachloroethylene, trichloroethane, and toluene. In addition to their potential toxicity, biorefractive compounds can cause taste and odor problems in water. They are not completely removed by biological treatment, and water contaminated with these compounds must be treated by physical and chemical means, including air stripping, solvent extraction, ozonation, and carbon adsorption.

Methyl *tert*-butyl ether (MTBE),

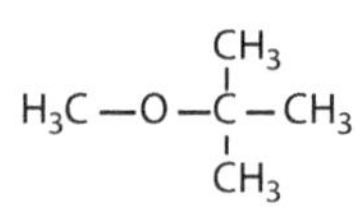

was once used as a gasoline octane booster, but was phased out after it appeared as a low-level water pollutant in the United States. Levels of this chemical in recreational lakes and reservoirs were attributed largely to emissions of unburned fuel from recreational motorboats and personal watercraft having two-cycle engines that discharge their exhausts directly to the water.

Perfluorinated organic compounds constitute a unique class of POPs. They occur as completely fluorinated hydrocarbon derivatives, such as CF_4, in the atmosphere, where they are regarded as atmospheric pollutants and potential greenhouse gases (see Chapter 2, Section 2.11). Other perfluorinated compounds that are organic acids or their salts have been encountered as water pollutants. Most commonly cited are salts of perfluorooctane sulfonic acid; others include salts of perfluorinated carboxylic acids, such as perfluorohexanoic acid:

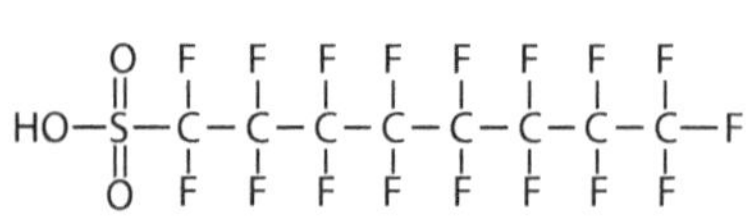

Perfluorooctane sulfonic acid

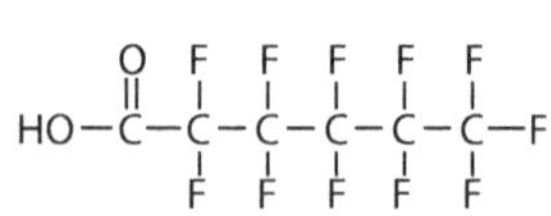

Perfluorohexanoic acid

Perfluorocarbons have been used commercially since the 1950s, primarily as coatings to resist soil and grease in paper products, fabrics, carpet material, and leather. Scotchgard fabric protector, once manufactured by 3M Corporation, contained perfluorooctane sulfonates, but this use has been discontinued. Perfluorocarbons have also been used as surfactants in oil drilling fluids and firefighting foams. Other applications have included alkaline cleaners, floor polish formulations, etching baths, and even denture cleaners. Perfluorocarbons have been detected in water, fish blood and liver, and human blood.

Brominated compounds have been recognized as significant environmental and water pollutants in recent years and have even been found in mothers' milk in some countries. These compounds have been manufactured as flame retardants, largely for use in polymers and textiles. The most common of the brominated compounds likely to be encountered as pollutants are polybrominated diphenyl ethers and tetrabromobisphenol:

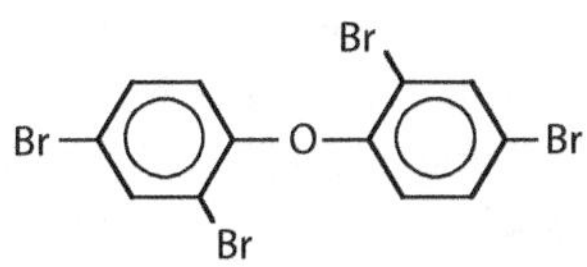

2,2′,4,4′ tetrabromodiphenyl ether

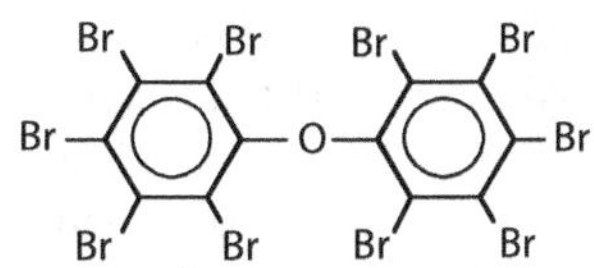

Decabromodiphenyl ether

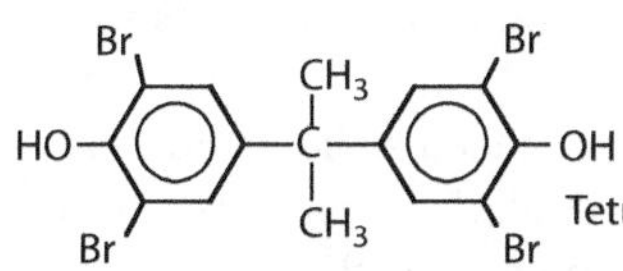

Tetrabromobisphenol A

Benzotriazole and tolyltriazoles (structural formulas below) are complexing agents for metals that are widely used as anticorrosive additives by forming a thin complexing film on metal surfaces, thereby protecting the metal from corrosion. These compounds are used in a variety of products including hydraulic fluids, cooling fluids, antifreeze formulations, and aircraft deicer fluids, and for silver protection in dishwasher detergents. Because of these uses, the triazoles are commonly encountered as "down-the-drain" chemicals that get into wastewaters. Their widespread use, high water solubility, and poor biodegradability make them among the most widely encountered chemicals in waters receiving treated wastewater.[9]

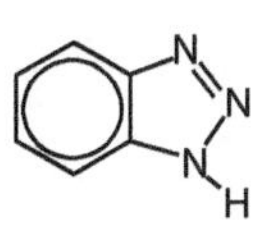

Benzotriazole

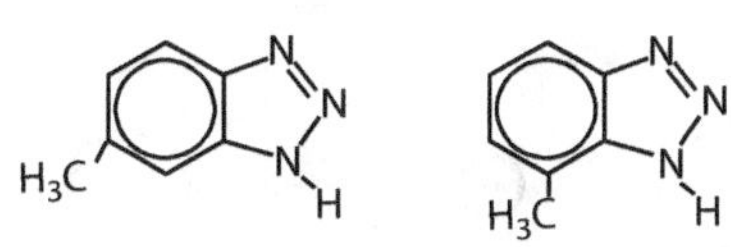

Tolyltriazoles

Naphthenic acid is a complex and variable mixture of carboxylic acids of molar masses in the approximate range of 180–350 and typically containing one $-CO_2H$ group and one 5- or 6-membered ring per molecule that is a by-product of petroleum refining. Naphthenic acids recovered from petroleum refining are used for solvents, lubricants, corrosion inhibitors, metal naphthenate synthesis, fuel additives, deicing, dust control, wood preservation, and road stabilization. Water pollution from naphthenic acids is most severe in the tar sands processing area of Alberta, Canada, where caustic hot water is used to wash heavy hydrocarbon crude oil from sand, leaving huge quantities of tailings of clay, sand, and water contaminated with 80–120 mg/L naphthenic acids. The acids are toxic to aquatic organisms and are endocrine disrupting substances.

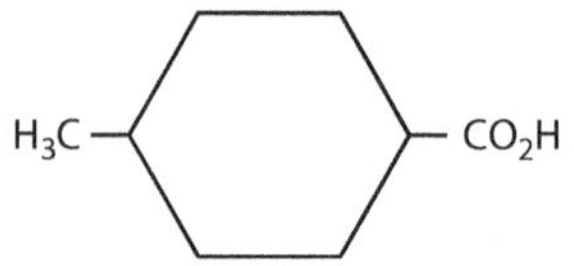

4-Methyl-1-cyclohexane carboxylic acid, a typical low-molar-mass naphthenic acid

7.14 RADIONUCLIDES IN THE AQUATIC ENVIRONMENT

The massive production of *radionuclides* (radioactive isotopes) by weapons and nuclear reactors since World War II has been accompanied by increasing concern about the effects of radioactivity on health and the environment. Radionuclides are produced as fission products of heavy nuclei of elements such as uranium or plutonium. They are also produced by the reaction of neutrons with stable nuclei. These phenomena are illustrated in Figure 7.17 and specific examples are given in Table 7.6. Radionuclides are formed in large quantities as waste products in nuclear power generation. Their ultimate disposal is a problem that has caused much controversy regarding the widespread use of nuclear power. Artificially produced radionuclides are also widely used in industrial and medical applications, particularly as "tracers." With so many possible sources of radionuclides, it is impossible to entirely eliminate radioactive contamination of aquatic systems. Furthermore, radionuclides may enter aquatic systems from natural sources. Therefore, the transport, reactions and biological concentration of

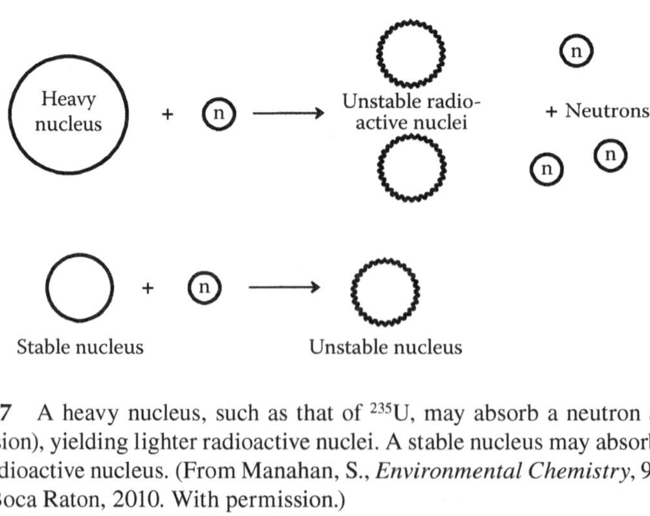

FIGURE 7.17 A heavy nucleus, such as that of ^{235}U, may absorb a neutron and break up (undergo fission), yielding lighter radioactive nuclei. A stable nucleus may absorb a neutron to produce a radioactive nucleus. (From Manahan, S., *Environmental Chemistry*, 9th ed., Taylor & Francis, Boca Raton, 2010. With permission.)

TABLE 7.6
Radionuclides in Water

Radionuclide	Half-Life	Nuclear Reaction, Description, Source
		Naturally Occurring and from Cosmic Reactions
Carbon-14	5730 y[a]	$^{14}N(n,p)^{14}C$,[b] thermal neutrons from cosmic or nuclear-weapon sources reacting with N_2
Silicon-32	~300 y	$^{40}Ar(p,x)^{32}Si$, nuclear spallation (splitting of the nucleus) of atmospheric argon by cosmic-ray protons
Potassium-40	~1.4 × 10^9 y	0.0119% of natural potassium including potassium in the body
		Naturally Occurring from ^{238}U Series
Radium-226	1620 y	Diffusion from sediments, atmosphere
Lead-210	21 y	$^{226}Ra \rightarrow$ 6 steps $\rightarrow$ ^{210}Pb
Thorium-230	75,200 y	$^{238}U \rightarrow$ 3 steps $\rightarrow$ ^{230}Th produced *in situ*
Thorium-234	24 d	$^{238}U \rightarrow$ ^{234}Th produced *in situ*
		From Reactor and Weapons Fission[c]
Strontium-90 (28 y) Iodine-131 (8 d) Cesium-137 (30 y)		
Barium-140 (13 d) > Zirconium-95 (65 d) > Cerium-141 (33 d) > Strontium-89 (51 d) > Ruthenium-103 (40 d) > Krypton-85 (10.3 y)		
		From Nonfission Sources
Cobalt-60	5.25 y	From nonfission neutron reactions in reactors
Manganese-54	310 d	From nonfission neutron reactions in reactors
Iron-55	2.7 y	$^{56}Fe(n,2n)^{55}Fe$, from high-energy neutrons acting on iron in weapons hardware
Plutonium-239	24,300 y	$^{238}U(n,g)^{239}Pu$, neutron capture by uranium

Source: From Manahan, S., *Environmental Chemistry*, 9th ed., Taylor & Francis, Boca Raton, 2010. With permission.

[a] Abbreviations: y, years; d, days.

[b] This notation shows the isotope nitrogen-14 reacting with a neutron, *n*, giving off a proton, *p*, and forming the carbon-14 isotope; other nuclear reactions can be deduced from this notation where *x* represents nuclear fragments from spallation.

[c] The first three fission-product radioisotopes listed below as products of reactor and weapons fission are of most significance because of their high yields and biological activity. The other fission products are listed in generally decreasing order of yield.

radionuclides in aquatic ecosystems are of great importance to the environmental chemist.

Radionuclides differ from other nuclei in that they emit *ionizing radiation*—alpha particles, beta particles, and gamma rays. The most massive of these emissions is the *alpha particle*, a helium nucleus of atomic mass 4, consisting of two neutrons and two protons. The symbol for an alpha particle is ${}^4_2\alpha$. An example of alpha production is found in the radioactive decay of uranium-238:

$$ {}^{238}_{92}U \rightarrow {}^{234}_{90}Th + {}^{4}_{2}\alpha \quad (7.12) $$

This transformation occurs when a uranium nucleus, atomic number 92 and atomic mass 238, loses an alpha particle, atomic number 2 and atomic mass 4, to yield a thorium nucleus, atomic number 90 and atomic mass 234.

Beta radiation consists of either highly energetic, negative electrons, which are designated ${}_{-1}^{0}\beta$, or positive electrons, called positrons, and designated ${}_{1}^{0}\beta$. A typical beta emitter, chlorine-38, may be produced by irradiating chlorine with neutrons. The chlorine-37 nucleus, natural abundance 24.5%, absorbs a neutron to produce chlorine-38 and gamma radiation:

$$ {}_{17}^{37}\mathrm{Cl} + {}_{0}^{1}\mathrm{n} \rightarrow {}_{17}^{38}\mathrm{Cl} + \gamma \qquad (7.13) $$

The chlorine-38 nucleus is radioactive and loses a negative *beta particle* to become an argon-38 nucleus:

$$ {}_{17}^{38}\mathrm{Cl} \rightarrow {}_{18}^{38}\mathrm{Ar} + {}_{-1}^{0}\beta \qquad (7.14) $$

Since the negative beta particle has essentially no mass and a −1 charge, the stable product isotope, argon-38, has the same mass and a charge 1 greater than chlorine-38.

Gamma rays are electromagnetic radiation similar to x-rays, though more energetic. Since the energy of gamma radiation is often a well-defined property of the emitting nucleus, it may be used in some cases for the qualitative and quantitative analysis of radionuclides.

The primary effect of alpha particles, beta particles, and gamma rays on materials is the production of ions; therefore, they are called *ionizing radiation*. Due to their large size, alpha particles do not penetrate matter deeply, but cause an enormous amount of ionization along their short path of penetration. Therefore, alpha particles present little hazard outside the body, but are very dangerous when ingested. Although beta particles are more penetrating than alpha particles, they produce much less ionization per unit path length. Gamma rays are much more penetrating than particulate radiation, but cause much less ionization. Their degree of penetration is proportional to their energy.

The *decay* of a specific radionuclide follows first-order kinetics; that is, the number of nuclei disintegrating in a short time interval is directly proportional to the number of radioactive nuclei present. The rate of decay, $-\mathrm{d}N/\mathrm{d}t$, is given by the equation

$$ \text{Decay rate} = \frac{\mathrm{d}N}{\mathrm{d}t} = \lambda N \qquad (7.15) $$

where N is the number of radioactive nuclei present and λ is the rate constant, which has units of reciprocal time. Since the exact number of disintegrations per second is difficult to determine in the laboratory, radioactive decay is often described in terms of the measured *activity*, A, which is proportional to the absolute rate of decay. The first-order decay equation may be expressed in terms of A:

$$ A = A_0 e^{-\lambda t} \qquad (7.16) $$

where A is the activity at time t, A_0 is the activity when t is zero, and e is the natural logarithm base. The *half-life*, $t_{1/2}$, is generally used instead of λ to characterize a radionuclide:

$$t_{1/2} = \frac{0.693}{\lambda} \tag{7.17}$$

As the term implies, a half-life is the period of time during which half of a given number of atoms of a specific kind of radionuclide decay. Ten half-lives are required for the loss of 99.9% of the activity of a radionuclide.

Radiation damages living organisms by initiating harmful chemical reactions in tissues. For example, bonds are broken in the macromolecules that carry out life processes. In cases of acute radiation poisoning, bone marrow, which produces red blood cells, is destroyed and the concentration of red blood cells is diminished. Radiation-induced genetic damage is of great concern. Such damage may not become apparent until many years after exposure. As humans have learned more about the effects of ionizing radiation, the dosage level considered to be safe has steadily diminished. For example, the U.S. Nuclear Regulatory Commission has dropped the maximum permissible concentration of some radioisotopes to levels of less than one ten-thousandth of those considered safe in the early 1950s. Although it is possible that even the slightest exposure to ionizing radiation entails some damage, some radiation is unavoidably received from natural sources including the radioactive ^{40}K found in all humans. For the majority of the population, exposure to natural radiation exceeds that from artificial sources.

The study of the ecological and health effects of radionuclides involves consideration of many factors. Among these are the type and energy of radiation emitter and the half-life of the source. In addition, the degree to which the particular element is absorbed by living species and the chemical interactions and transport of the element in aquatic ecosystems are important factors. Radionuclides having very short half-lives may be hazardous when produced but decay too rapidly to affect the environment into which they are introduced. Radionuclides with very long half-lives may be quite persistent in the environment but of such low activity that little environmental damage is caused. Therefore, in general, radionuclides with intermediate half-lives are the most dangerous. They persist long enough to enter living systems while still retaining a high activity. Because they may be incorporated within living tissue, radionuclides of "life elements" are particularly dangerous. Much concern has been expressed over strontium-90, a common waste product of nuclear testing. This element is interchangeable with calcium in bone. Strontium-90 fallout drops onto pasture and cropland and is ingested by cattle. Eventually, it enters the bodies of infants and children by way of cow's milk.

Some radionuclides found in water, primarily radium and potassium-40, originate from natural sources, particularly leaching from minerals. Others come from pollutant sources, primarily nuclear power plants and testing of nuclear weapons. The levels of radionuclides found in water typically are measured in units of pCi/L (picocuries per liter), where a curie is 3.7×10^{10} disintegrations per second, and a pCi is 1×10^{-12} that amount, or 3.7×10^{-2} disintegrations per second (2.2 disintegrations per minute).

The radionuclide of most concern in drinking water is *radium*, Ra. Areas in the United States where significant radium contamination of water has been observed include the uranium-producing regions of the western United States, Iowa, Illinois, Wisconsin, Missouri, Minnesota, Florida, North Carolina, Virginia, and the New England states.

The U.S. Environmental Protection Agency specifies maximum contaminant levels (MCL) for total radium ($^{226}Ra + ^{228}Ra$) in drinking water in units of pCi/L. In the past, perhaps as many as several hundred municipal water supplies in the United States have exceeded permissible levels, which has required finding alternative sources or additional treatment to remove radium. Fortunately, conventional water softening processes, which are designed to take out excessive levels of calcium, are relatively efficient in removing radium from water.

The possible contamination of water by fission-product radioisotopes from nuclear power production is of some concern. (If nations continue to refrain from testing nuclear weapons above ground, it is hoped that radioisotopes from this source will contribute only minor amounts of radioactivity to water.) Table 7.6 summarizes the major natural and artificial radionuclides likely to be encountered in water.

Transuranic elements are also of concern in the oceanic environment. These alpha emitters are long-lived and highly toxic. Included among these elements are various isotopes of neptunium, plutonium, americium, and curium. Specific isotopes, with half-lives in years given in parentheses, are Np-237 (2.14×10^6), Pu-236 (2.85), Pu-238 (87.8), Pu-239 (2.44×10^4), Pu-240 (6.54×10^3), Pu-241 (15), Pu-242 (3.87×10^5), Am-241 (433), Am-243 (7.37×10^6), Cm-242 (0.22), and Cm-244 (17.9).

LITERATURE CITED

1. Grove, R. A., C. J. Henny, and J. L. Kaiser, Worldwide sentinel species for assessing and monitoring environmental contamination in rivers, lakes, reservoirs, and estuaries, *Journal of Toxicology and Environmental Health, Part B: Critical Reviews*, 12, 25–44, 2008.
2. Arnot, J. A. and Frank A. P. C. Gobas, A review of bioconcentration factor (BCF) and bioaccumulation factor (BAF) assessments for organic chemicals in aquatic organisms, *Environmental Reviews*, 14, 257–297, 2008.
3. Vetter, W., Marine halogenated natural products of environmental relevance, *Reviews of Environmental Contamination and Toxicology*, 188, 1–57, 2006.
4. Gosselin, R. E., R. P. Smith, and H. C. Hodge, Paraquat, in *Clinical Toxicology of Commercial Products*, 5th ed., pp. III-328–III-336, Williams and Wilkins, Baltimore/London, 1984.
5. Pieper, D. H. and M. Seeger, Bacterial metabolism of polychlorinated biphenyls, *Journal of Molecular Microbiology and Biotechnology*, 15, 121–138, 2008.
6. Focazio, M. J., D. W. Kolpin, K. K. Barnes, E. T. Furlong, M. T. Meyer, S. D. Zaugg, L. B. Barber, and M. E. Thurman, A national reconnaissance of pharmaceuticals and other organic wastewater contaminants in the United States, *Science of the Total Environment*, 402, 192–216, 2008.
7. Khetan, S. K. and T. J. Collins, Human pharmaceuticals in the aquatic environment: A challenge to green chemistry, *Chemical Reviews*, 107, 2319–2364, 2007.
8. Lindström, A., I. J. Buerge, T. Poiger, P.-A. Bergqvist, M. D. Müller, and H.-R. Buser, Occurrence and environmental behavior of the bactericide triclosan and its methyl derivative in surface waters and in wastewater, *Environmental Science and Technology*, 36, 2322–2329, 2002.

9. Giger, W., Christian S., and H.-P. Kohler, Benzotriazole and tolyltriazole as aquatic contaminants. 1. Input and occurrence in rivers and lakes, *Environmental Science and Technology*, 40, 7186–7192, 2006.
10. Manahan, S., *Environmental Chemistry*, 9th ed., Taylor & Francis, Boca Raton, 2010.

SUPPLEMENTARY REFERENCES

Alley, E. R., *Water Quality Control Handbook*, 2nd ed., McGraw-Hill, New York, 2007.

Burk, A. R., Ed., *Water Pollution: New Research*, Nova Science Publishers, New York, 2005.

Calhoun, Y., Ed., *Water Pollution*, Chelsea House Publishers, Philadelphia, 2005.

Eckenfelder, W. W., D. L. Ford, and A. J. Englande, *Industrial Water Quality*, 4th ed., McGraw-Hill, New York, 2009.

Fellenberg, G. and A. Wier, *The Chemistry of Pollution*, 3rd ed., Wiley, Hoboken, NJ, 2000.

Gilliom, R. J., *Pesticides in the Nation's Streams and Ground Water, 1992–2001: The Quality of Our Nation's Waters*, U.S. Geological Survey, Reston, VA, 2006.

Hamilton, D. and S. Crossley, Eds, *Pesticide Residues in Food and Drinking Water: Human Exposure and Risks*, Wiley, New York, 2004.

Howd, R. A. and A. M. Fan, *Risk Assessment for Chemicals in Drinking Water*, Wiley, Hoboken, NJ, 2007.

Kaluarachchi, J. J., *Groundwater Contamination by Organic Pollutants: Analysis and Remediation*, American Society of Civil Engineers, Reston, VA, 2001.

Knepper, T. P., D. Barcelo, and P. de Voogt, *Analysis and Fate of Surfactants in the Aquatic Environment*, Elsevier, Amsterdam, 2003.

Laws, E. A. *Aquatic Pollution: An Introductory Text*, 3rd ed., Wiley, New York, 2000.

Lewinsky, A. A., Ed., *Hazardous Materials and Wastewater: Treatment, Removal and Analysis*, Nova Science Publishers, New York, 2007.

Lipnick, R. L., D. C. G. Muir, K. C. Jones, J. L. M. Hermens, and D. Mackay, Eds, *Persistent, Bioaccumulative, and Toxic Chemicals II: Assessment and Emerging New Chemicals*, American Chemical Society, Washington, DC, 2000.

Livingston, J. V., Ed., *Focus on Water Pollution Research*, Nova Science Publishers, New York, 2006.

Mason, C. F., *Biology of Freshwater Pollution*, 4th ed., Prentice Hall College Division, Upper Saddle River, NJ, 2002.

Raven, P. H., L. R. Berg, and D. M. Hassenzahl, *Environment*, 6th ed., Wiley, Hoboken, NJ, 2008.

Ravenscroft, P., H. Brammer, and K. Richards, *Arsenic Pollution: A Global Synthesis*, Blackwell, Malden, MA, 2009.

Research Council Committee on Drinking Water Contaminants, *Classifying Drinking Water Contaminants for Regulatory Consideration*, National Academy Press, Washington, DC, 2001.

Rico, D. P., C. A. Brebbia, and Y. Villacampa Esteve, Eds, *Water Pollution IX*, WIT Press, Southampton, UK, 2008.

Ritter, W. F. and A. Shirmohammadi, Eds, *Agricultural Nonpoint Source Pollution: Watershed Management and Hydrology*, CRC Press/Lewis Publishers, Boca Raton, FL, 2000.

Stollenwerk, K. G. and A. H. Welch, Eds, *Arsenic in Ground Water: Geochemistry and Occurrence*, Kluwer Academic Publishers, Hingham, MA, 2003.

Sullivan, P. J., F. J. Agardy, and J. J. J. Clark, *The Environmental Science of Drinking Water*, Elsevier Butterworth-Heinemann, Burlington, MA, 2005.

Viessman, W. and M. J. Hammer, *Water Supply and Pollution Control*, 7th ed., Pearson Prentice Hall, Upper Saddle River, NJ, 2005.

Water Environment Research, a research publication of the Water Environment Federation, Water Environment Federation, Alexandria, VA. This journal contains many articles of interest to water science; the annual reviews are especially informative.

Wheeler, W. B., *Pesticides in Agriculture and the Environment*, Marcel Dekker, New York, 2002.
Whitacre, D. M., Ed., *Reviews of Environmental Contamination and Toxicology*, 196, Springer-Verlag, New York, 2008 (published annually).
Xie, Y., *Disinfection Byproducts in Drinking Water: Form, Analysis, and Control*, CRC Press, Boca Raton, FL, 2002.

QUESTIONS AND PROBLEMS

1. Which of the following statements are true regarding chromium in water: (a) chromium(III) is suspected of being carcinogenic, (b) chromium(III) is less likely to be found in a soluble form than chromium(VI), (c) the toxicity of chromium(III) in electroplating wastewaters is decreased by oxidation to chromium(VI), (d) chromium is not an essential trace element, (e) chromium is known to form methylated species analogous to methylmercury compounds.
2. What do mercury and arsenic have in common in regard to their interactions with bacteria in sediments?
3. What are some characteristics of radionuclides that make them especially hazardous to humans?
4. To what class do pesticides containing the following group belong?

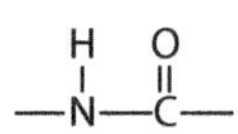

5. Consider the following compound:

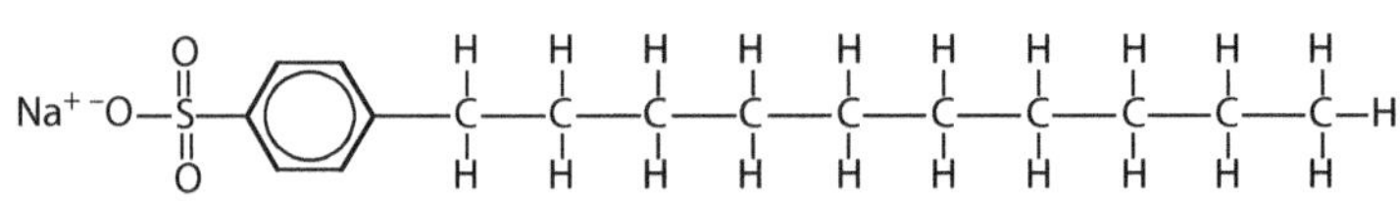

Which of the following characteristics is not possessed by the compound: (a) one end of the molecule is hydrophilic and the other end in hydrophobic, (b) surface-active qualities, (c) the ability to lower surface tension of water, (d) good biodegradability, (e) tendency to cause foaming in sewage treatment plants.

6. A certain pesticide is fatal to fish fingerlings at a level of 0.50 ppm in water. A leaking metal can containing 5.00 kg of the pesticide was dumped into a stream with a flow of 10.0 L/s moving at 1 km/h. The container leaks pesticide at a constant rate of 5 mg/s. For what distance (in km) downstream is the water contaminated by fatal levels of the pesticide by the time the container is empty?
7. Give a reason why Na_3PO_4 would not function well as a detergent builder, whereas $Na_3P_3O_{10}$ is satisfactory, though it is a source of pollutant phosphate.
8. Of the compounds $CH_3(CH_2)_{10}CO_2H$, $(CH_3)_3C(CH_2)_2CO_2H$, $CH_3(CH_2)_{10}CH_3$, and ϕ-$(CH_2)_{10}CH_3$ (where ϕ represents a benzene ring), which is the most readily biodegradable?

9. A pesticide sprayer got stuck while trying to ford a stream flowing at a rate of 136 L/s. Pesticide leaked into the stream for exactly 1 h and at a rate that contaminated the stream at a uniform 0.25 ppm of methoxychlor. How much pesticide was lost from the sprayer during this time?
10. A sample of water contaminated by the accidental discharge of a radionuclide used for medicinal purposes showed an activity of 12,436 counts per second at the time of sampling and 8966 counts per second exactly 30 days later. What is the half-life of the radionuclide?
11. What are the two reasons that soap is environmentally less harmful than ABS surfactant used in detergents?
12. What is the exact chemical formula of the specific compound designated as PCB?
13. Match each compound designated by a letter with the description corresponding to it designated by a number.

(a) CdS (b) $(CH_3)_2AsH$ (c)

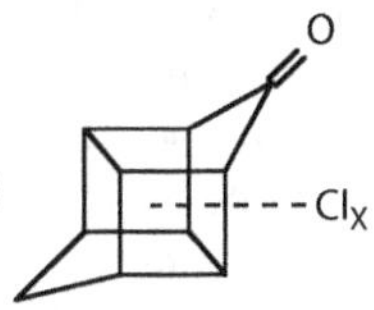

(d)

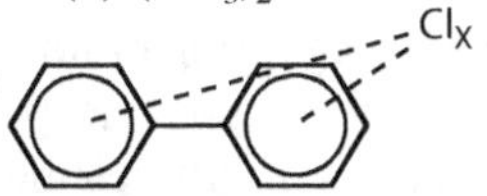

1. Pollutant released to a U.S. stream by a poorly controlled manufacturing process.
2. Insoluble form of a toxic trace element likely to be found in anoxic sediments.
3. Common environmental pollutant formerly used as a transformer coolant.
4. Chemical species thought to be produced by bacterial action.

14. A radioisotope has a nuclear half-life of 24 h and a biological half-life of 16 h (half of the element is eliminated from the body in 16 h). A person accidentally swallowed sufficient quantities of this isotope to give an initial "whole body" count rate of 1000 counts per minute. What was the count rate after 16 h?
15. What is the primary detrimental effect on organisms of salinity in water arising from dissolved NaCl and Na_2SO_4?
16. Give a specific example of each of the following general classes of water pollutants: (a) trace elements, (b) metal–organic combinations, and (c) pesticides.
17. A polluted water sample is suspected of being contaminated with one of the following: soap, ABS surfactant, or LAS surfactant. The sample has a very low BOD (biochemical oxygen demand) relative to its TOC (total organic carbon content). Which is the contaminant?
18. Of the following, the one that is *not* a cause of or associated with eutrophication is (A) eventual depletion of oxygen in the water, (B) excessive phosphate, (C) excessive algal growth, (D) excessive nutrients, and (E) excessive O_2.

19. Match the pollutants on the left with effects or other significant aspects on the right, below:

A. Salinity	1. Excessive productivity
B. Alkalinity	2. Can enter water from pyrite or from the atmosphere
C. Acidity	3. Osmotic effects on organisms
D. Nitrate	4. From soil and mineral strata

20. Of the following heavy metals, choose the one most likely to have microorganisms involved in its mobilization in water and explain why this is so: (A) lead, (B) mercury, (C) cadmium, (D) chromium, and (E) zinc.
21. Of the following, choose the true statement: (A) eutrophication results from the direct discharge of toxic pollutants into water, (B) treatment of a lake with phosphates is a process used to deter eutrophication, (C) alkalinity is the most frequent limiting nutrient in eutrophication, (D) eutrophication results from excessive plant or algal growth, (E) eutrophication is generally a beneficial phenomenon because it produces oxygen.
22. Of the following, the statement that is *untrue* regarding radionuclides in the aquatic environment is (A) they emit ionizing radiation, (B) they invariably come from human activities, (C) radionuclides of "life elements," such as iodine-131, are particularly dangerous, (D) normally the radionuclide of the most concern in drinking water is radium, (E) they may originate from the fission of uranium nuclei.
23. From the formulas below match the following: (A) lowers surface tension of water, (B) a carbamate, (C) a herbicide, (D) a noncarbamate insecticide

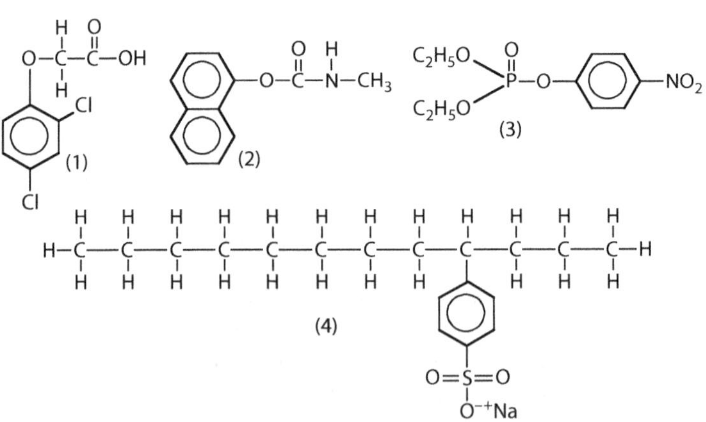

24. PCBs (A) consist of over 200 congeners with different numbers of chlorine atoms, (B) are noted for their biological instability and, therefore, toxicity, (C) occur primarily as localized pollutants, (D) are not known to undergo any biodegradation processes, (E) had no common uses, but were produced as manufacturing by-products.

8 Water Treatment

8.1 WATER TREATMENT AND WATER USE

The treatment of water may be divided into three major categories:

- Purification for domestic use
- Treatment for specialized industrial applications
- Treatment of wastewater to make it acceptable for release or reuse

The type and degree of treatment are strongly dependent upon the source and intended use of the water. Water for domestic use must be thoroughly disinfected to eliminate disease-causing microorganisms, but may contain appreciable levels of dissolved calcium and magnesium (hardness). Water to be used in boilers may contain bacteria but must be quite soft to prevent scale formation. Wastewater being discharged into a large river may require less rigorous treatment than water to be reused in an arid region. As world demand for limited water resources grows, more sophisticated and extensive means will have to be employed to treat water.

Most physical and chemical processes used to treat water involve similar phenomena, regardless of their application to the three main categories of water treatment listed above. Therefore, after introductions to water treatment for municipal use, industrial use, and disposal, each major kind of treatment process is discussed as it applies to all of these applications.

8.2 MUNICIPAL WATER TREATMENT

The modern water treatment plant is often called upon to perform wonders with the water fed to it. The clear, safe, even tasteful water that comes from a faucet may have started as a murky liquid pumped from a polluted river laden with mud and swarming with bacteria. Or, its source may have been well water, much too hard for domestic use and containing high levels of stain-producing dissolved iron and manganese. The water treatment plant operator's job is to make sure that the water plant product presents no hazards to the consumer.

A schematic diagram of a typical municipal water treatment plant is shown in Figure 8.1. This particular facility treats water containing excessive hardness and a high level of iron. The raw water taken from wells first goes to an aerator. Contact of the water with air removes volatile solutes such as hydrogen sulfide, carbon dioxide, methane, and volatile odorous substances such as methane thiol (CH_3SH) and bacterial metabolites. Contact with oxygen also aids iron removal by oxidizing soluble iron(II) to insoluble iron(III). The addition of lime as CaO or $Ca(OH)_2$ after aeration raises the pH and results in the formation of precipitates containing the hardness ions Ca^{2+} and Mg^{2+}. These precipitates settle from the water in a primary basin. Much of the solid material remains in suspension and requires the addition of coagulants [such as iron(III) and

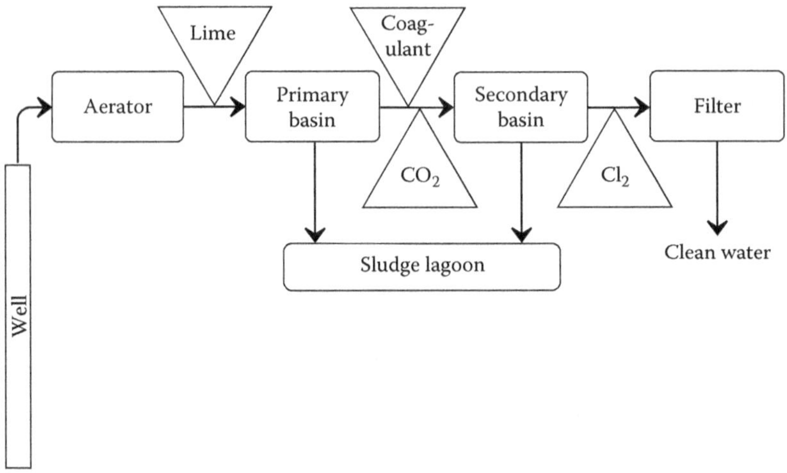

FIGURE 8.1 Schematic of a municipal water treatment plant. (From Manahan, S., *Environmental Chemistry*, 9th ed., Taylor & Francis, Boca Raton, 2010. With permission.)

aluminum sulfates, which form gelatinous metal hydroxides] to settle the colloidal particles. Activated silica or synthetic polyelectrolytes may also be added to stimulate coagulation or flocculation. The settling occurs in a secondary basin after the addition of carbon dioxide to lower the pH. Sludge from both the primary and secondary basins is pumped to a sludge lagoon. The water is finally chlorinated, filtered, and pumped to the city water mains.

8.3 TREATMENT OF WATER FOR INDUSTRIAL USE

Water is widely used in various process applications in industry. Other major industrial uses are boiler feedwater and cooling water. The kind and degree of treatment of water in these applications depends upon the end use. As examples, cooling water may require only minimal treatment, removal of corrosive substances and scale-forming solutes is essential for boiler feedwater, and water used in food processing must be free of pathogens and toxic substances. Improper treatment of water for industrial use can cause problems such as corrosion, scale formation, reduced heat transfer in heat exchangers, reduced water flow, and product contamination. These effects may cause reduced equipment performance or equipment failure, increased energy costs due to inefficient heat utilization or cooling, increased costs for pumping water, and product deterioration. Obviously, the effective treatment of water at minimum cost for industrial use is a very important area of water treatment.

Numerous factors must be taken into consideration in designing and operating an industrial water treatment facility. These include the following:

- Water requirement
- Quantity and quality of available water sources
- Sequential use of water (successive uses for applications requiring progressively lower water quality)

- Water recycle
- Discharge standards

The various specific processes employed to treat water for industrial use are discussed in later sections of this chapter. *External treatment*, usually applied to the plant's entire water supply, uses processes such as aeration, filtration, and clarification to remove material that may cause problems from water. Such substances include suspended or dissolved solids, hardness, and dissolved gases. Following this basic treatment, the water may be divided into different streams, some to be used without further treatment, and the rest to be treated for specific applications.

Internal treatment is designed to modify the properties of water for specific applications. Examples of internal treatment include the following:

- Reaction of dissolved oxygen (DO) with hydrazine or sulfite
- Addition of chelating agents to react with dissolved Ca^{2+} and prevent formation of calcium deposits
- Addition of precipitants, such as phosphate used for calcium removal
- Treatment with dispersants to inhibit scale
- Addition of inhibitors to prevent corrosion
- Adjustment of pH
- Disinfection for food processing uses or to prevent bacterial growth in cooling water

An important aspect of industrial process water treatment is the use of antiscalants and dispersants. The former prevent scale from materials such as $CaCO_3$ from building up and dispersants keep scale particles from adhering to surfaces by keeping them dispersed in water. One of the most effective agents for this purpose is polyacrylate polymer formed by the polymerization of acrylic acid and treatment with base. This polymer binds with scale-forming substances and keeps them dispersed in water by virtue of the negative charge of the polyacrylate. This same dispersant quality is useful in detergent formulations, some of which are about 5% polyacrylate. Polyacrylate is not biodegradable and accumulates with sludge residues from water treatment processes.

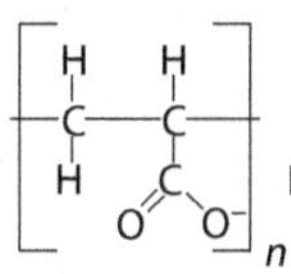

Polyacrylate polymer

8.4 SEWAGE TREATMENT

Typical municipal sewage contains oxygen-demanding materials, sediments, grease, oil, scum, pathogenic bacteria, viruses, salts, algal nutrients, pesticides, refractory organic compounds, heavy metals, and an astonishing variety of flotsam ranging from children's socks to sponges. It is the job of the waste treatment plant to remove as much of this material as possible.

Several characteristics are used to describe sewage. These include turbidity (international turbidity units), suspended solids (ppm), total dissolved solids (ppm), acidity (H^+ ion concentration or pH), and DO (in ppm O_2). BOD is used as a measure of oxygen-demanding substances.

Current processes for the treatment of wastewater may be divided into three main categories of primary treatment, secondary treatment, and tertiary treatment, each of which is discussed separately. Also discussed are total wastewater treatment systems, based largely upon physical and chemical processes.

Waste from a municipal water system is normally treated in a *publicly owned treatment works, POTW.* In the United States, these systems are allowed to discharge only effluents that have attained a certain level of treatment, as mandated by Federal law.

8.4.1 Primary Waste Treatment

Primary treatment of wastewater consists of the removal of insoluble matter such as grit, grease, and scum from water. The first step in primary treatment normally is screening. Screening removes or reduces the size of trash and large solids that get into the sewage system. These solids are collected on screens and scraped off for subsequent disposal. Most screens are cleaned with power rakes. Comminuting devices shred and grind solids in the sewage. Particle size may be reduced to the extent that the particles can be returned to the sewage flow.

Grit in wastewater consists of such materials as sand and coffee grounds which do not biodegrade well and generally have a high settling velocity. *Grit removal* is practiced to prevent its accumulation in other parts of the treatment system, to reduce clogging of pipes and other parts, and to protect moving parts from abrasion and wear. Grit is normally allowed to settle in a tank under conditions of low flow velocity, and it is then scraped mechanically from the bottom of the tank.

Primary sedimentation removes both settleable and floatable solids. During primary sedimentation there is a tendency for flocculant particles to aggregate for better settling, a process that may be aided by the addition of chemicals. The material that floats in the primary settling basin is known collectively as grease. In addition to fatty substances, the grease consists of oils, waxes, free fatty acids, and insoluble soaps containing calcium and magnesium. Normally, some of the grease settles with the sludge and some floats to the surface, where it may be removed by a skimming device.

8.4.2 Secondary Waste Treatment by Biological Processes

The most obvious harmful effect of biodegradable organic matter in wastewater is BOD, consisting of a biochemical oxygen demand for dissolved oxygen by microorganism-mediated degradation of the organic matter. *Secondary wastewater treatment* is designed to remove BOD, usually by taking advantage of the same kind of biological processes that would otherwise consume oxygen in water receiving the wastewater. Secondary treatment by biological processes takes many forms but consists basically of the action of microorganisms provided with added oxygen degrading organic material in solution or in suspension until the BOD of the waste has been reduced to acceptable levels.[1] The waste is oxidized biologically under conditions

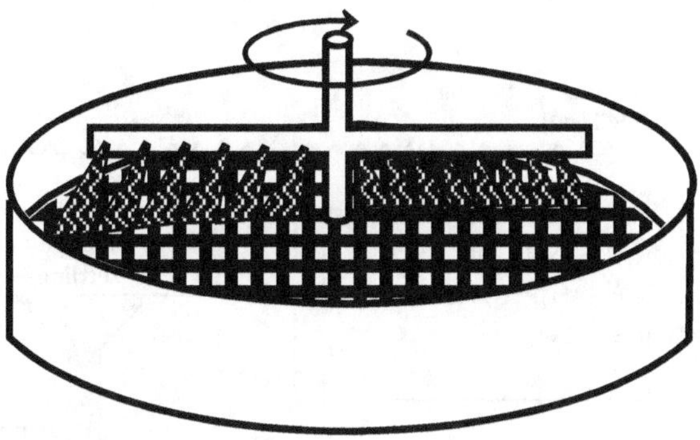

FIGURE 8.2 Trickling filter for secondary waste treatment. (From Manahan, S., *Environmental Chemistry*, 9th ed., Taylor & Francis, Boca Raton, 2010. With permission.)

controlled for optimum bacterial growth, and at a site where this growth does not influence the environment.

One of the simplest biological waste treatment processes is the *trickling filter* (Figure 8.2) in which wastewater is sprayed over rocks or other solid support material covered with microorganisms. The structure of the trickling filter is such that contact of the wastewater with air is allowed and degradation of organic matter occurs by the action of the microorganisms.

Rotating biological reactors (*contactors*), another type of treatment system, consist of groups of large plastic discs mounted close together on a rotating shaft. The device is positioned such that at any particular instant half of each disc is immersed in wastewater and half exposed to air. The shaft rotates constantly, so that the submerged portion of the discs is always changing. The discs, usually made of high-density polyethylene or polystyrene, accumulate thin layers of attached biomass, which degrades organic matter in the sewage. Oxygen is absorbed by the biomass and by the layer of wastewater adhering to it during the time that the biomass is exposed to air.

Both trickling filters and rotating biological reactors are examples of fixed-film biological (FFB) or attached growth processes. The greatest advantage of these processes is their low energy consumption. The energy consumption is minimal because it is not necessary to pump air or oxygen into the water, as is the case with the popular activated sludge process described below. The trickling filter has long been a standard means of wastewater treatment, and a number of wastewater treatment plants use trickling filters at present.

The *activated sludge process*, Figure 8.3, is probably the most versatile and effective of all wastewater treatment processes. Microorganisms in the aeration tank convert organic material in wastewater to microbial biomass and CO_2. Organic nitrogen is converted to ammonium ion or nitrate. Organic phosphorus is converted to orthophosphate. The microbial cell matter formed as part of the waste degradation processes is normally kept in the aeration tank until the microorganisms are past the log phase of growth (Section 6.7), at which point the cells flocculate relatively well to form settleable solids. These solids settle out in a settler and a fraction of them is discarded. Part of the

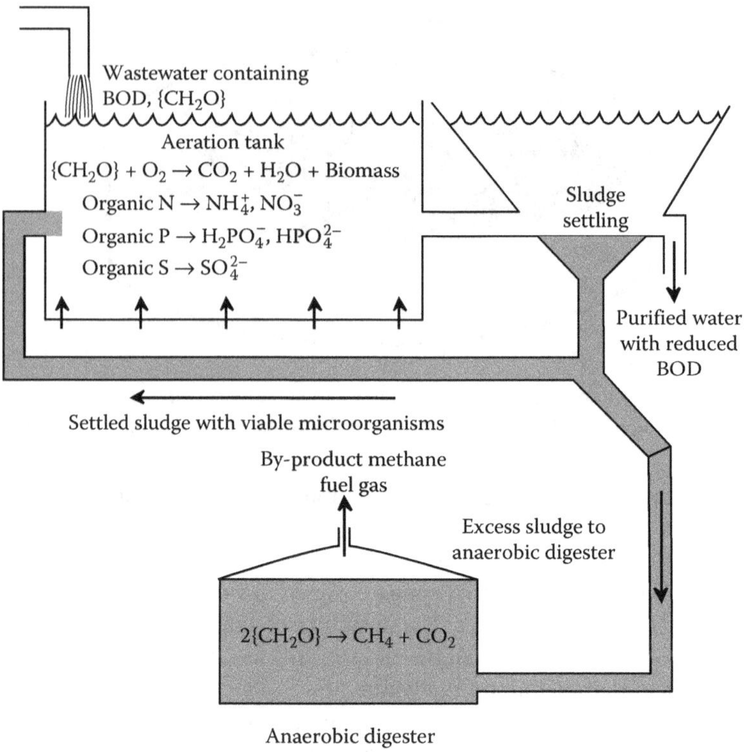

FIGURE 8.3 Activated sludge process. (From Manahan, S., *Environmental Chemistry*, 9th ed., Taylor & Francis, Boca Raton, 2010. With permission.)

solids, the return sludge, is recycled to the head of the aeration tank and comes into contact with fresh sewage. The combination of a high concentration of "hungry" cells in the return sludge and a rich food source in the influent sewage provides optimum conditions for the rapid degradation of organic matter. The degradation of organic matter that occurs in an activated sludge facility also occurs in streams and other aquatic environments. However, in general, when a degradable waste is put into a stream, it encounters only a relatively small population of microorganisms capable of carrying out the degradation process. Thus, several days may be required for the buildup of a sufficient population of organisms to degrade the waste. In the activated sludge process, continual recycling of active organisms provides the optimum conditions for waste degradation, and a waste may be degraded within the very few hours that it is present in the aeration tank.

The activated sludge process provides two pathways for the removal of BOD, as illustrated schematically in Figure 8.4. BOD may be removed by (1) oxidation of organic matter to provide energy for the metabolic processes of the microorganisms and (2) synthesis, incorporation of the organic matter into cell mass. In the first pathway, carbon is removed in the gaseous form as CO_2. The second pathway provides for removal of carbon as a solid in biomass. That portion of the carbon converted to CO_2

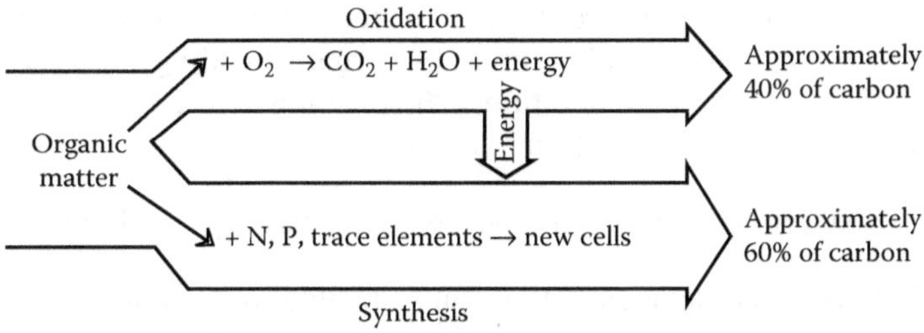

FIGURE 8.4 Pathways for the removal of BOD in biological wastewater treatment. (From Manahan, S., *Environmental Chemistry*, 9th ed., Taylor & Francis, Boca Raton, 2010. With permission.)

is vented to the atmosphere and does not present a disposal problem. The disposal of waste sludge, however, is a problem, primarily because it is only about 1% solids and contains many undesirable components. Normally, partial water removal is accomplished by drying on sand filters, vacuum filtration, or centrifugation. The dewatered sludge may be incinerated or used as landfill. To a certain extent, sewage sludge may be digested in the absence of oxygen by methane-producing anaerobic bacteria to produce methane and carbon dioxide, a process that reduces both the volatile-matter content and the volume of the sludge by about 60%. A carefully designed plant may produce enough methane to provide for all of its power needs:

$$2\{CH_2O\} \rightarrow CH_4 + CO_2 \tag{8.1}$$

One of the most desirable means of sludge disposal is to use it to fertilize and condition soil. However, care has to be taken that excessive levels of heavy metals are not applied to the soil as sludge contaminants. Problems with various kinds of sludges resulting from water treatment are discussed further in Section 8.10.

Activated sludge wastewater treatment is the most common example of an oxic suspended culture process. Many factors must be considered in the design and operation of an activated sludge wastewater treatment system. These include parameters involved with the process modeling and kinetics. The microbiology of the system must be considered. In addition to BOD removal, phosphorus and nitrogen removal must also be taken into account. Oxygen transfer and solids separation are important. Industrial wastes and the fates and effects of industrial chemicals (xenobiotics) must also be considered.

Nitrification (the microbially mediated conversion of ammonium nitrogen to nitrate; see Section 6.11) is a significant process that occurs during biological waste treatment. Ammonium ion is normally the first inorganic nitrogen species produced in the biodegradation of nitrogenous organic compounds. It is oxidized, under the appropriate conditions, first to nitrite by *Nitrosomonas* bacteria, then to nitrate by *Nitrobacter*:

$$2NH_4^+ + 3O_2 \rightarrow 4H^+ + 2NO_2^- + 2H_2O \tag{8.2}$$

$$2NO_2^- + O_2 \rightarrow 2NO_3^- \tag{8.3}$$

These reactions occur in the aeration tank of the activated sludge plant and are favored in general by long retention times, low organic loadings, large amounts of suspended solids, and high temperatures. Nitrification can reduce sludge settling efficiency because the denitrification reaction

$$4NO_3^- + 5\{CH_2O\} + 4H^+ \rightarrow 2N_2(g) + 5CO_2(g) + 7H_2O \quad (8.4)$$

occurring in the oxygen-deficient settler causes bubbles of N_2 to form on the sludge floc (aggregated sludge particles), making it so buoyant that it floats to the top. This prevents settling of the sludge and increases the organic load in the receiving waters. Under the appropriate conditions, however, advantage can be taken of this phenomenon to remove nutrient nitrogen from water (see Section 8.9).

8.4.3 Membrane Bioreactor

A problem with the activated sludge process is the difficulty encountered in settling the suspended biomass. Incomplete separation of the suspended solids in the sludge settling unit can result in solids contamination of the effluent and a dilute sludge that lacks sufficient biomass of active organisms required for effective waste biodegradation. These problems can be overcome with a *membrane bioreactor* in which a suspension of active biomass is maintained in an aeration tank and treated water is withdrawn through a membrane filter (see Section 8.6). In some configurations the membrane is immersed in the aeration chamber and the treated effluent drawn through the membrane filter under vacuum whereas in others the treated effluent is pumped through the filter under pressure.

8.4.4 Tertiary Waste Treatment

Unpleasant as the thought may be, many people drink used water—water that has been discharged from a municipal sewage treatment plant or from some industrial process. This raises serious questions about the presence of pathogenic organisms or toxic substances in such water. Because of high population density and heavy industrial development, the problem is especially acute in Europe where some municipalities process 50% or more of their water from "used" sources. Obviously, there is a great need to treat wastewater in a manner that makes it amenable to reuse. This requires treatment beyond the secondary processes.

Tertiary waste treatment (sometimes called *advanced waste treatment*) is a term used to describe a variety of processes performed on the effluent from secondary waste treatment.[1] The contaminants removed by tertiary waste treatment fall into the general categories of (1) suspended solids, (2) dissolved inorganic materials, and (3) dissolved organic compounds, including the important class of algal nutrients. Low levels of substances and their metabolites, such as pharmaceuticals, synthetic and natural hormones, and personal care products that are discharged in sewage, pose challenges in advanced wastewater treatment. Suspended solids are primarily responsible for residual biological oxygen demand in secondary sewage effluent waters. The dissolved organics are the most hazardous from the standpoint of potential toxicity. The major problem with dissolved inorganic materials is that presented by algal nutrients, primarily nitrates, and phosphates. In addition, potentially hazardous toxic metals may be found among the dissolved inorganics.

In addition to these chemical contaminants, secondary sewage effluent often contains a number of disease-causing microorganisms, requiring disinfection in cases where humans may later come into contact with the water. Among the bacteria that may be found in secondary sewage effluent are organisms causing tuberculosis, dysenteric bacteria (*Bacillus dysenteriae*, *Shigella dysenteriae*, *Shigella paradysenteriae*, *Proteus vulgaris*), cholera bacteria (*Vibriocholerae*), bacteria causing mud fever (*Leptospiraicterohemorrhagiae*), and bacteria causing typhoid fever (*Salmonella typhosa*, *Salmonella paratyphi*). In addition, viruses causing diarrhea, eye infections, infectious hepatitis, and polio may be encountered. Ingestion of sewage still causes disease, even in more developed nations.

8.4.5 Physical–Chemical Treatment of Municipal Wastewater

Complete physical–chemical wastewater treatment systems offer both advantages and disadvantages relative to biological treatment systems. The capital costs of physical–chemical facilities can be less than those of biological treatment facilities, and they usually require less land. They are better able to cope with toxic materials and overloads. However, they require careful operator control and consume relatively large amounts of energy.

Basically, a physical–chemical treatment process involves:

- Removal of scum and solid objects
- Clarification, generally with addition of a coagulant, and frequently with the addition of other chemicals (such as lime for phosphorus removal)
- Filtration to remove filterable solids
- Activated carbon adsorption
- Disinfection

The basic steps of a complete physical–chemical wastewater treatment facility are shown in Figure 8.5.

During the early 1970s, it appeared likely that physical–chemical treatment would largely replace biological treatment. However, higher chemical and energy costs since then have slowed the development of physical–chemical facilities.

8.5 INDUSTRIAL WASTEWATER TREATMENT

Before treatment, industrial wastewater should be characterized fully and the biodegradability of wastewater constituents determined. The options available for the treatment of wastewater are summarized briefly in this section and discussed in greater detail in later sections.

One of two major ways of removing organic wastes is biological treatment by an activated sludge or related process (see Section 8.4 and Figure 8.3). It may be necessary to acclimate microorganisms to the degradation of constituents that are not normally biodegradable. Consideration needs to be given to possible hazards of

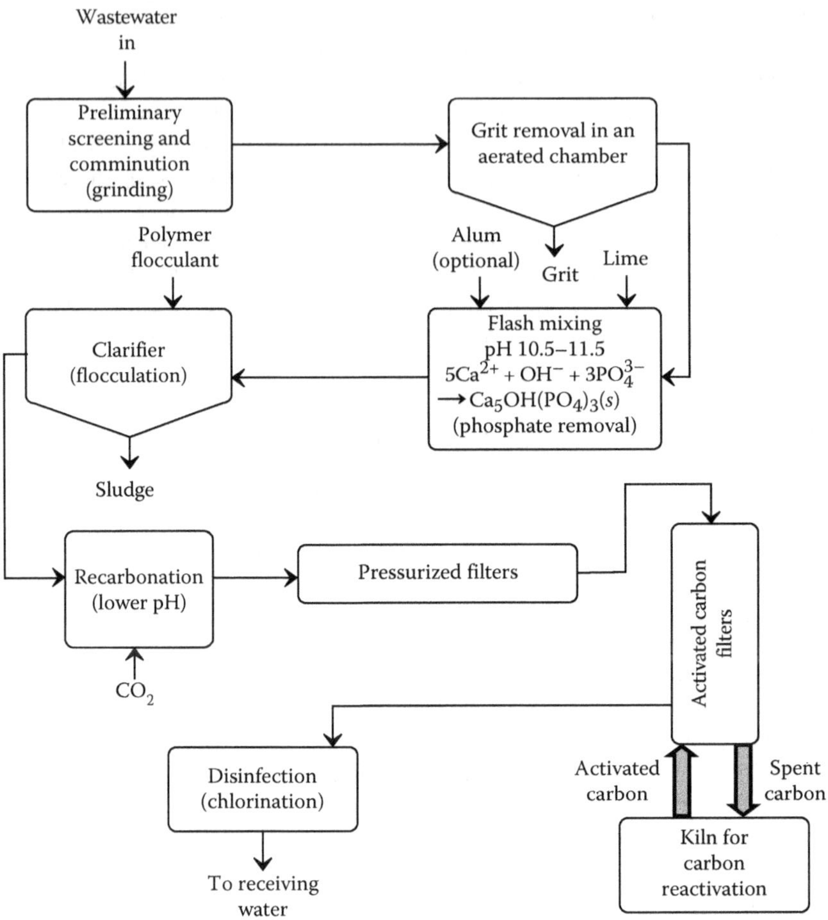

FIGURE 8.5 Major components of a complete physical–chemical treatment facility for municipal wastewater. (From Manahan, S., *Environmental Chemistry*, 9th ed., Taylor & Francis, Boca Raton, 2010. With permission.)

biotreatment sludges, such as those containing excessive levels of heavy metal ions. The other major process for the removal of organics from wastewater is sorption by activated carbon (see Section 8.8), usually in columns of granular activated carbon. Activated carbon and biological treatment can be combined with the use of powdered activated carbon in the activated sludge process. The powdered activated carbon sorbs some constituents that may be toxic to microorganisms and is collected with the sludge. A major consideration with the use of activated carbon to treat wastewater is the hazard that spent activated carbon may present from the wastes it retains. These hazards may include those of toxicity or reactivity, such as those hazards posed by wastes from the manufacture of explosives sorbed to activated carbon. Regeneration of the carbon is expensive and can be hazardous in some cases.

Wastewater can be treated by a variety of chemical processes, including acid/base neutralization, precipitation, and oxidation/reduction. Sometimes these steps must precede biological treatment; for example, acidic or alkaline wastewater must be neutralized in order for microorganisms to thrive in it. Cyanide in the wastewater may be oxidized with chlorine and organics with ozone, hydrogen peroxide promoted with ultraviolet radiation, or DO at high temperatures and pressures. Heavy metals may be precipitated with base, carbonate, or sulfide.

Wastewater can be treated by several physical processes. In some cases, simple density separation and sedimentation can be used to remove water-immiscible liquids and solids. Filtration is frequently required, and flotation by gas bubbles generated on particle surfaces may be useful. Wastewater solutes can be concentrated by evaporation, distillation, and membrane processes, including reverse osmosis, hyperfiltration, and ultrafiltration. Organic constituents can be removed by solvent extraction, air stripping, or steam stripping.

Synthetic resins are useful for removing some pollutant solutes from wastewater. Organophilic resins have proven useful for the removal of alcohols; aldehydes; ketones; hydrocarbons; chlorinated alkanes, alkenes, and aryl compounds; esters, including phthalate esters; and pesticides. Cation exchange resins are effective for the removal of heavy metals.

8.6 REMOVAL OF SOLIDS

Relatively large solid particles are removed from water by simple *settling* and *filtration*. The removal of colloidal solids from water usually requires *coagulation*.[2] Salts of aluminum and iron are the coagulants most often used in water treatment. Of these, alum or filter alum is the most commonly used. This substance is a hydrated aluminum sulfate, $Al_2(SO_4)_3 \cdot 18H_2O$. When this salt is added to water, the aluminum ion hydrolyzes by reactions that consume alkalinity in the water, such as

$$Al(H_2O)_6^{3+} + 3HCO_3^- \rightarrow Al(OH)_3(s) + 3CO_2 + 6H_2O \tag{8.5}$$

The gelatinous hydroxide thus formed carries suspended material with it as it settles. Furthermore, it is likely that positively charged hydroxyl-bridged dimers such as

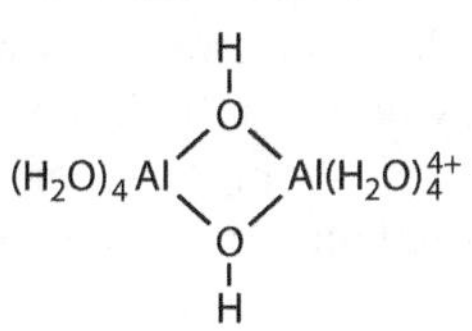

and higher polymers are formed which interact specifically with colloidal particles, bringing about coagulation. Sodium silicate partially neutralized by acid aids coagulation, particularly when used with alum. Metal ions in coagulants also react with virus proteins and destroy viruses in water.

Anhydrous iron(III) sulfate added to water forms iron(III) hydroxide in a reaction analogous to Reaction 8.5. An advantage of iron(III) sulfate is that it works over a wide pH range of approximately 4–11. Hydrated iron(II) sulfate, or copperas, $FeSO_4 \cdot 7H_2O$, is also commonly used as a coagulant. It forms a gelatinous precipitate of hydrated iron(III) oxide; in order to function, it must be oxidized to iron(III) by DO in the water at a pH higher than 8.5, or by chlorine, which can oxidize iron(II) at lower pH values.

Natural and synthetic polyelectrolytes are used in flocculating particles. Among the natural compounds so used are starch and cellulose derivatives, proteinaceous materials, and gums composed of polysaccharides. More recently, selected synthetic polymers, including neutral polymers and both anionic and cationic polyelectrolytes, that are effective flocculants have come into use.

Coagulation–filtration is a much more effective procedure than filtration alone for the removal of suspended material from water. As the term implies, the process consists of the addition of coagulants that aggregate the particles into larger size particles, followed by filtration. Either alum or lime, often with added polyelectrolytes, is most commonly employed for coagulation.

The filtration step of coagulation–filtration is usually performed on a medium such as sand or anthracite coal. Often, to reduce clogging, several media with progressively smaller interstitial spaces are used. One example is the *rapid sand filter*, which consists of a layer of sand supported by layers of gravel particles, the particles becoming progressively larger with increasing depth. The substance that actually filters the water is coagulated material that collects in the sand. As more material is removed, the buildup of coagulated material eventually clogs the filter and must be removed by back-flushing.

An important class of solids that must be removed from wastewater consists of suspended solids in secondary sewage effluent that arise primarily from sludge that was not removed in the settling process. These solids account for a large part of the BOD in the effluent and may interfere with other aspects of tertiary waste treatment, such as by clogging membranes in reverse osmosis water treatment processes. The quantity of material involved may be rather high. Processes designed to remove suspended solids often will remove 10–20 mg/L of organic material from secondary sewage effluent. In addition, a small amount of the inorganic material is removed.

An important class of solids that must be removed from wastewater consists of suspended solids in secondary sewage effluent that arise primarily from sludge that was not removed in the settling process. These solids account for a large part of the BOD in the effluent and may interfere with other aspects of tertiary waste treatment, such as by clogging membranes in reverse osmosis water treatment processes. The quantity of material involved may be rather high. Processes designed to remove suspended solids often will remove 10–20 mg/L of organic material from secondary sewage effluent. In addition, a small amount of the inorganic material is removed.

8.6.1 Dissolved Air Flotation

Many of the particles found in water have low densities close to or even less than that of water. Particles less dense than water have a tendency to rise to the surface from which they can be skimmed off, but this is often a slow and incomplete process. The removal of such particles can be aided by *dissolved air flotation* in which small air bubbles are formed that attach to particles causing them to float. As shown in

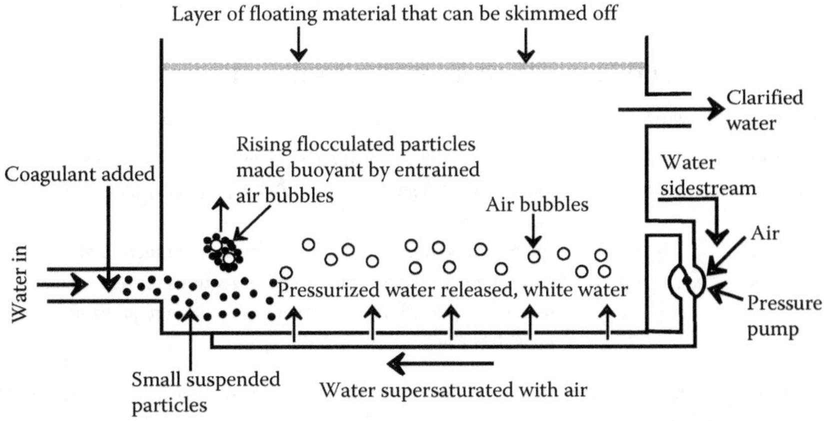

FIGURE 8.6 Illustration of dissolved air flotation in which pressurized water supersaturated with air is released into the bottom of a tank of water producing an abundance of small bubbles giving the water a white or milky appearance. Flocculated particles that entrain air bubbles are buoyant and rise to the top where the flocculated material may be skimmed off. (From Manahan, S., *Environmental Chemistry*, 9th ed., Taylor & Francis, Boca Raton, 2010. With permission.)

Figure 8.6, flotation of particles with air can be accompanied by coagulation of the particles as aided by a coagulant. Water supersaturated with air under pressure is released to the bottom of a tank where bubbles are formed in a layer of milky (white) water. Bubble formation accompanied by flocculation of the particles entrains bubbles in the floc, which floats to the surface where it is skimmed off.

8.6.2 Membrane Filtration Processes

Filtration through membranes under pressure is an especially effective means of removing solids and impurities from water.[3] The purified water that goes through a membrane is the *permeate* and the smaller amount of material that does not go through the membrane is a *retentate*. Membranes normally operate by size exclusion. As membranes with smaller openings are used, smaller particles and even molecules and ions are excluded, but higher pressures are required and more energy is consumed. In order of decreasing pore size, the common membrane processes are *microfiltration* > *ultrafiltration* > *nanofiltration* > and *hyperfiltration*. Microfiltration membranes have pores of 0.1–2 μm in size, and the other processes use membranes with progressively smaller pores. The main types of membrane processes and their uses are summarized in Table 8.1 and reverse osmosis is discussed in Section 8.9.

A problem common to all membrane processes is that posed by the retentate, which is concentrated in the substances that are removed from water. In some cases, this material may be discharged with wastewater, and retentate from the reverse osmosis desalination of seawater may be returned to the sea. Other options, depending upon the source of water treated, include evaporation of water and incineration of the residue, reclamation of materials from some industrial wastewaters, and disposal in deep saline water aquifers.

TABLE 8.1
Major Membrane Processes Used for Water Treatment

Process	Pressure (atm)	Contaminants Removed
Microfiltration	<5	Suspended solids, emulsified components, bacteria, protozoa
Ultrafiltration	2–8	Macromolecules above 5000–100,000 molecular mass (a function of pore size)
Nanofiltration	5–15	Molecules above 200–500 molecular mass (a function of pore size)
Hyperfiltration	15–100	Most solutes and ions; brackish water requires reverse osmosis pressures up to 15 bar; seawater desalination requires pressures up to 100 bar

Source: From Manahan, S., *Environmental Chemistry*, 9th ed., Taylor & Francis, Boca Raton, 2010. With permission.

8.7 REMOVAL OF CALCIUM AND OTHER METALS

Calcium and magnesium salts, which generally are present in water as bicarbonates or sulfates, cause water hardness. One of the most common manifestations of water hardness is the insoluble "curd" formed by the reaction of soap with calcium or magnesium ions. The formation of these insoluble soap salts is discussed in Section 7.10. Although ions that cause water hardness do not form insoluble products with detergents, they do adversely affect detergent performance. Therefore, calcium and magnesium must be complexed or removed from water in order for detergents to function properly.

Another problem caused by hard water is the formation of mineral deposits. For example, when water containing calcium and bicarbonate ions is heated, insoluble calcium carbonate is formed:

$$Ca^{2+} + 2HCO_3^- \rightarrow CaCO_3(s) + CO_2(g) + H_2O \tag{8.6}$$

This product coats the surfaces of hot water systems, clogging pipes, and reducing heating efficiency. Dissolved salts such as calcium and magnesium bicarbonates and sulfates can be especially damaging in boiler feedwater. Clearly, the removal of water hardness is essential for many uses of water.

Several processes are used for softening water. On a large scale, such as in community water-softening operations, the lime-soda process is used. This process involves the treatment of water with lime, $Ca(OH)_2$, and soda ash, Na_2CO_3. Calcium is precipitated as $CaCO_3$ and magnesium as $Mg(OH)_2$. When the calcium is present primarily as "bicarbonate hardness," it can be removed by the addition of $Ca(OH)_2$ alone:

$$Ca^{2+} + 2HCO_3^- + Ca(OH)_2 \rightarrow 2CaCO_3(s) + 2H_2O \tag{8.7}$$

When bicarbonate ion is not present at substantial levels, a source of CO_3^{2-} must be provided at a high enough pH to prevent conversion of most of the CO_3^{2-} to HCO_3^-.

These conditions are obtained by the addition of Na_2CO_3. For example, calcium present as the chloride can be removed from water by the addition of soda ash:

$$Ca^{2+} + 2Cl^- + 2Na^+ + CO_3^{2-} \rightarrow CaCO_3(s) + 2Cl^- + 2Na^+ \quad (8.8)$$

Note that the removal of bicarbonate hardness results in a net removal of soluble salts from solution, whereas removal of nonbicarbonate hardness involves the addition of at least as many equivalents of ionic material as are removed.

The precipitation of magnesium as the hydroxide requires a higher pH than the precipitation of calcium as the carbonate:

$$Mg^{2+} + 2OH^- \rightarrow Mg(OH)_2(s) \quad (8.9)$$

The high pH required may be provided by the basic carbonate ion from soda ash:

$$CO_3^{2-} + H_2O \rightarrow HCO_3^- + OH^- \quad (8.10)$$

Some large-scale lime-soda softening plants make use of the precipitated calcium carbonate product as a source of additional lime. The calcium carbonate is first heated to at least 825°C to produce quicklime, CaO:

$$CaCO_3 + \text{heat} \rightarrow CaO + CO_2(g) \quad (8.11)$$

The quicklime is then slaked with water to produce calcium hydroxide:

$$CaO + H_2O \rightarrow Ca(OH)_2 \quad (8.12)$$

The water softened by lime-soda softening plants usually suffers from two defects. First, because of supersaturation effects, some $CaCO_3$ and $Mg(OH)_2$ usually remain in solution. If not removed, these compounds will precipitate at a later time and cause harmful deposits or undesirable cloudiness in water. The second problem results from the use of highly basic sodium carbonate, which gives the product water an excessively high pH, up to pH 11. To overcome these problems, the water is recarbonated by bubbling CO_2 into it. The carbon dioxide converts the slightly soluble calcium carbonate and magnesium hydroxide to their soluble bicarbonate forms:

$$CaCO_3(s) + CO_2 + H_2O \rightarrow Ca^{2+} + 2HCO_3^- \quad (8.13)$$

$$Mg(OH)_2(s) + 2CO_2 \rightarrow Mg^{2+} + 2HCO_3^- \quad (8.14)$$

The CO_2 also neutralizes excess hydroxide ion:

$$OH^- + CO_2 \rightarrow HCO_3^- \quad (8.15)$$

The pH generally is brought within the range 7.5–8.5 by recarbonation. The source of CO_2 used in the recarbonation process may be from the combustion of carbonaceous fuel. Scrubbed stack gas from a power plant frequently is utilized. Water adjusted to a pH, alkalinity, and Ca^{2+} concentration very close to $CaCO_3$ saturation is labeled *chemically stabilized.* It neither precipitates $CaCO_3$ in water mains, which can clog the pipes, nor dissolves protective $CaCO_3$ coatings from the pipe surfaces. Water with Ca^{2+} concentration much below $CaCO_3$ saturation is called *aggressive* water.

Calcium may be removed from water very efficiently by the addition of orthophosphate:

$$5Ca^{2+} + 3PO_4^{3-} + OH^- \rightarrow Ca_5OH(PO_4)_3(s) \tag{8.16}$$

It should be pointed out that the chemical formation of a slightly soluble product for the removal of undesired solutes such as hardness ions, phosphate, iron, and manganese must be followed by sedimentation in a suitable apparatus. Frequently, coagulants must be added, and filtration employed for complete removal of these sediments.

Water may be purified by ion exchange, the reversible transfer of ions between aquatic solution and a solid material capable of bonding ions. The removal of NaCl from solution by two ion exchange reactions is a good illustration of this process. First the water is passed over a solid cation exchanger in the hydrogen form, represented by $H^{+-}\{Cat(s)\}$:

$$H^{+-}\{Cat(s)\} + Na^+ + Cl^- \rightarrow Na^{+-}\{Cat(s)\} + H^+ + Cl^- \tag{8.17}$$

Next, the water is passed over an anion exchanger in the hydroxide ion form, represented by $OH^{-+}\{An(s)\}$:

$$OH^{-+}\{An(s)\} + H^+ + Cl^- \rightarrow Cl^{-+}\{An(s)\} + H_2O \tag{8.18}$$

Thus, the cations in solution are replaced by hydrogen ion and the anions by hydroxide ion, yielding water as the product.

The softening of water by ion exchange does not require the removal of all ionic solutes, just those cations responsible for water hardness. Generally, therefore, only a cation exchanger is necessary. Furthermore, the sodium rather than the hydrogen form of the cation exchanger is used, and the divalent cations are replaced by sodium ion. Sodium ion at low concentrations is harmless in water to be used for most purposes, and sodium chloride is a cheap and convenient substance with which to recharge the cation exchangers.

A number of materials have ion-exchanging properties. Among the minerals especially noted for their ion exchange properties are the aluminum silicate minerals, or *zeolites.* An example of a zeolite which has been used commercially in water softening is glauconite, $K_2(MgFe)_2Al_6(Si_4O_{10})_3(OH)_{12}$. Synthetic zeolites have been prepared by drying and crushing the white gel produced by mixing solutions of sodium silicate and sodium aluminate.

The discovery in the mid-1930s of synthetic ion exchange resins composed of organic polymers with attached functional groups marked the beginning of modern ion exchange technology. Structural formulas of typical synthetic ion exchangers are shown in Figures 8.7 and 8.8. The cation exchanger shown in Figure 8.7 is called a *strongly acidic cation exchanger* because the parent $-SO_3^-H^+$ group is a strong acid.

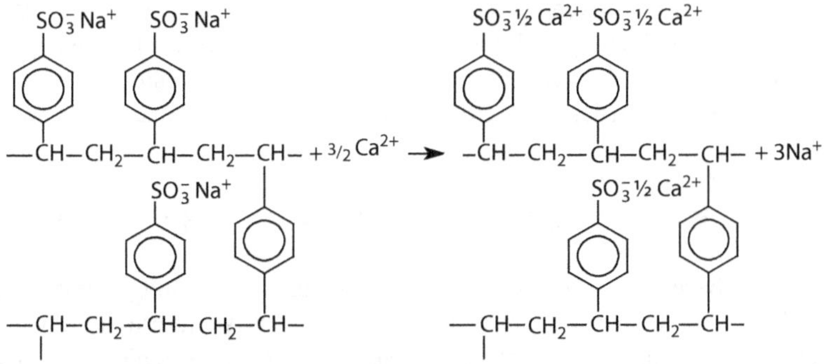

FIGURE 8.7 Strongly acidic cation exchanger. Sodium exchange for calcium in water is shown. (From Manahan, S., *Environmental Chemistry*, 9th ed., Taylor & Francis, Boca Raton, 2010. With permission.)

When the functional group binding the cation is the $-CO_2^-$ group, the exchange resin is called a *weakly acidic cation exchanger*, because the $-CO_2H$ group is a weak acid. Figure 8.8 shows a *strongly basic anion exchanger* in which the functional group is a quaternary ammonium group, $-N^+(CH_3)_3$. In the hydroxide form, $-N^+(CH_3)_3OH^-$, the hydroxide ion is readily released, so the exchanger is classified as *strongly basic*.

The water-softening capability of a cation exchanger is shown in Figure 8.7, where sodium ion on the exchanger is exchanged for calcium ion in solution. The same reaction occurs with magnesium ion. Water softening by cation exchange is a widely used, effective, and economical process. In many areas having a low water flow, however, widespread home water softening by ion exchange causes deterioration of water quality arising from the contamination of wastewater by sodium chloride. Such contamination results from the periodic need to regenerate a water softener with sodium chloride in order to displace calcium and magnesium ions from the resin and replace these hardness ions with sodium ions:

$$Ca^{2+-}\{Cat(s)\}_2 + 2Na^+ + 2Cl^- \rightarrow 2Na^{+-}\{Cat(s)\} + Ca^{2+} + 2Cl^- \quad (8.19)$$

During the regeneration process, a large excess of sodium chloride must be used—several pounds for a home water softener. Appreciable amounts of dissolved sodium chloride can be introduced into sewage by this route.

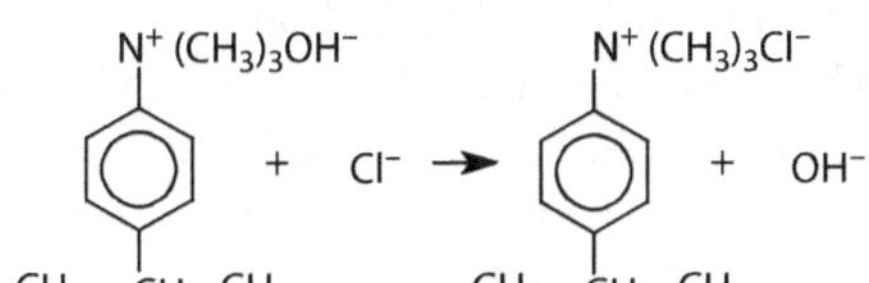

FIGURE 8.8 Strongly basic anion exchanger. Chloride exchange for hydroxide ion is shown. (From Manahan, S., *Environmental Chemistry*, 9th ed., Taylor & Francis, Boca Raton, 2010. With permission.)

Strongly acidic cation exchangers are used for the removal of water hardness. Weakly acidic cation exchangers having the $-CO_2H$ group as a functional group are useful for removing alkalinity. Alkalinity generally is manifested by bicarbonate ion, a species that is a sufficiently strong base to neutralize the acid of a weak acid cation exchanger:

$$2\text{R-CO}_2\text{H} + \text{Ca}^{2+} + 2\text{HCO}_3^- \rightarrow [\text{R-CO}_2^-]_2\text{Ca}^{2+} + 2\text{H}_2\text{O} + 2\text{CO}_2. \qquad (8.20)$$

However, weak bases such as sulfate ion or chloride ion are not strong enough to remove hydrogen ion from the carboxylic acid exchanger. An additional advantage of these exchangers is that they may be regenerated almost stoichiometrically with dilute strong acids, thus avoiding the potential pollution problem caused by the use of excess sodium chloride to regenerate strongly acidic cation changers.

Chelation or, as it is sometimes known, *sequestration*, is an effective method of softening water without actually having to remove calcium and magnesium from solution. A complexing agent is added which greatly reduces the concentrations of free hydrated cations, as shown by some of the example calculations in Chapter 3. For example, chelating calcium ion with excess EDTA anion (Y^{4-}),

$$\text{Ca}^{2+} + \text{Y}^{4-} \rightarrow \text{CaY}^{2-} \qquad (8.21)$$

reduces the concentration of hydrated calcium ion, preventing the precipitation of calcium carbonate:

$$\text{Ca}^{2+} + \text{CO}_3^{2-} \rightarrow \text{CaCO}_3(s) \qquad (8.22)$$

Polyphosphate salts, EDTA, and NTA (see Chapter 3) are chelating agents commonly used for water softening. Polysilicates are used to complex iron.

8.7.1 Removal of Iron and Manganese

Soluble iron and manganese are found in many groundwaters because of reducing conditions which favor the soluble +2 oxidation state of these metals (see Chapter 4). Iron is the more commonly encountered of the two metals. In groundwater, the level of iron seldom exceeds 10 mg/L, and that of manganese is rarely higher than 2 mg/L. The basic method for removing both of these metals depends upon oxidation to higher insoluble oxidation states. The oxidation is generally accomplished by aeration. The rate of oxidation is pH-dependent in both cases, with a high pH favoring more rapid oxidation. The oxidation of soluble Mn(II) to insoluble MnO_2 is a complicated process. It appears to be catalyzed by solid MnO_2, which is known to adsorb Mn(II). This adsorbed Mn(II) is slowly oxidized on the MnO_2 surface.

Chlorine and potassium permanganate are sometimes employed as oxidizing agents for iron and manganese. There is some evidence that organic chelating agents with reducing properties hold iron(II) in a soluble form in water. In such cases, chlorine is effective because it destroys the organic compounds and enables the oxidation of iron(II).

In water with a high level of carbonate, $FeCO_3$ and $MnCO_3$ may be precipitated directly by raising the pH above 8.5 by the addition of sodium carbonate or lime. This approach is less popular than oxidation, however.

Relatively high levels of insoluble iron(III) and manganese(IV) frequently are found in water as colloidal material which is difficult to remove. These metals may be associated with humic colloids or "peptizing" organic material that binds to colloidal metal oxides, stabilizing the colloid.

Heavy metals such as copper, cadmium, mercury, and lead are found in wastewaters from a number of industrial processes. Because of the toxicity of many heavy metals, their concentrations must be reduced to very low levels prior to release of the wastewater. A number of approaches are used in heavy metals removal.

Lime treatment, discussed earlier in this section for calcium removal, precipitates heavy metals as insoluble hydroxides, basic salts, or coprecipitated with calcium carbonate or iron(III) hydroxide. This process does not completely remove mercury, cadmium, or lead, so their removal is aided by addition of sulfide (most heavy metals are sulfide-seekers):

$$Cd^{2+} + S^{2-} \rightarrow CdS\,(s) \tag{8.23}$$

Heavy chlorination is frequently necessary to break down metal-solubilizing ligands. Lime precipitation does not normally permit recovery of metals and is sometimes undesirable from the economic viewpoint.

Electrodeposition (reduction of metal ions to metal by electrons at an electrode), *reverse osmosis* (see Section 8.9), and *ion exchange* are frequently employed for metal removal. Solvent extraction using organic-soluble chelating substances is also effective in removing many metals. *Cementation*, a process by which a metal deposits by reaction of its ion with a more readily oxidized metal, may be employed:

$$Cu^{2+} + Fe \text{ (iron scrap)} \rightarrow Fe^{2+} + Cu \tag{8.24}$$

Activated carbon adsorption effectively removes some metals from water at the part per million level. Sometimes a chelating agent is sorbed to the charcoal to increase metal removal.

Even when not specifically designed for the removal of heavy metals, most waste treatment processes remove appreciable quantities of the more troublesome heavy metals encountered in wastewater. Biological waste treatment effectively removes metals from water. These metals accumulate in the sludge from biological treatment, so sludge disposal must be given careful consideration.

Various physical–chemical treatment processes effectively remove heavy metals from wastewaters. One such treatment is lime precipitation followed by activated-carbon filtration. Activated-carbon filtration may also be preceded by treatment with iron(III) chloride to form an iron(III) hydroxide floc, which is an effective heavy metals scavenger. Similarly, alum, which forms aluminum hydroxide, may be added prior to activated-carbon filtration.

The form of the heavy metal has a strong effect upon the efficiency of metal removal. For instance, chromium(VI) is normally more difficult to remove than chromium(III). Chelation may prevent metal removal by solubilizing metals (see Chapter 3).

In the past, removal of heavy metals has been largely a fringe benefit of wastewater treatment processes. Currently, however, more consideration is being given to design and operating parameters that specifically enhance heavy-metals removal as part of wastewater treatment.

8.8 REMOVAL OF DISSOLVED ORGANICS

Very low levels of exotic organic compounds in drinking water are suspected of contributing to cancer and other maladies. Water disinfection processes, which by their nature involve chemically rather severe conditions, particularly of oxidation, have a tendency to produce *disinfection by-products*. Some of these are chlorinated organic compounds produced by chlorination of organics in water, especially humic substances. Removal of organics to very low levels prior to chlorination has been found to be effective in preventing THM formation. Another major class of disinfection by-products consists of organooxygen compounds such as aldehydes, carboxylic acids, and oxoacids.

A variety of organic compounds survive, or are produced by, secondary wastewater treatment and should be considered as factors in discharge or reuse of the treated water. Almost half of these are humic substances (see Section 3.13) with a molecular-mass range of 1000–5000. Among the remainder are found ether-extractable materials, carbohydrates, proteins, detergents, tannins, and lignins. Because of their high molecular mass and anionic character, the humic compounds influence some of the physical and chemical aspects of waste treatment. The ether extractables contain many of the compounds that are resistant to biodegradation and are of particular concern regarding potential toxicity, carcinogenicity, and mutagenicity. In the ether extract are found many fatty acids, hydrocarbons of the *n*-alkane class, naphthalene, diphenylmethane, diphenyl, methylnaphthalene, isopropylbenzene, dodecylbenzene, phenol, phthalates, and triethylphosphate.

The standard method for the removal of dissolved organic material is adsorption on activated carbon, a product that is produced from a variety of carbonaceous materials including wood, pulp-mill char, peat, and lignite. The carbon is produced by charring the raw material anoxically below 600°C, followed by an activation step consisting of partial oxidation. Carbon dioxide may be employed as an oxidizing agent at 600–700°C.

$$CO_2 + C \rightarrow 2CO \tag{8.25}$$

or the carbon may be oxidized by steam at 800–900°C:

$$H_2O + C \rightarrow H_2 + CO \tag{8.26}$$

These processes develop porosity, increase the surface area, and leave the C atoms in arrangements that have affinities for organic compounds.

Activated carbon comes in two general types: granulated activated carbon, consisting of particles 0.1–1 mm in diameter, and powdered activated carbon, in which most of the particles are 50–100 μm in diameter.

The exact mechanism by which activated carbon holds organic materials is not known. However, one reason for the effectiveness of this material as an adsorbent is its tremendous surface area. A solid cubic foot of carbon particles may have a combined pore and surface area of approximately 10 square miles!

Although interest is increasing in the use of powdered activated carbon for water treatment, currently granular carbon is more widely used. It may be employed in a fixed bed, through which water flows downward. Accumulation of particulate matter requires periodic backwashing. An expanded bed in which particles are kept slightly separated by water flowing upward may be used with less chance of clogging.

Economics require regeneration of the carbon, which is accomplished by heating it to 950°C in a steam–air atmosphere. This process oxidizes adsorbed organics and regenerates the carbon surface, with an approximately 10% loss of carbon.

Removal of organics may also be accomplished by adsorbent synthetic polymers. Such polymers as Amberlite XAD-4 have hydrophobic surfaces and strongly attract relatively insoluble organic compounds, such as chlorinated pesticides. The porosity of these polymers is up to 50% by volume, and the surface area may be as high as 850 m^2/g. They are readily regenerated by solvents such as isopropanol and acetone. Under appropriate operating conditions, these polymers remove virtually all nonionic organic solutes; for example, phenol at 250 mg/L is reduced to less than 0.1 mg/L by appropriate treatment with Amberlite XAD-4.

Oxidation of dissolved organics holds some promise for their removal. Ozone, hydrogen peroxide, molecular oxygen (with or without catalysts), chlorine and its derivatives, permanganate, or ferrate [iron(VI)] can be used. Electrochemical oxidation may be possible in some cases. High-energy electron beams produced by high-voltage electron accelerators also have the potential to destroy organic compounds.

8.8.1 Removal of Herbicides

Because of their widespread application and persistence, herbicides have proven to be particularly troublesome in drinking water sources. Their levels vary with season related to times that they are applied to control weeds. The more soluble ones are most likely to enter drinking water sources. One of the most troublesome is atrazine, which is often manifested by its metabolite desethylatrazine. Activated carbon treatment is the best means of removing herbicides and their metabolites from drinking water sources. A problem with activated carbon is that of *preloading*, in which natural organic matter in the water loads up the carbon and hinders uptake of pollutant organics such as herbicides. Pretreatment to remove such organic matter, such as flocculation and precipitation of humic substances, can significantly increase the efficacy of activated carbon for the removal of herbicides and other organics.

8.9 REMOVAL OF DISSOLVED INORGANICS

In order for complete water recycling to be feasible, inorganic-solute removal is essential. The effluent from secondary waste treatment generally contains 300–400 mg/L more dissolved inorganic material than does the municipal water supply. It is obvious, therefore, that 100% water recycling without removal of inorganics would cause the accumulation of an intolerable level of dissolved material. Even when water is not destined for immediate reuse, the removal of the inorganic nutrients phosphorus and nitrogen is highly desirable to reduce eutrophication downstream. In some cases the removal of toxic trace metals is needed.

One of the most obvious methods for removing inorganics from water is distillation. However, distillation generally uses a lot of energy adding substantially to the cost. Furthermore, volatile materials such as ammonia and odorous compounds are carried over to a large extent in the distillation process unless special preventative measures are taken. Freezing produces a very pure water, but is considered uneconomical with present technology. Ion exchange and membrane processes are the most cost-effective means of removing inorganic materials from water. These processes are discussed below.

8.9.1 Ion Exchange

The ion exchange method for softening water is described in detail in Section 8.7. The ion exchange process used for removal of inorganics consists of passing the water successively over a solid cation exchanger and a solid anion exchanger, which replace cations and anions by hydrogen ion and hydroxide ion, respectively, so that each equivalent of salt is replaced by a mole of water. For the hypothetical ionic salt MX, the reactions are the following where $^{-}\{\text{Cat}(s)\}$ represents the solid cation exchanger and $^{+}\{\text{An}(s)\}$ represents the solid anion exchanger:

$$H^{+-}\{\text{Cat}(s)\} + M^{+} + X^{-} \rightarrow M^{+-}\{\text{Cat}(s)\} + H^{+} + X^{-} \tag{8.27}$$

$$OH^{-+}\{\text{An}(s)\} + H^{+} + X^{-} \rightarrow X^{-+}\{\text{An}(s)\} + H_2O \tag{8.28}$$

The cation exchanger is regenerated with strong acid and the anion exchanger with strong base.

Demineralization by ion exchange generally produces water of a very high quality. Unfortunately, some organic compounds in wastewater foul ion exchangers, and microbial growth on the exchangers can diminish their efficiency. In addition, regeneration of the resins is expensive, and the concentrated wastes from regeneration require disposal in a manner that will not damage the environment.

8.9.2 Electrodialysis

Electrodialysis consists of applying a direct current across a body of water separated into layers by membranes alternately permeable to cations and anions. Cations migrate toward the cathode and anions toward the anode. Cations and anions both enter one layer of water, and both leave the adjacent layer. Thus, layers of water enriched in salts alternate with those from which salts have been removed. The water in the brine-enriched layers is recirculated to a certain extent to prevent excessive accumulation of brine. The principles involved in electrodialysis treatment are shown in Figure 8.9.

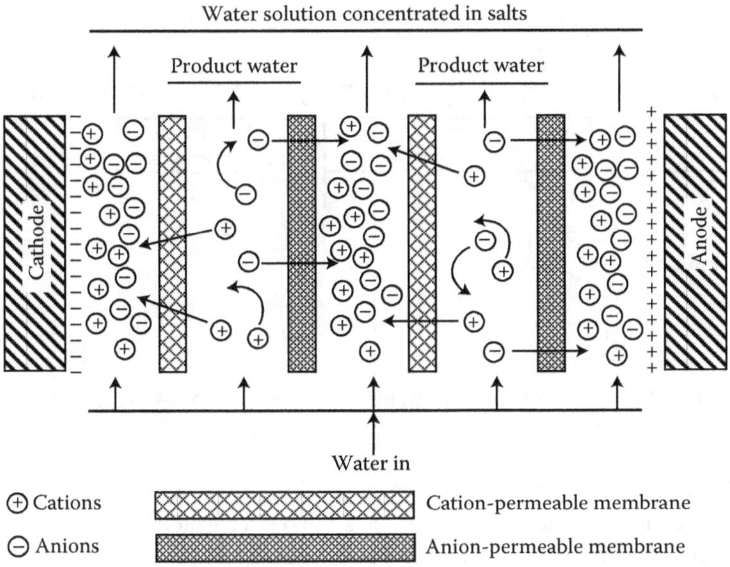

FIGURE 8.9 Electrodialysis apparatus for the removal of ionic material from water. (From Manahan, S., *Environmental Chemistry*, 9th ed., Taylor & Francis, Boca Raton, 2010. With permission.)

Fouling caused by various materials can cause problems with reverse osmosis treatment of water. Although the relatively small ions constituting the salts dissolved in wastewater readily pass through the membranes, large organic ions (proteins, for example) and charged colloids migrate to the membrane surfaces, often fouling or plugging the membranes and reducing efficiency. In addition, growth of microorganisms on the membranes can cause fouling.

Experience with pilot plants indicates that electrodialysis has the potential to be a practical and economical method to remove up to 50% of the dissolved inorganics from secondary sewage effluent after pretreatment to eliminate fouling substances. Such a level of efficiency would permit repeated recycling of water without dissolved inorganic materials reaching unacceptably high levels.

8.9.3 Reverse Osmosis

Reverse osmosis is one of several pressure-driven membrane processes for water purification which also include nonfiltration, ultrafiltration, and microfiltration (Section 8.6). Reverse osmosis is a very useful and well-developed technique for the purification and desalination of water. As illustrated in Figure 8.10, it consists of forcing pure water through a semipermeable membrane that allows the passage of water but not of other material. This process, which is not simply sieve separation or ultrafiltration, depends on the preferential sorption of water on the surface of a porous cellulose acetate or polyamide membrane. Pure water from the sorbed layer is forced through pores in the membrane under pressure. If the thickness of the sorbed water

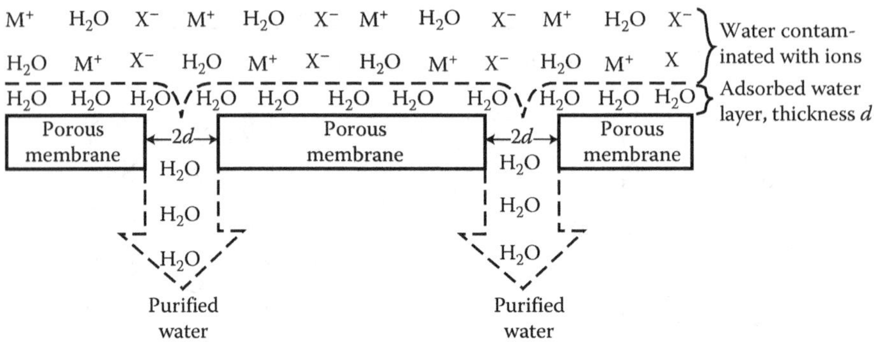

FIGURE 8.10 Solute removal from water by reverse osmosis. (From Manahan, S., *Environmental Chemistry*, 9th ed., Taylor & Francis, Boca Raton, 2010. With permission.)

layer is *d*, the pore diameter for optimum separation should be 2*d*. The optimum pore diameter depends upon the thickness of the sorbed pure water layer and may be several times the diameters of the solute and solvent molecules.

In an example of recycling that is in keeping with the practice of green chemistry, spent reverse osmosis membranes have been treated with potassium permanganate, which removes substances that clog the membranes, but also removes the salt-rejecting surface layer, drastically reducing their salt rejection properties. The treated membranes are then used for filtration to remove as much as 94% of suspended solids from water.

Nanofiltration is a pressurized membrane filtration process that does not remove singly charged salt ions that are removed by osmosis, but may be effective in removing hardness (Ca^{2+}). Nanofiltration operates at lower pressures than reverse osmosis, thus requiring less energy and less expense. Recently it has gained popularity as a drinking water treatment process.

8.9.4 Phosphorus Removal

Advanced waste treatment normally requires removal of phosphorus to reduce algal growth. Algae may grow at PO_4^{3-} levels as low as 0.05 mg/L. Growth inhibition requires levels well below 0.5 mg/L. Since municipal wastes typically contain approximately 25 mg/L of phosphate (as orthophosphates, polyphosphates, and insoluble phosphates), the efficiency of phosphate removal must be quite high to prevent algal growth. This removal may occur in the sewage treatment process (1) in the primary settler, (2) in the aeration chamber of the activated sludge unit, or (3) after secondary waste treatment.

Activated sludge treatment removes about 20% of the phosphorus from sewage. Thus, an appreciable fraction of largely biological phosphorus is removed with the sludge. Detergents and other sources contribute significant amounts of phosphorus to domestic sewage and considerable phosphate ion remains in the effluent. However, some wastes, such as carbohydrate wastes from sugar refineries, are so deficient in phosphorus that supplementation of the waste with inorganic phosphorus is required for proper growth of the microorganisms degrading the wastes.

Under some sewage plant operating conditions, much greater than normal phosphorus removal has been observed. In such plants, characterized by high DO and high pH levels in the aeration tank, removal of 60–90% of the phosphorus has been attained, yielding two or three times the normal level of phosphorus in the sludge. In a conventionally operated aeration tank of an activated sludge plant, the CO_2 level is relatively high because of the release of the gas by the degradation of organic material. A high CO_2 level results in a relatively low pH, because of the weakly acidic nature of dissolved CO_2 in water. The aeration rate is generally not maintained at a very high level because oxygen is transferred relatively more efficiently from air when the DO levels in water are relatively low. Therefore, the aeration rate normally is not high enough to sweep out sufficient dissolved carbon dioxide to bring its concentration down to low levels. Thus, the pH generally is low enough that phosphate is maintained primarily in the form of the $H_2PO_4^-$ ion. However, at a higher rate of aeration in a relatively hard water, the CO_2 is swept out, the pH rises, and reactions such as the following occur:

$$5Ca^{2+} + 3HPO_4^{2-} + 2H_2O \rightarrow Ca_5OH(PO_4)_3(s) + 4H^+ \quad (8.29)$$

The precipitated hydroxyapatite or other form of calcium phosphate is incorporated in the sludge floc. Reaction 8.29 is strongly hydrogen ion-dependent, and an increase in the hydrogen ion concentration drives the equilibrium back to the left. Thus, under anoxic conditions when the sludge medium becomes more acidic due to higher CO_2 levels, the phosphate returns to solution.

Chemically, phosphate is most commonly removed by precipitation. Some common precipitants and their products are shown in Table 8.2. Precipitation processes are capable of at least 90–95% phosphorus removal at reasonable cost. Lime has the advantages of low cost and ease of regeneration. The efficiency with which phosphorus is removed by lime is not as high as would be predicted by the low solubility of hydroxyapatite, $Ca_5OH(PO_4)_3$. Some of the possible reasons for this are slow precipitation of $Ca_5OH(PO_4)_3$, formation of nonsettling colloids; precipitation of calcium as $CaCO_3$ in certain pH ranges, and the fact that phosphate may be present as condensed phosphates (polyphosphates) which form soluble complexes with calcium ion.

TABLE 8.2
Chemical Precipitants for Phosphate and Their Products

Precipitant(s)	Product
$Ca(OH)_2$	$Ca_5OH(PO_4)_3$ (hydroxyapatite)
$Ca(OH)_2 + NaF$	$Ca_5F(PO_4)_3$ (fluorapatite)
$Al_2(SO_4)_3$	$AlPO_4$
$FeCl_3$	$FePO_4$
$MgSO_4$	$MgNH_4PO_4$

Source: From Manahan, S., *Environmental Chemistry*, 9th ed., Taylor & Francis, Boca Raton, 2010. With permission.

Lime, $Ca(OH)_2$, is the chemical most commonly used for phosphorus removal:

$$5Ca(OH)_2 + 3HPO_4^{2-} \rightarrow Ca_5OH(PO_4)_3(s) + 3H_2O + 6OH^- \qquad (8.30)$$

Phosphate can be removed from solution by adsorption on some solids, particularly activated alumina, Al_2O_3. Removals of up to 99.9% of orthophosphate have been achieved with this method.

8.9.5 Nitrogen Removal

Next to phosphorus, nitrogen is the algal nutrient most commonly removed as part of advanced wastewater treatment. Numerous chemical, biological, ion exchange, and membrane techniques can be used to remove nitrogen from water. Mineralization of organic nitrogen in wastewater treatment first produces ammonium ion, NH_4^+. The most common chemical process for nitrogen removal from wastewater is reaction of nitrogen in the form of ammonium ion with hypochlorite (from chlorine) to produce elemental nitrogen that is evolved as a gas:

$$NH_4^+ + HCOl \rightarrow NH_2Cl + H_2O + H^+ \qquad (8.31)$$

$$2NH_2Cl + HOCl \rightarrow N_2(g) + 3H^+ + 3Cl^- + H_2O \qquad (8.32)$$

Nitrification followed by denitrification is the most common biological means of removing inorganic nitrogen from wastewater and arguably the most effective technique. The first step is an essentially complete conversion of ammonia and organic nitrogen to nitrate under strongly aerobic conditions, achieved by more extensive than normal aeration of the sewage:

$$NH_4^+ + 2O_2 \text{ (nitrifying bacteria)} \rightarrow NO_3^- + 2H^+ + H_2O \qquad (8.33)$$

The second step is the reduction of nitrate to nitrogen gas. This reaction is also bacterially catalyzed and requires a carbon source and a reducing agent such as methanol, CH_3OH.

$$6NO_3^- + 5CH_3OH + 6H^+ \text{ (denitrifying bacteria)} \rightarrow 3N_2(g) + 5CO_2 + 13H_2O \qquad (8.34)$$

The denitrification process may be carried out either in a tank or on a carbon column. In pilot plant operation, conversions of 95% of the ammonia to nitrate and 86% of the nitrate to nitrogen have been achieved.

8.10 SLUDGE

Perhaps the most pressing water treatment problem at this time has to do with sludge collected or produced during water treatment. Finding a safe place to put the sludge or a use for it has proven troublesome, and the problem is aggravated by the large number of water treatment systems.

Some sludge is present in wastewater prior to treatment and may be collected from it. Such sludge includes human wastes, garbage grindings, organic wastes, and inorganic silt and grit from storm water runoff, and organic and inorganic wastes from commercial and industrial sources. There are two major kinds of sludge generated in a waste treatment plant. The first of these is organic sludge from activated sludge, trickling filter, or rotating biological reactors. The second is inorganic sludge from the addition of chemicals, such as in phosphorus removal (see Section 8.9).

Most commonly, sewage sludge is subjected to anaerobic digestion in a digester designed to allow bacterial action to occur in the absence of air. This reduces the mass and volume of sludge and ideally results in the formation of a stabilized humus material. Disease agents are also destroyed in the process.

Following digestion, sludge is generally conditioned and thickened to concentrate and stabilize it and make it more amenable to removal of water. Relatively inexpensive processes, such as gravity thickening, may be employed to get the moisture content down to about 95%. Sludge may be further conditioned chemically by the addition of iron or aluminum salts, lime, or polymers.

Sludge dewatering is employed to convert the sludge from an essentially liquid material to a damp solid containing not more than about 85% water. This may be accomplished on sludge drying beds consisting of layers of sand and gravel. Mechanical devices may also be employed, including vacuum filtration, centrifugation, and filter presses. Heat may be used to aid the drying process.

Ultimately, disposal of the sludge is required. Until banned by the U.S. Congress in the late 1980s, ocean disposal of sewage sludge was commonly practiced. Currently, the two main alternatives for sludge disposal are land spreading and incineration.

Rich in nutrients, waste sewage sludge contains around 5% N, 3% P, and 0.5% K on a dry-mass basis and can be used to fertilize and condition soil. The humic material in the sludge improves the physical properties and CEC of the soil. Possible accumulation of heavy metals is of some concern insofar as the use of sludge on cropland is concerned. Sewage sludge is an efficient heavy metals scavenger and may contain elevated levels of zinc, copper, nickel, and cadmium. These and other metals tend to remain immobilized in soil by chelation with organic matter, adsorption on clay minerals, and precipitation as insoluble compounds such as oxides or carbonates. However, increased application of sludge on cropland has caused distinctly elevated levels of zinc and cadmium in both leaves and grain of corn. Therefore, caution has been advised in heavy or prolonged application of sewage sludge to soil. Prior control of heavy metal contamination from industrial sources has greatly reduced the heavy metal content of sludge and enabled it to be used more extensively on soil.

A significant problem in sewage treatment can arise from sludge sidestreams. These consist of water removed from sludge by various treatment processes. Sewage treatment processes can be divided into mainstream treatment processes (primary clarification, trickling filter, activated sludge, and rotating biological reactor) and sidestream processes. During sidestream treatment, sludge is dewatered, degraded, and disinfected by a variety of processes, including gravity thickening, dissolved air flotation, anaerobic (anoxic) digestion, oxic digestion, vacuum filtration, centrifugation, belt-filter press filtration, sand-drying-bed treatment, sludge-lagoon settling, wet air oxidation, and pressure filtration. Each of these produces a liquid by-product sidestream which is circulated back to the mainstream. These add to the BOD and suspended solids of the mainstream.

A variety of chemical sludges are produced by various water treatment and industrial processes. Among the most abundant of such sludges is alum sludge produced by the hydrolysis of Al(III) salts used in the treatment of water, which creates gelatinous aluminum hydroxide:

$$Al^{3+} + 3OH^-(aq) \rightarrow Al(OH)_3(s) \tag{8.35}$$

Alum sludges containing aluminum hydroxide normally are 98% or more water and are very difficult to dewater.

Both iron(II) and iron(III) compounds are used for the removal of impurities from wastewater by precipitation of $Fe(OH)_3$. The sludge contains $Fe(OH)_3$ in the form of soft, fluffy precipitates that are difficult to dewater beyond 10% or 12% solids.

The addition of either lime, $Ca(OH)_2$, or quicklime, CaO, to water is used to raise the pH to about 11.5 and cause the precipitation of $CaCO_3$, along with metal hydroxides and phosphates. Calcium carbonate is readily recovered from lime sludges and can be recalcined to produce CaO, which can be recycled through the system.

Metal hydroxide sludges are produced in the removal of metals such as lead, chromium, nickel, and zinc from wastewater by raising the pH to such a level that the corresponding hydroxides or hydrated metal oxides are precipitated. The disposal of these sludges is a substantial problem because of their toxic heavy metal content. Reclamation of the metals is a possible green alternative for these sludges.

Pathogenic (disease-causing) microorganisms may persist in the sludge left from the treatment of sewage. These include bacteria such as *Enterobacter* and *Shigella*, viruses including hepatitis and enteroviruses, protozoa such as *Entamoeba* and *Giardia*, and helminth worms such as *Ascaris* and *Toxocara*. Many of these organisms present potential health hazards, and there is risk of public exposure when the sludge is applied to soil. Therefore, it is necessary both to be aware of pathogenic microorganisms in municipal wastewater treatment sludge and to find a means of reducing the hazards caused by their presence.

The most significant organisms in municipal sewage sludge include (1) indicators of fecal pollution, including fecal and total coliform, (2) pathogenic bacteria, including *Salmonellae* and *Shigellae*, (3) enteric (intestinal) viruses, including enterovirus and poliovirus, and (4) parasites, such as *Entamoeba histolytica* and *Ascaris lumbricoides.*

Several ways are recommended to significantly reduce levels of pathogens in sewage sludge. Oxic digestion involves air agitation of the sludge for periods of 40–60 days (longer times are employed with low sludge temperatures). Air drying involves draining and/or drying of the liquid sludge for at least 3 months in a layer 20–25 cm thick. This operation may be performed on underdrained sand beds or in basins. Anaerobic digestion involves maintenance of the sludge in an anaerobic state for periods of time ranging from 60 days at 20°C to 15 days at temperatures exceeding 35°C. Composting involves mixing dewatered sludge cake with bulking agents subject to decay, such as wood chips or shredded municipal refuse, and allowing the action of bacteria to promote decay at temperatures ranging up to 45–65°C. The higher temperatures tend to kill pathogenic bacteria. Finally, pathogenic organisms may be destroyed by lime stabilization in which sufficient lime is added to raise the pH of the sludge to 12 or higher.

8.11 WATER DISINFECTION

Chlorine is the most commonly used disinfectant employed for killing bacteria in water. When chlorine is added to water, it rapidly hydrolyzes according to the reaction

$$Cl_2 + H_2O \rightarrow H^+ + Cl^- + HOCl \tag{8.36}$$

which has the following equilibrium constant:

$$K = \frac{[H^+][Cl^-][HOCl]}{[Cl_2]} = 4.5 \times 10^{-4}. \tag{8.37}$$

Hypochlorous acid, HOCl, is a weak acid that dissociates according to the reaction,

$$HOCl \rightleftarrows H^+ + OCl^- \tag{8.38}$$

with an ionization constant of 2.7×10^{-8}. From the above it can be calculated that the concentration of elemental Cl_2 is negligible at equilibrium above pH 3 when chlorine is added to water at levels below 1.0 g/L.

Sometimes, hypochlorite salts are substituted for chlorine gas as a disinfectant. Calcium hypochlorite, $Ca(OCl)_2$, is commonly used. The hypochlorites are safer to handle than gaseous chlorine.

The two chemical species formed by chlorine in water, HOCl and OCl^-, are known as *free available chlorine*. Free available chlorine is very effective in killing bacteria. In the presence of ammonia, monochloramine, dichloramine, and trichloramine are formed:

$$NH_4^+ + HOCl \rightarrow NH_2Cl \text{ (monochloramine)} + H_2O + H^+ \tag{8.39}$$

$$NH_2Cl + HOCl \rightarrow NHCl_2 \text{ (dichloramine)} + H_2O \tag{8.40}$$

$$NHCl_2 + HOCl \rightarrow NCl_3 \text{ (trichloramine)} + H_2O \tag{8.41}$$

The chloramines are called *combined available chlorine*. Chlorination practice frequently provides for formation of combined available chlorine which, although a weaker disinfectant than free available chlorine, is more readily retained as a disinfectant throughout the water distribution system. Too much ammonia in water is considered undesirable because it exerts excess demand for chlorine.

At sufficiently high Cl:N ratios in water containing ammonia, some HOCl and OCl^- remain unreacted, and a small quantity of NCl_3 is formed. The ratio at which this occurs is called the *breakpoint*. Chlorination beyond the breakpoint ensures disinfection. It has the additional advantage of destroying the more common materials that cause odor and taste in water.

At moderate levels of NH_3-N (approximately 20 mg/L), when the pH is between 5.0 and 8.0, chlorination with a minimum 8:1 mass ratio of Cl to NH_3-nitrogen produces efficient denitrification:

$$NH_4^+ + HOCl \rightarrow NH_2Cl + H_2O + H^+ \tag{8.39}$$

$$2NH_2Cl + HOCl \rightarrow N_2\,(g) + 3H^+ + 3Cl^- + H_2O \tag{8.42}$$

This reaction is used to remove pollutant ammonia from wastewater. However, problems can arise from chlorination of organic wastes. Typical of such by-products is chloroform, produced by the chlorination of humic substances in water.

Chlorine is used to treat water other than drinking water. It is employed to disinfect effluent from sewage treatment plants, as an additive to the water in electric power plant cooling towers, and to control microorganisms in food processing.

8.11.1 Chlorine Dioxide

Chlorine dioxide, ClO_2, is an effective water disinfectant that is of particular interest because, in the absence of impurity Cl_2, it does not produce impurity THMs in water treatment. In acidic and neutral water, respectively, the two half-reactions for ClO_2 acting as an oxidant are the following:

$$ClO_2 + 4H^+ + 5e^- \rightleftarrows Cl^- + 2H_2O \tag{8.43}$$

$$ClO_2 + e^- \rightleftarrows ClO_2^- \tag{8.44}$$

In the neutral pH range, chlorine dioxide in water remains largely as molecular ClO_2 until it contacts a reducing agent with which to react. Chlorine dioxide is a gas that is violently reactive with organic matter and explosive when exposed to light. For these reasons, it is not shipped, but is generated on-site by processes such as the reaction of chlorine gas with solid sodium hypochlorite:

$$2NaClO_2\,(s) + Cl_2\,(g) \rightleftarrows 2ClO_2\,(g) + 2NaCl\,(s) \tag{8.45}$$

A high content of elemental chlorine in the product may require its purification to prevent unwanted side reactions from Cl_2.

As a water disinfectant, chlorine dioxide does not chlorinate or oxidize ammonia or other nitrogen-containing compounds. Some concern has been raised over possible health effects of its main degradation by-products, ClO_2^- and ClO_3^-.

8.11.2 Ozone and Other Oxidants

Ozone is sometimes used as a disinfectant in place of chlorine, particularly in Europe. Figure 8.11 shows the main components of an ozone water treatment system.

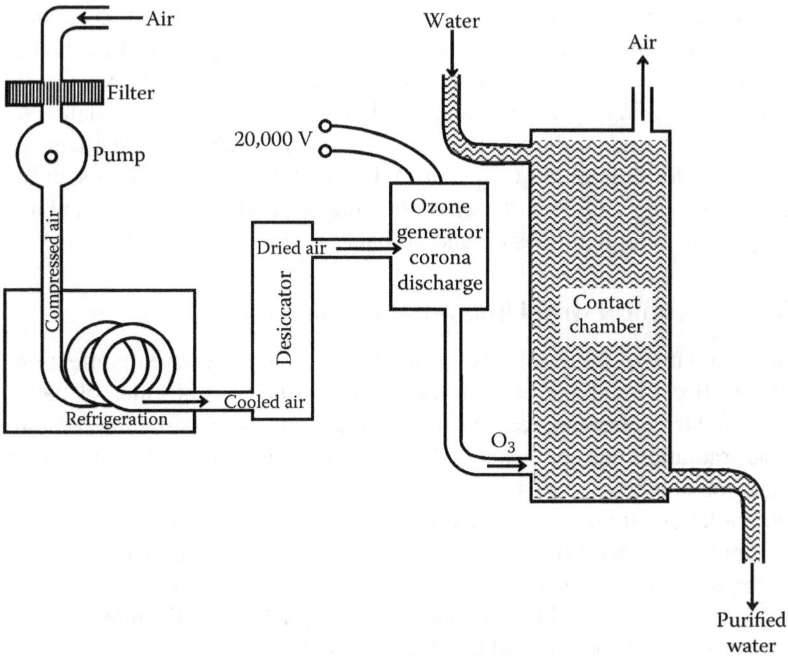

FIGURE 8.11 Schematic diagram of a typical ozone water-treatment system. (From Manahan, S., *Environmental Chemistry*, 9th ed., Taylor & Francis, Boca Raton, 2010. With permission.)

Basically, air is filtered, cooled, dried, and pressurized, then subjected to an electrical discharge of approximately 20,000 V. The ozone produced is then pumped into a contact chamber where water contacts the ozone for 10–15 min. Concern over possible production of toxic organochlorine compounds by water chlorination processes has increased interest in ozonation. Furthermore, ozone is more destructive to viruses than is chlorine. Unfortunately, the solubility of ozone in water is relatively low, which limits its disinfective power.

Ozone oxidizes water contaminants directly through the reaction of O_3 and indirectly by generating hydroxyl radical, HO•, a reactive strong oxidant. A major consideration with ozone is the rate at which it decomposes spontaneously in water, according to the overall reaction,

$$2O_3 \rightarrow 3O_2\,(g) \tag{8.46}$$

Because of the decomposition of ozone in water, some chlorine must be added to maintain disinfectant throughout the water distribution system.

Iron(VI) in the form of ferrate ion, FeO_4^{2-}, is a strong oxidizing agent with excellent disinfectant properties. It has the additional advantage of removing heavy metals, viruses, and phosphate. It may well find increased application for disinfection in the future.[4]

Another oxidant that is becoming widely used as a cleaning and bleaching agent in detergents and that, because of its oxidizing ability, probably has disinfecting abilities as well, is sodium percarbonate. This solid, commonly called *percarbonate*, is produced by mixing hydrogen peroxide, H_2O_2, with sodium carbonate, Na_2CO_3, along with $MgSO_4$ and Na_2SiO_3 stabilizers to produce a stable solid with an approximate empirical formula of $Na_2CO_3 \cdot 3H_2O_2$ in which the H_2O_2 may be stabilized by hydrogen bonding to CO_3^{2-} ion. In water, this material releases oxygen and acts as a bleaching agent, detergent booster, and probably bactericide as well.

8.11.3 Disinfection with Ultraviolet Radiation

Ultraviolet radiation, commonly represented as $h\nu$, is an effective disinfection agent that acts by direct breakage of chemical bonds in the biomolecules of pathogenic organisms and by generating reactive ions and radicals that destroy microorganisms. Ultraviolet radiation may be generated by high-powered mercury vapor lamps and solar radiation can be used as a source of ultraviolet radiation. Because it is reagent-free, ultraviolet radiation is a particularly green means of water disinfection. It works well in combination with other disinfection agents, such as chlorine or ozone, and reduces the quantities of such agents required. Studies are underway on the use for water disinfection of titanium dioxide, TiO_2, a photocatalytic material that can destroy pathogens when activated by ultraviolet radiation.

8.12 NATURAL WATER PURIFICATION PROCESSES

Virtually all of the materials that waste treatment processes are designed to eliminate may be absorbed by soil or degraded in soil.[5] In fact, most of these materials are essential for soil fertility. Wastewater may provide the water that is essential to plant growth, in addition to the nutrients—phosphorus, nitrogen, and potassium—usually provided by fertilizers. Wastewater also contains essential trace elements and vitamins. Stretching the point a bit, the degradation of organic wastes provides the CO_2 essential for photosynthetic production of plant biomass.

Soil may be viewed as a natural filter for wastes. Most organic matter is readily degraded in soil and, in principle, soil constitutes an excellent primary, secondary, and tertiary treatment system for water. Soil has physical, chemical, and biological characteristics that can enable wastewater detoxification, biodegradation, chemical decomposition, and physical and chemical fixation. A number of soil characteristics are important in determining its use for land treatment of wastes. These characteristics include physical form, ability to retain water, aeration, organic content, acid–base characteristics, and oxidation–reduction behavior. Soil is a natural medium for a number of living organisms that may have an effect upon biodegradation of wastewaters, including those that contain industrial wastes. Of these, the most important are bacteria, including those from the genera *Agrobacterium*, *Arthrobacteri*, *Bacillus*, *Flavobacterium*, and *Pseudomonas*. Actinomycetes and fungi are important in decay of vegetable matter and may be involved in biodegradation of wastes. Other unicellular organisms that may be present in or on soil are protozoa and algae. Soil animals, such as earthworms, affect soil parameters such as soil texture. The growth of plants in soil may have an influence on its waste treatment potential in such aspects as uptake of soluble wastes and erosion control.

Early civilizations, such as the Chinese, used human organic wastes to increase soil fertility, and the practice continues today. The ability of soil to purify water was noted well over a century ago. In 1850 and 1852, J. Thomas Way, a consulting chemist to the Royal Agricultural Society in England, presented two papers to the Society entitled "Power of Soils to Absorb Manure." Way's experiments showed that soil is an ion exchanger. Much practical and theoretical information on the ion exchange process resulted from his work.

If soil treatment systems are not properly designed and operated, odor can become an overpowering problem. The author of this book is reminded of driving into a small town, recalled from some years before as a very pleasant place, and being assaulted with a virtually intolerable odor. The disgruntled residents pointed to a large spray irrigation system on a field in the distance—unfortunately upwind—spraying liquified pig manure as part of an experimental feedlot waste treatment operation. The experiment was not deemed a success and was discontinued by the investigators, presumably before they met with violence from the local residents.

8.12.1 Industrial Wastewater Treatment by Soil

Wastes that are amenable to land treatment are biodegradable organic substances, particularly those contained in municipal sewage and in wastewater from some industrial operations, such as food processing. However, through acclimation over a long period of time, soil bacterial cultures may develop that are effective in degrading normally recalcitrant compounds that occur in industrial wastewater. Acclimated microorganisms are found particularly at contaminated sites, such as those where soil has been exposed to crude oil for many years.

Soil used for land treatment must meet a number of criteria. It should be at least a meter thick because wastes that reach underlying rock strata tend to degrade poorly. The soil slope should be moderate to reduce runoff. The water table should lie at a sufficient depth to avoid leaching into groundwater. Treatment sites should be located at appropriate distances from dwellings or other anthrospheric constructs that could be adversely affected by exposure to wastes. Shallow cultivation of soil to which wastes have been applied promotes degradation by exposure to atmospheric oxygen.

Land treatment is most used for petroleum refining wastes and is applicable to the treatment of fuels and wastes from leaking underground storage tanks. It can also be applied to biodegradable organic chemical wastes, including some organohalide compounds. Land treatment is not suitable for the treatment of wastes containing acids, bases, toxic inorganic compounds, salts, heavy metals, and organic compounds that are excessively soluble, volatile, or flammable.

8.13 GREEN WATER

Much of the world including parts of the United States suffers from chronic shortages of water. Throughout history, water used by humans has come from freshwater groundwater and surface sources, relying on nature's hydrologic cycle for purification to remove dissolved salt. Largely untapped is the more than 97% of earth's water that is too saline to use, most of which is present in the oceans, but including significant amounts in underground saline aquifers. Assuming sufficient availability of affordable energy, these waters can be subjected to *desalination* to provide freshwater suitable for human consumption and irrigation.

The two main means of water desalination are distillation and reverse osmosis. Distillation normally is carried out in several stages in which brine feedstock is heated to progressively higher temperatures as it is used to condense steam from each stage, the highest of which is heated by an external source. At each step, the heated brine is subjected to a lower pressure enabling more water to evaporate or "flash" to produce steam that condenses in a heat exchanger cooled by incoming brine feed. This approach enables maximum efficiency of energy use. Because of the energy intensity of this multistage distillation, it is usually combined with a power plant using otherwise waste heat from the power plant steam turbine outlet. An interesting possibility for small-scale water desalination is the use of solar power to heat the desalination units.

The world's largest seawater distillation plant is the multistage flash distillation unit in Shoaibi, Saudi Arabia, a nation that gets about 70% of its municipal water supplies from water desalination. This operation is combined with an oil-fired steam power plant, which uses the desalination plant for cooling and provides the heat needed for desalination.

In addition to distillation, the other significant alternative for providing freshwater is reverse osmosis discussed in Section 8.9. Some communities in arid lands close to a seacoast get their freshwater from this source.

Both distillation and reverse osmosis produce concentrated brine by-products (retentate) that requires disposal. This material can be pumped back into the ocean or, in some cases, to deep underground saltwater aquifers. Some of it can also be evaporated, in some cases using solar energy, to yield a salt product with commercial value.

Some inland areas have large resources of saline groundwater (brackish water) that can be desalinated. An advantage of this source is that the salt level is often less than that of seawater, which makes the energy requirements relatively lower. Residue from purification of saline groundwater can be pumped back into the aquifer at a depth or location that does not contaminate the source. Utilization of saline groundwater could vastly increase the availability of water in many water-deficient areas.

8.13.1 Reuse and Recycling of Wastewater

The most readily available water source is wastewater discharged by municipalities, industrial concerns, and even irrigation. Wastewater from municipal and industrial use usually requires treatment prior to discharge and it is relatively straightforward and inexpensive to remove inorganic and organic solutes from the water to make it suitable for use. It has an advantage over saline seawater or groundwater in that its dissolved salt content is usually much less, lowering the cost of desalination.

Water reuse and recycling are becoming much more common as demands for water exceed supply. *Unplanned reuse* occurs as the result of waste effluents entering receiving waters or groundwater and subsequently being taken into a water distribution system. A typical example of unplanned water reuse occurs in London, which withdraws water from the Thames River that may have been through other water systems at least once, and which uses groundwater sources unintentionally recharged with sewage effluents from a number of municipalities. *Planned reuse* utilizes wastewater treatment systems deliberately designed to bring water up to standards required for subsequent applications. The term *direct reuse* refers to water that has retained its identity from a previous application; reuse of water that has lost its identity is termed *indirect reuse*. The distinction also needs to be made between recycling and reuse. *Recycling* occurs internally before water is ever discharged. An example is condensation of steam in a steam power plant followed by return of the steam to boilers. *Reuse* occurs, for example, when water discharged by one user is taken as a water source by another user.

Reuse of water continues to grow because of two major factors. The first of these is lack of supply of water. The second is that widespread deployment of modern water treatment processes significantly enhances the quality of water available for reuse. These two factors come into play in semiarid regions in countries with advanced technological bases. For example, Israel, which is dependent upon irrigation for essentially all its agriculture, reuses about two-third of the country's sewage effluent for irrigation, whereas the United States, where water is relatively more available, uses only about 2–3% of its water for this purpose.

Since drinking water and water used for food processing require the highest quality of all large applications, intentional reuse for potable water is relatively less desirable, though widely practiced unintentionally or out of necessity. This leaves three applications with the greatest potential for reuse:

- Irrigation for cropland, golf courses, and other applications requiring water for plant and grass growth. This is the largest potential application for reused water and one that can take advantage of plant nutrients, particularly nitrogen and phosphorus, in water.
- Cooling and process water in industrial applications. For some industrial applications, relatively low quality water can be used and secondary sewage effluent is a suitable source.
- *Groundwater recharge*. Groundwater can be recharged with reused water either by direct injection into an aquifer or by applying the water to land, followed by percolation into the aquifer. The latter, especially, takes advantage of biodegradation and chemical sorption processes to further purify the water.

The major concern with recycled water is the potential presence of harmful pollutants. Water is readily disinfected to eliminate disease-causing bacteria. Viruses are more persistent, but can also be eliminated. Organic contaminants at very low levels may cause harm and can be very difficult to eliminate. As discussed in Section 7.13, these contaminants can include pharmaceuticals and their metabolic products such as synthetic 17α-ethinylestradiol and the natural hormone 17β-estradiol, used in oral contraceptives; other endocrine disrupting species may be present as well.

It is inevitable that water recycling and reuse will continue to grow. This trend will increase the demand for water treatment, both qualitatively and quantitatively. In addition, it will require more careful consideration of the original uses of water to minimize water deterioration and enhance its suitability for reuse.

Figure 8.12 shows a system for reuse of water. Most of the water treatment processes discussed in this chapter may be utilized in purifying water for reuse and employed in an operation such as the one shown (see especially the physical/chemical wastewater purification system shown in Figure 8.5). The system shown in Figure 8.12 shows some of the main operations employed in a system for total water recycle and the applications of the purified water product. It takes secondary wastewater effluent from which most of the BOD has been removed by the activated sludge process and discharges it into wetlands where plants and algae grow profusely in the nutrient-rich water. This biomass has the potential to be harvested and utilized as a fuel or raw material using thermochemical gasification or related measures. Depending upon geological conditions, significant amounts of water in the

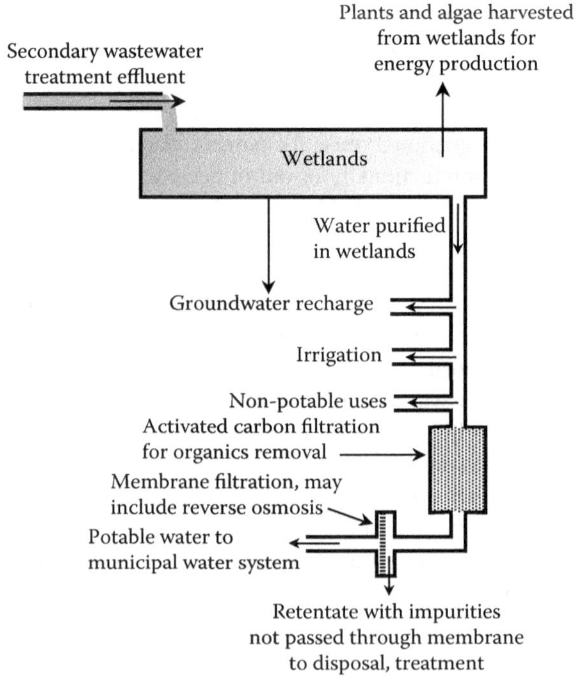

FIGURE 8.12 System in which wastewater effluent from secondary sewage treatment is upgraded to water that can be used for a number of purposes, including even recirculation back into the municipal water system. (From Manahan, S., *Environmental Chemistry*, 9th ed., Taylor & Francis, Boca Raton, 2010. With permission.)

wetlands may infiltrate into groundwater, where it is purified further by natural processes underground. Water exiting the wetlands may be run over a bed of activated carbon to remove organics. The water can also be subjected to highly efficient membrane filtration processes including even reverse osmosis for removal of all salts. The solute-enriched retentate from treatment of water by membrane processes can be pumped into deep underground saline aquifers in some areas, discarded to the ocean in coastal regions (often with mixing with wastewater or cooling water to dilute the impurities in the retentate), or evaporated to produce a solid or sludge residue that can be disposed.

The world's largest water plant devoted to bringing sewage effluent up to a standard permitting potable uses began operation in the Orange County Water District in southern California in late 2007. Known as the Groundwater Replenishment System this $481 million project is fed with sewage treated by secondary wastewater treatment processes and pumps its clean water product into underground aquifers to serve as future municipal water sources and to prevent saltwater intrusion from the nearby ocean. Around 275 million liters per day of sewage effluent is first forced through microfilters to remove solids, then treated by reverse osmosis to remove dissolved salts, and finally subjected to intense ultraviolet radiation to break down organic contaminants, especially pharmaceuticals and their metabolites before being discharged to underground aquifers. It is expected to serve as a model for similar installations around the world.

In coastal areas and on islands, a major advantage of upgrading wastewater to a standard that it can be injected into underground aquifers is the prevention of seawater intrusion into the aquifers (Figure 8.13). As aquifers are pumped down, the lowered water level in the aquifer can enable infiltration of saline water, thus ruining the groundwater quality. Injection of treated wastewater into the aquifer can raise the water table in the aquifer to a level that prevents seawater intrusion.

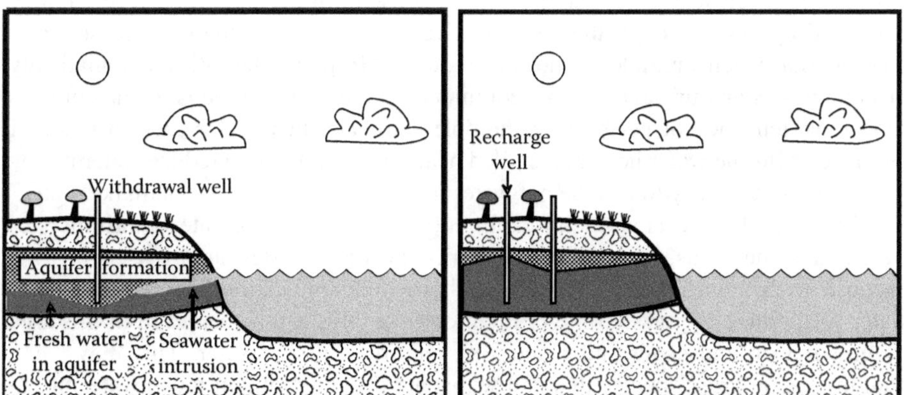

FIGURE 8.13 Injection of reclaimed water into underground aquifers in coastal areas can prevent seawater intrusion into the aquifer formation. (From Manahan, S., *Environmental Chemistry*, 9th ed., Taylor & Francis, Boca Raton, 2010. With permission.)

8.14 WATER CONSERVATION

Water conservation, simply using less water, is one of the most effective, and clearly the fastest, means to ensure water supply. Water conservation falls into several major categories:

1. Indoor and household water conservation practices
2. Water-conserving devices and appliances
3. Outdoor and landscaping water conservation
4. Efficient irrigation practices
5. Conservation in agricultural operations other than irrigation
6. Water-efficient manufacturing

Each of these categories is summarized briefly below.

Indoor and household water conservation practices are generally those simple measures that water consumers may implement to reduce use of water. Examples include sequential uses of water such as saving water from rinsing dishes to use for household cleaning, ensuring that there are no leaking faucets or toilet appliances, taking showers by first lathering one's wet body without water flowing followed by rinsing with water for the minimum length of time, minimizing the number and water volume of toilet flushes, minimizing use of kitchen sink disposal devices for food waste disposal, operating clothes washers and dishwashers only with full loads, and other common sense measures to reduce water use. Implementation of these measures is largely a matter of education and good environmental citizenship. In cases of real water shortage, force may be applied through the painful mechanism of sharply higher water utility rates.

A number of water-saving devices and appliances are available. These can be quite simple and cheap, such as aerators with flow constrictors on faucets or showerheads, or they may be large and expensive such as low-water-consumption clothes washers that function well with cold water. Toilets that use minimal water per flush are now mandated by law, although the less effective of these may actually increase water consumption when multiple flushes are required. In public facilities, automatically flushing toilets and urinals are now commonplace, although some have the counterproductive tendency of performing multiple automatic flushes for each instance of use. Instant in-line water heaters installed near faucet outlets can reduce water use by making it unnecessary to run the hot water faucet for some time to obtain hot water.

Water used for lawns and landscaping often is one of the greatest consumers of water. In some areas, such as public parks, purified wastewater is used for this purpose. In principle, water used in some household applications can be recycled to water yard plants or grass. Some of the greater savings can be achieved in this area by growing ground cover and plants that require minimal water. This may mean planting native grass species that can withstand drought without much watering. In extreme cases, yards have been converted to gravel cover. Mulching with materials such as tree bark or grass clippings can reduce water use significantly.

Irrigation of agricultural crops is one of the largest consumers of water and one that is amenable to conservation. Spray irrigation devices lose large quantities of water to evaporation and should be replaced by systems that do not spray the water into the air. The ultimate in irrigation efficiency is drip irrigation that applies minimum quantities of water directly to the plant roots. Application of excess water in irrigation should be minimized. However, particularly in areas that rarely receive enough rain to flush the soil, application of minimal amounts of irrigation water can result in harmful accumulation of salts in the soil.

Livestock production and processing agricultural products use large amounts of water. The production of 1 kg of beef requires around 20,000 L of water, including the water required to grow the grain that the animal eats. About 6000 L of water are required to produce 1 kg of chicken meat. Significantly less water is needed to produce an equivalent amount of vegetable protein. The production of fuel ethanol from the fermentation of sugar from corn requires relatively large amounts of water. Choices made in the kinds of agricultural products produced for market have important implications for water conservation.

Manufacturing can be very water-intensive. The manufacture of a single automobile requires around 150,000 L of water. Refining crude oil to fuel and other petroleum products requires approximately 44 times the volume of water as the gasoline produced. The conservation of water in manufacturing and commercial activities is an important aspect of industrial ecology (see Section 1.5).

8.15 PROTECTING WATER SUPPLIES FROM ATTACK

The fact that drinking water is distributed centrally to large numbers of people makes water systems a concern to terrorist threats. History abounds with examples of large populations devastated by contaminated water supplies that caused diseases such as cholera and typhoid. In modern times, contaminated water in developing countries causes outbreaks of debilitating and often fatal dysentery, a leading cause of mortality in infants and children in some areas. Even in industrialized countries in modern times, drinking water contamination can cause illness and even fatalities. For example, in 1993 contamination of the Milwaukee water system by the protozoan *Cryptosporidium parvum* caused illness in over 400,000 people, more than 50 of whom died as a result.

Deliberate chemical contamination of an entire water system is an unlikely, but not impossible, event. Incidents have been recorded in which terrorist groups attempted to obtain cyanide salts, presumably with the objective of poisoning water supplies. Toxins from microorganisms are likely weapons of groups wishing to poison water. The most toxic of these is the botulinus toxin, an astoundingly poisonous substance produced by *Botulinus* bacteria. Direct contamination by bacteria is possible. The most likely candidates are *Bacillus anthracis*, *Shigella dysenteriae*, *Vibriocholerae*, and *Yersiniapestis*. Waterborne strains of *Escherichia coli* bacteria that produce the shiga toxin (the toxin produced by *Shigelladysenteriae*) continue to cause fatal infections in the United States from time to time. Shiga-producing *E. coli* in the municipal water supply of Walkerton, Ontario, Canada, caused approximately 3000 people to become ill and seven to die in May 2000. In an incident suggestive of the potential for terrorist attack, 12 staff of a laboratory of a large medical center developed acute diarrheal illness requiring hospitalization of four of the victims due to infection by *Shigella dysenteriae*. Cultures taken from the victims were found to be identical to those found on doughnuts or muffins that they had eaten in the facility break room and to a culture of *Shigella dysenteriae* maintained in the laboratory, a portion of which was missing. This led to the conclusion that the contamination was deliberate.[6]

Fortunately, water supplies are not particularly vulnerable to attack. It would be rather difficult to infiltrate a water treatment system, and injecting a toxic substance into the water distribution system beyond the treatment plant is unlikely. Vulnerability to infectious microbial agents is greatly reduced by maintaining a disinfectant chlorine residual in the system. Some of the likely chemical agents hydrolyze in water or are destroyed by the action of chlorine, and their effects are reduced by the great dilution provided by water systems that distribute huge quantities of water. Even in the event of contamination of a water supply, provision of relatively small quantities of bottled water for drinking would reduce the hazard to very low levels. Prudence dictates that water supply, treatment, and distribution systems be maintained in a secure state and monitored for likely agents of attack, but water systems need not be the subject of undue concern.

LITERATURE CITED

1. Wiesmann, U., I. S. Choi, and E.-M. Dombrowski, *Fundamentals of Biological Wastewater Treatment*, Wiley-VCH, Hoboken, NJ, 2007.
2. Bratby, J., *Coagulation and Flocculation in Water and Wastewater Treatment*, 2nd ed., IWA Publishing, Seattle, 2006.
3. Hager, L. S., *Membrane Systems for Wastewater Treatment*, McGraw-Hill, New York, 2006.
4. Sharma, V., Ed., *Ferrates: Synthesis, Properties, and Applications in Water and Wastewater Treatment*, American Chemical Society, Washington, DC, 2008.
5. Crites, R. W., E. J. Middlebrooks, and S. C. Reed, *Natural Wastewater Treatment Systems*, CRC/Taylor & Francis, Boca Raton, FL, 2006.
6. Kolavic, S. A., A. Kimura, S. L. Simons, L. Slutsker, S. Barth, and C. E. Haley, An outbreak of *Shigella dysenteriae* type 2 among laboratory workers due to intentional food contamination, *Journal of the American Medical Association*, 278, 396–398, 1997.
7. Manahan, S., *Environmental Chemistry*, 9th ed., Taylor & Francis, Boca Raton, 2010.

SUPPLEMENTARY REFERENCES

American Water Works Association, *Reverse Osmosis and Nanofiltration*, 2nd ed., American Water Works Association, Denver, CO, 2007.

American Water Works Association, *Water Chlorination/Chloramination Practices and Principles*, 2nd ed., American Water Works Association, Denver, CO, 2006.

Ash, M. and I. Ash, *Handbook of Water Treatment Chemicals*, Gower, Aldershot, England, 1996.

Brennan, M.B. Waterworks: Research accelerates on advanced water-treatment technologies as their use in purification grows, *Chemical and Engineering News*, 79, 32–38, 2001.

Crittenden, J. C., *Water Treatment Principles and Design*, 2nd ed., Wiley, New York, 2005.

Faust, S. D. and O. M. Aly, Eds., *Chemistry of Water Treatment*, 2nd ed., American Water Works Association, Denver, CO, 1997.

Hammer, M. J., *Water and Wastewater Technology*, 6th ed., Prentice Hall, Upper Saddle River, NJ, 2008.

Rakness, K. L., *Ozone in Drinking Water Treatment: Process Design, Operation, and Optimization*, American Water Works Association, Denver, 2005.

Rimer, A. E., *The Impacts of Membrane Process Residuals on Wastewater Treatment: A Guidance Manual*, WateReuse Foundation, Alexandria, VA, 2008.

Russell, D. L., *Practical Wastewater Treatment*, Wiley-Interscience, Hoboken, NJ, 2006.

Spellman, F. R., *Handbook of Water and Wastewater Treatment Plant Operations*, 2nd ed., Taylor & Francis, Boca Raton, FL, 2008.

Sullivan, P. J., F. J. Agardy, and J. J. J. Clark, *The Environmental Science of Drinking Water*, Elsevier Butterworth-Heinemann, Burlington, MA, 2005.

Water Environment Federation, *Operation of Municipal Wastewater Treatment Plants*, 6th ed., McGraw-Hill, New York, 2008.

QUESTIONS AND PROBLEMS

The use of Internet resources is assumed in answering any of the questions. These would include such things as constants and conversion factors as well as additional information needed to complete an answer.

1. Consider the equilibrium reactions and expressions covered in Chapter 3. How many moles of NTA should be added to 1000 L of water having a pH of 9 and containing CO_3^{2-} at 1.00×10^{-4} M to prevent precipitation of $CaCO_3$? Assume a total calcium level of 40 mg/L.
2. What is the purpose of the return sludge step in the activated sludge process?
3. What are the two processes by which the activated sludge process removes soluble carbonaceous material from sewage?
4. Why might hard water be desirable as a medium if phosphorus is to be removed by an activated sludge plant operated under conditions of high aeration?
5. How does reverse osmosis differ from a simple sieve separation or ultrafiltration process?
6. How many liters of methanol would be required daily to remove the nitrogen from a 200,000-L/day sewage treatment plant producing an effluent containing 50 mg/L of nitrogen? Assume that the nitrogen has been converted to NO_3^- in the plant. The denitrifying reaction is Reaction 8.34.
7. Discuss some of the advantages of physical–chemical treatment of sewage as opposed to biological wastewater treatment. What are some disadvantages?
8. Why is recarbonation necessary when water is softened by the lime-soda process?
9. Assume that a waste contains 300 mg/L of biodegradable $\{CH_2O\}$ and is processed through a 200,000-L/day sewage treatment plant which converts 40% of the waste to CO_2 and H_2O. Calculate the volume of air (at 25°C, 1 atm) required for this conversion. Assume that the O_2 is transferred to the water with 20% efficiency.
10. If all of the $\{CH_2O\}$ in the plant described in Question 9 could be converted to methane by anoxic digestion, how many liters of methane (STP) could be produced daily?
11. Assuming that aeration of water does not result in the precipitation of calcium carbonate, of the following, which one that would not be removed by

aeration: hydrogen sulfide, carbon dioxide, volatile odorous bacterial metabolites, alkalinity, iron?

12. In which of the following water supplies would moderately high water hardness be most detrimental: municipal water, irrigation water, boiler feedwater, drinking water (in regard to potential toxicity).
13. Which solute in water is commonly removed by the addition of sulfite or hydrazine?
14. A wastewater containing dissolved Cu^{2+} ion is to be treated to remove copper. Which of the following processes would *not* remove copper in an insoluble form: lime precipitation, cementation, treatment with NTA, ion exchange, reaction with metallic Fe.
15. Match each water contaminant in the left column with its preferred method of removal in the right column.

A. Mn^{2+}	1. Activated carbon
B. Ca^{2+} and HCO_3^-	2. Raise pH by addition of Na_2CO_3
C. THM compounds	3. Addition of lime
D. Mg^{2+}	4. Oxidation

16. A cementation reaction employs iron to remove Cd^{2+} present at a level of 350 mg/L from a wastewater stream. Given that the atomic mass of Cd is 112.4 and that of Fe is 55.8, how many kilograms of Fe are consumed in removing all the Cd from 4.50×10^6 L of water?
17. Consider municipal drinking water from two different kinds of sources, one a flowing, well-aerated stream with a heavy load of particulate matter, and the other an anoxic groundwater. Describe possible differences in the water treatment strategies for these two sources of water.
18. In treating water for industrial use, consideration is often given to "sequential use of the water." What is meant by this term? Give some plausible examples of sequential use of water.
19. Active biomass is used in the secondary treatment of municipal wastewater. Describe three ways of supporting a growth of the biomass, contacting it with wastewater, and exposing it to air.
20. Using appropriate chemical reactions for illustration, show how calcium present as the dissolved HCO_3^- salt in water is easier to remove than other forms of hardness, such as dissolved $CaCl_2$.
21. Match the pollutant or impurity on the left with a reagent used to treat it from the list on the right, below:

A. Bacteria	1. Ozone
B. PCB	2. $Al_2(SO_4)_3$
C. H_2S	3. Activated carbon
D. Colloidal matter	4. Air

22. Match the water treatment step or process on the right, below, with the substance that it treats or removes on the left.

A. Oxygen	1. Live microorganisms

B. Calcium, Ca^{2+}	2. Phosphate, PO_4^{3-}
C. $\{CH_2O\}$	3. Hydrazine (H_4N_2) or sulfite (SO_3^{2-})
D. Colloidal solids	4. $Al_2(SO_4)_3 \cdot 18H_2O$

23. Regarding secondary wastewater treatment, the true statement of the following is (A) the activated sludge process is predominantly a physical/chemical process, (B) trickling filters make use of a mass of biological sludge that is continuously pumped over the filter, (C) excess sludge from activated sludge treatment is likely to undergo the process represented by $2\{CH_2O\} \rightarrow CH_4 + CO_2$, (D) the activated sludge process gets rid of all of the sludge as soon as it is made, (E) the trickling filter is an anaerobic (oxygen-free) treatment process.
24. Of the following, the true statement pertaining to water disinfection is (A) disinfection with ozone is particularly desirable because the ozone persists throughout the water distribution system, (B) disinfection with ozone is noted for producing toxic organic by-products, (C) organic substances are left in water to be treated by chlorine to retain disinfecting capacity in the water, (D) the main disadvantage of chlorination of water is that Cl_2 gas is the only effective chlorinating agent, (E) chlorine dioxide, ClO_2, is an effective water disinfectant that does not produce impurity THMs in water treatment.
25. Of the following, designate the technique *least likely* to significantly lower the concentration of heavy metals in water and explain your choice: (A) cementation, (B) anion exchange, (C) addition of S^{2-}, (D) biological waste treatment, (E) lime.
26. Of the following, designate the *least likely* to be effective in reducing total levels of dissolved ions (salts) from water and explain your choice: (A) reverse osmosis, (B) cation exchange followed by anion exchange, (C) distillation, (D) ion exchange as practiced for water softening, (E) electrodialysis.
27. Using internet resources, attempt to find instances in which deliberate poisoning of water supplies has been used as a tactic for warfare or terrorism.
28. Bottled water is a product with enormous commercial value. What are the advantages of using this source of drinking water? What are the environmental disadvantages of this product?

9 Chemical Analysis of Water and Wastewater

9.1 GENERAL ASPECTS OF ENVIRONMENTAL CHEMICAL ANALYSIS

Scientists' understanding of the environment can only be as good as their knowledge of the identities and quantities of pollutants and other chemical species in water, air, soil, and biological systems. Therefore, proven, state-of-the-art techniques of chemical analysis, properly employed, are essential to environmental chemistry. The present time is a very exciting period in the evolution of analytical chemistry, characterized by the development of new and improved analysis techniques that enable detection of much lower levels of chemical species and a vastly increased data throughput. These developments pose some challenges. Because of the lower detection limits of some instruments, it is now possible to see quantities of pollutants that would have escaped detection previously, resulting in difficult questions regarding the setting of maximum allowable limits of various pollutants. The increased output of data from automated instruments has in many cases overwhelmed human capacity to assimilate and understand it.

Challenging problems still remain in developing and utilizing techniques of environmental chemical analysis. Not the least of these problems is knowing which species should be measured, or even whether or not an analysis should be performed at all. The quality and choice of analyses is much more important than the number of analyses performed. Indeed, a persuasive argument can be made that, given modern capabilities in analytical chemistry, too many analyses of environmental samples are performed, whereas fewer, more carefully planned analyses would yield more useful information.

In addition to a discussion of water analysis, this chapter covers some of the general aspects of environmental chemical analysis and the major techniques that are used to determine a wide range of analytes (species measured). Many techniques are common to water, air, soil, and biological sample analyses, and reference is made to them in chapters that follow.

9.1.1 Error and Quality Control

A crucial aspect of any chemical analysis is the validity and quality of the data that it produces. All measurements are subject to error, which may be *systematic* (of the same magnitude and same direction) or *random* (varying in both magnitude and direction). Systematic errors cause the measured values to vary consistently from the true values; this variation is known as the *bias*. The degree to which a measured value comes close to the actual value of an analytical measurement is called the *accuracy* of the measurement, reflecting both systematic and random errors. It is essential for the analyst to determine these error components in the measurement of environmental samples, including water samples. The identification and control of systematic and random errors falls in the category of *quality control (QC)* procedures. It is beyond the scope of this chapter to go into any detail on these crucial procedures for which the reader is referred to a work on standard methods for the analysis of water.[1]

In order for the results from a laboratory to be meaningful, the laboratory needs a *quality assurance plan* specifying measures taken to produce data of known quality. An important aspect of such a plan is the use of laboratory control standards consisting of samples with very accurately known analyte levels in a carefully controlled matrix. Such standard reference materials are available in the United States from the National Institute of Standards and Technology (NIST) for many kinds of samples.

Many environmental analytes are present at very low levels, which challenge the ability of the method used to detect and accurately quantify them. (Typically, pharmaceuticals and their metabolites are present at nanogram to low-picogram per liter levels in wastewater.) Therefore, the *detection limit* of a method of analysis is quite important. Defining detection limit has long been a controversial topic in chemical analysis. Every analytical method has a certain degree of noise. The detection limit is an expression of the lowest concentration of analytes that can be measured above the noise level with a specified degree of confidence in an analytical procedure. In the detection of analytes, two kinds of errors can be defined. A Type I error occurs when the measurement finds an analyte present when it is actually absent. A Type II error occurs when the measurement finds an analyte absent when it is actually present.

Detection limits can be further categorized into several different subcategories. The *instrument detection limit (IDL)* is the analyte concentration capable of producing a signal 3 times the standard deviation of the noise. The *lower level of detection (LLD)* is the quantity of analyte that will produce a measurable signal 99% of the time; it is about 2 times the IDL. The *method detection limit* (*MDL*) is measured like the LLD, except that the analyte is taken through the whole analytical procedure, including steps such as extraction and sample cleanup; it is about 4 times the IDL. Finally, the *practical quantitation limit (PQL)*, which is about 20 times the IDL, is the lowest level achievable in routine analysis in laboratories.

9.1.2 Water Analysis Methods

Methods of analysis are published for a huge number of water constituents and contaminants. It is not possible to cover them in a comprehensive manner in one short chapter. For analytical procedures, the reader is referred to sources of methods. The most comprehensive of these is the classic *Standard Methods for the Examination of Water and Wastewater.*[2] The U.S. EPA publishes methods for water analysis that are listed in an index of methods[3] and available from the National Technical Information Service.[4] Another useful source of methods is available on a CD ROM from Genium Publishing Corp.[5] Current issues in water analysis are reviewed periodically in the journal *Analytical Chemistry.*[6]

9.2 CLASSICAL METHODS

Before sophisticated instrumentation became available, most important water quality parameters and some air pollutant analyses were done by *classical methods*, which require only chemicals, balances to measure masses, burettes, volumetric flasks and pipettes to measure volumes, and other simple laboratory glassware. The two major classical methods are *volumetric analysis*, in which volumes of reagents are measured, and *gravimetric analysis*, in which masses are measured. Some of these methods are still used today, and many have been adapted to instrumental and automated procedures.

The most common classical methods for pollutant analysis are titrations, largely used for water analysis. Some of the titration procedures used are discussed in this section.

Acidity (see Section 3.3) is determined simply by titrating hydrogen ion with base. Titration to the methyl orange endpoint (pH 4.5) yields the "free acidity" due to strong acids (HCl and H_2SO_4). Carbon dioxide does not, of course, appear in this category. Titration to the phenolphthalein endpoint (pH 8.3) yields total acidity and accounts for all acids except those weaker than HCO_3^-.

Alkalinity may be determined by titration with H_2SO_4 to pH 8.3 to neutralize bases as strong as, or stronger than, carbonate ion,

$$CO_3^{2-} + H^+ \rightarrow HCO_3^- \tag{9.1}$$

or by titration to pH 4.5 to neutralize bases weaker than CO_3^{2-}, but as strong as, or stronger than, HCO_3^-:

$$HCO_3^- + H^+ \rightarrow H_2O + CO_2(g) \tag{9.2}$$

Titration to the lower pH yields total alkalinity.

The calcium and magnesium ions responsible for water hardness are readily titrated at pH 10 with a solution of EDTA, a chelating agent discussed in Sections 3.6 and 3.9. The titration reaction is

$$Ca^{2+}(\text{or } Mg^{2+}) + H_2Y^{2-} \rightarrow CaY^{2-}(\text{or } MgY^{2-}) + 2H^+ \tag{9.3}$$

where H_2Y^{2-} is the partially ionized EDTA chelating agent. Eriochrome Black T is used as an indicator, and it requires the presence of magnesium, with which it forms a wine red complex.

Several oxidation–reduction titrations can be used for environmental chemical analysis. Oxygen is determined in water by the Winkler titration. The first reaction in the Winkler method is the oxidation of manganese(II) to manganese(IV) by the oxygen analyte in a basic medium; this reaction is followed by acidification of the brown hydrated MnO_2 in the presence of I^- ion to release free I_2, and then titration of the liberated iodine with standard thiosulfate, using starch as an endpoint indicator:

$$Mn^{2+} + 2OH^- + \tfrac{1}{2}O_2 \rightarrow MnO_2(s) + H_2O \tag{9.4}$$

$$MnO_2(s) + 2I^- + 4H^+ \rightarrow Mn^{2+} + I_2 + 2H_2O \tag{9.5}$$

$$I_2 + 2S_2O_3^{2-} \rightarrow S_4O_6^{2-} + 2I^- \tag{9.6}$$

A back-calculation from the amount of thiosulfate required yields the original quantity of dissolved oxygen (DO) present. Biochemical oxygen demand (BOD) (see Section 7.9), is determined by adding a microbial "seed" to the diluted sample, saturating with air, incubating for five days, and determining the oxygen remaining. The results are calculated to show BOD as mg/L of O_2. A BOD of 80 mg/L, for example, means that biodegradation of the organic matter in a liter of the sample would consume 80 mg of oxygen.

9.3 SPECTROPHOTOMETRIC METHODS

9.3.1 Absorption Spectrophotometry

Absorption spectrophotometry of light-absorbing species in solution, historically called colorimetry when visible light is absorbed, is still used for the analysis of some water and air pollutants. Basically, absorption spectrophotometry consists of measuring the percent transmittance (%T) of monochromatic light passing through a light-absorbing solution as compared to the amount of light passing through a blank solution containing everything in the medium but the sought-for constituent (100%). The absorbance (*A*) is defined as

$$A = \log\frac{100}{\%\mathrm{T}} \tag{9.7}$$

The relationship between *A* and the concentration (*C*) of the absorbing substance is given by Beer's law:

$$A = abC \tag{9.8}$$

where *a* is the absorptivity, a wavelength-dependent parameter characteristic of the absorbing substance; *b* is the path length of the light through the absorbing solution; and *C* is the concentration of the absorbing substance. A linear relationship between *A* and *C* at constant path length indicates adherence to Beer's law. In many cases, analyses may be performed even when Beer's law is not obeyed, if a suitable calibration curve is prepared. A color-developing step is usually required in which the sought-for substance reacts to form a colored species, and in some cases a colored species is extracted into a nonaqueous solvent to provide a more intense color and a more concentrated solution.

A number of solution spectrophotometric methods have been used for the determination of water and air pollutants. Some of these are summarized in Table 9.1.

9.3.2 Atomic Absorption and Emission Analyses

Atomic absorption analysis is commonly used for the determination of metals in environmental samples. This technique is based on the absorption of monochromatic light by a cloud of atoms of the analyte metal. The monochromatic light can be produced by a source composed of the same atoms as those being analyzed. The source produces intense electromagnetic radiation with a wavelength exactly the same as

TABLE 9.1
Solution Spectrophotometric (Colorimetric) Methods of Analysis

Analyte	Reagent and Method
Arsenic	Reaction of arsine, AsH_3, with silver diethylthiocarbamate in pyridine, forming a red complex
Boron	Reaction with curcumin, forming red rosocyanine
Bromide	Reaction of hypobromite with phenol red to form bromphenol blue-type indicator
Chlorine	Development of color with orthotolidine
Cyanide	Formation of a blue dye from the reaction of cyanogen chloride, CNCl, with pyridine-pyrazolone reagent, measured at 620 nm
Fluoride	Decolorization of a zirconium-dye colloidal precipitate ("lake") by the formation of colorless zirconium fluoride and free dye
Nitrate and nitrite	Nitrate is reduced to nitrite, which is diazotized with sulfanilamide and coupled with *N*-(l-naphthyl)-ethylenediamine dihydrochloride to produce a highly colored azo dye measured at 540 nm
Nitrogen, Kjeldahl method	Digestion in sulfuric acid to NH_4^+ followed by treatment with alkaline phenol reagent and sodium hypochlorite to form blue indo-phenate phenol measured at 630 nm
Phenols	Reaction with 4-aminoantipyrine at pH 10 in the presence of potassium ferricyanide, forming an antipyrine dye which is extracted into pyridine and measured at 460 nm
Phosphate	Reaction with molybdate ion to form a phosphomolybdate, which is selectively reduced to intensely colored molybdenum blue
Selenium	Reaction with diaminobenzidine, forming colored species absorbing at 420 nm
Silica	Formation of molybdosilicic acid with molybdate, followed by reduction to a heteropoly blue measured at 650 or 815 nm
Sulfide	Formation of methylene blue
Surfactants	Reaction with methylene blue to form blue salt
Tannin and lignin	Blue color from tungstophosphoric and molybdophosphoric acids

Source: From Manahan, S., *Environmental Chemistry*, 9th ed., Taylor & Francis, Boca Raton, 2010. With permission.

that absorbed by the atoms, resulting in extremely high selectivity. The basic components of an atomic absorption instrument are shown in Figure 9.1. The key element is the hollow cathode lamp in which atoms of the analyte metal are energized such that they become electronically excited and emit radiation with a very narrow wavelength band characteristic of the metal. This radiation is guided by the appropriate optics through a flame into which the sample is aspirated. In the flame, most metallic compounds are decomposed, and the metal is reduced to the elemental state, forming a cloud of atoms. These atoms absorb a fraction of radiation in the flame. The fraction of radiation absorbed increases with the concentration of the sought-for element in the sample according to the Beer's law relationship (Equation 9.8). The attenuated light beam next goes to a monochromator to eliminate extraneous light resulting from the flame, and then to a detector.

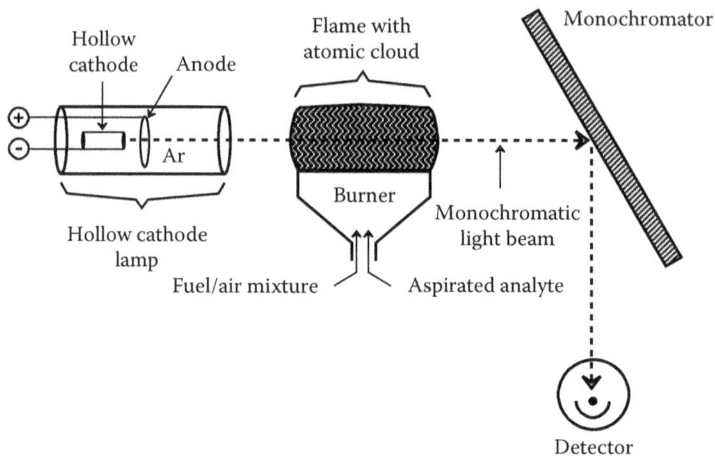

FIGURE 9.1 Basic components of a flame atomic absorption spectrophotometer. (From Manahan, S., *Environmental Chemistry*, 9th ed., Taylor & Francis, Boca Raton, 2010. With permission.)

Atomizers other than a flame can be used. The most common of these is the graphite furnace, an electrothermal atomization device that consists of a hollow graphite cylinder placed so that the light beam passes through it. A small sample of up to 100 μL is inserted through a hole on the upper side of the horizontally oriented tube. An electric current is passed through the tube to heat it—gradually at first to dry the sample, then rapidly to vaporize and excite the metal analyte. The absorption of metal atoms in the hollow portion of the tube is measured and recorded as a spike-shaped signal. A diagram of a graphite furnace with a typical output signal is shown in Figure 9.2. The major advantage of the graphite furnace is that it gives detection limits up to 1000 times lower than those of conventional flame devices.

A special technique for the flameless atomic absorption analysis of mercury involves room temperature reduction of mercury to the elemental state by tin(II) chloride in solution, followed by sweeping the mercury into an absorption cell with air. Nanogram (10^{-9} g) quantities of mercury can be determined by measuring mercury absorption at 253.7 nm.

9.3.3 Atomic Emission Techniques

Metals may be determined in water, atmospheric particulate matter, and biological samples very well by observing the spectral lines emitted when they are heated to a very high temperature. An especially useful atomic emission technique is inductively coupled plasma atomic emission spectroscopy (ICP/AES). The "flame" in which analyte atoms are excited in plasma emission consists of an incandescent plasma (ionized gas) of argon heated inductively by radiofrequency energy at 4–50 MHz and 2–5 kW (Figure 9.3). The energy is transferred to a stream of argon through an induction coil, producing temperatures up to 10,000 K. The sample atoms are subjected to temperatures around 7000 K, twice those of the hottest conventional

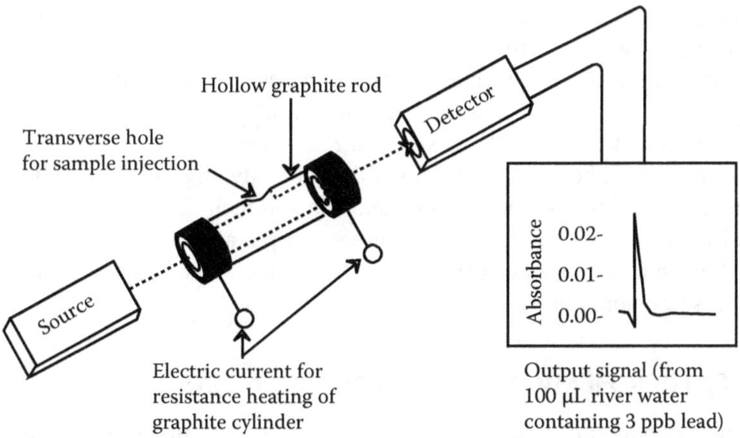

FIGURE 9.2 Graphite furnace for atomic absorption analysis and typical output signal. (From Manahan, S., *Environmental Chemistry*, 9th ed., Taylor & Francis, Boca Raton, 2010. With permission.)

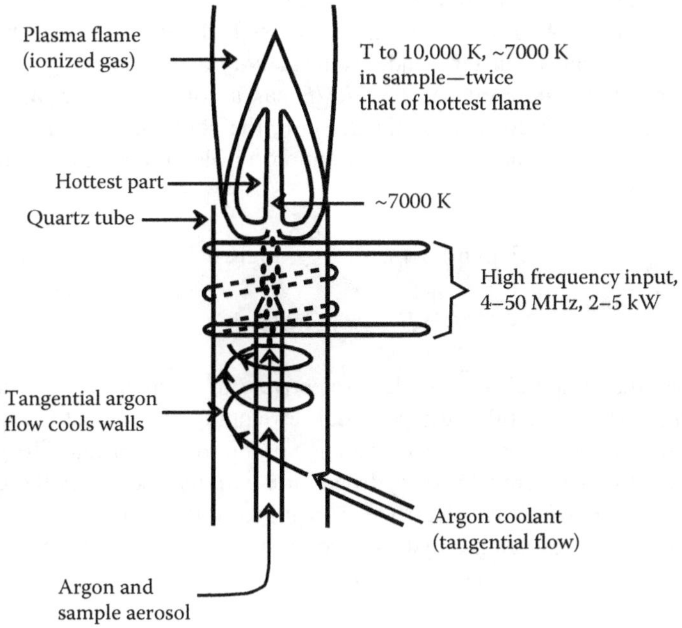

FIGURE 9.3 Schematic diagram showing inductively coupled plasma used for optical emission spectroscopy. (From Manahan, S., *Environmental Chemistry*, 9th ed., Taylor & Francis, Boca Raton, 2010. With permission.)

flames (e.g., nitrous oxide–acetylene operates at 3200 K). Since emission of light increases exponentially with temperature, lower detection limits are obtained. Furthermore, the technique enables emission analysis of some of the environmentally important metalloids such as arsenic, boron, and selenium. Interfering chemical reactions and interactions in the plasma are minimized as compared to flames. Of greatest significance, however, is the capability of analyzing as many as 30 elements simultaneously, enabling a true multielement analysis technique. Plasma atomization combined with mass spectrometric measurement of analyte elements (ICP/MS) has developed as an especially powerful means for multielement analysis that can even be used for some nonmetals.

9.4 ELECTROCHEMICAL METHODS OF ANALYSIS

Potentiometric, voltammetric, and amperometric water analyses utilize electrochemical sensors. Potentiometry is based on the general principle that the relationship between the potential of a measuring electrode and that of a reference electrode is a function of the log of the activity of an ion in solution. For a measuring electrode responding selectively to a particular ion, this relationship is given by the Nernst equation,

$$E = E^0 + \frac{2.303RT}{zF}\log(a_z), \tag{9.9}$$

where E is the measured potential; E^0 is the standard electrode potential; R is the gas constant; T is the absolute temperature; z is the signed charge (+ for cations, – for anions); F is the Faraday constant; and a is the activity of the ion being measured. At a given temperature, the quantity $2.303RT/F$ has a constant value; at 25°C, it is 0.0592 V (59.2 mV). At constant ionic strength, the activity, a, is directly proportional to concentration, and the Nernst equation for electrodes responding to Cd^{2+} and F^- may be written as

$$E(\text{in mV}) = E^0 + \frac{59.2}{2}\log[\mathrm{Cd}^{2+}], \tag{9.10}$$

$$E = E^0 - 59.2\log[\mathrm{F}^-] \tag{9.11}$$

Electrodes that respond more or less selectively to various ions are called *ion-selective electrodes*. Generally, the potential-developing component is a membrane of some kind that allows for selective exchange of the sought-for ion. The glass electrode used for the measurement of hydrogen-ion activity and pH is the oldest and most widely used ion-selective electrode. The potential is developed at a glass membrane that selectively exchanges hydrogen ion in preference to other cations, giving a Nernstian response to hydrogen ion activity, a_{H^+}:

$$E = E^0 + 59.2\log(a_{\mathrm{H}^+}), \quad E = E^0 - 59.2\ \mathrm{pH} \tag{9.12}$$

The pH meters used with glass electrodes are calibrated directly in pH units using standard buffer solutions.

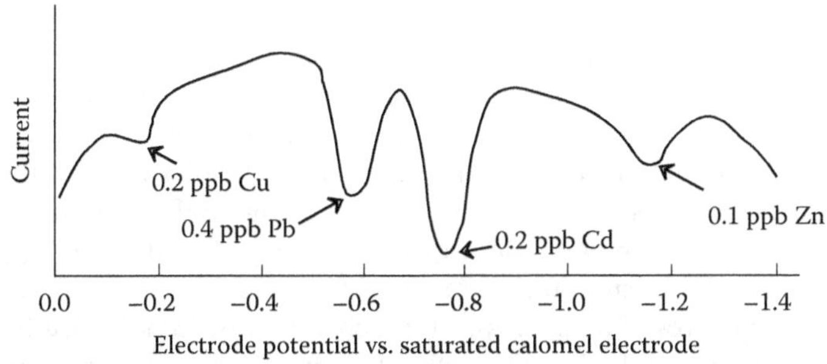

FIGURE 9.4 Differential-pulse anodic-stripping voltammogram of tap water at a mercury-plated, wax-impregnated graphite electrode. (From Manahan, S., *Environmental Chemistry*, 9th ed., Taylor & Francis, Boca Raton, 2010. With permission.)

Of the ion-selective electrodes other than glass electrodes, the fluoride electrode is the most successful. It is well behaved, relatively free of interferences, and has an adequately low detection limit and a long range of linear response. Like all ion-selective electrodes, its electrical output is in the form of a potential signal that is proportional to log of concentration. Because of the logarithmic response of potential to concentration, small errors in E lead to relatively high concentration errors.

Voltammetric techniques, the measurement of current resulting from potential applied to a microelectrode, have found some applications in water analysis. One such technique is differential-pulse polarography, in which the potential is applied to the microelectrode in the form of small pulses superimposed on a linearly increasing potential. The current is read near the end of the voltage pulse and compared to the current just before the pulse was applied. It has the advantage of minimizing the capacitive current from charging the microelectrode surface, which sometimes obscures the current due to the reduction or oxidation of the species being analyzed. Anodic-stripping voltammetry involves deposition of metals on an electrode surface over a period of several minutes followed by stripping them off very rapidly using a linear anodic sweep. The electrodeposition concentrates the metals on the electrode surface, and increased sensitivity results. An even better technique is to strip the metals off using a differential-pulse signal. A differential-pulse anodic-stripping voltammogram of copper, lead, cadmium, and zinc in tap water is shown in Figure 9.4.

9.5 CHROMATOGRAPHY

First described in the literature in the early 1950s, gas chromatography (GC) has played an essential role in the analysis of organic materials. GC is both a qualitative and a quantitative technique; for some analytical applications of environmental importance, it is remarkably sensitive and selective. GC is based on the principle that when a mixture of volatile materials transported by a carrier gas is passed through a column containing an adsorbent solid phase or, more commonly, an absorbing liquid phase coated on a solid material, each volatile component will be partitioned between the carrier gas and the solid or liquid. The length of time required for the volatile component to traverse the column is proportional to the degree to which it is retained by the nongaseous phase. Since different components may be retained to different degrees, they will emerge from the end of the column at different times. If a suitable detector is available, the time at which the component emerges from the column and the quantity of the component are both measured. A recorder trace of the detector response appears as peaks of different sizes, depending on the quantity of material producing the detector response. Both quantitative and (within limits) qualitative analyses of the sought-for substances are obtained.

The essential features of a gas chromatograph are shown schematically in Figure 9.5. The carrier gas generally is argon, helium, hydrogen, or nitrogen. The sample is injected as a single compact plug into the carrier gas stream immediately ahead of the column entrance. If the sample is liquid, the injection chamber is heated to vaporize the liquid rapidly. The separation column may consist of a metal or glass tube packed with an inert solid of high surface area covered with a liquid phase, or it may consist of an active solid, which enables the separation to occur. More commonly, capillary columns are employed, which consist of very small diameter, very long tubes in which the liquid phase is coated on the inside of the column.

The component that primarily determines the sensitivity of gas chromatographic analysis and, for some classes of compounds, the selectivity as well, is the detector. One such device is the thermal conductivity detector, which responds to changes in the thermal conductivity of gases passing over it. The electron-capture detector, which is especially useful for halogenated hydrocarbons and phosphorus compounds, operates through the capture of electrons emitted by a β-particle source. The flame-ionization gas chromatographic detector is very sensitive for the detection of organic compounds. It is based on the phenomenon by which organic compounds form highly conducting fragments, such as C^+, in a flame. Application of a potential gradient across the flame results in a small current that may be readily measured. The mass spectrometer, described in Section 9.6, may be used as a detector for a GC. A combined GC/MS instrument is an especially powerful analytical tool for organic compounds.

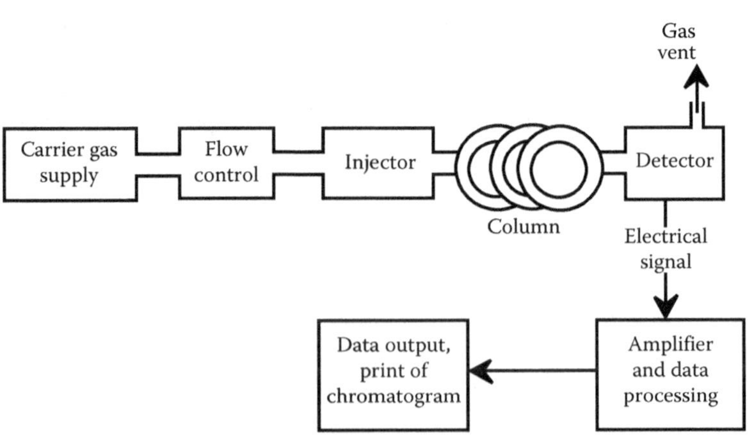

FIGURE 9.5 Schematic diagram of the essential features of a GC. (From Manahan, S., *Environmental Chemistry*, 9th ed., Taylor & Francis, Boca Raton, 2010. With permission.)

Chromatographic analysis requires that a compound exhibit at least a few mm of vapor pressure at the highest temperature at which it is stable. In many cases, organic compounds that cannot be chromatographed directly may be converted to derivatives that are amenable to gas chromatographic analysis. It is seldom possible to analyze organic compounds in water by direct injection of the water into the GC; higher concentration is usually required. Two techniques commonly employed to remove volatile compounds from water and concentrate them are extraction with solvents and purging volatile compounds with a gas, such as helium; concentrating the purged gases on a short column; and driving them off by heat into the chromatograph.

9.5.1 High-Performance Liquid Chromatography

A liquid mobile phase used with very small column-packing particles enables high-resolution chromatographic separation of materials in the liquid phase. Very high pressures up to several thousand psi are required to get a reasonable flow rate in such systems. Analysis using such devices is called *high-performance liquid chromatography (HPLC)* and offers an enormous advantage in that the materials analyzed need not be changed to the vapor phase, a step that often requires preparation of a volatile derivative or results in decomposition of the sample. The basic features of a high-performance liquid chromatograph are the same as those of a gas chromatograph shown in Figure 9.5, except that a solvent reservoir and high-pressure pump are substituted for the carrier gas source and regulator. A hypothetical HPLC chromatogram is shown in Figure 9.6. Refractive index and ultraviolet detectors are both used for the detection of peaks coming from the liquid chromatograph column. Fluorescence detection can be especially sensitive for some classes of compounds.

Mass spectrometric detection of HPLC effluents has led to the development of LC/MS analysis. This powerful tool enables measurement of highly polar water contaminants without the necessity of forming volatile derivatives of analytes to be separated by GC and can be used to measure analytes down to the nanogram per liter levels. LC/MS is especially useful for the determination of a variety of emerging contaminants including pharmaceuticals and their metabolites, hormones, and endocrine-disrupting species.

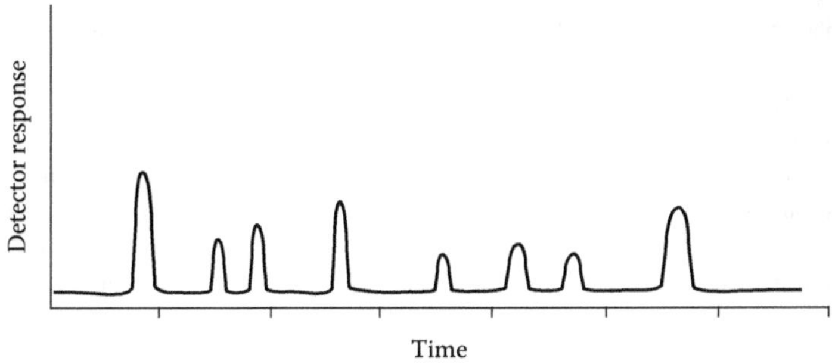

FIGURE 9.6 Hypothetical HPLC chromatogram. (From Manahan, S., *Environmental Chemistry*, 9th ed., Taylor & Francis, Boca Raton, 2010. With permission.)

9.5.2 Chromatographic Analysis of Water Pollutants

The U.S. EPA has developed a number of chromatography-based standard methods for determining water pollutants. Some of these methods use the purge-and-trap technique, bubbling gas through a column of water to purge volatile organics from the water followed by trapping the organics on solid sorbents, whereas others use solvent extraction to isolate and concentrate the organics. These methods are summarized in Table 9.2.

9.5.3 Ion Chromatography

The liquid chromatographic determination of ions, particularly anions, by *ion chromatography* has enabled the measurement of species that used to be very

TABLE 9.2
Chromatography-Based EPA Methods for Organic Compounds in Water

	Method Number			
Class of Compounds	**GC**	**GC/MS**	**HPLC**	**Example Analytes**
Purgeable halocarbons	601			Carbon tetrachloride
Purgeable aromatics	602			Toluene
Acrolein and acrylonitrile	603			Acrolein
Phenols	604			Phenol and chlorophenols
Benzidines			605	Benzidine
Phthalate esters	606			Bis(2-ethylhexylphthalate)
Nitrosamines	607			*N*-nitroso-*N*-dimethylamine
Organochlorine pesticides and PCBs	608			Heptachlor, PCB 1016
Nitroaromatics and isophorone	609			Nitrobenzene
PAHs	610		610	Benzo[*a*]pyrene
Haloethers	611			Bis(2-chloroethyl) ether
Chlorinated hydrocarbons	612			1,3-Dichlorobenzene
2,3,7,8-TCDD		613		2,3,7,8-TCDD
Organophosphorus pesticides	614			Malathion
Chlorinated herbicides	615			Dinoseb
Triazine pesticides	619			Atrazine
Purgeable organics		624		Ethylbenzene
Base/neutrals and acids		625		More than 70 organic compounds
Dinitro aromatic pesticides		646		Basalin (Fluchloralin)
Volatile organic compounds		1624		Vinyl chloride

Source: From Manahan, S., *Environmental Chemistry*, 9th ed., Taylor & Francis, Boca Raton, 2010. With permission.

troublesome for water chemists. The development of this technique has been facilitated by special detection techniques using suppressors to enable detection of analyte ions in the chromatographic effluent. Ion chromatography has been developed for the determination of most of the common anions, including arsenate, arsenite, borate, carbonate, chlorate, chlorite, cyanide, the halides, hypochlorite, hypophosphite, nitrate, nitrite, phosphate, phosphite, pyrophosphate, selenate, selenite, sulfate, sulfite, sulfide, trimetaphosphate, and tripolyphosphate. Cations, including the common metal ions, can also be determined by ion chromatography.

9.6 MASS SPECTROMETRY

Mass spectrometry (MS) is particularly useful for the identification of specific organic pollutants. It depends on the production of ions by an electrical discharge or chemical process, followed by separation based on the charge-to-mass ratio and measurement of the ions produced. The output of a mass spectrometer is a mass spectrum, such as the one shown in Figure 9.7. A mass spectrum is characteristic of a compound and serves to identify it. Computerized data banks for mass spectra have been established and are stored in computers interfaced with mass spectrometers. Identification of a mass spectrum depends on the purity of the compound from which the spectrum is taken. Prior separation by GC with continual sampling of the column effluent by a mass spectrometer, commonly called GC-MS, is particularly effective in the analysis of organic pollutants.

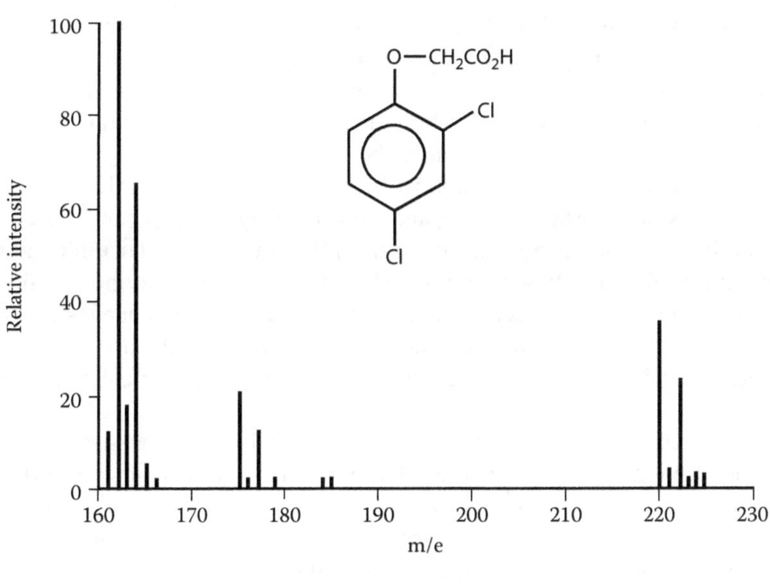

FIGURE 9.7 Partial mass spectrum of the herbicide 2,4-dichlorophenoxyacetic acid (2,4-D), a common water pollutant. (From Manahan, S., *Environmental Chemistry*, 9th ed., Taylor & Francis, Boca Raton, 2010. With permission.)

9.7 ANALYSIS OF WATER SAMPLES

The preceding sections of this chapter have covered the major kinds of analysis techniques that are used on water. In this section, several specific aspects of water analysis are addressed.

9.7.1 Physical Properties Measured in Water

The commonly determined physical properties of water are color, residue (solids), odor, temperature, specific conductance, and turbidity. Most of these terms are self-explanatory and will not be discussed in detail. All of these properties either influence or reflect the chemistry of the water. Solids, for example, arise from chemical substances either suspended or dissolved in the water and are classified physically as total, filterable, nonfilterable, or volatile. Specific conductance is a measure of the degree to which water conducts alternating current and reflects, therefore, the total concentration of dissolved ionic material. By necessity, some physical properties must be measured in the water without sampling (see discussion of water sampling below).

9.7.2 Water Sampling

It is beyond the scope of this text to describe water sampling procedures in detail. It must be emphasized, however, that the acquisition of meaningful data demands that correct sampling and storage procedures be used. These procedures may be quite different for various species in water. In general, separate samples must be collected for chemical and biological analysis because the sampling and preservation techniques differ significantly. Usually, the shorter the time interval between sample collection and analysis, the more accurate the analysis will be. Indeed, some analyses must be performed in the field within minutes of sample collection. Others, such as the determination of temperature, must be performed on the body of water itself. Within a few minutes after collection, water pH may change, dissolved gases (oxygen, carbon dioxide, hydrogen sulfide, and chlorine) may be lost, or other gases (oxygen and carbon dioxide) may be absorbed from the atmosphere. Therefore, analyses of temperature, pH, and dissolved gases should always be performed in the field. Furthermore, precipitation of calcium carbonate accompanies changes in the pH-alkalinity-calcium carbonate relationship following sample collection. Analysis of a sample after standing may thus give erroneously low values for calcium and total hardness.

Oxidation–reduction reactions may cause substantial errors in analysis. For example, soluble iron(II) and manganese(II) are oxidized to insoluble iron(III) and manganese(IV) compounds when an anoxic water sample is exposed to atmospheric oxygen. Microbial activity may decrease phenol or biochemical oxygen demand (BOD) values, change the nitrate–nitrite–ammonia balance, or alter the relative proportions of sulfate and sulfide. Iodide and cyanide are frequently oxidized. Chromium(VI) in solution may be reduced to insoluble chromium(III). Sodium, silicate, and boron are leached from glass container walls.

Samples can be divided into two major categories. *Grab samples* are taken at a single time and in a single place. Therefore, they are very specific with respect to time and location. *Composite samples* are collected over an extended time and may encompass different locations as well. In principle, the average results from a large number of grab samples give the same information as a composite sample. A composite sample has the advantage of providing an overall picture from only one analysis. On the other hand, it may miss extreme concentrations and important variations that occur over time and space.

9.7.2.1 Extractors

The ease and effectiveness of various kinds of solid-phase devices for water sampling is steadily increasing their use in water analysis. Based on size and physical configuration, several categories of such devices are available. One of these is the conventional solid-phase extractor (SPE) containing an extracting solid in a column. Activated carbon has been used for decades for this purpose, but synthetic materials, such as those composed of long hydrocarbon chains (C18) bound to solids have been found to be quite useful. A typical procedure uses a polymer-divinylstyrene extraction column to remove pesticides from water. The pesticide analytes are eluted from the SPE with ethyl acetate and measured by GC.

Solid-phase microextraction (SPME) devices constitute a second kind of SPE. These make use of very small diameter tubes in which analytes are bonded directly to the extractor walls, and then eluted directly into a chromatograph. The use of SPME devices for the determination of haloethers in water has been described.

A third kind of device that is simple, convenient, and available for a number of classes of compounds consists of disks composed of substances that bind with and remove analytes from water when the water is filtered through them. As an example, solid-phase extraction disks can be used to remove and concentrate radionuclides from water, including ^{99}Tc, ^{137}Cs, ^{90}Sr, and ^{238}Pu. Organic materials sampled from water with such disks include haloacetic acids and acidic and neutral herbicides.

Liquid–liquid extraction of organic-soluble analytes has long been used to extract these materials from water and isolate them in a solvent that can be reduced to a small volume in which analytes can be further concentrated by a Kuderna–Danish evaporator or nitrogen blowing. Relatively high consumption of solvent (creating a disposal problem) and emulsion formation can be problems with liquid–liquid extraction. As an alternative for volatile and semivolatile analytes, membrane extraction can be used. This procedure entails running the water sample over one side of a membrane through which the analytes migrate to another solvent on the other side. Water can even be used as a collector solvent. In the case of ionizable organics, for example, the pH on the sample side can be adjusted such that the analyte species are neutral, in which form they can migrate through an organophilic membrane, and be collected in an aqueous solution that is at a pH adjusted such that the analyte species are ionized and hence cannot permeate back through the membrane. A procedure exists for the determination of semivolatile compounds in water in which the extracting solvent traverses the interior of hollow fibers and collects analytes from the water sample flowing over the outer surface of the fibers. By pressurizing the solvent, a significant fraction of it is forced through the membrane into the water thereby concentrating the organic-soluble analytes in the solvent phase.

9.7.3 Water Sample Preservation

It is not possible to completely protect a water sample from changes in composition. However, various additives and treatment techniques can be employed to minimize sample deterioration. These methods are summarized in Table 9.3.

The most general method of sample preservation is refrigeration to 4°C. Freezing normally should be avoided because physical changes including precipitate formation and gas loss may adversely affect sample composition. Acidification is commonly applied to metal samples to prevent their precipitation, and it also slows microbial action. In the case of metals, the samples should be filtered before adding acid to enable determination of dissolved metals. Sample holding times vary from zero for parameters such as temperature or DO measured by a probe to 6 months for metals. Many different kinds of samples, including those to be analyzed for acidity, alkalinity, and various forms of nitrogen or phosphorus, should not be held for more than 24 h. Details on water sample preservation are to be found in standard references on water analysis listed at the end of this chapter.

9.7.4 Total Organic Carbon in Water

It is important to measure dissolved organic carbon (Chapter 7) in water because it exerts an oxygen demand in water, often is in the form of toxic substances, and is a general indicator of water pollution. The measurement of total organic carbon, TOC, in water is performed by methods that, for the most part, totally oxidize the dissolved

TABLE 9.3
Preservatives and Preservation Methods Used with Water Samples

Preservative or Technique Used	Effect on Sample	Type of Samples for Which the Method is Employed
Nitric acid	Keeps metals in solution	Metal-containing samples
Sulfuric acid	Bactericide	Biodegradable samples containing organic carbon, oil, or grease
	Formation of sulfates with volatile bases	Samples containing amines or ammonia
Sodium hydroxide	Formation of sodium salts from volatile acids or cyanides	Samples containing volatile organic acids
Chemical	Fix a particular reaction constituent	Samples to be analyzed for DO using the Winkler method

Source: From Manahan, S., *Environmental Chemistry*, 9th ed., Taylor & Francis, Boca Raton, 2010. With permission.

organic material to produce carbon dioxide. The amount of carbon dioxide evolved is taken as a measure of TOC.

TOC can be determined by a technique that uses a dissolved oxidizing agent, commonly potassium peroxydisulfate, $K_2S_2O_8$, promoted by ultraviolet light to oxidize the organic carbon to CO_2. Phosphoric acid is added to the sample, which is sparged with nitrogen to drive off inorganic CO_2. After sparging, the sample is pumped to a chamber containing a lamp emitting ultraviolet radiation of 184.9 nm. This radiation produces reactive free radical species such as the hydroxyl radical, HO•, (see Chapter 1, Section 1.4). These active species bring about the rapid oxidation of dissolved organic compounds as shown in the following general reaction:

$$\text{Organics} + \text{HO}\bullet \rightarrow CO_2 + H_2O \tag{9.13}$$

After the completion of oxidation, the CO_2 is sparged from the system and measured with a gas chromatographic detector or by absorption in ultrapure water followed by a conductivity measurement. Figure 9.8 is a schematic of a TOC analyzer.

9.7.5 Measurement of Radioactivity in Water

There are several potential sources of radioactive materials that may contaminate water (see Section 7.13). Radioactive contamination of water is normally detected by measurements of gross β and gross α activity, a procedure that is simpler than detecting individual isotopes. The measurement is made from a sample formed by evaporating water to a very thin layer on a small pan, which is then inserted inside an internal proportional counter. This setup is necessary because β particles can penetrate only very thin detector windows, and α particles have essentially no penetrating power. More detailed information can be obtained for radionuclides that emit γ rays by the use of γ spectrum analysis. This technique employs solid-state detectors to

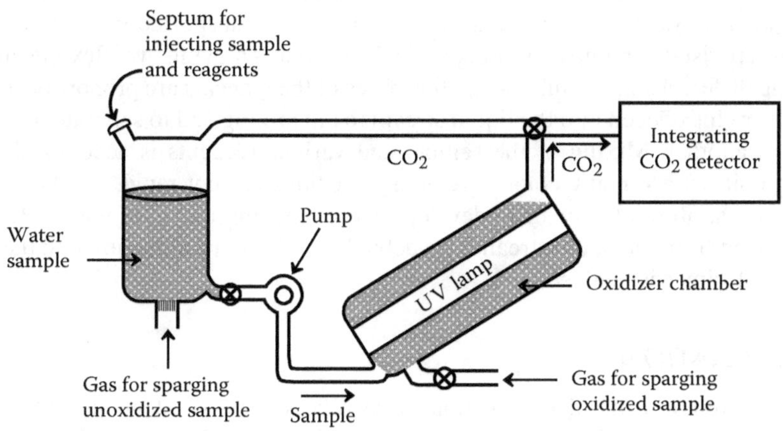

FIGURE 9.8 TOC analyzer employing UV-promoted sample oxidation. (From Manahan, S., *Environmental Chemistry*, 9th ed., Taylor & Francis, Boca Raton, 2010. With permission.)

resolve rather closely spaced γ peaks in the sample's spectra. In conjunction with multichannel spectrometric data analysis, it is possible to determine a number of radionuclides in the same sample without chemical separation. This method requires minimal sample preparation.

9.7.6 Biological Toxins

Toxic substances produced by microorganisms are of some concern in water. Photosynthetic cyanobacteria and some kinds of algae growing in water produce potentially troublesome toxic substances. An HPLC/MS method of analysis for such toxins has been described.[7]

9.7.7 Summary of Water Analysis Procedures

The main chemical parameters commonly determined in water are summarized in Table 9.4. In addition to these, a number of other solutes, especially specific organic pollutants, may be determined in connection with specific health hazards or incidents of pollution.

9.8 AUTOMATED WATER ANALYSES

Huge numbers of water analyses must often be performed in order to get meaningful results and for reasons of economics. This has resulted in the development of a number of automated procedures in which the samples are introduced through a sampler and the analyses performed and results posted without manual manipulation of reagents and apparatus. Such procedures have been developed and instruments marketed for the determination of a number of analytes, including alkalinity, sulfate, ammonia, nitrate/nitrite, and metals. Colorimetric procedures are popular for such automated analytical instruments, using simple, rugged colorimeters for absorbance measurements. Figure 9.9 shows an automated analytical system for the determination of alkalinity. The reagents and sample liquids are transported through the analyzer by a peristaltic pump consisting basically of rollers moving over flexible tubing. By using different sizes of tubing, the flow rates of the reagents are proportioned. Air bubbles are introduced into the liquid stream to aid mixing and to separate one sample from another. Mixing of the sample and various reagents is accomplished in mixing coils. Since many color-developing reactions are not rapid, a delay coil is provided that allows the color to develop before reaching the colorimeter. Bubbles are removed from the liquid stream by a debubbler prior to introduction into the flow cell for colorimetric analysis.

9.9 SPECIATION

Speciation refers to the determination of specific species of elements rather than simply analysis of the elements. It applies especially to metals and metalloids, which may be relatively easy to measure as the element, but much more difficult to determine as specific species of the element. For example, it is a relatively straightforward

TABLE 9.4
Chemical Parameters Commonly Determined in Water

Chemical Species	Significance in Water	Methods of Analysis
Acidity	Indicative of industrial pollution or acid mine drainage	Titration
Alkalinity	Water treatment, buffering, algal productivity	Titration
Aluminum	Water treatment, buffering	AA[a], ICP[b]
Ammonia	Algal productivity, pollutant	Spectrophotometry
Arsenic	Toxic pollutant	Spectrophotometry, AA, ICP
Barium	Toxic pollutant	AA, ICP
Beryllium	Toxic pollutant	AA, ICP, fluorimetry
Boron	Toxic to plants	Spectrophotometry, ICP
Bromide	Seawater intrusion, industrial waste	Spectrophotometry, potentiometry, ion chromatography
Cadmium	Toxic pollutant	AA, ICP
Calcium	Hardness, productivity, treatment	AA, ICP, titration
Carbon dioxide	Bacterial action, corrosion	Titration, calculation
Chloride	Saline water contamination	Titration, electrochemical, ion chromatography
Chlorine	Water treatment	Spectrophotometry
Chromium	Toxic pollutant (hexavalent Cr)	AA, ICP, colorimetry
Copper	Plant growth	AA, ICP
Cyanide	Toxic pollutant	Spectrophotometry, potentiometry, ion chromatography
Fluoride	Water treatment, toxic at high levels	Spectrophotometry, potentiometry, ion chromatography
Hardness	Water quality, water treatment	AA, titration
Iodide	Seawater intrusion, industrial waste	Catalytic effect, potentiometry, ion chromatography
Iron	Water quality, water treatment	AA, ICP, colorimetry
Lead	Toxic pollutant	AA, ICP, voltammetry
Lithium	May indicate some pollution	AA, ICP, flame photometry
Magnesium	Hardness	AA, ICP
Manganese	Water quality (staining)	AA, ICP
Mercury	Toxic pollutant	Flameless atomic absorption
Methane	Anaerobic bacterial action	Combustible-gas indicator
Nitrate	Algal productivity, toxicity	Spectrophotometry, ion chromatography
Nitrite	Toxic pollutant	Spectrophotometry, ion chromatography
Nitrogen (albuminoid)	Proteinaceous material	Spectrophotometry
(organic)	Organic pollution indicator	Spectrophotometry
Oil and grease	Industrial pollution	Gravimetry
Organic carbon	Organic pollution indicator	Oxidation-CO_2 measurement
Organic contaminants	Organic pollution indicator	Activated carbon adsorption
Oxygen	Water quality	Titration, electrochemical

TABLE 9.4 (continued)
Chemical Parameters Commonly Determined in Water

Chemical Species	Significance in Water	Methods of Analysis
Oxygen demand		
(biochemical)	Water quality and pollution	Microbiological titration
(chemical)	Water quality and pollution	Chemical oxidation-titration
Ozone	Water treatment	Titration
Pesticides	Water pollution	GC
pH	Water quality and pollution	Potentiometry
Phenols	Water pollution	Distillation-colorimetry
Phosphate	Productivity, pollution	Spectrophotometry
Phosphorus	Water quality and pollution (hydrolyzable)	Spectrophotometry
Potassium	Productivity, pollution	AA, ICP, flame photometry
Selenium	Toxic pollutant	Spectrophotometry, ICP, neutron activation
Silica	Water quality	Spectrophotometry, ICP
Silver	Water pollution	AA, ICP
Sodium	Water quality, saltwater intrusion	AA, ICP, flame photometry
Strontium	Water quality	AA, ICP, flame photometry
Sulfate	Water quality, water pollution	Ion chromatography
Sulfide	Water quality, water pollution	Spectrophotometry, titration, chromatography
Sulfite	Water pollution, oxygen scavenger	Titration, ion chromatography
Surfactants	Water pollution	Spectrophotometry
Tannin, lignin	Water quality, water pollution	Spectrophotometry
Vanadium	Water quality, water pollution	ICP
Zinc	Water quality, water pollution	AA, ICP

Source: From Manahan, S., *Environmental Chemistry*, 9th ed., Taylor & Francis, Boca Raton, 2010. With permission.

[a] AA denotes atomic absorption.

[b] ICP stands for inductively coupled plasma techniques in which the atoms in the atomized plasma are detected by atomic emission or by MS.

process to measure total tin in water, but much more challenging (and important) to determine organotin species that might be present as pollutants. Arsenic provides an important example of significant elemental species in water. Inorganic arsenic can be present as either As(III) or As(V), of which the former is generally more toxic. Organoarsenic compounds can be produced microbially and include monomethylarsonate and dimethylarsinate. Organoarsenic compounds are also used as additives to livestock feed and get into water supplies as runoff from feeding operations; roxarsone is one such compound.

The most generally applicable means of measuring elemental species is separation by chromatography followed by measurement of elemental levels in the

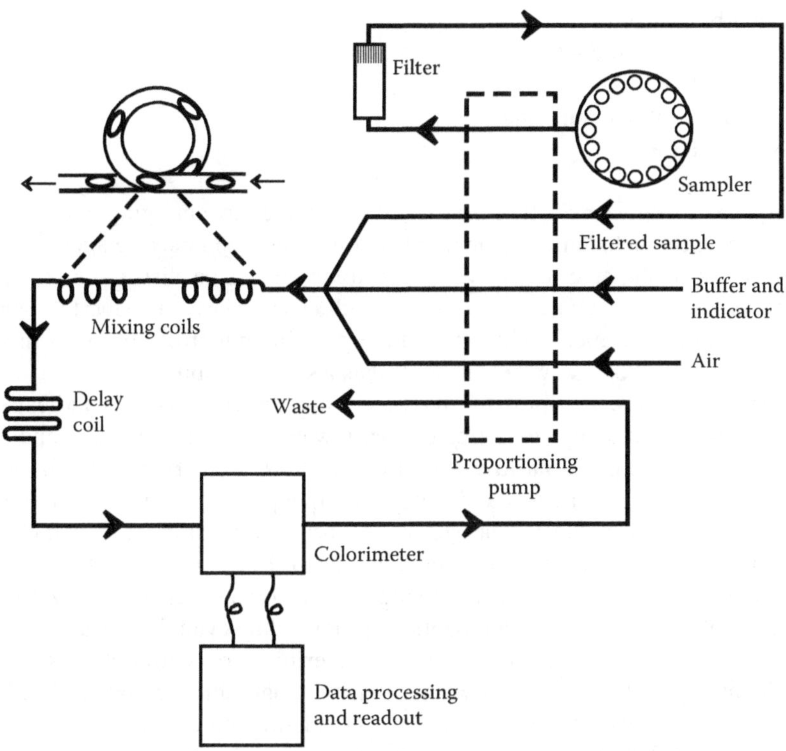

FIGURE 9.9 Automated analyzer system for the determination of total alkalinity in water. Addition of a water sample to a methyl orange solution buffered to pH 3.1 causes a loss of color in proportion to the alkalinity in the sample. (From Manahan, S., *Environmental Chemistry*, 9th ed., Taylor & Francis, Boca Raton, 2010. With permission.)

effluent. Specifically for arsenic, separation by anion exchange chromatography followed by measurement of arsenic with inductively coupled plasma-MS is a good technique.

9.10 EMERGING CONTAMINANTS IN WATER ANALYSIS

Emerging contaminants in water analysis refers to species that have been recognized as significant water contaminants relatively recently, the long-term effects of which are not very well known. Approximately 3000 different substances are now used as pharmaceutical ingredients including antibiotics, painkillers, antidiabetics, contraceptives, antidepressants, impotence drugs, and α blockers and these substances and their metabolites get into water at very low levels. Major classes of emerging contaminants include the following:

- Pharmaceuticals and their metabolites
- Endocrine disruptors
- Polybrominated compounds

- Microbial toxins
- Personal care products
- Disinfection products
- Organometallic compounds
- Nanomaterials

Some specific substances that have been recognized as emerging contaminants relatively recently include the following: benzotriazoles, complexing agents used as anticorrosives; naphthenic acids, which are toxic endocrine disruptors remaining from the extraction of crude oil from tar sands, particularly in Alberta, Canada; ethylenedibromide, formerly added to leaded gasoline but still a problem in some groundwaters; and dioxane, widely produced as a solvent stabilizer.

Emerging contaminants in drinking water supplies are generally indicative of contamination by incompletely treated wastewater. Pharmaceuticals and their metabolites have the potential to cause direct health effects. Endocrine disruptors may affect health as well as sexual development in aquatic organisms. The organometallic compounds of most concern are the organotin compounds, which have industrial uses and have been used as biocides. Microbial toxins include substances produced by photosynthetic bacteria and algae growing in water supplies. An example consists of microcystins, cyclic peptides produced by cyanobacteria.

A variety of analytical methods are used to determine concentrations of emerging contaminants in water. Especially useful are chromatographic methods coupled with MS or tandem MS (two mass spectrometers in tandem). Some of the specific emerging contaminants that can be determined by solid-phase extraction of water contaminants measured by liquid chromatography/tandem mass spectrometry (LC/MS/MS) using electrospray ionization include meprobamate, ethynylestradiol, progesterone, caffeine, testosterone, androstenedione, acetaminophen, erythromycin, TCEP, hydrocodone, oxybenzone, DEET, carbamazepine, diazepam, dilantin, sulfamethoxazole, trimethoprim, atrazine, triclosan, pentoxifylline, diclofenac, ibuprofen, naproxen, gemfibrozil, fluoxetine, and iopromide. Including pharmaceuticals, steroids, personal care products, and pesticides, this list gives an idea of compounds included among emerging contaminants.

9.11 CHIRAL CONTAMINANTS

Chiral molecules are those in which four different groups are arranged in three dimensions around an atom, usually of carbon, that constitutes a chiral center (see alanine below) such that the molecule cannot be directly superimposed over its mirror image. Two chiral molecules of the same compound, called *enantiomers*, have identical melting points, boiling points, and solubilities as well as generally identical chemical properties. They differ, often markedly, in their biochemical properties because of the exact fit required between enzyme active sites and the substrate molecules upon which the enzymes act. Therefore, chiral isomers may exhibit completely different environmental and toxicological behaviors. For example, one enantiomer of a herbicide may undergo biodegradation readily whereas the other is resistant to it. Biochemical differences between enantiomers may be especially pronounced for pesticides. For example, one enantiomer of a herbicide may kill weeds very effectively whereas the other has no effect. An example is the *R* enantiomer of herbicidal mecoprop (Figure 7.13), which is now marketed as a herbicide in place of the racemic mixture with the *S* enantiomer.

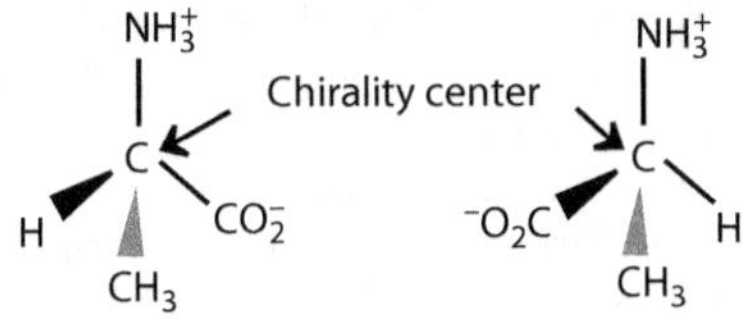

The two enantiomers of the amino acid alanine

Although important, the ability to determine specific enantiomers of contaminants in water is relatively new. Most commonly, the compounds are separated by LC and measured by MS. This requires a separation medium that is selective for particular enantiomers. Cyclodextrins are most commonly used for this purpose.

LITERATURE CITED

1. Eaton, A. D., L. S. Clesceri, E. W. Rice, A. E. Greenberg, and M. A. H. Franson, Eds, Data quality, Section 1030 in *Standard Methods for the Examination of Water and Wastewater*, 21st ed., American Public Health Association, Washington, DC, pp. 1-13–1-22, 2005.
2. Eaton, A. D., L. S. Clesceri, E. W. Rice, A. E. Greenberg, and M. A. H. Franson, Eds, *Standard Methods for the Examination of Water and Wastewater*, 21st ed., American Public Health Association, Washington, DC, 2005.
3. *Index to EPA Test Methods*, April 2003 revised, http://www.epa.gov/ne/info/testmethods/
4. *Methods and Guidance for the Analysis of Water*, Version 2.0 on CD ROM including EPA Series, 500, 600, and 1600 Methods, National Technical Information Service (NTIS), U.S. Department of Commerce, Springfield, VA, 2003.
5. *Understanding Environmental Methods*, Version 5.0 on CD ROM, Genium Publishing Corp., Amsterdam, NY, 2009.
6. Richardson, S. D., Water analysis: Emerging contaminants and current issues, *Analytical Chemistry*, 79, 4295–4324, 2007.
7. Hedmana, C. J., W. R. Kricka, D. A. Karner Perkinsa, E. A. Harrahyb, and W. C. Sonzognia, New measurements of cyanobacterial toxins in natural waters using high performance liquid chromatography coupled to tandem mass spectrometry, *Journal of Environmental Quality*, 37, 1817–1824, 2008.
8. Manahan, S., *Environmental Chemistry*, 9th ed., Taylor & Francis, Boca Raton, 2010.

SUPPLEMENTARY REFERENCES

AWWA Staff, *Simplified Procedures for Water Examination*, 5th ed., American Water Works Association, Denver, CO, 2002.

AWWA Staff, *Basic Laboratory Procedures for Wastewater Examination*, 4th ed., Water Environment Federation, Alexandria, VA, 2002.

Crompton, T. R., *Determination of Anions in Natural and Treated Waters*, Taylor & Francis, New York, 2002.

Crompton, T. R., *Chromatography of Natural and Treated Waters*, Taylor & Francis, New York, 2005.

Crompton, T. R., *Determination of Metals in Natural and Treated Waters*, Taylor & Francis, New York, 2002.

Dieken, F. P., *Methods Manual for Chemical Analysis of Water and Wastes*, Alberta Environmental Centre, Vergeville, Alberta, Canada, 1996.

Eaton, A. D., L. S. Clesceri, E. W. Rice, A. E. Greenberg, and M. A. H. Franson, Eds, *Standard Methods for the Examination of Water and Wastewater*, 21st ed., American Public Health Association, Washington, DC, 2005.

Kolle, W., *Wasseranalysen*, Wiley, New York, 2002.

Lewinsky, A. A., Ed., *Hazardous Materials and Wastewater: Treatment, Removal and Analysis*, Nova Science Publishers, New York, 2007.

Meyers, R. A., Ed., *The Encyclopedia of Environmental Analysis and Remediation*, Wiley, New York, 1998.

Nollet, L. M. L., Ed., *Handbook of Water Analysis*, 2nd ed., Taylor & Francis/CRC Press, Boca Raton, FL, 2007.

Patnaik, P., *Handbook of Environmental Analysis: Chemical Pollutants in Air, Water, Soil, and Solid Wastes*, CRC Press/Lewis Publishers, Boca Raton, FL, 1997.

Quevauviller, P. and C. Thompson, *Analytical Methods for Drinking Water: Advances in Sampling and Analysis*, Wiley, Hoboken, NJ, 2006.

Rump, H. H., *Laboratory Manual for the Examination of Water*, 3rd ed., Wiley, New York, 2001.

Tomar, M., *Laboratory Manual for the Quality Assessment of Water and Wastewater*, CRC Press/Lewis Publishers, Boca Raton, FL, 1999.

U.S. Environmental Protection Agency, Links to EPA and non-EPA methods of analysis for drinking water, www.epa.gov/safewater/methods/methods.html

U.S. Environmental Protection Agency, Links to methods developed by the EPA's Office of Ground Water and Drinking Water, www.epa.gov/safewater/methods/sourcalt.html

U.S. Environmental Protection Agency, Links to methods for the determination of microorganisms in water, www.epa.gov/nerlcwww

U.S. Environmental Protection Agency, Links to drinking water and marine water methods recently developed by the EPA's Office of Research and Development, www.epa.gov/nerlcwww/ordmeth.htm

U.S. Environmental Protection Agency, Information regarding local drinking water, www.epa.gov/safewater/dwinfo

QUESTIONS AND PROBLEMS

1. A soluble water pollutant forms ions in solution and absorbs light at 535 nm. What are the two physical properties of water influenced by the presence of this pollutant?
2. A sample was taken from the bottom of a deep, stagnant lake. Upon standing, bubbles evolved from the sample; the pH went up; and a white precipitate formed. From these observations, what may be said about the dissolved CO_2 and hardness in the water?
3. For which of the following analytes may nitric acid be used as a water sample preservative: H_2S; CO_2; metals; coliform bacteria; and cyanide?
4. In the form of what compound is oxygen fixed in the Winkler analysis of O_2?
5. Of the following analytical techniques, the water analysis techniques that would best distinguish between the hydrated $Ag(H_2O)_6^+$ ion and the complex

$Ag(NH_3)_2^+$ ion by direct measurement of the uncomplexed ion are (a) neutron-activation analysis, (b) atomic absorption, (c) inductively coupled plasma atomic emission spectroscopy, (d) potentiometry, and (e) flame emission.

6. A water sample was run through the colorimetric procedure for the analysis of nitrate, giving 55.0% transmittance. A sample containing 1.00 ppm nitrate run through the exactly identical procedure gave 24.6% transmittance. What was the concentration of nitrate in the first sample?
7. What is the molar concentration of HCl in a water sample containing HCl as the only contaminant and having a pH of 3.80?
8. A 200-mL sample of water required 9.12 mL of 0.0200 N standard H_2SO_4 for titration to the methyl orange endpoint, pH 4.5. What was the total alkalinity of the original sample?
9. Analysis of a lead-containing sample by graphite-furnace atomic absorption analysis gave a peak of 0.075 absorbance units when 50 μL of pure sample was injected. Lead was added to the sample such that the added concentration of lead was 6.0 μg/L. Injection of 50 μL of "spiked" sample gave an absorbance of 0.115 absorbance units. What was the concentration of lead in the original sample?
10. In a 2.63×10^{-4} M standard fluoride solution, a fluoride electrode read −0.100 V versus a reference electrode, and it read −0.118 V in an appropriately processed fluoride sample. What was the concentration of fluoride in the sample?
11. The activity of iodine-131 ($t_{1/2} = 8$ days) in a water sample 24 days after collection was 520 pCi/L. What was the activity on the day of collection?
12. Neutron irradiation of exactly 2.00 mL of a standard solution containing 1.00 mg/L of unknown heavy metal "X" for exactly 30 s gave an activity of 1257 counts per minute, when counted exactly 33.5 min after the irradiation, measured for a radionuclide product of "X" having a half-life of 33.5 min. Irradiation of an unknown water sample under identical conditions (2.00 mL, 30.0 s, and the same neutron flux) gave 1813 counts per minute when counted 67.0 min after irradiation. What was the concentration of "X" in the unknown sample?
13. Why is magnesium-EDTA chelate added to a magnesium-free water sample before it is to be titrated with EDTA for Ca^{2+}?
14. For what type of sample is the flame-ionization detector most useful?
15. Manganese from a standard solution was oxidized to MnO_4^- and diluted such that the final solution contained 1.00 mg/L of Mn. This solution had an absorbance of 0.316. A 10.00 mL wastewater sample was treated to develop the MnO_4 color and diluted to 250.0 mL. The diluted sample had an absorbance of 0.296. What was the concentration of Mn in the original wastewater sample?

10 Sustainable Energy
The Key to Everything

10.1 THE ENERGY CHALLENGE

This chapter on energy is included in a book about water chemistry because of the very close relationship between water and energy. Any meaningful consideration of water availability and quality must consider energy. And the provision of sustainable energy requires consideration of the availability of water and the effect that providing energy has upon water resources. One of the greater uses of water is for cooling electric power plants. Greenhouse warming largely results from carbon dioxide emitted to the atmosphere in the combustion of fossil fuels. Sustainable biomass energy sources are dependent upon water to grow plants; algae can be grown in water as a renewable biomass energy resource. With abundant sustainable energy, water supplies can be vastly extended by means including purification of municipal wastewater to drinking water standards, desalination of brackish groundwater, and even desalination of seawater. Many other correlations can be drawn between water and energy.

It is a rather bold claim that sustainable energy is "the key to everything." But a convincing argument can be made that most problems in the physical environment can be solved to at least a large degree if enough energy is available, if it is inexpensive enough, and if it can be used without doing irreparable environmental harm. Consider the following environmental and sustainability problems that can be solved at least in part with sufficient sustainable energy:

- Water: With enough energy, wastewater can be reclaimed to drinking water standards by reverse osmosis and other energy-consuming technologies and seawater can be desalinated.
- Food: With enough energy, marginal land can be reclaimed by measures such as leveling, terracing, and rock removal and irrigation water can be pumped from long distances or desalinated to grow food. Greenhouses can be heated even during winter months to grow high-value specialty foods.
- Wastes: The disposal of hazardous organic wastes in landfills, though widely practiced, is not a good idea. With sufficient energy, such wastes can be converted to forms that cannot do harm.
- Transportation: With sufficient sustainable energy, transportation problems can be solved by technologies such as electrified railways.
- Fuels: Biomass sources of fixed carbon can be converted to hydrocarbon fuels for applications for which there are no viable alternatives (such as aircraft) without adding any net amounts of greenhouse gas carbon dioxide to the atmosphere.

The list above can be extended to many other areas and to a large number of sustainability and environmental problems. The great challenge is, of course, that systems of energy utilization have developed that are unsustainable. One of the most obvious sustainability challenges is that humankind is running out of the sources of energy around which its economic systems have become based. This is especially true of petroleum. Peak U.S. petroleum production was passed several years ago, and it is likely that peak world production will be reached within a few years of 2010. Exorbitantly high petroleum prices in the first part of 2008 followed by a precipitous fall in price as world economies collapsed at the end of the year emphasized the economic uncertainties of reliance on petroleum as an energy source, especially by nations that do not have domestic sources. There are still abundant resources of coal, but their utilization to provide for energy with current technology almost certainly will cause unacceptable global warming. Therefore, the great challenge facing humankind during the next several decades is to develop sources of energy that will meet energy needs and that will not ruin Earth and its climate.

There are energy alternatives to fossil fuels that can be developed, that are environmentally safe (or can be made so), and that, taken in total, can be adequate to supply energy needs. These include wind, solar, biomass, geothermal, and nuclear energy sources. Some other miscellaneous sources, such as tidal energy, may contribute as well. Fossil fuels will continue to be used and may contribute sustainably for decades with sequestration of greenhouse gas carbon dioxide. And, of course, energy conservation and greatly enhanced efficiency of energy use will make substantial contributions. This chapter discusses the energy alternatives listed above with emphasis on energy sustainability.

10.2 NATURE OF ENERGY

Energy is the capacity to do work (basically, to move matter around) or *heat* in the form of the movement of atoms and molecules. *Kinetic energy* is contained in moving objects. One such example is the energy contained in a rapidly spinning flywheel, a device of potentially great importance for energy storage to even out the energy flow from intermittent solar and wind sources. *Potential energy* is stored energy, such as in an elevated reservoir of water used as a means of storing hydroelectric energy for later use that can be run through a hydroelectric turbine to generate electricity as needed.

A very important form of potential energy is *chemical energy* stored in the bonds of molecules and released, usually as heat, when chemical reactions occur. For example, in the case of methane, CH_4, in natural gas, when the methane burns,

$$CH_4 + 2O_2 \rightarrow CO_2 + 2H_2O \tag{10.1}$$

the difference between the bond energies in the CO_2 and H_2O products and the CH_4 and O_2 reactants is released, primarily in the form of heat. If the heat is released by combustion of methane in a gas turbine, part of the heat energy can be converted to *mechanical energy* in the form of the rapidly spinning turbine and an electrical generator to which it is attached. The generator, in turn, converts the mechanical energy to *electrical energy*.

The standard unit of energy is the *joule*, abbreviated *J*. A total of 4.184 J of heat energy will raise the temperature of 1 g of liquid water by 1°C. This amount of heat is equal to 1 *calorie* of energy (1 cal = 4.184 J), the unit of energy formerly used in scientific work. A joule is a small unit, and the kilojoule, kJ, equal to 1000 J is widely used in describing chemical processes. The "calorie" commonly used to express the energy value of food (and its potential to produce fat) is actually a kilocalorie, kcal, equal to 1000 cal.

Power refers to energy generated, transmitted, or used per unit time. The unit of power is the *watt*, equal to an energy flux of 1 J/s). A compact fluorescent light bulb adequate to illuminate a desk area might have a rating of 21 W. A large power plant may put out electricity at a power level of 1000 *megawatts* (MW, where 1 MW is equal to 1 million watts). Power on a national or global scale is often expressed in *gigawatts*, each one of which is equal to a billion watts or even *terawatts*, where a terawatt is equal to a trillion watts.

The science that deals with energy in its various forms and with work is *thermodynamics*. There are some important laws of thermodynamics. The *first law of thermodynamics* states that energy is neither created nor destroyed. This law is also known as the *law of conservation of energy*. The first law of thermodynamics must always be kept in mind in the practice of green technology, the best practice of which requires the most efficient use of energy. Thermodynamics enables calculation of the amount of usable energy. As described by the laws of thermodynamics, only a relatively small amount of the potential energy in fuel can be converted to mechanical energy or electrical energy with the remaining energy from the combustion of the fuel dissipated as heat. With the application of green technology, much of this heat is salvaged for applications such as district heating of homes.

Although energy is neither created nor destroyed, the amount of useful energy that can be obtained from a system can be dissipated. Useful mechanical energy can be produced, for example, in a heat engine by harnessing part of the energy flow from a hot to a cooler part of the system (Section 10.4 and Equation 10.5). The amount of this useful energy is called *exergy* and it can go to zero in a system that has reached equilibrium.

10.3 SOURCES OF ENERGY USED IN THE ANTHROSPHERE

Before the 1800s, most of the energy used in the anthrosphere came from biomass produced during plant photosynthesis. Houses were heated with wood. Soil was cultivated and goods and people moved by the power of animals or by humans themselves, obtaining their energy from food biomass. Wind drove sailing ships and windmills and falling water moved waterwheels. These sources were renewable and sustainable, with solar energy captured by photosynthesis to generate biomass, wind produced by temperature and pressure differences in solar-heated masses of atmospheric air, and flowing water moved as part of the solar-driven hydrologic cycle.

Although coal from readily accessible deposits had long been used in small quantities for home heating, exploitation of this energy source grew rapidly after the invention of the steam engine around 1800. During the 1800s, coal became the predominant source of energy in the United States, England, Europe, and other countries that had readily accessible coal resources, a major shift from renewable biomass, wind, and water to a depletable resource that had to be dug from the ground. By 1900, petroleum had become a significant source of energy and by 1950 had surpassed coal as the United States' leading energy supply. Lagging behind petroleum, natural gas had become a significant energy supplier by 1950. By 1950, hydroelectric power was providing a large fraction of energy used in the anthrosphere and still is. By around 1975, nuclear energy was supplying significant amounts of electricity and has maintained an appreciable share worldwide until the present. Miscellaneous renewable sources including geothermal and, more recently, solar and wind energy are making increasing contributions to total energy supply. Biomass still contributes significantly to the total of the sources of energy used.

Figure 10.1 shows U.S. and world energy sources used annually as of the year 2000. The predominance of *fossil fuel* petroleum, natural gas, and coal are obvious. Estimates of the amounts of fossil fuels available differ. Those of the quantities of recoverable fossil fuels in the world before 1800 as estimated around 1970 are given in Figure 10.2. By far, the greatest recoverable fossil fuel is in the form of coal and lignite.

Although world coal resources are enormous and potentially can fill energy needs for a century or two, their utilization would become intolerable to earth's environment because of environmental disruption from mining and emissions of carbon dioxide long before coal resources were exhausted. Assuming only uranium-235 as a fission fuel source, total recoverable reserves of nuclear fuel are roughly about the same as fossil fuel reserves. These are several orders of magnitude higher if the use of breeder reactors that convert normally unfissionable uranium-238 to fissionable plutonium-239 is assumed. Extraction of only 2% of the deuterium present in the earth's oceans would yield about a billion times as much energy by controlled nuclear fusion as was originally present in fossil fuels! This prospect is tempered by the lack of success in developing a controlled nuclear fusion reactor. Geothermal power, currently utilized in northern California, Italy, Iceland, and New Zealand, has the potential for providing a significant percentage of energy worldwide. The same limited potential is

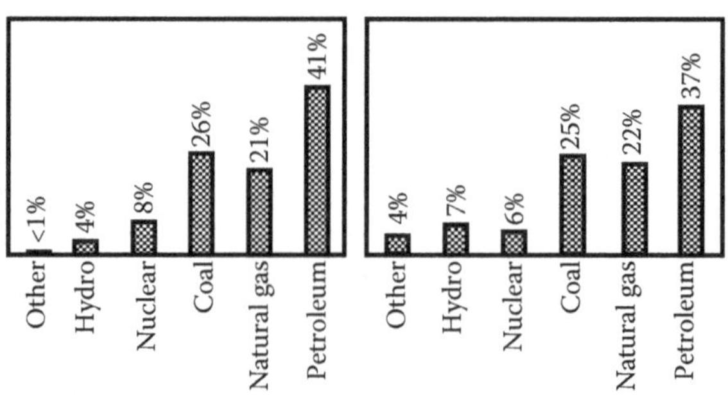

FIGURE 10.1 U.S. (left) and world (right) sources of energy. Percentages of total rounded to the nearest 1%. (From Manahan, S., *Environmental Chemistry*, 9th ed., Taylor & Francis, Boca Raton, 2010. With permission.)

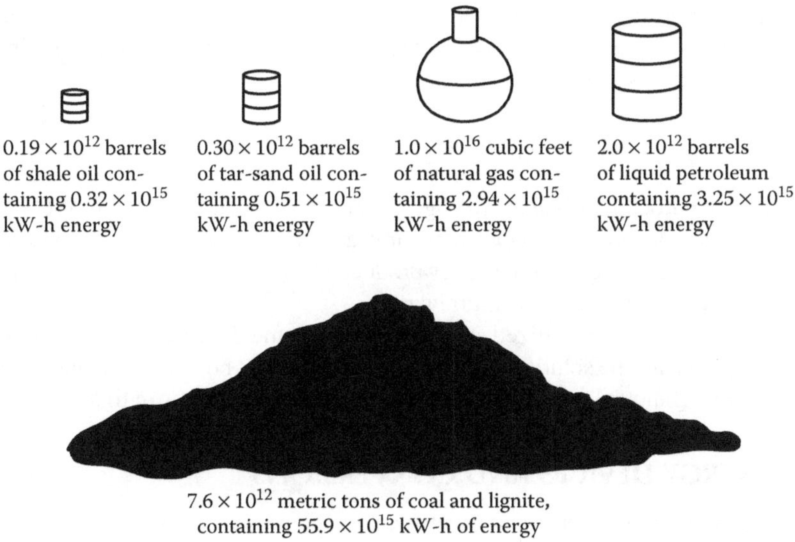

FIGURE 10.2 Original amounts of the world's recoverable fossil fuels. (Quantities in thermal kilowatt hours of energy based on data taken from M. K. Hubbert, "The energy resources of the earth," in *Energy and Power*, W. H. Freeman and Co., San Francisco, 1971.) (From Manahan, S., *Environmental Chemistry*, 9th ed., Taylor & Francis, Boca Raton, 2010. With permission.)

characteristic of several renewable energy resources, including hydroelectric energy, tidal energy, and especially wind power. All of these will continue to contribute significant amounts of energy. Renewable, nonpolluting solar energy comes close to being an ideal energy source and it almost certainly has a bright future.

The contributions of different fossil fuels to greenhouse gas carbon dioxide emissions vary with the chemical nature of the fuel with greater contributions from those that have relatively less hydrogen. For example, the chemical reaction for the combustion of methane, CH_4,

$$CH_4 + 2O_2 \rightarrow CO_2 + 2H_2O + \text{energy} \tag{10.2}$$

shows that 2 molecules of H_2O are produced for each molecule of CO_2. Since the conversion of chemically bound hydrogen to H_2O produces a large amount of heat, a relatively smaller amount of CO_2 is released per unit of heat generated. Petroleum hydrocarbons, such as those in gasoline or diesel fuel contain essentially only 2 atoms of H per atom of C. The combustion of such a molecule showing the conversion of 1 atom of C to CO_2 is represented as

$$CH_2 + \tfrac{3}{2}O_2 \rightarrow CO_2 + H_2O + \text{energy} \tag{10.3}$$

demonstrating that only half as much hydrocarbon-bound H is burned per molecule of CO_2 produced, so significantly less energy is produced per C atom than in the combustion of natural gas. Coal is even worse. Coal is a black hydrocarbon with an approximate simple formula of $CH_{0.8}$, so the combustion of an atom of carbon in coal can be represented as the following:

$$CH_{0.8} + 1.2O_2 \rightarrow CO_2 + 0.4H_2O + \text{energy} \quad (10.4)$$

Much less hydrocarbon-bound hydrogen is available to burn per atom of C in coal compared to petroleum or, especially, natural gas, so the amount of carbon dioxide emitted to the atmosphere per unit energy produced from coal is higher than with petroleum and much higher than with natural gas.

The problem that industrialized societies have become dependent on unsustainable fossil energy is clear, the solution less so. Alternatives must be developed and the transition to them will not be easy. The alternatives are discussed later in this chapter.

10.4 ENERGY DEVICES AND CONVERSIONS

Energy occurs in many forms and its utilization requires conversion to other forms. Many devices exist for the utilization of energy and its conversion to other forms. The most common of these are shown in Figure 10.3. The types of energy available, the forms in which it is utilized, and the processes by which it is converted to other forms have a number of implications for green technology and sustainability. For example, the wind turbine shown in Figure 10.3(1), once in place, continues to pump electricity into the power grid with virtually no harm to the environment (although some people regard them as unsightly while others think they are picturesque), whereas the steam power plant shown in Figure 10.3(2) requires mining of depletable coal, combustion of the fossil fuel with its potential for air pollution, control of air pollutants, and means for cooling the steam exiting the turbine, with its potential for thermal pollution of waterways.

An important aspect of energy utilization is its conversion to usable forms. For example, gasoline burned in automobile engines comes originally from petroleum pumped from underground, the petroleum constituents are separated, molecules with properties suitable for engine fuel are produced chemically, the gasoline product is burned in an internal combustion engine converting chemical to mechanical energy, and the mechanical energy is transmitted to the wheels of the automobile in the form of kinetic energy that moves the automobile. Significantly, less than half of the energy in the gasoline is actually converted to mechanical energy of the automobile's motion; the rest is dissipated as waste heat through the engine's cooling system.

Figure 10.4 illustrates major forms of energy and conversions between them. A significant point of this illustration is the very large ranges of energy conversion efficiencies from just a few percent or less to almost 100%. These differences suggest areas in which improvements may be sought. One of the most striking efficiencies is the less than 0.5% conversion of light energy to chemical energy by photosynthesis. Despite such a low conversion efficiency, photosynthesis has generated the fossil fuels, from which industrialized societies now get their energy and provides a significant fraction of energy in areas where wood and agricultural wastes are used. Doubling photosynthesis efficiency with genetically engineered plants

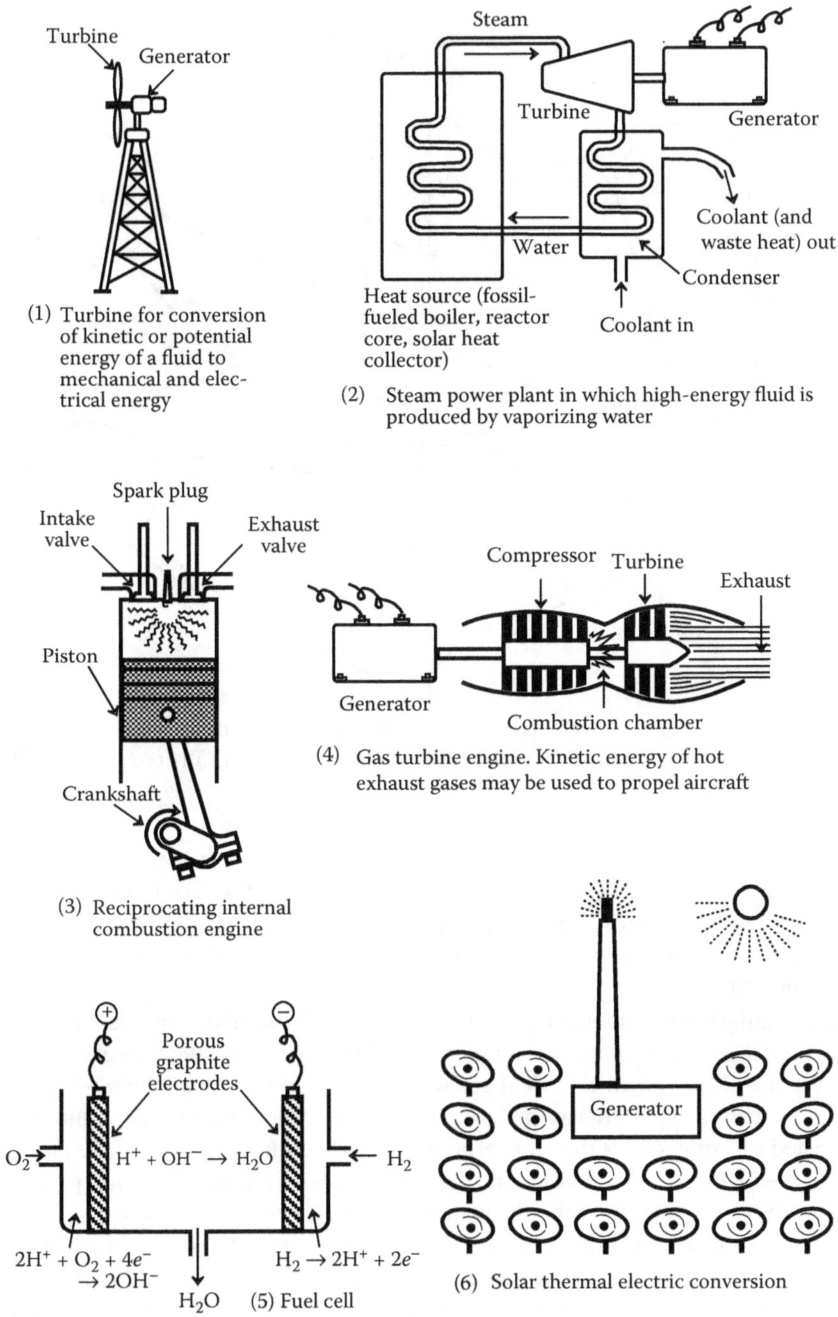

FIGURE 10.3 Examples of many devices for the collection of energy and its conversion to other forms. (From Manahan, S., *Environmental Chemistry*, 9th ed., Taylor & Francis, Boca Raton, 2010. With permission.)

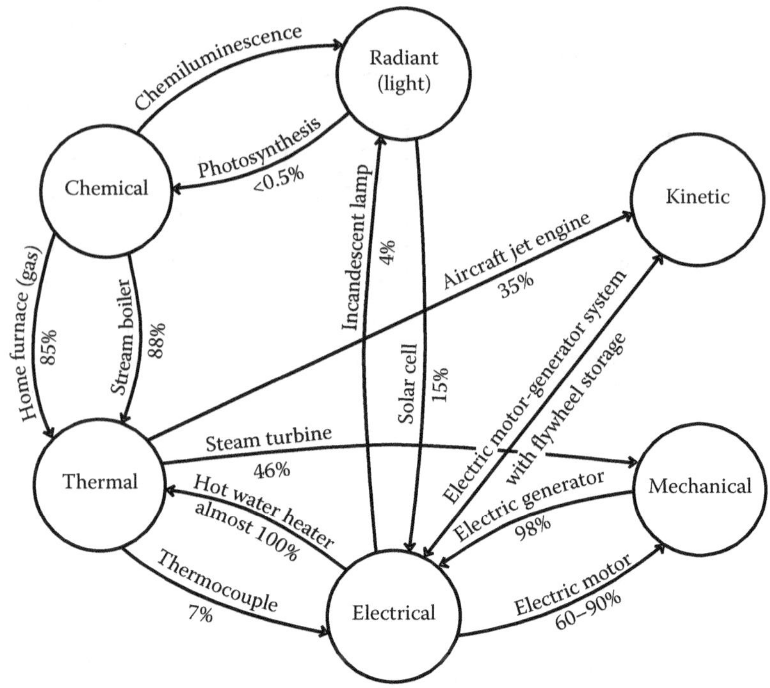

FIGURE 10.4 Major types of energy and conversions between them showing conversion efficiencies. (From Manahan, S., *Environmental Chemistry*, 9th ed., Taylor & Francis, Boca Raton, 2010. With permission.)

could be a major factor in making biomass a more desirable energy source. Replacement of woefully inefficient incandescent light bulbs with fluorescent bulbs that are 5–6 times more efficient in converting electrical energy to light can save large amounts of energy and will be mandated in the United States by legislation passed in 2007.

A particularly important energy conversion carried out in the anthrosphere is that of heat, such as from chemical combustion of fuel, to mechanical energy used to propel a vehicle or run an electrical generator. This occurs, for example, when gasoline in a gasoline engine burns, generating hot gases that move pistons in the engine connected to a crankshaft that converts the up-and-down movement of the piston to rotary motion that drives a vehicle's wheels. It also occurs when hot steam generated at high pressure in a boiler flows through a turbine connected directly to an electrical generator. A device, such as a steam turbine, in which heat energy is converted to mechanical energy is called a heat engine. Unfortunately, the laws of thermodynamics dictate that the conversion of heat to mechanical energy is always much less than 100% efficient. The efficiency of this conversion is given by the Carnot equation,

$$\text{Percent efficiency} = \frac{T_1 - T_2}{T_1} \times 100 \tag{10.5}$$

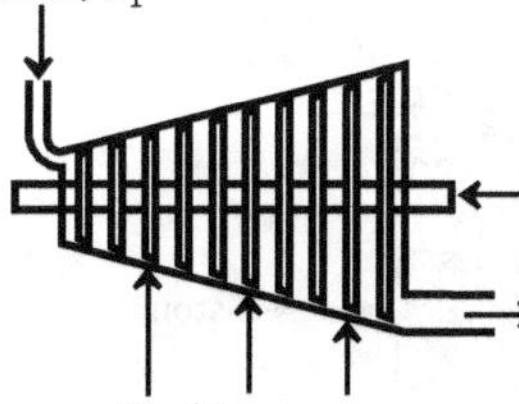

FIGURE 10.5 In a steam turbine, superheated steam impinges on vanes attached to a shaft to produce mechanical energy. For generation of electricity, the shaft is coupled to an electrical generator. (From Manahan, S., *Environmental Chemistry*, 9th ed., Taylor & Francis, Boca Raton, 2010. With permission.)

in which T_1 is the inlet temperature (e.g., of steam into a steam turbine) and T_2 is the outlet temperature, both expressed in Kelvin (°C + 273). Consider a steam turbine as shown in Figure 10.5. Substitution into the Carnot equation of 875 K for T_1 and 335 K for T_2 gives a maximum theoretical efficiency of 62%. However, it is not possible to introduce all the steam at the highest temperature and friction losses occur so that the energy conversion efficiency of most modern steam turbines is just below 50%. About 80% of the chemical energy released by combustion of fossil fuel in a boiler is actually transferred to water to produce steam so that the net efficiency for conversion of chemical energy in fossil fuels to mechanical energy to produce electricity is about 40%. The overall conversion of chemical energy to electricity is essentially the same because an electrical generator converts virtually all of the energy of a rotating turbine to electricity. Because nuclear reactor peak temperatures are limited for safety reasons, their conversion of nuclear energy to electricity is only about 30%.

A particularly important machine for converting chemical to mechanical energy is the internal combustion piston engine shown in Figure 10.6. Most internal combustion engines operate on a cycle of four strokes. In the first of these, the piston moves downward drawing air or an air/fuel mixture into the cylinder. Next, with both valves closed, the air or air/fuel mixture is compressed as the piston moves upward. With the piston near the top of the cylinder (a point at which fuel may be injected if only air is compressed), ignition occurs and the burning fuel creates a mass of highly pressurized combustion gas in the cylinder, which drives the piston down in the third stroke. The exhaust valve then opens and the exhaust gas is expelled during the exhaust stroke.

The efficiency of the internal combustion engine increases with the peak temperature reached by the burning fuel, which increases with the degree of compression during the compression stroke (around 20:1 for a modern diesel engine). This temperature is highest for the diesel engine in which the compression is so high that fuel injected into the combustion chamber ignites without a spark plug ignition source. Whereas a standard gasoline engine is typically about 25% efficient in converting chemical energy in fuel to mechanical energy, a diesel engine is typically 37% efficient, with some reaching higher values.

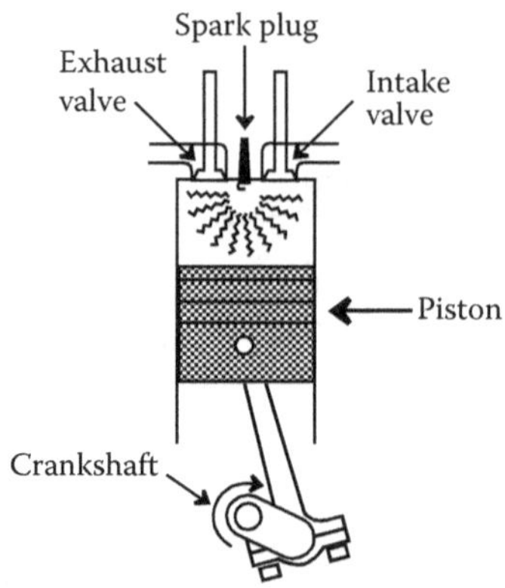

FIGURE 10.6 An internal combustion piston engine in which a very rapidly burning mixture of air and fuel drives a piston downward during the power stroke and this motion is converted to rotary mechanical motion by the crankshaft. (From Manahan, S., *Environmental Chemistry*, 9th ed., Taylor & Francis, Boca Raton, 2010. With permission.)

Although highly superior from the standpoint of efficiency, diesel engines do have some disadvantages with respect to emissions. The first of these is that the combustion zone is not homogeneous because the fuel is injected into the highly compressed air at the top of the compression stroke resulting in incomplete combustion and production of carbon particles; improperly adjusted diesel engines are a major source of particle air pollution in urban areas. In addition, because of their very high combustion temperatures and high ignition pressures, diesel engines tend to produce elevated levels of air pollutant nitrogen oxides. Recent advances in diesel engine design, computerized control, and exhaust pollutant control devices have greatly reduced diesel engine emissions.

10.4.1 Fuel Cells

Fuel cells are devices that convert the energy released by electrochemical reactions directly to electricity without going through a combustion process and electricity generator. Fuel cells are the primary means for utilizing hydrogen fuel and are becoming more common as electrical generators. A fuel cell has an anode at which elemental hydrogen is oxidized, releasing electrons to an external circuit, and a cathode at which elemental oxygen is reduced by electrons introduced from the external circuit, as shown by the half-reactions in Figure 10.7. The H^+ ions generated at the anode migrate to the cathode through a solid membrane permeable to protons. The net reaction is

$$2H_2 + O_2 \rightarrow 2H_2O + \text{electrical energy} \tag{10.6}$$

and the only product of the fuel cell reactions is water.

Although elemental hydrogen is the ultimate fuel for fuel cells, it may be produced by the chemical breakdown of hydrogen-rich fuels, such as methane, methanol,

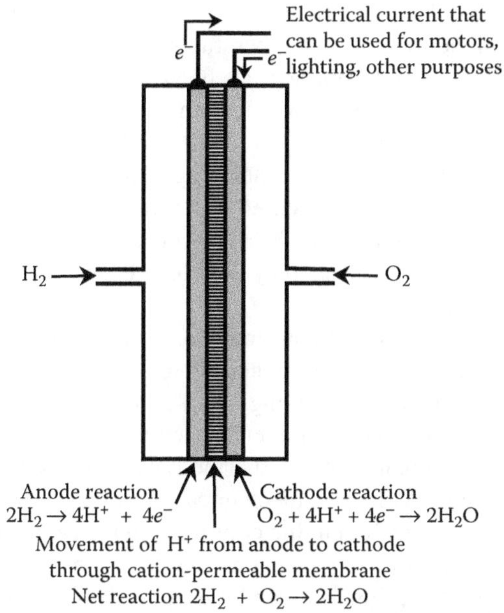

FIGURE 10.7 Cross-sectional diagram of a fuel cell in which elemental hydrogen can be reacted with elemental oxygen to produce electricity directly with water as the only chemical product. (From Manahan, S., *Environmental Chemistry*, 9th ed., Taylor & Francis, Boca Raton, 2010. With permission.)

or even gasoline, a process that also generates carbon dioxide. Tubular-style solid-oxide fuel cells, such as those manufactured by Siemens Westinghouse, operate at an elevated temperature of about 1000°C and produce an exhaust that is hot enough to drive a turbine or even to cogenerate steam. Such systems may be able to develop overall efficiencies of up to 80%.

10.5 GREEN TECHNOLOGY AND ENERGY CONVERSION EFFICIENCY

One of the best ways to conserve fuel resources is through increasing the efficiency of energy conversion including that of chemical to mechanical energy with the intermediate step of production of heat energy. Many advances have been made in this area since the late 1800s. Part of the increase in conversion of fuel energy to electricity going from around 4% conversion in 1900 to more than 40% at present resulted from increasing the input temperature (T_1 in the Carnot equation) in the heat engines driving electrical generators. Energy use efficiency increased by more than fourfold when picturesque steam engines were replaced by diesel/electric locomotives during the 1940s and 1950s. Substitution of diesel engines for gasoline engines in trucks and farm and construction equipment have resulted in gains in energy efficiency.

Much of the increased efficiency in fuel utilization has come from improved materials that allow higher operating temperatures. In addition to high-temperature-tolerant metals in engines, a contribution has been made by lubricating oils that do not break down at high temperatures. Much of the progress has been achieved with better engineering, now greatly aided by computerized design, evaluation, and manufacturing of engines. Engineers of a century ago had never heard of green technology, and probably would not have cared had they known about it. But they did understand costs of fuel (which, on the basis of constant value currency, were often higher then than they are now) and they welcomed the greater efficiencies they achieved on the basis of costs.

A key aspect of the most efficient conversion of chemical to mechanical energy in engines is the precise control of operational aspects such as ignition timing, valve timing, and fuel injection. In modern engines, key operation parameters are controlled by computers leading to optimum efficiency in engine operation.

As an inevitable consequence of the thermodynamics described by the Carnot equation, engines that convert heat to mechanical energy cannot utilize much of the heat, which is carried away by an engine cooling system. Typically, a small portion of this heat is used in automotive heaters on cold days. On a broader scale, such as municipal electrical systems, this heat can be used for heating buildings. Such efficiencies are discussed under "Combined Power Cycles" in Section 10.17.

10.6 ENERGY CONSERVATION AND RENEWABLE ENERGY SOURCES

Any consideration of energy needs and production must take energy conservation into consideration. This does not have to mean cold classrooms with thermostats set at 60°F in mid-winter, nor swelteringly hot homes with no air-conditioning, nor total reliance on the bicycle for transportation, although these and even more severe conditions are routine in many countries. The fact remains that the United States and several other industrialized nations have wasted energy at a deplorable rate. For example, U.S. energy consumption is higher per capita than that of some other countries that have equal, or significantly better, living standards. Obviously, a great deal of potential exists for energy conservation that will ease the energy problem.

Efficient use of energy can in fact correlate positively with higher economic standards. Figure 10.8 shows a plot of the ratio of energy use per unit of gross domestic product in developed industrialized nations and illustrates a steady and favorable decrease of energy required relative to economic output. Whereas in 2000, 1.7 barrels of oil equivalents were required per $1000 gross domestic product in developed nations, the corresponding figure for developing nations, which tend to lack advanced means of using energy efficiently, was 5.2 barrels, or 3 times as much. These figures indicate the substantial potential for decreased energy consumption by energy conscious development of the less industrially advanced nations as well as the additional conservation that can be achieved if citizens of industrialized nations can be persuaded to forego wasteful energy practices such as using excessively large and inefficient vehicles and living in overly large dwellings.

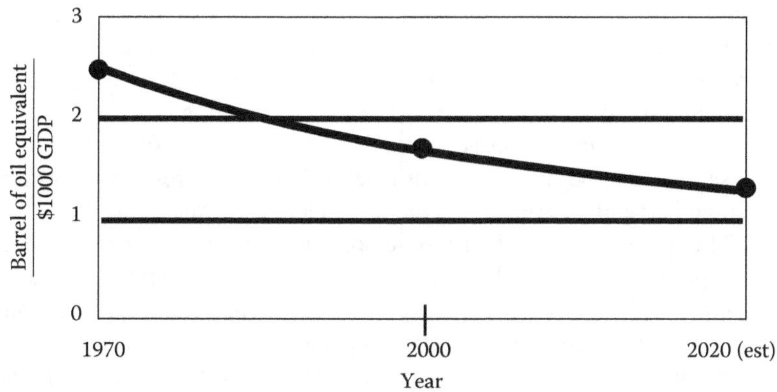

FIGURE 10.8 Plot of barrel of oil equivalent required per $1000 gross domestic product (GDP) as a function of year in industrialized nations. (From Manahan, S., *Environmental Chemistry*, 9th ed., Taylor & Francis, Boca Raton, 2010. With permission.)

Transportation is the economic sector with the greatest potential for increased efficiencies. The private auto and airplane are only about one-third as efficient as buses or trains for transportation. Transportation of freight by a truck requires about 3800 Btu/ton-mile, compared to only 670 Btu/ton-mile for a train. Truck transport is terribly inefficient (as well as dangerous, labor-intensive, and environmentally disruptive) compared to rail transport. Major shifts in current modes of transportation in the United States will not come without anguish, but energy conservation dictates that they be made.

Figure 10.9 shows the trend in U.S. automobile fuel economy during recent decades. The gains through about 1990 were very impressive, then dropped off as less fuel efficient vehicles became more popular. If the same trends from this period would have been maintained, the U.S. automobile fleet would by now average close to 40 miles per

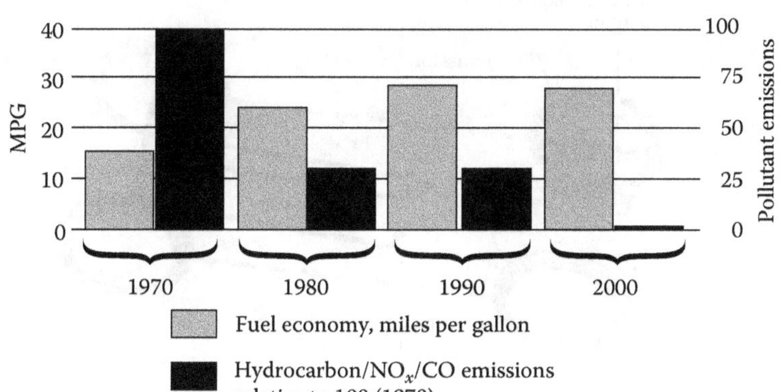

FIGURE 10.9 U.S. auto fleet fuel economy and emissions over three decades. Fuel economy has improved markedly while emissions have been greatly reduced. (From Manahan, S., *Environmental Chemistry*, 9th ed., Taylor & Francis, Boca Raton, 2010. With permission.)

gallon (MPG). Such a figure is readily achievable without seriously compromising safety or comfort and, as is obvious from the figure, with much lower emissions from pollutants today as compared to in 1970. In 2010, revised U.S. fuel standards were announced mandating fleet average mileage of 35.5 miles per gallon in 2016.

Household and commercial uses of energy are relatively efficient. Here again, appreciable savings can be made. The all-electric home requires much more energy (considering the percentage wasted in generating electricity) than a home heated with fossil fuels. The sprawling ranch-house style home uses much more energy per person than does an apartment unit, row house, or even a home of comparable floor area built in a compact format (more like a square box). Improved insulation, sealing around the windows, and other measures can conserve a great deal of energy. Electricity generating plants centrally located in cities can provide waste heat for commercial and residential heating and cooling and, with proper pollution control, can use municipal refuse for part of their fuel, thus reducing quantities of solid wastes requiring disposal.

One of the greatest contributions to energy conservation and energy use efficiency in very recent years has been the *hybrid vehicle* that uses an internal combustion engine to produce electricity that is stored for propulsion of the vehicle in a nickel-metal-hydride battery (Figure 10.10). Although not greatly more efficient for prolonged driving at highway speeds, these vehicles have achieved improvements up to 50% in stop-and-go driving in traffic. For routine operation, the internal combustion engine supplies all the power needed plus additional power, if required, to run the generator to recharge a battery (larger than the battery in a conventional automobile but significantly smaller than the battery in an all-electric vehicle). When a surge of power is required, electricity from the storage battery drives the electric motor to produce the additional power. The braking system also generates electricity that is stored in the battery. When the vehicle is stopped, the internal combustion engine does not run, which also saves fuel.

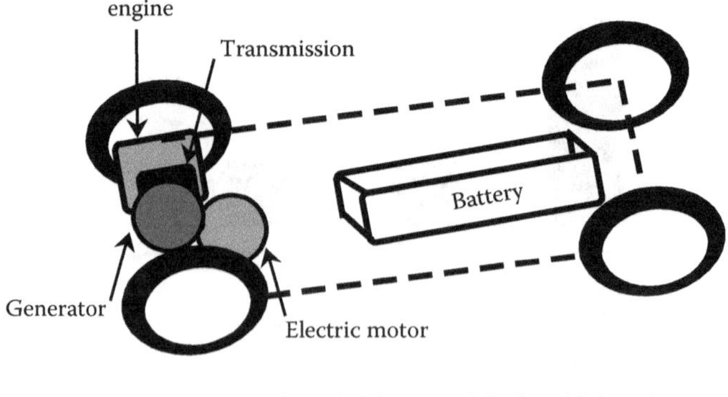

FIGURE 10.10 Major components of a hybrid automobile in which an internal combustion engine drives a generator to produce electricity to drive a vehicle. Electrical energy is stored in a battery. Electrical energy is also salvaged for storage from braking the vehicle. (From Manahan, S., *Environmental Chemistry*, 9th ed., Taylor & Francis, Boca Raton, 2010. With permission.)

Vastly reduced fuel consumption can be achieved by hybrid vehicles now under development that have batteries that can be charged by an external electrical source as well as the onboard engine. Such vehicles could give around 40–50 km driving range from an overnight charge alone, sufficient for the majority of routine driving. As the battery runs down, the internal combustion engine would recharge it as is the case with current hybrids. As of 2009, batteries were under development that could be recharged efficiently by both an external source of electricity and by an internal combustion engine coupled to a generator.

Although gasoline engines are now employed in hybrid vehicles, the ultimate in fuel economy could be achieved with an inherently more efficient diesel engine as the internal combustion engine component. By allowing the diesel engine to run at a generally steady rate, the output of exhaust pollutants, which are produced at higher levels by diesel engines as the engine speed is changed, could be greatly reduced. Furthermore, diesel engines idle with remarkably little fuel consumption, so that the diesel engine would not need to be turned off when the vehicle is stopped, thus staying hot and further reducing emissions when it is brought up to speed.

As scientists and engineers undertake the crucial task of developing alternative energy sources to replace dwindling petroleum and natural gas supplies, energy conservation must receive proper emphasis. In fact, zero energy-use growth, at least on a per capita basis, is a worthwhile and achievable goal. Such a policy would go a long way toward solving many environmental problems. With ingenuity, planning, and proper management, it could be achieved while increasing the standard of living and quality of life.

Closely related to energy conservation is the concept of *renewable energy* from sources that do not run out. Essentially all of these depend on energy from the sun including direct solar energy, wind driven by the solar heating of air masses, falling water from the solar-powered hydrologic cycle, and biomass formed from photosynthesis. For most of its lifetime on Earth, humankind has depended entirely on renewable sources of energy, and most countries are again emphasizing these sources.

Enlightened sustainable energy policies are being implemented in some developing countries. China implemented a new law on renewable energy at the beginning of 2006. This policy encourages renewable energy alternatives including wind power, biomass energy, and biomethane generation. Long known for its utilization of wastes (including even use of human wastes as fertilizer for growing vegetables), China has constructed many waste-to-methane generators in rural areas with 17 million families served by such facilities by 2006. Experimental biopower projects burning crop biomass by-products have been undertaken. As of 2006, China had 80 million square meters of solar collectors to heat water, equivalent to the energy from 10 million tons per year of coal. The total renewable energy capacity of China in 2006 was 7% of China's energy use, equivalent to 160 million tons of coal per year.

10.7 PETROLEUM AND NATURAL GAS

Liquid *petroleum* occurs in rock formations ranging in porosity from 10% to 30%. Up to half of the pore space is occupied by water. The oil in these formations must flow over long distances to an approximately 15-cm diameter well from which it is pumped. The rate of flow depends on the permeability of the rock formation, the viscosity of the oil, the driving pressure behind the oil, and other factors. Because of limitations in these factors, *primary recovery* of oil yields an average of about 30% of the oil in the formation, although it is sometimes as little as 15%. More oil can be obtained using *secondary recovery* techniques, which involve forcing water under pressure into the oil-bearing formation to drive the oil out. Primary and secondary recovery together typically extract somewhat less than 50% of the oil from a formation. Finally, *tertiary recovery* can be used to extract even more oil, normally through the injection of pressurized carbon dioxide, which forms a mobile solution with the oil and allows it to flow more easily to the well. Other chemicals, such as detergents, may be used to aid in tertiary recovery. Currently, about 300 billion barrels of U.S. oil are not available through primary recovery alone. A recovery efficiency of 60% through secondary or tertiary techniques could double the amount of available petroleum. Much of this would come from fields that have already been abandoned or essentially exhausted using primary recovery techniques.

The year 2008 was a very interesting one for world oil production and consumption. In July of that year, prices reached record levels of more than $150/barrel and many experts forecast that prices would continue to increase with gasoline going well over the $5/gallon reached in parts of the United States. With the world economic collapse that took place in the fall of 2008, petroleum prices fell precipitously, reaching levels of around $40/barrel. The year also saw the first drop in world oil consumption since 1983 down to 85.8 million barrels per day and U.S. consumption dropped 6.3% to 19.4 million barrels per day.

Shale oil is a possible substitute for liquid petroleum. Shale oil is a pyrolysis product of oil shale, a rock containing organic carbon in a complex structure of biological origin from eons past called kerogen. Oil shale is believed to contain approximately 1.8 trillion barrels of shale oil that could be recovered from deposits in Colorado, Wyoming, and Utah. In the Colorado Piceance Creek basin alone, more than 100 billion barrels of oil could be recovered from prime shale deposits. However, the environmental implications of recovering shale oil by heating oil shale include production of vast amounts of greenhouse gas carbon dioxide. Furthermore, potential water pollution problems from water-soluble salt residues left over from the pyrolysis of oil shale make it unlikely that this resource will ever be developed on a large scale.

Natural gas, consisting almost entirely of methane, is a very attractive fuel that produces few pollutants and less carbon dioxide per unit energy than any other fossil fuel. In addition to its use as a fuel, natural gas can be converted to many other hydrocarbon materials. It can be used as a raw material for the Fischer–Tropsch synthesis of gasoline. As of the early 2000s, increased demand for natural gas had led to tight supplies in the United States. Production of natural gas from coal seams in Wyoming has required pumping saline, alkaline water from the seams, which has caused water pollution problems. By 2009, exploitation of natural gas in "tight" shale formations opened by fracturing with water had resulted in increased supply of natural gas in the United States and some other countries.

10.8 COAL

From Civil War times until World War II, *coal* was the dominant energy source behind industrial expansion in most nations. However, after World War II, the greater convenience of lower cost petroleum resulted in a decrease in the use of coal for energy in the United States and in a number of other countries. Annual coal production in the United States fell by about one-third, reaching a low of approximately 400 million tons in 1958, but since then coal production for electricity generation has reached about 1 billion tons per year in the United States. About one-third of the world's energy and around 50% of electrical energy is provided by coal.

The general term *coal* describes a large range of solid fossil fuels derived from partial degradation of plants. Coal is differentiated largely by *coal rank* based on the percentage of fixed carbon, percentage of volatile matter, and heating value. The approximate average empirical formula of coal is $CH_{0.8}$ and coal typically contains from 1 to several percent sulfur, nitrogen, and oxygen. Of these elements, sulfur bound to the organic coal molecule and mixed with coal as mineral pyrite, FeS_2, presents major environmental problems because of production of air pollutant sulfur dioxide during combustion. Much of the FeS_2 can be removed physically from coal prior to combustion and sulfur dioxide can be removed from stack gas by various scrubbing processes.

10.8.1 Coal Conversion

As shown in Figure 10.11, coal can be converted to gaseous, liquid, or low-sulfur, low-ash solid fuels such as coal char (coke) or solvent-refined coal (SRC). Coal conversion is an old idea; a house belonging to William Murdock at Redruth, Cornwall, England, was illuminated with coal gas in 1792. The first municipal coal-gas system was employed to light Pall Mall in London in 1807. The coal-gas industry began in the United States in 1816. The early coal-gas plants used coal pyrolysis (heating in the absence of air) to produce a hydrocarbon-rich product particularly useful for illumination. Later in the 1800s, the water-gas process was developed, in which steam was impinged on hot coal coke to produce a mixture consisting primarily of H_2 and CO. It was necessary to add volatile hydrocarbons to this "carbureted" water-gas to bring its illuminating power up to that of gas prepared by coal pyrolysis. The United States had 11,000 coal gasifiers operating in the 1920s. At the peak of its use in 1947, the water-gas method accounted for 57% of U.S.-manufactured gas. The gas was made in low-pressure, low-capacity gasifiers that by today's standards would be inefficient and environmentally unacceptable (many sites of these old plants have been designated as hazardous waste sites because of residues of coal tar and other wastes). This was definitely not a green technology because of the high toxicity of carbon monoxide in the gas product, and many people were killed by release of CO in their houses. During World War II, Germany developed a major synthetic petroleum industry based on coal, which reached a peak capacity of 100,000 barrels per day in 1944. A synthetic petroleum plant operating in Sasol, South Africa, reached a capacity of several tens of thousands of tons of coal per day in the 1970s and

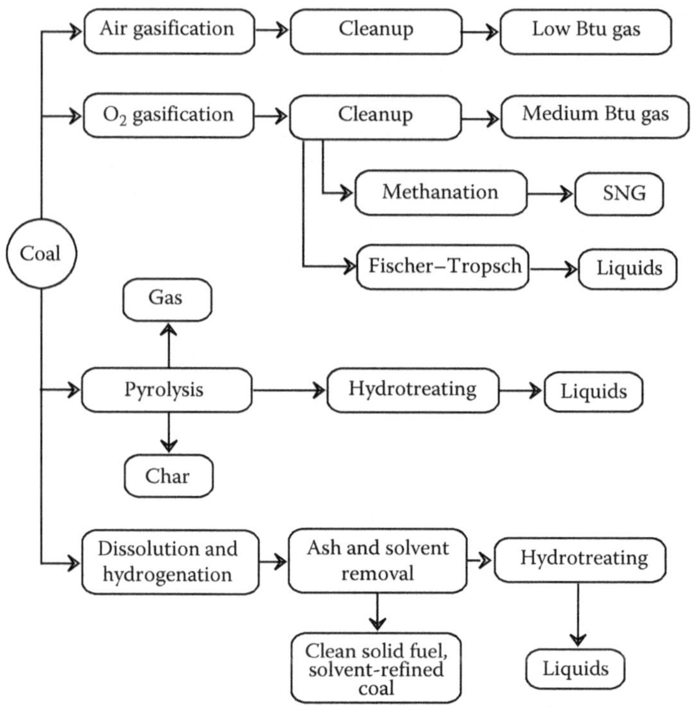

FIGURE 10.11 Routes to coal conversion. Btu refers to British thermal units, a measure of the heat energy that can be obtained from a fuel. Methanation means synthesis of CH_4 gas. Hydrogenation and hydrotreating refer to reaction with elemental H_2 gas. (From Manahan, S., *Environmental Chemistry*, 9th ed., Taylor & Francis, Boca Raton, 2010. With permission.)

currently produces hydrocarbons and feedstocks equivalent to about 150,000 barrels of petroleum per day.

A number of environmental implications are involved in the widespread use of coal conversion. These include strip mining, water consumption in arid regions, lower overall energy conversion compared to direct coal combustion, and increased output of atmospheric carbon dioxide. These plus economic factors have prevented coal conversion from being practiced on a very large scale. However, coal conversion does enable relatively facile carbon sequestration (see Section 10.9), which could enable much more sustainable coal utilization.

10.9 CARBON SEQUESTRATION FOR FOSSIL FUEL UTILIZATION

Carbon sequestration, which prevents carbon dioxide generated by fossil fuels from entering the atmosphere, holds the promise of enabling utilization of fossil fuels without contributing to greenhouse warming. Basically, the various schemes that have been proposed entail capturing carbon dioxide from a product or waste stream and sequestering it in a place where it cannot enter the atmosphere. Several approaches have been suggested or tried for capturing carbon dioxide and there are several possibilities for sequestration. The term "carbon sequestration" is also used in a more general sense to apply to the removal of carbon dioxide from the atmosphere, especially by photosynthesis.

There are several possible sinks in which carbon dioxide can be sequestered. The largest of these, a natural sink for the gas, is the ocean. Earth's oceans have an almost inexhaustible capacity for carbon dioxide. However, lowering the average pH of the oceans by as little as 0.1 pH unit from acidic carbon dioxide could have a serious adverse effect on ocean life and productivity; there is some evidence to suggest that shells of some small sea creatures have become thinner because of higher levels of dissolved carbon dioxide in ocean water. Deep saline formations also have a very high capacity for carbon dioxide sequestration. Depleted oil and gas reservoirs and unmineable coal seams have much lower, but still significant, carbon dioxide capacities.

Geological carbon dioxide sequestration can be accomplished by injecting the gas into porous sedimentary formations at depths exceeding approximately 1000 m. Experience in the petroleum industry with underground disposal of carbon dioxide and injection of the gas into oil-bearing formations for petroleum recovery have provided the technology required for geological carbon dioxide sequestration. The carbon dioxide injected into sedimentary formations rises and is confined by poorly permeable caprock. Breaches in caprock, such as those from abandoned oil wells, can result in carbon dioxide release. Eventually, the carbon dioxide dissolves in the generally saline pore waters in the sedimentary formation into which it is injected. Chemical reactions in the water and with the surrounding geological strata can result in long-term stability of the carbon dioxide. It is much more effective and less expensive to force carbon dioxide into unsaturated saline groundwater where it is dissolved than to compress the gas into a pressurized gas form underground.[1]

The first commercial application of carbon dioxide sequestration has operated since 1996 in the North Sea about 240 km from the Norwegian coast in a region known as the Sleipner oil and gas field. The natural gas product from this field is about 9% carbon dioxide, a value that must be reduced to 2.5% for commercial distribution of the gas. Whereas all other gas-producing operations simply discharge the carbon dioxide removed to the atmosphere, at Sleipner it is pumped under pressure into a 200-m thick layer of sandstone, the Utsira formation, which is about 1000 m below the seabed. A mixture of carbon dioxide and hydrogen sulfide removed from sour natural gas that is abundant in Alberta, Canada, is now being sequestered underground.[2]

The easiest sources of carbon dioxide to capture are those from industrial processes that generate the gas in high concentrations. An example of such a process is the fermentation of carbohydrates to produce ethanol for fuel or other uses. This source provides much of the carbon dioxide that is used commercially at present. The largest anthropogenic source of carbon dioxide discharged to the atmosphere is generated in power plants fueled with fossil fuels. These sources present a substantial challenge for carbon dioxide removal because they are so dilute. A power plant fueled with carbon-rich coal produces an exhaust stream that is 13–15% carbon dioxide, whereas one burning hydrogen-rich methane produces only 3–5% carbon dioxide. A third possibility is to capture carbon dioxide released from the gasification of fossil fuels, particularly coal (see Section 10.8). Normally, gasification is performed using oxygen as an oxidant, and the initial product consists of carbon dioxide and combustible H_2 and CO gases. Carbon monoxide in the synthesis gas product can be reacted with steam,

$$CO + H_2O \rightarrow H_2 + CO_2 \tag{10.7}$$

to produce nonpolluting elemental hydrogen fuel and carbon dioxide.

In late 2007, the U.S. Department of Energy announced funding for the construction of the first U.S. power plant with carbon dioxide sequestration, the FutureGen coal-powered power plant to be located at a site in Mattoon, Illinois. The project was abruptly canceled in early 2008 allegedly because of greatly increased costs, but a Government Accountability Office report released in March 2009 showed that the costs had been overestimated by $500 million because of an accounting cost mathematics error! The 275 MW power plant was to be large enough to supply electricity to 275,000 households and to sequester 1–2 million tons of carbon dioxide per year. Figure 10.12 is a schematic of a coal-fired power plant with carbon sequestration such as the FutureGen power plant. A gas turbine powered by hydrogen fuel produced by coal gasification generates electricity. Hot exhaust gas from the gas turbine is used to

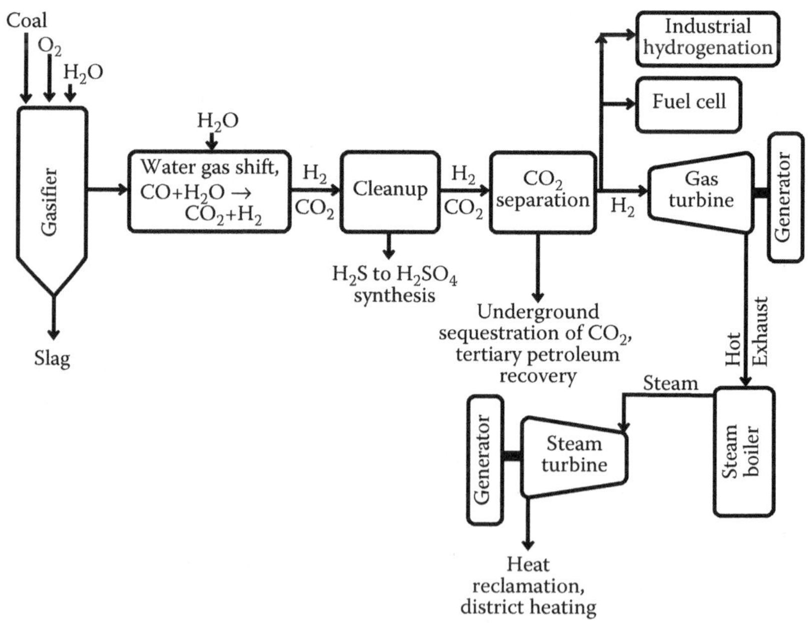

FIGURE 10.12 A coal-based carbon sequestration operation. Electricity is generated at generators connected both to the gas turbine and steam turbine. Reclaimed hydrogen sulfide is used to synthesize sulfuric acid. Hydrogen product can be employed for fuel cells and industrial hydrogenation as well as to power the gas turbine. Sequestered carbon dioxide can be used for petroleum recovery or simply disposed underground. Waste heat is reclaimed for district heating and industrial heating applications. (From Manahan, S., *Environmental Chemistry*, 9th ed., Taylor & Francis, Boca Raton, 2010. With permission.)

raise steam in a boiler and this steam powers a steam turbine that is also coupled to a generator. This combination results in very efficient electrical power generation. The production of intermediate hydrogen adds options for powering fuel cells or for using hydrogen in chemical synthesis or manufacture of synthetic hydrocarbon fuels. By-product hydrogen sulfide must be removed from the hydrogen product and can be used for making sulfuric acid, an important industrial chemical, or can be sequestered with the carbon dioxide. The carbon dioxide by-product is sequestered in mineral formations at depths up to around 2000 m. For power plants located near petroleum-bearing formations, carbon dioxide can be pumped into oil-bearing mineral formations for tertiary petroleum recovery.

10.10 INDUSTRIAL ECOLOGY FOR ENERGY AND CHEMICALS

An excellent example of a system of industrial ecology for sustainable utilization of fossil fuels is provided by the Great Plains Synfuel Plant located near Beulah, North Dakota. This complex is designed to take advantage of abundant deposits of lignite, a relatively low-heating-value, high-water-content form of coal that occurs in abundance in North Dakota. A schematic diagram of the system is shown in Figure 10.13. The heart of the plant consists of 14 Lurgi-type gasifiers that are 13 m high and 4 m in diameter processing 16,000 metric tons of lignite per day at temperatures up to 1200°C to produce a synthesis gas mixture of combustible H_2 and CO along with

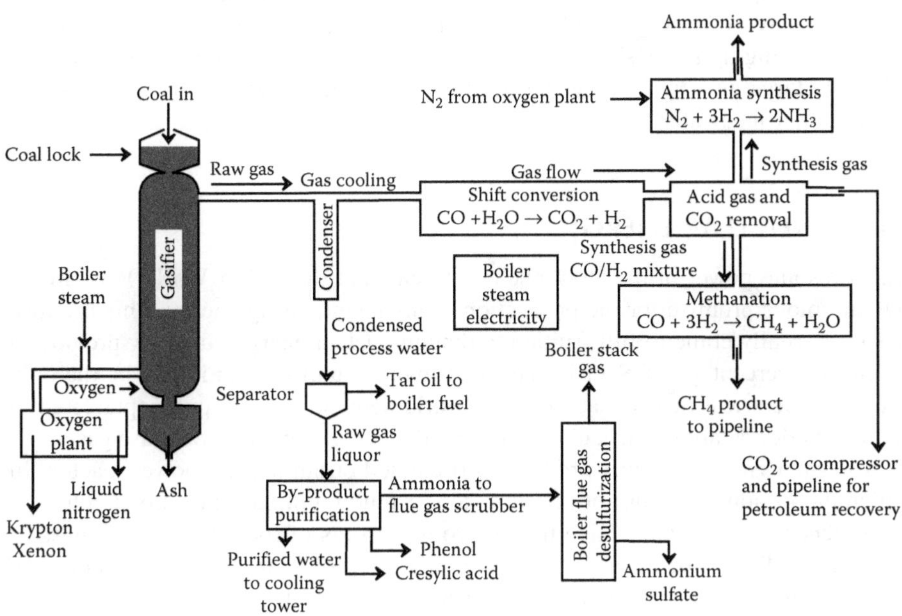

FIGURE 10.13 Schematic of the Great Plains lignite coal gasification complex. (From Manahan, S., *Environmental Chemistry*, 9th ed., Taylor & Francis, Boca Raton, 2010. With permission.)

by-products CO_2, water, and smaller quantities of tars, oils, phenolic compounds, ammonia, and H_2S. Useful fuel hydrocarbons, phenol, cresols, ammonia, and sulfur compounds are extracted from the water, which is used for cooling water. The gas mixture is subjected to a shift reaction that increases the ratio of H_2 to CO and this mixture is reacted to produce methane (CH_4) that is fed into a natural gas pipeline. Part of the synthesis gas stream is diverted to an ammonia synthesis plant to make this valuable fertilizer product. A key feature of the plant is the extraction of CO_2 from the synthesis gas. The carbon dioxide is compressed and sent through a pipeline to Canada where it is pumped underground to depleted petroleum-bearing strata for secondary recovery of crude oil.

The Great Plains Synfuels plant provides an excellent example of a diversified industrial ecology complex. It utilizes an abundant resource (lignite coal) in a way that significantly reduces the greenhouse gas emissions and other potential environmental impacts from this resource. Rather than having to transport large quantities of lignite by rail to distant power plants, the energy content of the lignite is converted to methane, the most environmentally friendly fossil fuel, which is moved with minimum environmental disruption through pipelines. An even higher value product, fertilizer ammonia, is synthesized at the site using elemental hydrogen made from gasifying the lignite and elemental nitrogen isolated from air in the air liquefaction operation required to prepare elemental oxygen required for gasification. The carbon dioxide by-product is not released to the atmosphere where it could aggravate the global warming problem but is pumped to oilfields where it is pumped underground to increase crude oil recovery. Commercially valuable ammonium sulfate by-product, a useful source of nitrogen and sulfur soil nutrients, is recovered from the water released during lignite gasification. In an area where water shortages can occur, the relatively large amount of water released by lignite gasification is used for cooling and as boiler feedwater.

10.11 NUCLEAR ENERGY

The awesome power of the atom nucleus revealed at the end of World War II held out enormous promise for the production of abundant, cheap energy. This promise has never really come to full fruition, although nuclear energy currently provides a significant percentage of electric energy in many countries, and it may be the only source of electrical power that can meet world demand without unacceptable environmental degradation, particularly through the generation of greenhouse gases.

Nuclear fission for power production is carried out in nuclear power reactors in which the fission (splitting) of uranium-235 or plutonium nuclei occurs. Each such event generates two radioactive fission product atoms of roughly half the mass of the nucleus fissioned, an average of 2.5 neutrons, plus an enormous amount of energy compared to normal chemical reactions. The neutrons, initially released as fast-moving, highly energetic particles, are slowed to thermal energies in a moderator medium. For a reactor operating at a steady state, exactly one of the neutron products from each fission is used to induce another fission reaction in a chain reaction (Figure 10.14).

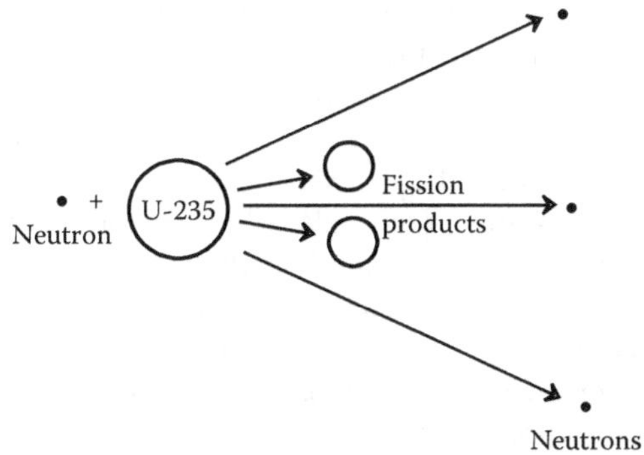

FIGURE 10.14 Fission of a uranium-235 nucleus. (From Manahan, S., *Environmental Chemistry*, 9th ed., Taylor & Francis, Boca Raton, 2010. With permission.)

The energy from these nuclear reactions is used to heat water in the reactor core and produce steam to drive a steam turbine, as shown in Figure 10.15. As noted in Section 10.4, temperature limitations make nuclear power less efficient in converting heat to mechanical energy and, therefore, to electricity, than fossil energy conversion processes.

A limitation of fission reactors is the fact that only 0.71% of natural uranium is fissionable uranium-235. This situation could be improved by the development of *breeder reactors*, which convert uranium-238 (natural abundance 99.28%) to fissionable plutonium-239.

A major consideration in the widespread use of nuclear fission power is the production of large quantities of highly radioactive waste products. These remain lethal

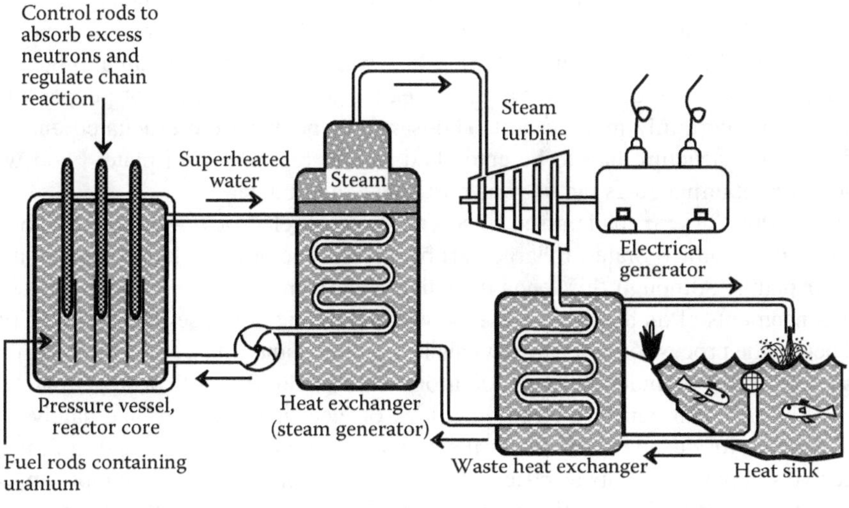

FIGURE 10.15 A typical nuclear fission power plant. (From Manahan, S., *Environmental Chemistry*, 9th ed., Taylor & Francis, Boca Raton, 2010. With permission.)

for thousands of years. They must either be stored in a safe place or be disposed of permanently in a safe manner. At the present time, spent fuel elements are being stored under water at the reactor sites. Under current regulations in most countries, the wastes from this fuel will eventually have to be buried. An alternative favored by many investigators is to process the material in the spent fuel elements to remove radioactive products from uranium fuel, isolate the relatively short-lived fission products that decay spontaneously within several hundred years, and bombard the longer-lived nuclear wastes with neutrons in nuclear reactors. The absorption of neutrons by the nuclei of the nuclear waste elements causes *transmutation* in which the elements are converted to other elements or fission products with shorter half lives resulting in relatively rapid production of stable isotopes. Radioactive waste elements for which transmutation is feasible include plutonium, americium, neptunium, curium, technetium-99, and iodine-129. Plutonium, americium, neptunium, and curium are heavy actinide elements that are fissionable and add fuel value in a nuclear reactor.

Another problem to be faced with nuclear fission reactors is their eventual decommissioning. There are three possible solutions. One is dismantling soon after shutdown, in which the fuel elements are removed, various components are flushed with cleaning fluids, and the reactor is cut up by remote control and buried. "Safe storage" involves letting the reactor stand 30–100 years to allow for radioactive decay, followed by dismantling. The third alternative is entombment, that is, encasing the reactor in a concrete structure.

The course of nuclear power development was altered drastically by two accidents. The first of these occurred on March 28, 1979, with a partial loss of coolant water from the Metropolitan Edison Company's nuclear reactor located on Three Mile Island in the Susquehanna River, 28 miles outside of Harrisburg, Pennsylvania. Known as the TMI incident, the result was a loss of control, overheating, and partial disintegration of the reactor core. Some radioactive xenon and krypton gases were released and some radioactive water was dumped into the Susquehanna River. Eventually the reactor building was sealed. A much worse accident occurred at Chernobyl in the Soviet Union in April 1986, when a reactor blew up and burned, spreading radioactive debris over a wide area and killing a number of people (officially 31, but certainly many more). Thousands of people were evacuated and the entire reactor structure had to be entombed in concrete and steel plate. Food was seriously contaminated as far away as northern Scandinavia.

As of 2006, 28 years had passed since a new nuclear electric power plant had been ordered in the United States, in large part because of the projected high costs of new nuclear plants. Although this tends to indicate hard times for the nuclear industry, pronouncements of its demise may be premature and by 2008, several U.S. utilities had begun the process of ordering new nuclear power plants and a significant number of such plants were under construction around the world. Properly designed nuclear fission reactors can generate large quantities of electricity reliably and safely and have done so for decades in U.S. naval submarines and carrier ships and in France, which gets about 80% of its electricity from nuclear sources. The single most important factor that may lead to the renaissance of nuclear energy is the threat to the atmosphere from greenhouse gases produced in large quantities by fossil fuels. It can be argued that nuclear energy is the only proven alternative that can provide the amounts of energy required within acceptable limits of cost, reliability, and environmental effects.

New designs for nuclear power plants can enable power reactors that are much safer and environmentally acceptable than those built with older technologies. The proposed new designs incorporate built-in passive safety features that work automatically in the event of problems that could lead to incidents such as TMI or Chernobyl with older reactors. These devices—which depend on passive phenomena such as gravity feeding of coolant, evaporation of water, or convection flow of fluids—give the reactor the desirable characteristics of *passive stability*. They have also enabled significant simplification of hardware, with only about half as many pumps, pipes, and heat exchangers as are contained in older power reactors.

10.11.1 Nuclear Fusion

The fusion of a deuterium nucleus and a tritium nucleus releases a lot of energy as shown below, where MeV stands for million electron volts, a unit of energy:

$$^{2}_{1}\mathrm{H} + {}^{3}_{1}\mathrm{H} \rightarrow {}^{4}_{2}\mathrm{He} + {}^{1}_{0}\mathrm{n} + 17.6\ \mathrm{MeV}\ \text{(energy released per fusion)} \qquad (10.8)$$

This reaction is responsible for the enormous explosive power of the "hydrogen bomb." So far it has eluded efforts at containment for a practical continuous source of energy. And since physicists have been trying to make it work on a practical basis for the last approximately 50 years, it will probably never be done. (Within about 15 years after the discovery of the phenomenon of nuclear fission, it was being used in a power reactor to power a nuclear submarine.) However, the tantalizing possibility of using the essentially limitless supply of deuterium, an isotope of hydrogen, from earth's oceans for nuclear fusion still give some investigators the hope of a practical nuclear fusion reactor and some research continues toward that end.

Nuclear fusion was the subject of one of the greatest scientific embarrassments of modern times that occurred in 1989. This incident came about when investigators at the University of Utah announced that they had accomplished the so-called cold fusion of deuterium during the electrolysis of deuterium oxide (heavy water). This resulted in an astonishing flurry of activity as scientists throughout the world sought to repeat the results, whereas others ridiculed the idea. Unfortunately, for the dream of a cheap and abundant source of energy, the skeptics were right, and the whole story of cold fusion stands as a lesson in the (temporary) triumph of wishful technological thinking over scientific good sense.

10.12 GEOTHERMAL ENERGY

Underground heat in the form of steam, hot water, or hot rock used to produce steam has been employed as an energy resource for about a century and can be regarded as largely renewable. This energy was first harnessed for the generation of electricity at Larderello, Italy, in 1904, and has since been developed in Japan, Russia, New Zealand, the Philippines, and at the Geysers in northern California.

Underground dry steam is relatively rare, but is the most desirable from the standpoint of power generation. More commonly, energy reaches the surface as superheated water and steam. In some cases, the water is so pure that it can be used for irrigation and livestock; in other cases, it is loaded with corrosive, scale-forming salts. Utilization of the heat from contaminated geothermal water generally requires that the water be reinjected into the hot formation after heat removal to prevent contamination of surface water.

The utilization of hot rocks for energy requires fracturing of the hot formation, followed by injection of water and withdrawal of steam. This technology is still in the experimental state, but promises approximately 10 times as much energy production as steam and hot water sources.

Land subsidence and seismic effects, such as the mini-earthquakes that occur when water is pumped under extreme pressure into hot rock formations that fracture as a consequence, are environmental factors that may hinder the development of geothermal power. However, this energy source holds considerable promise, and its development continues.

10.13 THE SUN: AN IDEAL, RENEWABLE ENERGY SOURCE

Solar power is an ideal source of energy that is unlimited in supply, widely available, and inexpensive. It does not add to the earth's total heat burden or produce chemical air and water pollutants. On a global basis, utilization of only a small fraction of solar energy reaching the earth could provide for all energy needs. In the United States, for example, with conversion efficiencies ranging from 10% to 30%, it would only require collectors ranging in area from one-tenth down to one-thirtieth that of the state of Arizona to satisfy present U.S. energy needs. (This is still an enormous amount of land, and there are economic and environmental problems related to the use of even a fraction of this amount of land for solar energy collection. Certainly, many residents of Arizona would not be pleased at having so much of the state covered by solar collectors, and some environmental groups would protest the resultant shading of rattlesnake habitat.)

There are vast land areas available in the United States and throughout the world that receive exceptional levels of solar energy suitable for power generation. Factors involved in evaluating areas for solar energy generation include proximity to the Equator, relatively high altitude, and consistent absence of cloud cover. The terrestrial area that has optimum conditions is in the Sahara Desert of southeast Niger in Africa, which receives an average of 6.78 kWh/m^2 each day, which is close to the amount of energy required to heat water for a typical U.S. household.

Solar power cells (photovoltaic cells) for the direct conversion of sunlight to electricity have been developed and are widely used for energy in space vehicles. With present technology, however, they are relatively expensive in most places for large-scale generation of electricity, although the economic gap is narrowing. Most schemes for the utilization of solar power depend on the collection of thermal energy followed by conversion to electrical energy. The simplest example of such an approach involves focusing sunlight on a steam-generating boiler (see Illustration 6 in Figure 10.3). Parabolic reflectors can be used to focus sunlight on pipes containing heat-transporting fluids. Selective coatings on these pipes ensure that most of the incident energy is absorbed.

The most efficient and reliable type of solar-powered heat engine to generate electricity is the *Stirling engine*, which has achieved close to 30% overall efficiency in the conversion of solar energy to electricity. The Stirling engine uses hydrogen gas as a working fluid in a sealed system in which pistons connected by rods in cylinders pump the gas back and forth in the system. The gas is heated at a heater head kept at a relatively high temperature by sunlight focused from an array of reflective mirrors that concentrate sunlight on the heater head. The resultant pressure forces one of the pistons down in a power stroke. The gas is then transferred to a piston on the cooled side of the engine through a regenerator consisting of a material that captures some of the heat to be used in the next cycle, thereby greatly increasing efficiency. For a system planned for installation in California's Mojave desert,[3] each Stirling unit is rated at 25 kW power, running at 1800 rpm, and generating electricity at 480 V and 60 Hz. A major advantage of this system is its modular character that enables scaling from a single unit to a very large number generating hundreds of MW of energy in a single location. As of 2008, it appeared that the Stirling system would become a leading contender for the utilization of solar power in areas such as California's Mojave Desert and Africa's Sahara that receive intense, consistent sunlight.

The direct conversion of energy in sunlight to electricity is accomplished by special solar voltaic cells. Most types of photovoltaic cells depend on the electronic properties of silicon atoms containing low levels of other elements. In a typical photovoltaic cell, the cell consists of two layers of silicon, a donor layer that is doped with about 1 ppm of arsenic atoms and an acceptor layer that is doped with about 1 ppm of boron. As noted in Chapter 11, Lewis symbols use dots to represent the outermost valence electrons of atoms, those that can be lost, gained, or shared in chemical bonds. Examination of the Lewis symbols of silicon, arsenic, and boron

$$\cdot\dot{\mathrm{Si}}\!: \qquad \cdot\underset{\cdot}{\dot{\mathrm{As}}}\!: \qquad \dot{\mathrm{B}}\!:$$

shows that substitution of an arsenic atom with its 5 valence electrons for a silicon atom with its 4 valence electrons in the donor layer gives a site with an excess of 1 electron, whereas substitution of a boron atom with only 3 electrons for a silicon atom in the acceptor layer gives a site "hole" that is deficient in 1 electron. The surface of a donor layer in contact with an acceptor layer contains electrons that are attracted to the acceptor layer. When light shines on this area, the energy of the photons of light can push these electrons back onto the donor layer, from which they can go through an external circuit back to the acceptor layer, as shown in Figure 10.16. This flow of electrons constitutes an electrical current that can be used for energy.

Solar voltaic cells based on crystalline silicon have operated with a 30% efficiency for experimental cells and 15–20% for commercial units available in 2008, at a cost of around 20–25 cents/kWh, compared to 4–7 cents/kWh for fossil fuel fired power plants and 6–9 cents for those fired by biomass. Part of the high cost results from the fact that the silicon used in the cells must be cut as small wafers from silicon

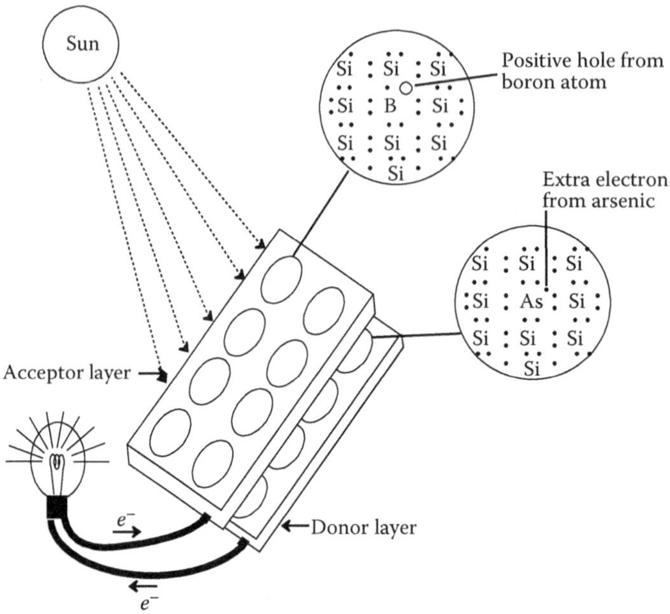

FIGURE 10.16 The operation of a photovoltaic cell. (From Manahan, S., *Environmental Chemistry*, 9th ed., Taylor & Francis, Boca Raton, 2010. With permission.)

crystals for mounting on the cell surfaces. Significant advances in costs and technology are being made with thin-film photovoltaics, which use an amorphous silicon alloy. These cells are only about half as efficient as those made with crystalline silicon, but cost only about 25% as much. A newer approach to the design and construction of amorphous silicon film photovoltaic devices uses three layers of amorphous silicon to absorb, successively, short wavelength ("blue"), intermediate wavelength ("green"), and long wavelength ("red") light, as shown in Figure 10.17. Thin-film solar panels constructed with this approach have achieved solar-to-electricity energy conversion efficiencies just over 10%, lower than those using crystalline silicon, but higher than other amorphous film devices. The low cost and relatively high conversion efficiencies of these solar panels should enable production of electricity at only about twice the cost of conventional electrical power, which would be competitive in some situations.

Systems have been developed in which solar cells lining glass tubes are used to generate electricity. Such a configuration is well adapted to rooftops because it captures light from all angles including even light from below reflected from a white roof. It has been estimated that covering all flat rooftops in the United States with such collectors could provide about 150 GW of electricity, 15% of U.S. consumption.

A major disadvantage of solar energy is its intermittent nature. However, flexibility inherent in an electric power grid would enable it to accept up to 15% of its total power input from solar energy units without special provision for energy storage.

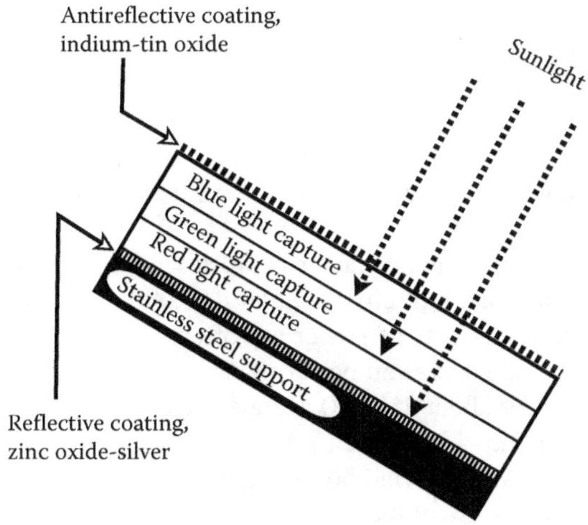

FIGURE 10.17 High-efficiency thin-film solar photovoltaic cell using amorphous silicon. (From Manahan, S., *Environmental Chemistry*, 9th ed., Taylor & Francis, Boca Raton, 2010. With permission.)

Existing hydroelectric facilities may be used for pumped-water energy storage in conjunction with solar electricity generation. Heat or cold can be stored in water, in a latent form in water (ice) or eutectic salts, or in beds of rock. Enormous amounts of heat can be stored in water as a supercritical fluid contained at high temperatures and very high pressures deep underground. Mechanical energy can be stored with compressed air or flywheels. Utilization of solar energy to produce elemental hydrogen as a means to store, transfer, and utilize energy, as discussed in Section 10.16 below will probably come into widespread use.

No really insurmountable barriers exist to block the development of solar energy, such as might be the case with fusion power. In fact, the installation of solar space and water heaters became widespread in the late 1970s, and research on solar energy was well supported in the United States until after 1980, when it became fashionable to believe that free-market forces had solved the "energy crisis." At present, shortages of energy and concern over the effects of global warming due to the use of fossil fuels is leading to much increased interest in solar energy. With the installation of more heating devices and the probable development of some cheap, direct solar electrical generating capacity, it is likely that during the coming century, solar energy will be providing an appreciable percentage of energy needs in areas receiving abundant sunlight.

10.14 ENERGY FROM MOVING AIR AND MOVING WATER

10.14.1 Surprising Success of Wind Power

Wind power using huge turbines mounted on high towers and coupled to electrical generators is emerging at a somewhat surprising rate as a source of renewable energy.

Although used for centuries with windmills that drove grain grinding and water pumping operations and during the early 1900s for small-scale electricity generation, especially in remote locations, modern large-scale wind-powered electrical generators emerged during the 1990s as economical means of generating electrical power. Wind power is completely renewable and nonpolluting. It is an indirect means of utilizing solar energy because winds are caused by the movement of air masses heated by the sun (Figure 10.18).

Wind power has become a major factor in world energy generation with an estimated capacity at the beginning of 2010 of about 158,000 megawatts. (A megawatt is abbreviated MW. Large amounts of power are often expressed in gigawatts, where 1 GW = 1000 MW. A 1000 megawatt power plant is considered to be quite large and can provide electricity for about 250,000 average U.S. homes. Typical modern large wind turbines generate 1.5–4 MW each.) As of the beginning of 2010, the United States had about 35,000 MW of wind power capacity installed with plants of significant size in about 30 states. As of the beginning of 2010, Texas was the leading state for wind energy production in the United States with an installed capacity of 9403 MW. The next eight states were Iowa (3604 MW), California (2798 MW), Washington (1849 MW), Minnesota (1810 MW), Oregon (1758 MW), Illinois (1547 MW), Colorado (1244 MW), and North Dakota (1203 MW). Wind energy in the United States is now sufficient to power 8.9 million average households eliminating as much as 57 million tons of carbon dioxide emissions per year.

Illustrating the importance of wind power globally, by the beginning of 2010 the European Union (EU) countries had around 75,000 MW of wind power capacity with eleven EU countries, about one-third of all EU nations, each having at least 1000 MW

FIGURE 10.18 Wind-powered electrical generators mounted on towers are becoming increasingly common sights in the world in areas where consistent wind makes this nonpolluting source of renewable energy practical. (From Manahan, S., *Environmental Chemistry*, 9th ed., Taylor & Francis, Boca Raton, 2010. With permission.)

installed capacity. At the beginning of 2010 the United States was the leading country in wind energy capacity with Germany and China tied for second, each with about 25,000 MW capacity. Next came Spain (19,100 MW) and India (10,900 MW). Though a small country, Denmark has the highest per capita wind power capacity with a rapid rate of growth. Because of its aggressive program for wind power development, China may well become number one in wind power within a few years.

As of 2008, Calgary, Canada, had in operation a new $140 million wind electrical power system that powers three-fourth of the city's municipal buildings and provides the energy for the municipal light rail transportation system. The system is equivalent to taking 30,000 automobiles off the road.

Wind power is especially attractive for some agricultural regions. One reason is that such regions often have low population densities so that there are fewer objections to wind power installations. A commercial wind turbine generating 1.8 MW is a formidable machine, typically mounted on towers around 80 m high to take advantage of higher, more consistent wind speeds at such heights, and may have blades 40 m long. However, the footprints of these structures are relatively small and do not occupy much farmland. The electricity is conveyed from the turbines in underground lines eliminating surface power lines. Adding to the potential attractiveness of wind power in agricultural regions is the fact that electricity generated by wind energy can be used to electrolyze water to produce elemental H_2 and O_2, an application not handicapped by wind's intermittent nature. The H_2 currently required to make NH_3 fertilizer is usually produced from steam reacting with methane (natural gas), a relatively expensive process, and its production from water by inexpensive wind energy could help keep the price of ammonia fertilizer at reasonable levels. Furthermore, both elemental H_2 and O_2 can be used to convert crop by-product biomass to hydrocarbon fuels, as discussed in Section 10.15.

Northern regions, including parts of Alaska, Canada, the Scandinavian countries, and Russia often have consistently strong wind conditions conducive to the generation of wind power. Isolation from other sources of energy makes wind power attractive for many of these regions. Severe climate conditions in these regions pose special challenges for wind generators. One problem can be the buildup of rime consisting of ice condensed directly on structures from supercooled fog in air. (In warmer regions, the remains of insects zapped by the rotating turbine blades have built up to the point of reducing the aerodynamic efficiency of the blades.)

Although wind turbines have been growing in size leading to much greater economy in power production, there is also a market for small wind turbines that can be used for an individual home or commercial building. Small wind turbines have been installed in a number of locations including Logan International Airport in Boston and the Brooklyn Naval yard. Although the cost of these turbines is much less than that of large commercial turbines, the cost of electricity produced by the small turbines is much higher.

10.14.2 Energy from Moving Water

Water flowing in contact with a device called a waterwheel is one of the oldest sources of power other than humans or animals. Grain mills driven by waterpower existed in ancient Greece and Rome, and large waterwheels developing up to 50 hp were constructed in the Middle Ages. In colonial North America, waterwheels drove grist mills and sawmills and were further applied to leather, textile, and machine shop operations. Because of problems with low water flow in the summer and ice formation in the winter, these operations were rapidly displaced when steam engines became available in the early 1800s.

With the development of electric power in the late 1800s, waterpower underwent a spectacular renaissance to drive electrical generators. The first practical hydroelectric plant went into operation on the Fox River near Appleton, Wisconsin, in 1882. Hydroelectric power grew rapidly as an energy source from that time and by 1980 accounted for about 25% of world electrical production and 5% of total world energy use. The potential to construct hydroelectric plants is favored by mountainous terrain with large river valleys and is distributed relatively evenly around the world. China has about one-tenth of the world's potential for hydroelectric power. About 99% of Norway's electric power is hydroelectric accounting for about 50% of that country's energy use.

The largest hydroelectric project in the world is the Three Gorges installation on the huge Yangtze River in China. Located at the end of a number of steep canyons, the dam spans 2.3 km across the river valley and reaches a height of 185 m. When filled, the reservoir formed by the dam will extend for 630 km with an average width of 1.3 km. When the dam was finished in 2009, it had 26 generating units each capable of generating 700 MW of hydroelectric power, a total capacity of 18.2 GW. Beginning in 2009, the construction of six more units in a subterranean powerhouse will bring the total generating capacity to 22.4 GW.

The sustainability and environmental acceptability of hydroelectric power present a mixed picture. In the modern age, construction of water impoundments tends to displace significant numbers of people; more than 1 million were displaced for the Chinese Three Gorges project. Altering the flow of rivers can change their aquatic ecology. Esthetics can be harmed by filling scenic river valleys with impounded water. In some cases in the United States, dams are being dismantled to restore river valleys to their former state. However, hydroelectric power prevents release of greenhouse gases from equivalent fossil energy-powered plants. Reservoirs can provide recreational facilities and serve as sources of fish and water used by municipalities, industries, and agriculture.

10.14.3 Energy from Moving Water without Dams

A promising approach to the utilization of energy from moving water without building dams is provided by *hydrokinetic* and *wave energy conversion* devices that tap the kinetic energy of moving water in river flows, tides, or ocean waves. These devices can be put in natural streams, tidal estuaries, ocean currents, and constructed waterways. One promising device consists of a turbine with large blades spaced relatively far apart and connected directly to a generator that can be placed in a water current, such as in a river. Such turbines can be attached directly to structures constructed on waterway beds or can be attached to bridge supports. Bidirectional turbines have been designed for use in tidal currents that flow back and forth.

10.15 BIOMASS ENERGY

Fossil fuels originally came from photosynthetic processes. Photosynthesis is a promising source of combustible chemicals to be used for energy production including transportation fuels[4] and could certainly produce all needed organic raw materials to substitute for petroleum in the current petrochemicals industry. It suffers from the disadvantage of being a very inefficient means of solar energy collection (a collection efficiency of only a fraction of a percent by photosynthesis is typical of most common plants). However, the overall energy conversion efficiency of several plants, such as sugarcane, is around 0.6%. Furthermore, some plants, such as *Euphorbia lathyrus* (gopher plant), a small bush growing wild in California, produce hydrocarbon emulsions directly. The fruit of the Philippine plant, *Pittsosporum reiniferum*, can be burned for illumination due to its high content of hydrocarbon terpenes, primarily α-pinene and myrcene. Conversion of agricultural plant residues to energy could be employed to provide much of the energy required for agricultural production. Indeed, until about 80 years ago, virtually all of the energy required in agriculture—hay and oats for horses, home-grown food for laborers, and wood for home heating—originated from plant materials produced on the land. (An interesting exercise is to calculate the number of horses required to provide the energy used for transportation at the present time in the Los Angeles basin. It can be shown that such a large number of horses would fill the entire basin with manure at a rate of several feet per day.)

Annual world production of biomass is estimated at 146 billion metric tons, mostly from uncontrolled plant growth. Many farm crops and trees can produce around 2 metric tons per acre per year of dry biomass, and some algae and grasses can produce significantly more. The heating value of this biomass is 5000–8000 Btu/lb about half of typical values for coal. However, biomass contains virtually no ash or sulfur, both problems with coal. Another sustainability advantage of biomass is that all of the carbon in it is taken from carbon dioxide in the atmosphere so that biomass combustion does not add any net quantities of carbon dioxide to the atmosphere. Indeed, use of biomass to produce hydrogen-rich methane or elemental hydrogen along with sequestration of by-product carbon dioxide would result in an overall loss of carbon dioxide from the atmosphere.

As it has been throughout history, biomass is significant as heating fuel, and in some parts of the world is the fuel most widely used for cooking. Scavenging wood for cooking fuel has been a major contributor to deforestation in some areas. About 15% of Finland's energy needs are provided by wood and wood products (including black liquor by-product from pulp and paper manufacture), about one-third of which is from solid wood. Despite the charm of a wood fire and the sometimes pleasant odor of wood smoke, air pollution from wood-burning stoves and furnaces is a significant problem in some areas. Currently, wood provides about 8% of world energy needs. This percentage could increase through the development of energy plantations consisting of trees grown solely for their energy content.

Biomass could be used to replace much of the 100 million metric tons of petroleum and natural gas currently consumed in the manufacture of primary chemicals in the world each year. Among the sources of biomass that could be used for chemical production are grains and sugar crops (for ethanol manufacture), oilseeds, animal by-products, manure, and sewage (the last two for methane generation). The biggest potential source of chemicals is the lignocellulose making up the bulk of most plant material (see below).

10.15.1 Ethanol Fuel

A major option for converting photosynthetically produced biochemical energy to forms suitable for internal combustion engines is the production of ethanol, C_2H_6O, by the fermentation of sugars from biomass. Suitably designed internal combustion engines can burn pure ethanol or a mixture of 85% ethanol and 15% gasoline called E85. More commonly, ethanol is blended in proportions of around 10% with gasoline to give *gasohol*, a fuel that can be used in existing internal combustion engines with little or no adjustment.

Gasohol boosts octane rating and reduces emissions of carbon monoxide. From a resource viewpoint, because of its photosynthetic origin, alcohol may be considered a renewable resource rather than a depletable fossil fuel. Ethanol is most commonly produced biochemically by the fermentation of carbohydrates. Brazil, a country that produces copious amounts of fermentable sugar from sugarcane, has been a leader in the manufacture of ethanol for fuel uses, with about 27 billion liters produced in 2006. All motor fuels in Brazil contain at least 24% ethanol and some fuel is essentially pure ethanol. A significant fraction of the gasoline consumed in the United States is supplemented with ethanol.

Although most of the ethanol that has been produced for fuel has been made from the fermentation of grain or sugar, there is legitimate concern that, considering the energy that goes into producing grain ethanol, there is no net energy gain. A potentially much more abundant and cheaper source of ethanol consists of biomass generated as a by-product of crop production including straw from wheat or rice production or cornstalks from growing corn. In the past, much rice straw from commercial production in the United States was simply burned to save the cost of cultivating it back into the soil. Straw cannot be fermented directly, but must be broken down to hexose and pentose sugars for fermentation. This has traditionally been done with acid treatment, which is expensive, although technologies exist for recycling acid. It is now generally agreed that production of ethanol from plant biomass by-products will require enzymatic hydrolysis with cellulase enzyme to produce the required sugars. The Canadian Iogen Corporation has a means for obtaining fermentable sugars from wheat straw and other plant materials and has attempted to develop a cost-effective commercial process.

10.15.2 Biodiesel Fuel

Biodiesel fuel is a growing source of renewable liquid hydrocarbon fuels. Unlike fuel ethanol, which must be transported in corrosion-resistant truck tanker trailers or railroad tankers, biodiesel fuel is readily moved through existing pipelines. Rudolf Diesel developed the high-compression, compression-ignited diesel engine in the late 1800s and first operated the engine in Augsburg, Germany, in 1893. He demonstrated his invention at the World Fair in Paris in 1900, receiving the "Grand Prix" (highest prize) for his invention. Interestingly, the fuel used in this and other demonstrations of Diesel's engine was peanut oil, and vegetable oils were the main source of fuel for diesel engines during the first two decades of their use.

Vegetable oils were eventually replaced by petroleum-based hydrocarbons. More recently, diesel fuels that are derivatives of the fatty acids in the oils have been synthesized from vegetable oil feedstocks. Vegetable oils from soybeans and other biological sources are used to make biodiesel fuel as discussed below.

Vegetable oils are fatty acid esters of glycerol, a three-carbon alcohol with 3-OH groups attached. To produce biodiesel fuel, the glycerol esters are hydrolyzed by strong base (NaOH) in the presence of methanol alcohol ($HOCH_3$) and the fatty acids are converted to their methyl esters, the molecules composing biodiesel fuel:

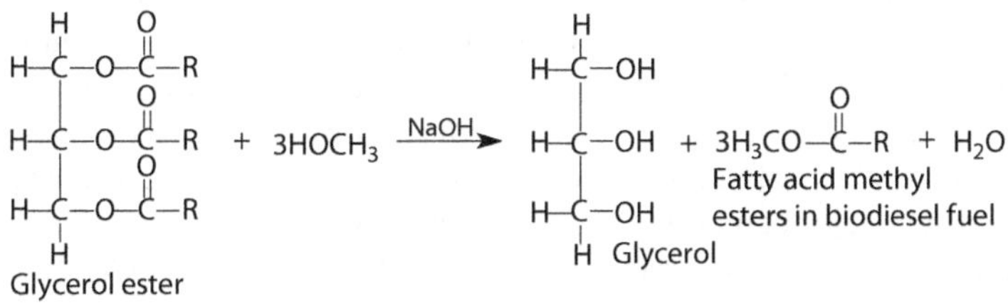

(10.9)

In this reaction, R stands for a long hydrocarbon chain in one of a number of fatty acids including stearic acid, linoleic acid, oleic acid, lauric acid, and behenic acid. For example, in stearic acid, R is a straight chain with 17 carbon atoms, $C_{17}H_{35}$.

Major oils that are used for biodiesel fuel production are rapeseed, sunflower, soybean, palm, coconut, and jatropha. Rapeseed, long grown for animal feed, is the largest source of oil for biodiesel manufacture and is widely produced in Europe, whereas soybean oil predominates in the United States. Both of these oils offer the advantage of providing a protein-rich animal feed after the oil has been squeezed from the seed. Palm oil, and to a lesser extent, coconut oil (from coconut trees) and jatropha (from *Jatropha curcus*, planted for hedges) are attractive because they are from perennial plants that thrive in the tropics. (In late 2008, Air New Zealand flew a Boeing 747 aircraft on which one engine was fueled with 50% jatropha-derived fuel and 50% conventional jet fuel; the aircraft had three other engines in case the experiment failed.) By 2009, the explosive growth of palm oil tree plantations in Malaysia and Indonesia was resulting in high levels of rain forest destruction.

An interesting possibility for biodiesel fuel production is algae that have an oil content exceeding 50%. Such algae can grow profusely in ponds fed with nutrient-rich effluent from wastewater treatment plants. Oil-producing algae have also been grown in the carbon dioxide-rich atmosphere provided by power plant stack gases. By growing algae in treated wastewater within a carbon-dioxide-rich atmosphere from power plants, nutrients can be removed from wastewater, thus reducing eutrophication in receiving waters, and some of the carbon dioxide can be removed from stack gas emissions.

The potential productivity of algae for biomass fuel production is spectacular. Whereas soybean and palm oil sources typically produce, respectively, up to 200 and 2500 L/acre annually of biodiesel fuel, algae have the potential to yield as much as 40,000 L/acre/year of the fuel. Furthermore, with adequate supplies of water, algae can be grown on desert lands that are unsuitable for growing soil-based crops. One experimental system grows algae in water that circulates through transparent plastic tubes. Some algae thrive in saltwater, which can come from the sea or underground formations bearing brackish water.

10.15.3 Unrealized Potential of Lignocellulose Fuels

Both grain- and sugar-based ethanol as well as vegetable oils used to make biodiesel fuel are not the best candidates for biomaterial fuels because they use only relatively small fractions of the plants consisting of the parts that have the most value for food, animal feed, and raw materials. A much more abundant source of fuel consists of the *lignocellulose* parts of plants, a material composed of large polymeric molecules with an approximate empirical formula of CH_2O that composes the structural members of plants including tree wood, stalks, straw, corncobs, and leaves. The three major components of lignocellulose and their approximate empirical formulas are cellulose/starch, $[C_6(H_2O)_5)]_n$; hemicellulose, $[C_5(H_2O)_4)]_n$; and lignin, $[C_{10}(H_{12}O)_3)]_n$.

Large amounts of crop by-product biomass are generated annually. Assuming conservatively that the amount of this material available in the United States is equal to the mass of corn grain produced, about 230 million tons of crop by-product biomass could be made available for fuel each year; the actual figure might be much higher.

The amount of biomass that could be generated from dedicated trees and grass is very high, an estimated 2240 million tons per year in the United States alone. A major advantage of this source is that it comes from perennial plants that can be grown on erodable land, much of which has been taken from agricultural production as the result of government programs. One of the plants that is remarkably productive of biomass consists of hybrid poplars, from the genus *Populis* that includes cottonwoods and aspens. These trees may grow more than 2 m/year and will establish new growth from the stumps of harvested trees, retaining their root systems that prevent soil erosion.

The grass most commonly considered for its biomass productivity is switchgrass. Native to North America, switchgrass is disease- and pest-resistant and requires little fertilizer. It tolerates both drought and flooding very well. Upland varieties of switchgrass grow up to 2 m tall in one growing season on well-drained soils. Lowland varieties can reach heights of 4 m and grow best on heavy soils in bottomlands. Improved varieties of switchgrass have been developed for animal forage and yield around 8 tons/acre of biomass each year. Average biomass yields of forests are only about half as high.

Another high-yielding grass native to swampy regions, such as the Florida everglades, is sawgrass (*Cladium jamaicense*), which gets its name from the saw-like serrations on its leaves. It is well adapted to cultivation in wet areas where other crops cannot be grown and has the additional advantage of providing good wildlife cover.

There are numerous pathways by which biomass and biowastes (including sewage and animal wastes) can be converted to energy, fuels, and feedstocks. These are the following:

- Direct combustion to produce heat and to raise steam to generate electricity
- Fermentation to produce ethanol (from sugar) or methane
- Pyrolysis to yield gaseous fuels (particularly methane), liquids (including hydrocarbons and oxygenated species), and solid carbon
- Thermochemical gasification to produce CO, H_2, CH_4, by-product liquids, and solid carbon
- Fischer–Tropsch synthesis of hydrocarbons from CO and H_2 derived from biomass
- Hydrogenation of oxygenated liquids from biomass to produce hydrocarbons
- Methyl esterification of oils to produce methyl ester biodiesel fuels

Sustainable preparation of biofuels from biomass makes it desirable to recover nutrients such as phosphorus and potassium from the biomass during processing.

Biomass can be used directly as a fuel with a heating value on a dry mass basis about half that of coal. It is extremely low in sulfur and ash and its mineral ash components do not contain toxic elements, such as the arsenic that occurs in some coals. The most direct way to use biomass fuel is direct combustion to produce heat. Biomass can be converted to other high-value fuels including hydrocarbons by gasification. One approach to biomass gasification begins with combustion of part of the biomass (represented by the formula $\{CH_2O\}$) with pure molecular oxygen oxidant (to avoid diluting the product gas with N_2 from air),

$$\{CH_2O\} + O_2 \rightarrow CO_2 + H_2O + \text{heat} \tag{10.10}$$

yielding heat required for the rest of the gasification process. Under the oxygen-deficient conditions through which gasification is carried out, part of the biomass is partially oxidized to combustible carbon monoxide, CO:

$$\{CH_2O\} + \tfrac{1}{2}O_2 \rightarrow CO + H_2O + \text{heat} \tag{10.11}$$

Part of the biomass is pyrolyzed by the heat produced by Reaction 10.10

$$\{CH_2O\} + \text{heat} \rightarrow C + H_2O \tag{10.12}$$

yielding hot carbon and combustible gaseous and liquid by-products. The hot carbon reacts with steam,

$$C + H_2O + \text{heat} \rightarrow CO + H_2 \tag{10.13}$$

yielding a *synthesis gas* mixture of CO and H_2. Biomass may also react as it is heated,

$$\{CH_2O\} + \text{heat} \rightarrow CO + H_2 \quad (10.14)$$

yielding synthesis gas. The carbon monoxide in synthesis gas can be subjected to the water-gas shift reaction,

$$CO + H_2O \rightarrow CO_2 + H_2 \quad (10.15)$$

giving elemental H_2 as the only gaseous product. The CO_2 generated in Reactions 10.10 and 10.15 can be separated and sequestered, such as by pumping into underground petroleum formations to enable recovery of petroleum, thereby preventing the release of this greenhouse-warming gas to the atmosphere.

Elemental hydrogen can be used as an end product of biomass gasification directly as a fuel in gas turbines and other heat engines or to generate electricity in fuel cells. Elemental hydrogen can be used to synthesize ammonia, NH_3, an important industrial chemical and fertilizer. A mixture of CO and H_2 can be reacted over a catalyst in a methanation reaction

$$CO + 3H_2 \rightarrow CH_4 + H_2O \quad (10.16)$$

to produce methane, which, made by this method, is called *synthetic natural gas* (SNG). With different proportions of hydrogen and CO reactants and a different catalyst, a mixture of CO and H_2 can react according to the Fischer–Tropsch reaction to yield a variety of hydrocarbons including gasoline, jet fuel, and diesel fuel as shown by the reaction below for the synthesis of octane, one of the hydrocarbons in gasoline:

$$8CO + 17H_2 \rightarrow C_8H_{18} + 8H_2O \quad (10.17)$$

A similar reaction can also be used to make ethanol and methanol, CH_3OH, which can be used as a fuel, gasoline additive, and to produce H_2 for fuel cells in vehicles.

An interesting possibility for increasing the amount of hydrocarbon fuel that can be obtained from biomass is to use H_2 and O_2 produced by the electrolysis of water with electricity generated by wind power,

$$2H_2O + \text{electrical energy} \rightarrow 2H_2(g) + O_2(g) \quad (10.18)$$

to react with biomass for gasification. The pure elemental oxygen can be used to produce energy from biomass as shown by Reaction 10.11 above without diluting the gas product with elemental nitrogen, N_2, as would be the case with air oxidant, and enabling sequestration of the CO_2 product. The elemental hydrogen generated by electrolysis of water can be reacted directly with biomass. Ideally, direct hydrogenation of biomass would produce predominantly hydrocarbons, such as methane generated by the following reaction:

$$\{CH_2O\} + 2H_2 \rightarrow CH_4 + H_2O \tag{10.19}$$

More commonly, hydrogenation produces a mixture of many liquids, most of which contain oxygen in compounds such as ethers, alcohols, and ketones. For fuel use, these compounds must be further treated with hydrogen to remove oxygen and leave hydrocarbon liquids. Such a process as outlined above provides a means for using the energy originally generated by windpower to make a high-energy hydrocarbon fuel that can be used for transportation or other purposes.

10.15.4 Biogas

A significant source of clean-burning methane can be obtained from the anoxic (oxygen-free) bacterial fermentation of biomass of a variety of kinds. Representing biomass as $\{CH_2O\}$, the biochemical reaction is the following:

$$2\{CH_2O\} \rightarrow CH_4 + CO_2 \tag{10.20}$$

This reaction has long been used in the anoxic digesters of sewage treatment plants to reduce the amount of degradable organic matter in excess sewage sludge. Such a well-balanced plant makes enough methane to provide for all its energy needs. Large livestock feeding operations may have digesters to produce methane from livestock manure and other biological wastes associated with the feeding operation. Another source of methane generated by anoxic fermentation is obtained by burying collector pipes in mounds of municipal solid wastes and collecting the off-gas.

An outstanding example of "grass roots" energy sustainability is provided by the widespread use of biogas generators in rural China, an area chronically short of fuel. The Chinese have built millions of anoxic digesters located just below the ground surface in which all kinds of biomass including animal wastes, human excrement, and vegetable and crop wastes are degraded in the absence of air to produce combustible methane gas. The digesters are usually located in proximity to dwellings and to sources of waste, including latrines and pigsties. Covered by a thin layer of soil, these facilities take advantage of solar heating to accelerate fermentation. In addition to producing fuel, anoxic digestion alleviates waste disposal and health problems through the destruction of pathogens in the waste by the fermentation process, which often takes place at a somewhat elevated temperature that kills most pathogenic organisms. An additional advantage is that the fertilizer value of the wastes is conserved so that residues from the digesters can be placed on soil to serve as fertilizer for crops without much concern over spreading pathogens. The contents of both nitrogen and phosphorus in the digester residues are much higher than in biomass that has been treated by mulching in an air environment.

The gas from the Chinese biogas facilities is used largely for cooking and lamps. In larger installations, internal combustion engines have been adapted to run in part on biogas. In the case of gasoline engines, biogas is fed into carburetors through an adjustable feed valve and in the case of diesel engines, it is fed into the air/fuel intake manifold to supplement the minimal amount of diesel fuel injected into the combustion chamber to initiate combustion during engine operation.

A major methane-generating system in the United States is overseen by Central Vermont Public Service in Vermont, utilizing manure from several large dairy farms. The largest dairy in this complex is the Pleasant Valley Farm where 1500 milking cows produce over 40 million pounds of milk a year. As a profitable by-product of this operation, anoxic fermentation for 3 weeks of manure from the cows produces methane used to generate 3.5 million kWh of electricity a year, enough to power 500 homes. The farm went from spending about $200/day for electricity to making about $1200/day from renewable electricity sales in 2008. The system cost about $2 million. As of late 2008, Central Vermont Public Service had five farms with a total of about 5000 cows in its "Cow Power" renewable energy program.

10.16 HYDROGEN AS A MEANS TO STORE AND UTILIZE ENERGY

Hydrogen gas, H_2, is an ideal chemical fuel in some respects that may serve as a storage medium for solar energy. There are several ways to generate elemental hydrogen.[5] As noted in the preceding section, solar-generated electricity can be used to electrolyze water:

$$2H_2O + \text{electrical energy} \rightarrow 2H_2(g) + O_2(g) \tag{10.21}$$

The hydrogen fuel product, and even elemental oxygen produced as an electrolysis by-product, can be piped some distance and the hydrogen burned without pollution, or it may be used in a fuel cell (Figure 10.7). This may, in fact, make possible a "hydrogen economy." Disadvantages of using hydrogen as a fuel include its low heating value per unit volume and the wide range of explosive mixtures it forms with air. Although not yet economical, photochemical processes can be used to split water to H_2 and O_2 that can be used to power fuel cells.

Fuel-cell-powered vehicles are now practical in some applications and Honda began production of the first commercial fuel-cell automobile in 2008. One of the greatest barriers to their widespread adoption has been their inability to carry sufficient hydrogen for an acceptable range. Several solutions to this problem are now being investigated. One is the potentially problematic use of very cold liquid hydrogen as a fuel source. Another is the use of very high pressure containers composed of multilayer cylinders wrapped with carbon composite and filled with hydrogen at pressures up to 10,000 psi (about 670 × atmospheric pressure!), reputed to contain sufficient hydrogen to propel an automobile up to 300 miles. Other systems use catalysts to break down liquid fuels, such as methanol or gasoline, to generate hydrogen for fuel cells.

Although there has been much enthusiasm in some quarters for hydrogen fuel and predictions of a new hydrogen economy, some of the more enthusiastic arguments for hydrogen fuel may be too optimistic. The most important point is that, unlike fossil fuels, hydrogen is not a primary source of energy and has to be made by processes such as the electrolysis of water (Reaction 10.18) that use other sources of energy.

Most of the 6 million tons of elemental hydrogen used in the United States each year is made from steam reforming of methane from natural gas:

$$CH_4 + H_2O \rightarrow 3H_2 + CO \tag{10.22}$$

The carbon monoxide product can be reacted with steam as noted in the preceding section,

$$CO + H_2O \rightarrow CO_2 + H_2 \tag{10.23}$$

to produce additional H_2 and the CO_2 product of this reaction can be sequestered as discussed in Section 10.9.

In principle, the process described above and the utilization of elemental hydrogen in fuel cells can provide a transport fuel that is pollution-free. However, methane gas is easier to store than elemental hydrogen and the modern internal combustion engine with associated emissions control equipment is virtually pollution-free. So the intermediate production of elemental hydrogen is unlikely to be the greenest approach. Production of elemental hydrogen by electrolysis of water using electricity from renewable sources, such as photovoltaics and windpower, is essentially nonpolluting, but it requires balancing the relatively inefficient electrolysis process against the value of the electricity that it consumes.

10.17 COMBINED POWER CYCLES

Combined power cycles enable much more efficient utilization of combustible fuels by first using the heat of combustion in a turbine coupled to an electrical generator, raising steam in a boiler with the hot exhaust gas from this turbine, using the steam to power a second turbine linked to a generator, and finally using the steam and hot water from the steam turbine for applications such as processing in the chemical industry, heating commercial buildings, or heating homes. A schematic diagram of a combined power cycle system is illustrated in Figure 10.19. The water condensed from the steam used for heating is pure and is recycled to the boiler, thus minimizing the amount of makeup boiler feedwater, which requires expensive treatment to make it suitable for use in boilers. The use of steam leaving a steam turbine for heating, a concept known as *district heating*, is commonly practiced in Europe (and many university campuses in the United States) and can save large amounts of fuel otherwise required for heating. Such a system as the one described is in keeping with the best practice of industrial ecology and should be employed whenever it is practical to do so.

10.18 A SYSTEM OF INDUSTRIAL ECOLOGY FOR METHANE PRODUCTION

Figure 10.20 illustrates an industrial ecosystem based on the production of methane fuel from renewable biomass and wind energy sources. Methane is the ideal

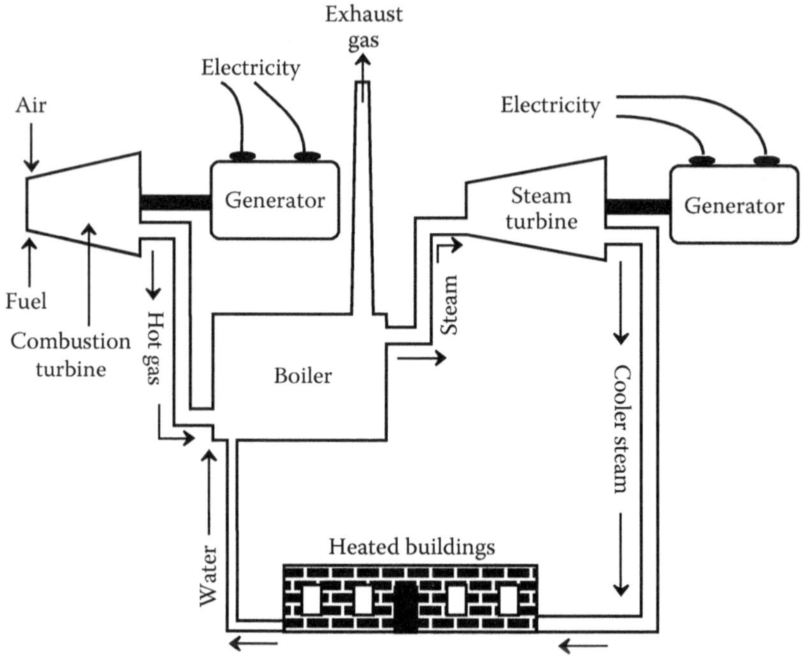

FIGURE 10.19 A combined power cycle in which combustible gas or oil is first used to fire a gas turbine connected to an electrical generator. The hot gases from this turbine are fed to a boiler to raise steam, which drives a steam turbine, also connected to a generator. The still hot exhaust steam from the steam turbine is used for process heat or conveyed to commercial or residential buildings for heating. The water condensed from the steam in this final application is returned to the power plant to generate more steam, thus conserving water and avoiding the necessity to treat more water to the high purity standards required by the boiler. (From Manahan, S., *Environmental Chemistry*, 9th ed., Taylor & Francis, Boca Raton, 2010. With permission.)

clean-burning fuel for virtually all applications that currently use fossil fuels except for aircraft. With hybrid automobile technology, methane can be used as motor fuel giving an acceptable range between refueling. With previously untapped sources coming on line in the United States and a number of other countries, methane is relatively available and inexpensive. Pipeline systems are in place for its distribution. And, when methane sources are eventually exhausted, the gas can be synthesized from biomass and elemental hydrogen produced from the electrolysis of water.

A system for producing methane renewably is outlined in Figure 10.20. Biomass is used as a source of fixed carbon that is reacted with elemental hydrogen made by electrolyzing water using renewable wind energy. Another source of methane in the system is anoxic fermentation of wastes, including sewage. Instead of municipal refuse in the community being disposed in landfill, it is first subjected to anoxic fermentation to generate methane after which the residues are thermochemically gasified to provide feedstock for chemical synthesis. Other aspects of the system are largely self-explanatory from the figure.

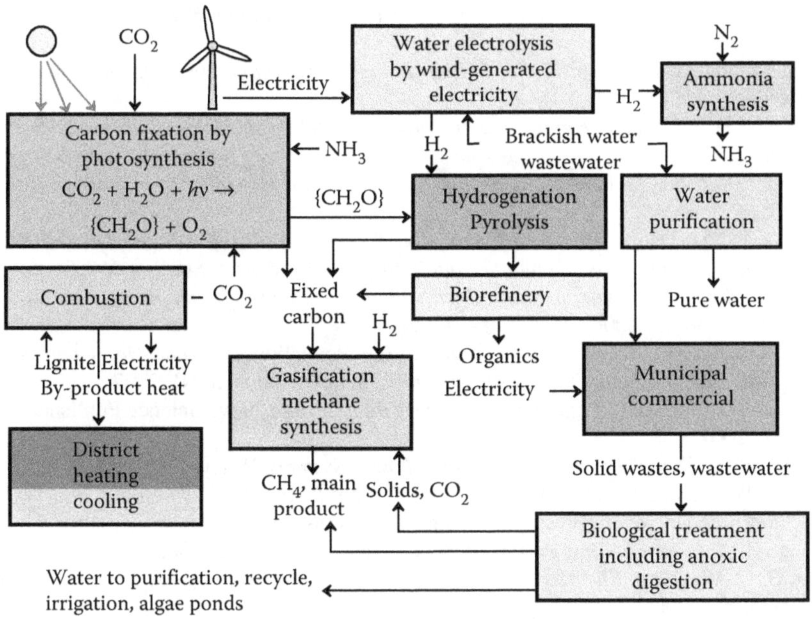

FIGURE 10.20 An industrial ecosystem based on the synthesis of methane and other fuels from biomass using wind-generated electrical power as the major energy source driving the system. (From Manahan, S., *Environmental Chemistry*, 9th ed., Taylor & Francis, Boca Raton, 2010. With permission.)

LITERATURE CITED

1. Leonenko, Y. and D. Keith, Reservoir engineering to accelerate the dissolution of CO_2 stored in aquifers, *Environmental Science and Technology*, 42, 2742–2747, 2008.
2. Bachu, S. and W. D. Gunter, Acid-gas injection in the Alberta Basin, Canada: A CO_2-storage experience, *Geological Society Special Publication*, 233, 225–234, 2004.
3. For an explanation of the planned system see: http://www.stirlingenergy.com/
4. Larson, E. D., A review of life-cycle analysis studies on liquid biofuel systems for the transport sector, *Energy for Sustainable Development*, 10, 109–126, 2006.
5. Holladay, J. D., J. Hu, D. L. King, and Y. Wang, An overview of hydrogen production technologies, *Catalysis Today*, 139, 244–260, 2009.
6. Manahan, S., *Environmental Chemistry*, 9th ed., Taylor & Francis, Boca Raton, 2010.

SUPPLEMENTARY REFERENCES

Amos, S., *Energy: A Historical Perspective and 21st Century Forecast*, The American Association of Petroleum Geologists, Tulsa, OK, 2005.

Brebbia, C. A. and V. Popov, Eds, *Energy and Sustainability*, WIT Press, Southampton, UK, 2007.

Brenes, M. D., Ed., *Biomass and Bioenergy: New Research*, Nova Science Publishers, New York, 2006.

Coley, D. A., *Energy and Climate Change: Creating a Sustainable Future*, Wiley, Hoboken, NJ, 2008.

Dickson, M. and M. Fanelli, Eds, *Geothermal Energy: Utilization and Technology*, Earthscan, Sterling, VA, 2005.

Fay, J. A. and D. S. Golomb, *Energy and the Environment*, Oxford University Press, New York, 2002.

Goetzberger, A. and V. U. Hoffmann, *Photovoltaic Solar Energy Generation*, Springer, Berlin, 2005.

Harvey, L. and D. Danny, Ed., *A Handbook on Low-Energy Buildings and District-Energy System: Fundamentals, Techniques and Examples*, Earthscan, Sterling, VA, 2006.

Hau, E., *Windturbines: Fundamentals, Technologies, Application, and Economics*, 2nd ed., Springer, Berlin, 2006.

Herbst, A. M. and G. W. Hopley, *Nuclear Energy Now: Why the Time Has Come for the World's Most Misunderstood Energy Source*, Wiley, Hoboken, NJ, 2007.

Hofman, K. A., Ed., *Energy Efficiency, Recovery and Storage*, Nova Science Publishers, New York, 2007.

Infield, D. and L. Freris, *Renewable Energy in Power Systems*, Wiley, Chichester, UK, 2008.

Jamasb, T., W. J. Nuttall, and M. G. Pollitt, Eds, *Future Electricity Technologies and Systems*, Cambridge University Press, Cambridge, UK, 2006.

Kanninen, B., Ed., *Atomic Energy*, Green Haven Press, San Diego, 2006.

Krauter, S. C. W., *Solar Electric Power Generation—Photovoltaic Energy Systems*, Springer, New York, 2006.

Kruger, P., *Alternative Energy Resources: The Quest for Sustainable Energy*, Wiley, Hoboken, NJ, 2006.

Larminie, J. and A. Dicks, *Fuel Cell Systems Explained*, 2nd ed., Wiley, New York, 2003.

Manwell, J. F., J. G. McGowan, and A. L. Rogers, *Wind Energy Explained: Theory, Design and Application*, Wiley, New York, 2002.

McGowan, T., *Biomass and Alternate Fuel Systems: An Engineering and Economic Guide*, Hoboken, NJ, 2009.

Nag, A., Ed., *Biofuels Refining and Performance*, McGraw-Hill, New York, 2008.

Nakaya, A., *Energy Alternatives*, ReferencePoint Press, San Diego, CA, 2008.

Nutall, W. J., *Nuclear Renaissance: Technologies and Policies for the Future of Nuclear Power*, Bristol, Philadelphia, 2005.

O'Hayre, R., *Fuel Cell Fundamentals*, Wiley, Hoboken, NJ, 2006.

Patel, M. R., *Wind and Solar Power Systems: Design, Analysis, and Operation*, 2nd ed., Taylor & Francis, London, 2006.

Peinke, J., *Wind Energy*, Springer, New York, 2006.

Romm, J. J., *The Hype about Hydrogen: Fact and Fiction in the Race to Save the Climate*, Island Press, Washington, 2004.

Savage, L., Ed., *Geothermal Power*, Greenhaven Press, Detroit, 2007.

Silveira, S., Ed., *Bioenergy, Realizing the Potential*, Elsevier, Amsterdam, 2005.

Soetaert, W. and E. Vandamme, *Biofuels*, Wiley, Hoboken, NJ, 2008.

Sørensen, B., *Renewable Energy Conversion, Transmission, and Storage*, Elsevier/Academic Press, Amsterdam, 2007.

Suppes, G. J. and T. S. Storvick, *Sustainable Nuclear Power*, Elsevier/Academic Press, Amsterdam, 2007.

Tabak, J., *Solar and Geothermal Energy*, Facts On File, New York, 2009.

Weiss, C. and W. B. Bonvillian, *Structuring an Energy Technology Revolution*, MIT Press, Cambridge, MA, 2009.

Worldwatch Institute, *Biofuels for Transport: Global Potential and Implications for Energy and Agriculture*, Earthscan, Sterling, VA, 2007.

QUESTIONS AND PROBLEMS

The use of Internet resources is assumed in answering any of the questions. These would include such things as constants and conversion factors as well as additional information needed to complete an answer.

1. The tail of a firefly glows, although it is not hot. Explain the kind of energy transformation that is most likely involved in the firefly's tail producing light.
2. What is the standard unit of energy? What unit did it replace? What is the relationship between these two units?
3. Which law states that energy is neither created nor destroyed?
4. What is the special significance of 1340 W?
5. What is the reaction in nature by which solar energy is converted to chemical energy?
6. In what respects is wind both one of the oldest, as well as one of the newest, sources of energy?
7. What are two major problems with reliance on coal and petroleum for energy?
8. Why does natural gas contribute less to greenhouse warming than does petroleum and much less than coal?
9. How might coal be utilized for energy without producing greenhouse gas carbon dioxide?
10. What is a large limiting factor in growing biomass for fuel, and in what respect does this limit hold hope for the eventual use of biomass fuel?
11. What relationship describes the limit to which heat energy can be converted to mechanical energy?
12. Why does a diesel-powered vehicle have significantly better fuel economy than a gasoline-powered vehicle of similar size?
13. Why is a nuclear power plant less efficient in converting heat energy to electricity than is a fossil-fueled power plant?
14. Instead of having a spark plug that ignites the fuel, a diesel engine has a glow plug that operates only during engine startup. Explain the operation of the glow plug.
15. Cite two examples of vastly increased efficiency of energy utilization that took place during the 1900s.
16. Describe a combined power cycle. How may it be tied with district heating?
17. What are three reactions used in biomass gasification?
18. What is a major proposed use of liquid methanol as a fuel for the future?
19. Describe a direct and an indirect way to produce electricity from solar energy.
20. What is the distinction between donor and acceptor layers in photovoltaic cells?
21. Using internet resources for information, list some possible means for storing energy generated from solar radiation?

22. What are the advantages of *Pittsosporum reiniferum* and *Euphorbia lathyrus* for the production of biomass energy?
23. Corn produces biomass in large quantities during its growing season. What are two potential sources of biomass fuel from corn, one that depends on the corn grain and the other that does not?
24. Does the use of biomass for fuel contribute to greenhouse gas carbon dioxide? Explain.
25. What fermentation process is used to generate a fuel from wastes, such as animal wastes?
26. What are two potential pollution problems that accompany the use of geothermal energy to generate electricity?
27. What basic phenomenon is responsible for nuclear energy? What keeps the process going?
28. What is the biggest problem with nuclear energy? Why is it not such a bad idea to store spent nuclear fuel at a reactor site for a number of years before moving it?
29. What is meant by passive stability in nuclear reactor design?
30. What is the status of thermonuclear fusion for power production?
31. Arrange the following energy conversion processes in order from the least to the most efficient: (a) electric hot water heater, (b) photosynthesis, (c) solar cell, (d) electric generator, (e) aircraft jet engine.
32. Considering the Carnot equation and common means for energy conversion, what might be the role of improved materials (metal alloys and ceramics) in increasing energy conversion efficiency?
33. As it is now used, what is the principle or basis for the production of energy from uranium by nuclear fission? Is this process actually used for energy production? What are some of its environmental disadvantages? What is one major advantage?
34. What would be at least two highly desirable features of nuclear fusion power if it could ever be achieved in a controllable fashion on a large scale?
35. Justify describing the sun as "an ideal energy source." What are two big disadvantages of solar energy?
36. What are some of the greater implications of the use of biomass for energy? How might such widespread use affect greenhouse warming? How might it affect agricultural production of food?

11 Fundamentals of Chemistry

11.1 THE SCIENCE OF MATTER

This chapter is designed to give those readers who have had little exposure to chemistry the basic knowledge needed to understand the material in the rest of the book. Although it is helpful for the reader to have had several courses in chemistry, including organic chemistry and quantitative chemical analysis, most of the material in this book can be understood with less. Indeed, a reader willing to do some independent study on the fundamentals of chemistry can understand much of the material in this book without ever having had any formal chemistry course work.

This chapter can serve two purposes. For the reader who has had no chemistry, it provides the concepts and terms basic to general chemistry. A larger category of reader consists of those who have had at least one chemistry course, but whose chemistry background, for various reasons, is inadequate. By learning the material in this chapter, plus the subject matter of Chapter 12, "Organic Chemistry," these readers can comprehend the rest of the material in the book. For a more complete coverage of basic chemistry, readers should consult one of a number of basic chemistry books, such as *Fundamentals of Sustainable Chemical Science*[1] and other supplementary references listed at the end of the chapter.

Chemistry is the science of matter. Therefore, it deals with all of the things that surround humankind, and with all aspects of the environment. Chemical properties and processes are central to environmental science. A vast variety of chemical reactions occur in water, for example, including acid–base reactions, precipitation reactions, and oxidation–reduction reactions largely mediated by microorganisms. Atmospheric chemical phenomena are largely determined by photochemical processes and chain reactions. A large number of organic chemical processes occur in the atmosphere. The geosphere, including soil, is the site of many chemical processes, particularly those that involve solids. The biosphere obviously is where the many biochemical processes crucial to the environment and to the toxic effects of chemicals occur.

This chapter emphasizes several aspects of chemistry. It begins with a discussion of the fundamental subatomic particles that make up all matter, and explains how these are assembled to produce atoms. In turn, atoms join together to make compounds. Chemical reactions and chemical equations that represent them are discussed. Solution chemistry is especially important to aquatic chemistry and is addressed in a separate section. The important, vast discipline of organic chemistry is crucial to all parts of the environment and is addressed in Chapter 12.

11.2 ELEMENTS

All substances are composed of only about a hundred fundamental kinds of matter called *elements*. Elements, themselves, may be of environmental concern. The "heavy metals," including lead, cadmium, and mercury, are well recognized as toxic substances in the environment. Elemental forms of otherwise essential elements may be very toxic or cause environmental damage. Oxygen in the form of ozone, O_3, is the agent most commonly associated with atmospheric smog pollution and is very toxic to plants and animals. Elemental white phosphorus is highly flammable and toxic.

Each element is made up of very small entities called *atoms*; all atoms of the same element behave identically chemically. The study of chemistry, therefore, can logically begin with elements and the atoms of which they are composed. Each element is designated by an *atomic number*, a name, and a *chemical symbol*, such as carbon, C; potassium, K (for its Latin name kalium); or cadmium, Cd. Each element has a characteristic *atomic mass* (atomic weight), which is the average mass of all atoms of the element. Atomic numbers of the elements are integrals ranging from 1 for hydrogen, H, to somewhat more than 100 for some of the transuranic elements (those beyond uranium). Atomic number is a unique, important way of designating each element, and it is equal to the number of protons in the nuclei of each atom of the element (see discussion of subatomic particles and atoms, below).

11.2.1 Subatomic Particles and Atoms

Figure 11.1 represents an atom of deuterium, a form of hydrogen. It is seen that such an atom is made up of even smaller subatomic particles—positively charged protons, negatively charged electrons, and uncharged (neutral) neutrons.

11.2.2 Subatomic Particles

The subatomic particles differ in mass and charge. Their masses are expressed by the atomic mass unit, u (also called the Dalton), which is also used to express the masses

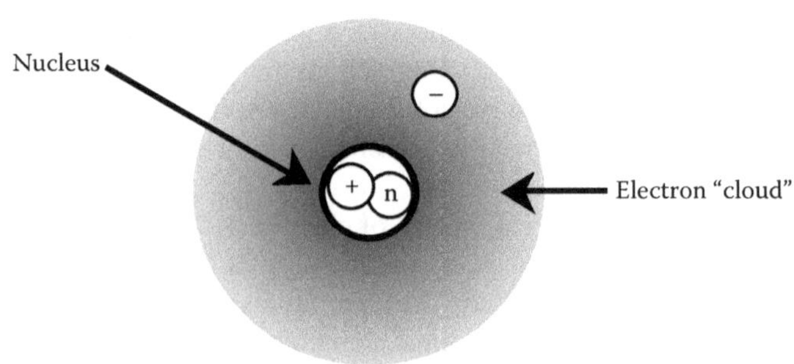

FIGURE 11.1 Representation of a deuterium atom. The nucleus contains one proton (+) and one neutron (n). The electron (–) is in constant, rapid motion around the nucleus, forming a cloud of negative electrical charge, the density of which drops off with increasing distance from the nucleus.

of individual atoms and molecules (aggregates of atoms). The atomic mass unit is defined as a mass equal to exactly 1/12 that of an atom of carbon-12, the isotope of carbon that contains six protons and six neutrons in its nucleus.

The proton, p, has a mass of 1.007277 u and a unit charge of +1. This charge is 1.6022×10^{-19} Coulombs, where a Coulomb is the amount of electrical charge involved in a flow of electrical current of 1 amp for 1 s. The neutron, n, has no electrical charge and a mass of 1.009665 u. The proton and neutron each have a mass of essentially 1 u and are said to have a *mass number* of 1. (Mass number is a useful concept expressing the total number of protons and neutrons, as well as the approximate mass, of a nucleus or subatomic particle.) The electron, *e*, has a unit electrical charge of −1. It is very light, however, with a mass of only 0.00054859 u, about 1/1840 that of the proton or neutron. Its mass number is 0. The properties of protons, neutrons, and electrons are summarized in Table 11.1.

Although it is convenient to think of the proton and neutron as having the same mass, and each is assigned a mass number of 1, it is seen in Table 11.1 that their exact masses differ slightly from each other. Furthermore, the mass of an atom is not exactly equal to the sum of the masses of subatomic particles composing the atom. This is because of the energy relationships involved in holding the subatomic particles together in atoms so that the masses of the atom's constituent subatomic particles do not add up to exactly the mass of the atom.

11.2.3 Atom Nucleus and Electron Cloud

Protons and neutrons, which have relatively high masses compared to electrons, are contained in the positively charged *nucleus* of the atom. The nucleus has essentially all of the mass, but occupies virtually none of the volume, of the atom. An uncharged atom has the same number of electrons as protons. The electrons in an atom are contained in a cloud of negative charge around the nucleus that occupies most of the volume of the atom. These concepts are emphasized in Figure 11.2.

11.2.4 Isotopes

Atoms with the *same* number of protons, but *different* numbers of neutrons in their nuclei are called *isotopes*. They are chemically identical atoms of the same element,

TABLE 11.1
Properties of Protons, Neutrons, and Electrons

Subatomic Particle	Symbol	Unit Charge	Mass Number	Mass in u	Mass in grams
Proton[a]	*p*	+1	1	1.007277	1.6726×10^{-24}
Neutron[a]	*n*	0	1	1.008665	1.6749×10^{-24}
Electron[a]	*e*	−1	0	0.000549	9.1096×10^{-28}

[a] The mass number and charge of each of these kinds of particles may be indicated by a superscipt and subscript, respectively, as in the symbols ${}^{1}_{1}\text{p}$, ${}^{1}_{0}\text{n}$, and ${}^{0}_{-1}\text{e}$.

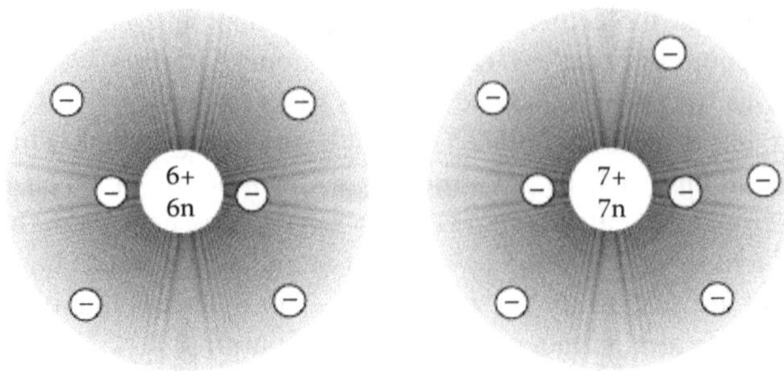

An atom of carbon, symbol C. Each C atom has 6 protons (+) in its nucleus, so the atomic number of C is 6. The atomic mass of C is 12.

An atom of nitrogen, symbol N. Each N atom has 7 protons (+) in its nucleus, so the atomic number of N is 7. The atomic mass of N is 14.

FIGURE 11.2 Atoms of carbon and nitrogen.

but have different masses and may differ in their nuclear properties. Some isotopes are *radioactive isotopes* or *radionuclides*, which have unstable nuclei that give off charged particles and gamma rays in the form of *radioactivity*. Radioactivity may have detrimental, or even fatal, health effects; a number of hazardous substances are radioactive and they can cause major environmental problems. The most striking example of such contamination resulted from a massive explosion and fire at a power reactor in the Ukrainian city of Chernobyl in 1986.

11.2.5 List of the Elements

A list of the elements is given in Table 11.2. Several unnamed elements above atomic number 109 have been made artificially with particle accelerators, but for the most part only a few radioactive atoms of each.

11.2.6 The Periodic Table

When elements are considered in order of increasing atomic number, it is observed that their properties are repeated in a periodic manner. For example, elements with atomic numbers 2, 10, and 18 are gases that do not undergo chemical reactions and consist of individual atoms, whereas those with atomic numbers larger by 1—elements with atomic numbers 3, 11, and 19—are unstable, highly reactive metals. An arrangement of the elements in a manner that reflects this recurring behavior is known as the *periodic table* (Figure 11.3). The periodic table is extremely useful in understanding chemistry and predicting chemical behavior. As shown in Figure 11.3, the entry for each element in the periodic table gives the element's atomic number, name, symbol, and atomic mass. More detailed versions of the table include other information as well.

11.2.6.1 Features of the Periodic Table

Groups of elements having similar chemical behavior are contained in vertical columns in the periodic table. *Main group* elements may be designated as A groups

TABLE 11.2
List of the Elements[a]

Atomic Number	Name	Atomic Symbol	Atomic Mass	Atomic Number	Name	Symbol	Atomic Mass
1	Hydrogen	H	1.00794	56	Barium	Ba	137.327
2	Helium	He	4.0026	57	Lanthanum	La	138.9055
3	Lithium	Li	6.941	58	Cerium	Ce	140.115
4	Beryllium	Be	9.01218	59	Praseodymium	Pr	140.9077
5	Boron	B	10.811	60	Neodymium	Nd	144.24
6	Carbon	C	12.011	61	Promethium	Pm	145
7	Nitrogen	N	14.0067	62	Samarium	Sm	150.36
8	Oxygen	O	15.9994	63	Europium	Eu	151.965
9	Fluorine	F	18.9984	64	Gadolinium	Gd	157.25
10	Neon	Ne	20.1797	65	Terbium	Tb	158.925
11	Sodium	Na	22.9898	66	Dysprosium	Dy	162.50
12	Magnesium	Mg	24.305	67	Holmium	Ho	164.9303
13	Aluminum	Al	26.98154	68	Erbium	Er	167.26
14	Silicon	Si	28.0855	69	Thulium	Tm	168.9342
15	Phosphorus	P	30.9738	70	Ytterbium	Yb	173.04
16	Sulfur	S	32.066	71	Lutetium	Lu	174.967
17	Chlorine	Cl	35.4527	72	Hafnium	Hf	178.49
18	Argon	Ar	39.948	73	Tantalum	Ta	180.9497
19	Potassium	K	39.0983	74	Tungsten	W	183.85
20	Calcium	Ca	40.078	75	Rhenium	Re	186.207

TABLE 11.2 (continued)
List of the Elements[a]

Atomic Number	Name	Atomic Symbol	Atomic Mass	Atomic Number	Name	Symbol	Atomic Mass
21	Scandium	Sc	44.9559	76	Osmium	Os	190.2
22	Titanium	Ti	47.88	77	Iridium	Ir	192.22
23	Vanadium	V	50.9415	78	Platinum	Pt	195.08
24	Chromium	Cr	51.9961	79	Gold	Au	196.9665
25	Manganese	Mn	54.938	80	Mercury	Hg	200.59
26	Iron	Fe	55.847	81	Thallium	Tl	204.383
27	Cobalt	Co	58.9332	82	Lead	Pb	207.2
28	Nickel	Ni	58.6934	83	Bismuth	Bi	208.98
29	Copper	Cu	63.546	84	Polonium	Po	209
30	Zinc	Zn	65.39	85	Astatine	At	210
31	Gallium	Ga	69.723	86	Radon	Rn	222
32	Germanium	Ge	72.61	87	Francium	Fr	223
33	Arsenic	As	74.9216	88	Radium	Ra	226.0254
34	Selenium	Se	78.96	89	Actinium	Ac	227.0278
35	Bromine	Br	79.904	90	Thorium	Th	232.038
36	Krypton	Kr	83.8	91	Protactinium	Pa	231.0359
37	Rubidium	Rb	85.4678	92	Uranium	U	238.0289

38	Strontium	Sr	87.62	93	Neptunium	Np	237.048
39	Yttrium	Y	88.9056	94	Plutonium	Pu	244
40	Zirconium	Zr	91.224	95	Americium	Am	243
41	Niobium	Nb	92.9064	96	Curium	Cm	247
42	Molybdenum	Mo	95.94	97	Berkelium	Bk	247
43	Technetium	Tc	98	98	Californium	Cf	251
44	Ruthenium	Ru	101.07	99	Einsteinium	Es	252
45	Rhodium	Rh	102.9055	100	Fermium	Fm	257.1
46	Palladium	Pd	106.42	101	Mendelevium	Md	258.1
47	Silver	Ag	107.8682	102	Nobelium	No	255
48	Cadmium	Cd	112.411	103	Lawrencium	Lr	260
49	Indium	In	114.82	104	Rutherfordium	Rf	261.11
50	Tin	Sn	118.710	105	Dubnium	Db	262.11
51	Antimony	Sb	121.757	106	Seaborgium	Sg	263.12
52	Tellurium	Te	127.60	107	Bohrium	Bh	262.12
53	Iodine	I	126.9045	108	Hassium	Hs	265
54	Xenon	Xe	131.29	109	Meitnerium	Mt	266
55	Cesium	Cs	132.9054				

[a] Elements above atomic number 92 have been made artificially.

Period	IA 1	IIA 2	IIIB 3	IVB 4	VB 5	VIB 6	VIIB 7	VIII 8	VIII 9	VIII 10	IB 11	IIB 12	IIIA 13	IVA 14	VA 15	VIA 16	VIIA 17	Noble gases 18
1	1 **H** 1.008																	2 **He** 4.003
2	3 **Li** 6.941	4 **Be** 9.012											5 **B** 10.81	6 **C** 12.01	7 **N** 14.01	8 **O** 16.00	9 **F** 19.00	10 **Ne** 20.18
3	11 **Na** 22.99	12 **Mg** 24.3											13 **Al** 26.98	14 **Si** 28.09	15 **P** 30.97	16 **S** 32.07	17 **Cl** 35.45	18 **Ar** 39.95
4	19 **K** 39.10	20 **Ca** 40.08	21 **Sc** 44.96	22 **Ti** 47.88	23 **V** 50.94	24 **Cr** 52.00	25 **Mn** 54.94	26 **Fe** 55.85	27 **Co** 58.93	28 **Ni** 58.69	29 **Cu** 63.55	30 **Zn** 65.39	31 **Ga** 69.72	32 **Ge** 72.59	33 **As** 74.92	34 **Se** 78.96	35 **Br** 79.9	36 **Kr** 83.8
5	37 **Rb** 85.47	38 **Sr** 87.62	39 **Y** 88.91	40 **Zr** 91.22	41 **Nb** 92.91	42 **Mo** 95.94	43 **Tc** 98.91	44 **Ru** 101.1	45 **Rh** 102.9	46 **Pd** 106.4	47 **Ag** 107.9	48 **Cd** 112.4	49 **In** 114.8	50 **Sn** 118.7	51 **Sb** 121.8	52 **Te** 127.6	53 **I** 126.9	54 **Xe** 131.3
6	55 **Cs** 132.9	56 **Ba** 137.3	57 **La** 138.9 *	72 **Hf** 178.5	73 **Ta** 180.9	74 **W** 183.8	75 **Re** 186.2	76 **Os** 190.2	77 **Ir** 192.2	78 **Pt** 195.1	79 **Au** 197.0	80 **Hg** 200.6	81 **Tl** 204.4	82 **Pb** 207.2	83 **Bi** 209.0	84 **Po** (210)	85 **At** (210)	86 **Rn** (222)
7	87 **Fr** (223)	88 **Ra** (226)	89 **Ac** (227) **	104 **Rf** (261)	105 **Ha** (262)	106 **Sg** (263)	107 **Ns** (262)	108 **Ha** (265)	109 **Mt** (266)									

Inner Transition Elements

Lanthanide series * 6	58 **Ce** 140.1	59 **Pr** 140.9	60 **Nd** 144.2	61 **Pm** 144.9	62 **Sm** 150.4	63 **Eu** 152.0	64 **Gd** 157.2	65 **Tb** 158.9	66 **Dy** 162.5	67 **Ho** 164.9	68 **Er** 167.3	69 **Tm** 168.9	70 **Yb** 173.0	71 **Lu** 175.0
Actinide series ** 7	90 **Th** 232.0	91 **Pa** 231.0	92 **U** 238.0	93 **Np** 237.0	94 **Pu** 239.1	95 **Am** 243.1	96 **Cm** 247.1	97 **Bk** 247.1	98 **Cf** 252.1	99 **Es** 252.1	100 **Fm** 257.1	101 **Md** 256.1	102 **No** 259.1	103 **Lr** 260.1

FIGURE 11.3 The periodic table of elements.

(1A and 2A on the left, 3A through 8A on the right). *Transition elements* are those between main groups 2A and 3A. *Noble gases* (group 8A), a group of gaseous elements that are virtually chemically unreactive, are in the far-right column. The chemical similarities of elements in the same group are especially pronounced for groups 1A, 2A, 7A, and 8A.

Horizontal rows of elements in the periodic table are called *periods*, the first of which consists of only hydrogen (H) and helium (He). The second period begins with atomic number 3 (lithium) and terminates with atomic number 10 (neon), whereas the third goes from atomic number 11 (sodium) through 18 (argon). The fourth period includes the first row of transition elements, whereas lanthanides and actinides are listed separately at the bottom of the table.

11.2.7 Electrons in Atoms

Although a detailed discussion of the placement of electrons in atoms determines how the atoms behave chemically and, therefore, the chemical properties of each element, it is beyond the scope of this chapter to discuss electronic structure in detail. Several key points pertaining to this subject are mentioned here.

Electrons in atoms are present in *orbitals* in which the electrons have different energies, orientations in space, and average distances from the nucleus. Each orbital may contain a maximum of two electrons. The placement of electrons in their orbitals determines the chemical behavior of an atom; in this respect, the outermost orbitals and the electrons contained in them are the most important. These *outer electrons* are the ones beyond those of the immediately preceding noble gas in the periodic table. They are of particular importance because they become involved in the sharing and transfer of electrons through which chemical bonding occurs that results in the formation of huge numbers of different substances from only a few elements.

Much of environmental chemistry and water chemistry are concerned with electrons in atoms. In the atmosphere, the absorption of electromagnetic radiation, primarily ultraviolet radiation, promotes electrons to higher energy levels, forming reactive excited species and reactive free radicals with unpaired electrons. These phenomena can result in *photochemical reactions* such as the formation of stratospheric ozone discussed in Chapter 1, Section 1.4. Atomic absorption and emission methods of elemental analysis (see Chapter 9, Section 9.3) involve transitions of electrons between energy levels.

11.2.7.1 Lewis Structures and Symbols of Atoms

Outer electrons are called *valence electrons* and are represented by dots in *Lewis symbols*, as shown for carbon and argon in Figure 11.4.

The four electrons shown for the carbon atom are those added beyond the electrons possessed by the noble gas that immediately precedes carbon in the periodic table (helium, atomic number 2). Eight electrons are shown around the symbol of argon. This is an especially stable electron configuration for noble gases known as an *octet*. (Helium is the exception among noble gases in that it has a stable shell of only two electrons.) When atoms interact through the sharing, loss, or gain of electrons to form molecules and chemical compounds (see Section 11.3) many attain an

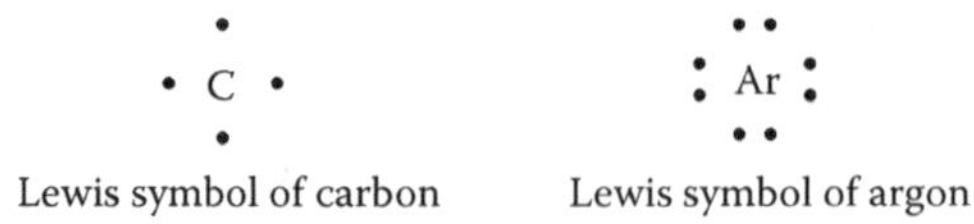

FIGURE 11.4 Lewis symbols of carbon and argon.

octet of outer shell electrons. This tendency is the basis of the *octet rule* of chemical bonding. (Two or three of the lightest elements, most notably hydrogen, attain stable helium-like electron configurations containing two electrons when they become chemically bonded.)

11.2.8 Metals, Nonmetals, and Metalloids

Elements are divided between metals and nonmetals; a few elements with intermediate character are called metalloids. *Metals* are elements that are generally solid, shiny in appearance, electrically conducting, and malleable—that is, they can be pounded into flat sheets without disintegrating. They tend to have only 1–3 outer electrons, which they may lose in forming chemical compounds. Examples of metals are iron, copper, and silver. Most metallic objects that are commonly encountered are not composed of just one kind of elemental metal, but are alloys consisting of homogeneous mixtures of two or more metals. *Nonmetals* often have a dull appearance, are not at all malleable, and frequently occur as gases or liquids. Colorless oxygen gas, green chlorine gas (transported and stored as a liquid under pressure), and brown bromine liquid are common nonmetals. Nonmetals tend to have close to a full octet of outer-shell electrons, and in forming chemical compounds they gain or share electrons. *Metalloids*, such as silicon or arsenic, are elements with properties intermediate between those of metals and nonmetals. Under some conditions, a metalloid may exhibit properties of metals, and under other conditions, properties of nonmetals.

11.3 CHEMICAL BONDING

Only a few elements, particularly the noble gases, exist as individual atoms; most atoms are joined by chemical bonds to other atoms. For example, elemental hydrogen

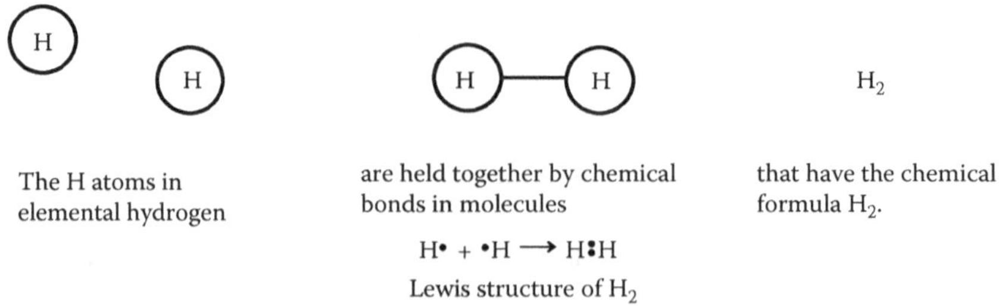

FIGURE 11.5 Molecule and Lewis formula of H_2.

exists as *molecules*, each consisting of 2 H atoms linked by a *chemical bond* as shown in Figure 11.5. Because hydrogen molecules contain 2 H atoms, they are said to be diatomic and are denoted by the *chemical formula*, H_2. The H atoms in the H_2 molecule are held together by a *covalent bond* made up of two electrons, each contributed by one of the H atoms, and shared between the atoms. (Bonds formed by transferring electrons between atoms are described later in this section.) The shared electrons in the covalent bonds holding the H_2 molecule together are represented by two dots between the H atoms in Figure 11.5. By analogy with Lewis symbols defined in the preceding section, such a representation of molecules showing outer-shell and bonding electrons as dots is called a *Lewis formula*.

11.3.1 Chemical Compounds

Most substances consist of two or more elements joined by chemical bonds. As an example consider the chemical combination of hydrogen and oxygen shown in Figure 11.6. Oxygen, chemical symbol O, has an atomic number of 8 and an atomic mass of 16.00; it exists in the elemental form as diatomic molecules of O_2. Hydrogen atoms combine with oxygen atoms to form molecules in which 2 H atoms are bonded to 1 O atom in a substance with a chemical formula of H_2O (water). A substance such as H_2O that consists of a chemically bonded combination of two or more elements is called a *chemical compound*. In the chemical formula for water the letters H and O are the chemical symbols of the two elements in the compound and the subscript 2 indicates that there are 2 H atoms per O atom. (The absence of a subscript after the O denotes the presence of just 1 O atom in the molecule.)

As shown in Figure 11.6, each of the hydrogen atoms in the water molecule is connected to the oxygen atom by a chemical bond composed of two electrons shared between the hydrogen and oxygen atoms. For each bond one electron is contributed by the hydrogen and one by oxygen. The two dots located between each H and O in the Lewis formula of H_2O represent the two electrons in the covalent bond joining these atoms. Four of the electrons in the octet of electrons surrounding O are involved

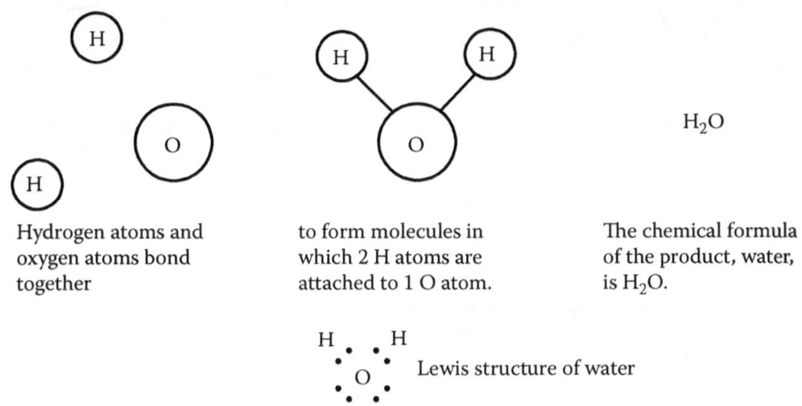

FIGURE 11.6 Formation and Lewis formula of a chemical compound, water.

in H–O bonds and are called bonding electrons. The other four electrons shown around the oxygen that are not shared with H are nonbonding outer electrons.

11.3.2 Molecular Structure

As implied by the representations of the water molecule in Figure 11.6, the atoms and bonds in H_2O form an angle somewhat greater than 90°. The shapes of molecules are referred to as their *molecular geometry*, which is crucial in determining the chemical and toxicological activity of a compound and structure–activity relationships.

11.3.3 Ionic Bonds

As shown in Figure 11.7, the transfer of electrons from one atom to another produces charged species called *ions*. Positively charged ions are called *cations* and negatively charged ions are called *anions*. Ions that make up a solid compound are held together by *ionic bonds* in a *crystalline lattice* consisting of an ordered arrangement of the ions in which each cation is largely surrounded by anions and each anion by cations. The attracting forces of the oppositely charged ions in the crystalline lattice constitute ionic bonds in the compound.

The formation of magnesium oxide is shown in Figure 11.7. In naming this compound, the cation is simply given the name of the element from which it was formed, magnesium. However, the ending of the name of the anion, *oxide*, is different from that of the element from which it was formed, ox*ygen*.

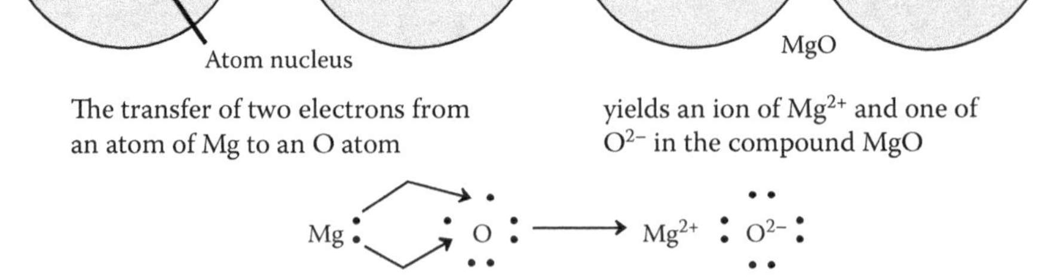

Formation of ionic MgO as shown by Lewis structures and symbols. In MgO, Mg has lost 2 electrons and is in the +2 oxidation state, Mg(II), and O has gained 2 electrons and is in the −2 oxidation state.

FIGURE 11.7 Ionic bonds are formed by the transfer of electrons and the mutual attraction of oppositely charged ions in a crystalline lattice.

Rather than individual atoms that have lost or gained electrons, many ions are groups of atoms bonded together covalently and having a net charge. A common example of such an ion is the ammonium ion, NH_4^+,

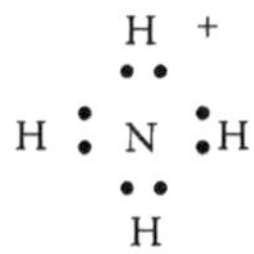

Ammonium ion, NH_4^+

which consists of four hydrogen atoms covalently bonded to a single nitrogen (N) atom and having a net electrical charge of +1 for the whole cation, as shown by its Lewis formula above.

11.3.4 Summary of Chemical Compounds and the Ionic Bond

The preceding several pages have just covered some material on chemical compounds and bonds that are essential to understand chemistry. To summarize, these are the following:

- Atoms of two or more different elements can form *chemical bonds* with each other to yield a product that is entirely different from the elements.
- Such a substance is called a *chemical compound.*
- The *formula* of a chemical compound gives the symbols of the elements and uses subscripts to show the relative numbers of atoms of each element in the compound.
- *Molecules* of some compounds are held together by *covalent bonds* consisting of shared electrons.
- Another kind of compound consists of *ions* composed of electrically charged atoms or groups of atoms held together by *ionic bonds* that exist because of the mutual attraction of oppositely charged ions.

11.3.5 Molecular Mass

The average mass of all molecules of a compound is its *molecular mass* (formerly called molecular weight). The molecular mass of a compound is calculated by multiplying the atomic mass of each element by the relative number of atoms of the element, then adding all the values obtained for each element in the compound. For example, the molecular mass of NH_3 is $14.0 + 3 \times 1.0 = 17.0$. As another example consider the following calculation of the molecular mass of ethylene, C_2H_4.

1. The chemical formula of the compound is C_2H_4
2. Each molecule of C_2H_4 consists of 2 C atoms and 4 H atoms
3. From the periodic table or Table 11.2, the atomic mass of C is 12.0 and that of H is 1.0

4. Therefore, the molecular mass of C_2H_4 is

$$\underbrace{12.0 + 12.0}_{\text{From 2 C atoms}} + \underbrace{1.0 + 1.0 + 1.0 + 1.0}_{\text{From 4 H atoms}} = 28.0$$

11.3.6 The Mole and Molar Mass

It is convenient to consider quantities of chemicals in compounds in units proportional to the number of molecules. Such a unit is the *mole*, which is expressed as the same number of grams of a compound as the molecular mass of the compound. Using this definition, a mole of C_2H_4, calculated above to have a molecular mass of 28.0, is 28.0 g and the *molar mass* of C_2H_4 is 28.0 g/mol. The molecular mass of H_2O is 18.0 and its molar mass is 18.0 g/mol. The definition applies equally well to ionic compounds that do not consist of distinct molecules. Sodium chloride, NaCl, consists of aggregates of equal numbers of Na^+ and Cl^- ions. The molecular mass of NaCl is 23.0 (from Na) + 35.5 (from Cl) = 58.5. Therefore, the molar mass of NaCl is 58.5 g/mol. In the case of elements that are composed of individual atoms, the molar mass is given by the atomic mass. The molar mass of helium, atomic mass 4.0, is 4.0 g/mol. Elemental hydrogen, atomic mass 1.0 consists of molecules each composed of two atoms (H_2) and has a molecular mass of 2.0. The molar mass of H_2 is 2.0 g/mol.

11.3.7 Oxidation State

The loss of two electrons from the magnesium atom as shown in Figure 11.7 is an example of *oxidation*, and the Mg^{2+} ion product is said to be in the +2 *oxidation state*. [A positive oxidation state or oxidation number is conventionally denoted by a Roman numeral within parentheses following the name or symbol of an element as in magnesium(II) and Mg(II)]. In gaining two negatively charged electrons in the reaction that produces magnesium oxide, the oxygen atom is *reduced* and is in the −2 oxidation state. (Unlike positive oxidation numbers, negative ones are not conventionally shown by Roman numerals in parentheses.) In chemical terms, an *oxidizer* is a species that takes electrons from a reducing agent in a chemical reaction. Many hazardous waste substances are oxidizers or strong reducers and oxidation/reduction is the driving force behind many dangerous chemical reactions. For example, the reducing tendencies of the carbon and hydrogen atoms in propane cause it to burn violently or explode in the presence of oxidizing oxygen in air. The oxidizing ability of concentrated nitric acid, HNO_3, enables it to react destructively with organic matter, such as cellulose or skin. As discussed in Chapter 4, oxidation–reduction phenomena are very important in aquatic chemical processes.

Covalently bonded atoms that have not actually lost or gained electrons to produce ions are also assigned oxidation states. This can be done because in covalent compounds electrons are not shared equally. Therefore, an atom of an element with a greater tendency to attract electrons is assigned a negative oxidation number compared to the positive oxidation number assigned to an element with a lesser tendency to attract electrons. For example, Cl atoms attract electrons more strongly than do H atoms so that in hydrogen chloride gas, HCl, the Cl atom is in the –1 oxidation state and the H atoms are in the +1 oxidation state. *Electronegativity* values are assigned to elements on the basis of their tendencies to attract electrons.

The oxidation state (oxidation number) of an element in a compound may have a strong influence on the hazards posed by the compound. For example, chromium from which each atom has lost three electrons to form a chemical compound, designated as chromium(III) or Cr(III), is not toxic, whereas chromium in the +6 oxidation state (CrO_4^{2-}, chromate) is regarded as a cancer-causing chemical when inhaled.

11.4 CHEMICAL REACTIONS AND EQUATIONS

Chemical reactions occur when substances are changed to other substances through the breaking and formation of chemical bonds. For example, water is produced by the chemical reaction of hydrogen and oxygen:

Hydrogen plus oxygen yields water

Chemical reactions are written as *chemical equations*. The chemical reaction between hydrogen and water is written as the *balanced chemical equation*

$$2H_2 + O_2 \rightarrow 2H_2O \tag{11.1}$$

in which the arrow is read as "yields" and separates the hydrogen and oxygen *reactants* from the water *product*. Note that because elemental hydrogen and elemental oxygen occur as *diatomic molecules* of H_2 and O_2, respectively, it is necessary to write the equation in a way that reflects these correct chemical formulas of the elemental form. All correctly written chemical equations are *balanced* in that *the same number of each kind of atom must be shown on both sides of the equation*. The equation above is balanced because of the following:

- On the left
  - There are 2 H_2 *molecules* each containing 2 H *atoms*, a total of 4 H atoms on the left.
  - There is 1 O_2 *molecule* containing 2 O *atoms* for a total of 2 O atoms on the left.

- On the right
  - There are 2 H_2O *molecules* each containing 2 H *atoms* and 1 O *atom* for a total of 4 H atoms and 2 O atoms on the right.

11.4.1 Reaction Rates

Most chemical reactions give off heat and are classified as exothermic reactions. The rate of a reaction may be calculated by the Arrhenius equation, which contains absolute temperature (K = °C + 273) in an exponential term. As a general rule, the speed of a reaction doubles for each 10°C increase in temperature. Reaction rate factors are important factors in fires or explosions involving hazardous chemicals.

11.5 SOLUTIONS

A liquid *solution* is formed when a substance in contact with a liquid becomes dispersed homogeneously throughout the liquid in a molecular form. The substance, called a *solute*, is said to *dissolve*. The liquid is called a *solvent*. There may be no readily visible evidence that a solute is present in the solvent; for example, a deadly poisonous solution of sodium cyanide in water looks like pure water. Some solutions have a strong color, as is the case for intensely purple solutions of potassium permanganate, $KMnO_4$. A solution may have a strong odor, such as that of ammonia, NH_3, dissolved in water. Solutions may consist of solids, liquids, or gases dissolved in a solvent. Technically, it is even possible to have solutions in which a solid is a solvent, although such solutions are not discussed in this book.

11.5.1 Solution Concentration

The quantity of solute relative to that of solvent or solution is called the *solution concentration*. Concentrations are expressed in numerous ways. Very high concentrations are often given as percent by weight. For example, commercial concentrated hydrochloric acid is 36% HCl, meaning that 36% of the weight has come from dissolved HCl and 64% from water solvent. Concentrations of very dilute solutions, such as those of hazardous waste leachate containing low levels of contaminants, are expressed as mass of solute per unit volume of solution. Common units are milligrams per liter (mg/L) or micrograms per liter (μg/L). Since a liter of water weighs essentially 1000 g, a concentration of 1 mg/L is equal to 1 part per million (ppm) and a concentration of 1 μg/L is equal to 1 part per billion (ppb).

Chemists often express concentrations in moles per liter, or *molarity*, M. Molarity is given by the relationship

$$M = \frac{\text{Number of moles of solute}}{\text{Number of liters of solution}} \tag{11.2}$$

The number of moles of a substance is its mass in grams divided by its molar mass. For example, the molecular mass of ammonia, NH_3, is 14 + 1 + 1 + 1, so a mole of ammonia has a mass of 17 g. Therefore, 17 g of NH_3 in 1 L of solution has a value of M equal to 1 mol/L.

11.5.2 Water as a Solvent

Most liquid wastes are solutions or suspensions of waste materials in water. Water has some unique properties as a solvent which arise from its molecular structure as represented by the Lewis structure of the water molecule below:

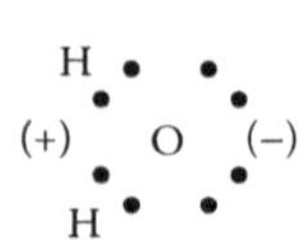

The H atoms are not on opposite sides of the O atom and the two H–O bonds form an angle of 105°. Furthermore, the O atom (–2 oxidation state) is able to attract electrons more strongly than the 2 H atoms (each in the +1 oxidation state) so that the molecule is *polar*, with the O atom having a partial negative charge and the end of the molecule with the 2 H atoms having a partial positive charge. This means that water molecules can cluster around ions with the positive ends of the water molecules attracted to negatively charged anions and the negative end to positively charged cations. This kind of interaction is part of the general phenomenon of *solvation*. It is specifically called *hydration* when water is the solvent and is partially responsible for water's excellent ability to dissolve ionic compounds including acids, bases, and salts.

Water molecules form a special kind of bond called a *hydrogen bond* with each other and with solute molecules that contain O, N, or S atoms. As its name implies, a hydrogen bond involves a hydrogen atom held between two other atoms of O, N, or S. Hydrogen bonding is partly responsible for water's ability to solvate and dissolve chemical compounds capable of hydrogen bonding.

As noted above, the water molecule is a polar species, which affects its ability to act as a solvent. Solutes may likewise have polar character. In general, solutes with polar molecules are more soluble in water than nonpolar ones. The polarity of an impurity solute in wastewater is a factor in determining how it may be removed from water. Nonpolar organic solutes are easier to take out of water by an adsorbent species such as activated carbon than are more polar solutes.

11.5.3 Solutions of Acids and Bases

11.5.3.1 Acid–Base Reactions

The reaction between H^+ ion from an acid and OH^- ion from a base is a *neutralization* reaction. As a specific example consider the reaction of H^+ from a solution of sulfuric acid, H_2SO_4, with OH^- from a solution of calcium hydroxide:

Acid, source of H^+ ion (→ H_2SO_4); Water (→ $2H_2O$)

$$H_2SO_4 + Ca(OH)_2 \longrightarrow 2H_2O + CaSO_4 \qquad (11.3)$$

Base, source of OH^- ion (→ $Ca(OH)_2$); Salt (→ $CaSO_4$)

In addition to water, which is always the product of a neutralization reaction, the other product is calcium sulfate, $CaSO_4$. This compound is a *salt* composed of Ca^{2+} ions and SO_4^{2-} ions held together by ionic bonds. A salt, consisting of a cation other than H^+ and an anion other than the OH^- ion, is the other product in addition to water produced when an acid and base react. Some salts are hazardous substances and environmental pollutants because of their dangerous or harmful properties. Some examples are the following:

- Ammonium perchlorate, NH_4ClO_4, (reactive oxidant)
- Barium cyanide, $Ba(CN)_2$ (toxic)

- Lead acetate, $Pb(C_2H_3O_2)_2$ (toxic)
- Thallium(I) carbonate, Tl_2CO_3 (toxic)

11.5.3.2 Concentration of H+ Ion and pH

Acids such as HCl and sulfuric acid (H_2SO_4) produce H^+ ion, whereas bases, such as sodium hydroxide and calcium hydroxide [NaOH and $Ca(OH)_2$, respectively], produce hydroxide ion, OH^-. Molar concentrations of water solutions of hydrogen ion, $[H^+]$, range over many orders of magnitude and are conveniently expressed by pH defined as

$$pH = -\log [H^+] \tag{11.4}$$

In absolutely pure water at 25°C, the value of $[H^+]$ is exactly 1×10^{-7} mol/L, the pH is 7.00, and the solution is *neutral* (neither acidic nor basic). *Acidic* solutions have pH values of less than 7, and *basic* solutions have pH values of greater than 7.

Strong acids and strong bases are *corrosive* substances that exhibit extremes of pH. They are destructive to materials and flesh. Strong acids can react with cyanide and sulfide compounds to release highly toxic hydrogen cyanide (HCN) or hydrogen sulfide (H_2S) gases, respectively. Bases liberate noxious ammonia gas (NH_3) from solid ammonium compounds.

11.5.3.3 Metal Ions Dissolved in Water

Metal ions dissolved in water have some unique characteristics that influence their properties as natural water constituents and heavy metal pollutants, and in biological systems. The formulas of metal ions are usually represented by the symbol for the ion followed by its charge. For example, iron(II) ion [from a compound such as iron(II) sulfate, $FeSO_4$] dissolved in water is represented as Fe^{2+}. Actually, in water solution each iron(II) ion is strongly solvated and bonded to water molecules, so that the formula is more correctly shown as $Fe(H_2O)_6^{2+}$. Many metal ions have a tendency to lose hydrogen ions from the solvating water molecules as shown by the following:

$$Fe(H_2O)_6^{2+} \rightarrow Fe(OH)(H_2O)_5^{+} + H^+ \tag{11.5}$$

Ions of the next higher oxidation state, iron(III), have such a tendency to lose H^+ ion in aqueous solution that, except in rather highly acidic solutions, they precipitate out as solid hydroxides, such as iron(III) hydroxide, $Fe(OH)_3$:

$$Fe(H_2O)_6^{3+} \rightarrow Fe(OH)_3(s) + 3H_2O + 3H^+ \tag{11.6}$$

11.5.3.4 Complex Ions Dissolved in Water

It was noted above that metal ions are solvated (hydrated) by binding to water molecules in aqueous solution. Some species in solution have a stronger tendency than water to bond to metal ions. An example of such a species is cyanide ion, CN^-, which displaces water molecules from some metal ions in solution as shown below:

$$Ni(H_2O)_4^{2+} + 4CN^- \rightleftarrows Ni(CN)_4^{2-} + 4H_2O \tag{11.7}$$

The species that bonds to the metal ion, cyanide in this case, is called a *ligand*, and the product of the reaction is a *complex ion* or metal complex. The overall process is called *complexation.*

11.5.4 Colloidal Suspensions

Very small particles on the order of 1 μm or less in size, called *colloidal particles*, may stay suspended in a liquid for an indefinite period of time. Such a mixture is a *colloidal suspension* and it behaves in many respects like a solution. Colloidal suspensions are used in many industrial applications. Many waste materials are colloidal and are often emulsions consisting of colloidal liquid droplets suspended in another liquid, usually wastewater. One of the challenges in dealing with colloidal wastes is to remove a relatively small quantity of colloidal material from a large quantity of water by precipitating the colloid. This process is called *coagulation* or *flocculation* and is often brought about by the addition of chemical agents.

11.5.5 Solution Equilibria

Many of the phenomena in aquatic chemistry (Chapters 3 through 8) and geochemistry (chemical phenomena in the geosphere) involve solution equilibrium. In a general sense, solution equilibrium deals with the extent to which *reversible* acid–base, solubilization–precipitation, complexation, or oxidation–reduction reactions proceed in a forward or backward direction. This is expressed for a generalized equilibrium reaction,

$$aA + bB \rightleftarrows cC + dD \tag{11.8}$$

by the following equilibrium constant expression:

$$\frac{[C]^c[C]^d}{[A]^a[B]^b} = K \tag{11.9}$$

where K is the equilibrium constant.

A reversible reaction may approach equilibrium from either direction. In the example above, if A were mixed with B, or C were mixed with D, the reaction would proceed in a forward or reverse direction such that the concentrations of species—[A], [B], [C], and [D]—substituted into the equilibrium expression gave a value equal to K.

As expressed by *Le Châtelier's principle*, a stress placed upon a system in equilibrium will shift the equilibrium to relieve the stress. For example, adding product "D" to a system in equilibrium will cause Reaction 11.8 to shift to the left, consuming "C" and producing "A" and "B," until the equilibrium constant expression is again satisfied. This *mass action effect* is the driving force behind many environmental chemical phenomena.

In most cases this book uses concentrations and pressures in equilibrium constant expression calculations. When this is done, K is not exactly constant with varying concentrations and pressures; it is an *approximate equilibrium constant* that applies only to limited conditions. *Thermodynamic equilibrium constants* are more exact forms derived from thermodynamic data that make use of *activities* in place of concentrations. At a specified temperature, the value of a thermodynamic equilibrium constant is applicable over a wide concentration range. The activity of a species, commonly denoted as a_X for species "X," expresses how effectively it interacts with its surroundings, such as other solutes or electrodes in solution. (The analogy may be drawn of an aquatic chemistry class with a "concentration" of 20 students per classroom. Their "activity" in relating to the subject is likely to be much higher on a cold, rainy day than on a balmy, sunny day in springtime.) Activities approach concentrations at low values of concentration. The thermodynamic equilibrium constant expression for Reaction 11.8 is expressed as the following in terms of activities:

$$\frac{a_C^c a_D^d}{a_A^a a_B^b} = K \tag{11.10}$$

There are several major kinds of equilibria in aqueous solution. One of these is acid–base equilibrium (see Chapter 3) as exemplified by the ionization of acetic acid, HAc,

$$HAc \rightleftarrows H^+ + Ac^- \tag{11.11}$$

for which the acid dissociation constant is

$$\frac{[H^+][Ac^-]}{[HAc]} = K = 1.75 \times 10^{-5} \tag{11.12}$$

Very similar expressions are obtained for the formation and dissociation of metal *complexes* or *complex ions* (Chapter 3) formed by the reaction of a metal ion in solution with a *complexing agent* or *ligand*, both of which are capable of independent existence in solution. This can be shown by the reaction of iron(III) ion and thiocyanate ligand

$$Fe^{3+} + SCN^- \rightleftarrows FeSCN^{2+} \tag{11.13}$$

for which the formation constant expression is:

$$\frac{[FeSCN^{2+}]}{[Fe^{3+}][SCN^-]} = K = 1.07 \times 10^3 \quad \text{(at 25°C)} \tag{11.14}$$

The bright red color of the $FeSCN^{2+}$ complex formed is used to test for the presence of iron(III) in acid mine water (Chapter 7).

An example of an *oxidation–reduction reaction*, which involves the transfer of electrons between species, is

$$MnO_4^- + 5Fe^{2+} + 8H^+ \rightleftarrows Mn^{2+} + 5Fe^{3+} + 4H_2O \tag{11.15}$$

for which the equilibrium expression is:

$$\frac{[Mn^{2+}][Fe^{3+}]^5}{[MnO_4^-][Fe^{2+}]^5[H^+]^8} = K = 3 \times 10^{62} \quad \text{(at 25°C)} \tag{11.16}$$

The value of K is calculated from the Nernst equation, as explained in Chapter 4.

11.5.5.1 Distribution between Phases

As discussed in Chapter 5, many important environmental chemical phenomena involve distribution of species between phases. This most commonly involves the equilibria between species in solution and in a solid phase. *Solubility equilibria* (see Chapter 5) deal with reactions such as

$$AgCl(s) \rightleftarrows Ag^+ + Cl^- \tag{11.17}$$

in which one of the participants is a solid that is slightly soluble (virtually insoluble). For the example shown above, the equilibrium constant is

$$[Ag^+][Cl^-] = K_{sp} = 1.82 \times 10^{-10} \quad \text{(at 25°C)} \tag{11.18}$$

a *solubility product*. Note that in the equilibrium constant expression there is not a value given for the solid AgCl. This is because the activity of a solid is constant at a specific temperature and is contained in the value of K_{sp}.

An important example of distribution between phases is that of a hazardous waste species partitioned between water and a body of immiscible organic liquid in a hazardous waste site. The equilibrium for such a reaction,

$$X(aq) \rightleftarrows X(org) \tag{11.19}$$

is described by the *distribution law* expressed by a *distribution coefficient* or *partition coefficient* in the following form:

$$\frac{[X(org)]}{[X(aq)]} = K_d \tag{11.20}$$

LITERATURE CITED

1. Manahan, S. E., *Fundamentals of Sustainable Chemical Science*, Taylor & Francis/CRC Press, Boca Raton, FL, 2009.

SUPPLEMENTARY REFERENCES

Brown, T. L., H. Eugene LeMay, and B. E. Bursten, *Chemistry: The Central Science*, 11th ed., Pearson Prentice-Hall, Upper Saddle River, NJ, 2008.

Chang, R. and J. Overby, *General Chemistry: The Essential Concepts*, 5th ed., McGraw-Hill, Boston, 2007.

Eubanks, L. P., C. H. Middlecamp, C. E. Heltzel, and S. W. Keller, *Chemistry in Context*, 6th ed., McGraw-Hill Higher Education, Boston, 2008.

Frost, L. D., T. S. Deal, and K. C. Timberlake, *General, Organic, and Biological Chemistry: An Integrated Approach*, Pearson Education, Upper Saddle River, NJ, 2011.

Goldberg, D., *Fundamentals of Chemistry*, 5th ed., McGraw-Hill, Dubuque, IA, 2007.

Hein, M. and S. Arena, *Foundations of College Chemistry*, 12th ed., Wiley, Hoboken, NJ, 2006.

Joesten, M., M. E. Castellion, and J. L. Hogg, *The World of Chemistry: Essentials*, 4th ed., Thomson Brooks/Cole, Belmont, CA, 2007.

McMurry, J., D. S. Ballantine, C. A. Hoeger, and V. E. Peterson, *Fundamentals of General, Organic, and Biological Chemistry*, 6th ed., Prentice-Hall, Upper Saddle River, NJ, 2009.

Petrucci, R. H. *General Chemistry: Principles and Modern Applications*, 9th ed., Pearson Education, Upper Saddle River, NJ, 2007.

Raymond, K. *General Organic and Biological Chemistry*, Wiley, Hoboken, NJ, 2010.

Silberberg, M. S., *Principles of General Chemistry*, McGraw-Hill, Boston, 2007.

Seager, S. L. M. R. Slabaugh, *Chemistry for Today: General, Organic, and Biochemistry*, 6th ed., Thomson Brooks/Cole, Belmont, CA, 2008.

Stoker, H. Stephen, *General, Organic, and Biological Chemistry*, 5th ed., Brooks Cole, Belmont, CA, 2008.

Whitten, K. W., *Chemistry*, 8th ed., Thomson Brooks/Cole, Belmont, CA, 2007.

QUESTIONS AND PROBLEMS

The use of Internet resources is assumed in answering any of the questions. These would include such things as constants and conversion factors as well as additional information needed to complete an answer.

1. What distinguishes a radioactive isotope from a "normal" stable isotope?
2. Why is the periodic table so named?
3. After examining Figure 11.7, consider what might happen when an atom of sodium (Na), atomic number 11, loses an electron to an atom of fluorine, (F), atomic number 9. What kinds of particles are formed by this transfer of a negatively charged electron? Is a chemical compound formed? If so, what is it called?
4. Match the following:
   1. O_2 (a) Element consisting of individual atoms
   2. NH_3 (b) Element consisting of chemically bonded atoms

3. Ar (c) Ionic compound
4. NaCl (d) Covalently bound compound

5. Consider the following atom:

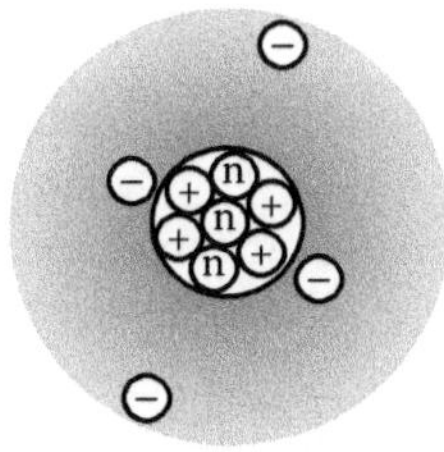

How many electrons, protons, and neutrons does it have? What is its atomic number? Give the name and chemical symbol of the element of which it is composed.

6. Give the chemical formula and molecular mass of the molecule represented below:

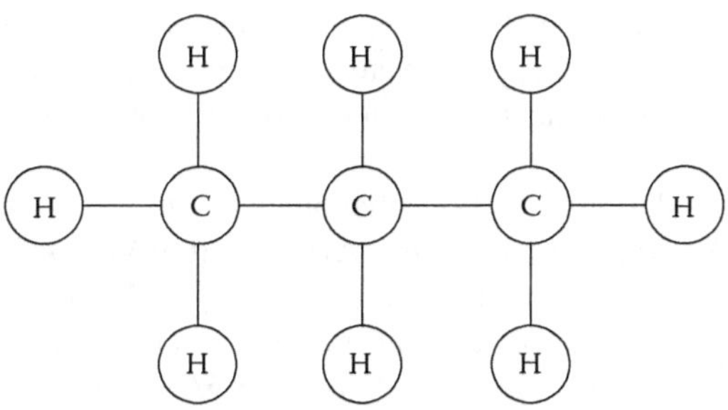

7. Calculate the molecular masses of (a) C_2H_2, (b) N_2H_4, (c) Na_2O, (d) O_3 (ozone), (e) PH_3, (f) CO_2.
8. Is the equation, $H_2 + O_2 \rightarrow H_2O$, a balanced chemical equation? Explain. Point out the reactants and products in the equation.
9. Define and distinguish the differences between environmental chemistry, environmental biochemistry, and toxicological chemistry.
10. An uncharged atom has the same number of ______________ as protons. The electrons in an atom are contained in a cloud of around the nucleus that occupies most of the ____________________ of the atom.
11. Match:

1. Argon (a) A halogen
2. Hydrogen (b) Fissionable element
3. Uranium (c) Noble gas
4. Chlorine (d) Has an isotope with a mass number of 2
5. Mercury (e) Toxic heavy metal

12. The entry for each element in the periodic table gives the element's ______________ and the periodic table is arranged horizontally in _________ and vertically in _________.

13. Electrons in atoms occupy __________ in which electrons have different ________________________. Each orbital may contain a maximum of ________ electrons.
14. The Lewis symbol of carbon is ________ in which each dot represents ________.
15. Elements that are generally solid, shiny in appearance, electrically conducting, and malleable are______________ whereas elements that tend to have a dull appearance, are not at all malleable, and frequently occur as gases or liquids are _______________. Elements with intermediate properties are called ____________________.
16. In the Lewis formula,

$$H:H$$

the two dots represent ___________________________.
17. Explain why, even though it consists of molecules containing chemical bonds, H_2 is not a chemical compound whereas H_2O is a chemical compound.
18. Using examples, distinguish between covalent and ionic bonds.
19. In terms of c, h, o and the appropriate atomic masses, write a mathematical formula for the molecular mass of a compound with a general formula of $C_cH_hO_o$.
20. Considering oxidation/reduction phenomena, when Al reacts with O_2 to produce Al_2O_3, which contains the Al^{3+} ion, the Al is said to have been ____________________ and is in the _______ oxidation state.
21. Calculate the concentration in moles per liter of (a) a solution that is 27.0% H_2SO_4 by mass and that has a density of 1198 g/L, and (b) of a solution that is 1 mg/L H_2SO_4 having a density of 1000 g/L.
22. Calculate the pH of the second solution described in the preceding problem, keeping in mind that each H_2SO_4 molecule yields two H^+ ions.
23. Write a balanced neutralization reaction between NaOH and H_2SO_4.
24. Distinguish between solutions and colloidal suspensions.
25. What is the nature of the Fe^{3+} ion? Why are solutions containing this ion acidic?
26. What kind of species is $Ni(CN)_4^{2-}$?
27. What is the solubility product expression for lead sulfate, $PbSO_4$, in terms of concentrations of Pb^{2+} ion and SO_4^{2-} ion?

12 Organic Chemistry

12.1 ORGANIC CHEMISTRY

Most carbon-containing compounds are *organic chemicals* and are addressed by the subject of *organic chemistry*. Organic chemistry is a vast, diverse discipline because of the enormous number of organic compounds that exist as a consequence of the versatile bonding capabilities of carbon. Such diversity is due to the ability of carbon atoms to bond to each other through single (two shared electrons) bonds, double (four shared electrons) bonds, and triple (six shared electrons) bonds in a limitless variety of straight chains, branched chains, and rings (Figure 12.1).

Among organic chemicals are included the majority of important industrial compounds, synthetic polymers, agricultural chemicals, biological materials, and most substances that are of concern because of their toxicities and other hazards. Pollution of the water, air, and soil environments by organic chemicals is an area of significant concern.

Chemically, most organic compounds can be divided among hydrocarbons, oxygen-containing compounds, nitrogen-containing compounds, sulfur-containing compounds, organohalides, phosphorus-containing compounds, or combinations of these kinds of compounds. Each of these classes of organic compounds is discussed briefly here.

All organic compounds, of course, contain carbon. Virtually all also contain hydrogen and have at least one C–H bond. The simplest organic compounds, and those easiest to understand, are those that contain only hydrogen and carbon. These compounds are called *hydrocarbons* and are addressed first among the organic compounds discussed in this chapter. Hydrocarbons are used here to illustrate some of the most fundamental points of organic chemistry, including organic formulas, structures, and names.

12.1.1 Molecular Geometry in Organic Chemistry

The three-dimensional shape of a molecule, that is, its molecular geometry, is particularly important in organic chemistry. This is because its molecular geometry determines, in part, the properties of an organic molecule, particularly its interactions with biological systems and how it is metabolized by organisms. Shapes of molecules are represented in drawings by lines of normal uniform thickness for bonds in the plane of the paper; broken lines for bonds extending away from the viewer; and heavy lines for bonds extending toward the viewer. These conventions are shown by the example of dichloromethane, CH_2Cl_2, an important organochloride solvent and extractant illustrated in Figure 12.2.

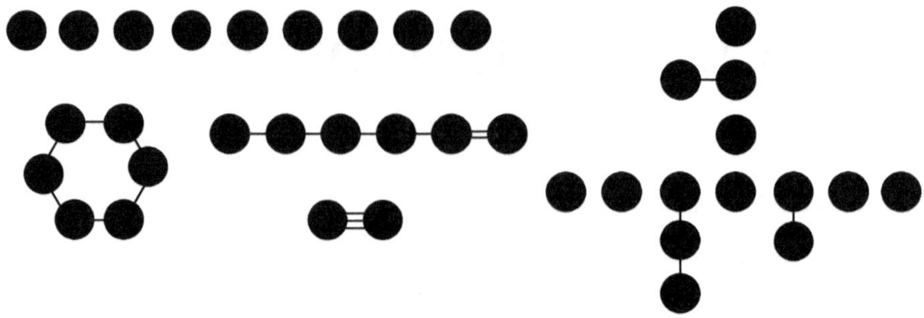

FIGURE 12.1 Carbon atoms in organic compounds bond with each other in straight chains, branched chains and rings. In addition to single bonds, carbon atoms may be joined by double, and even triple, bonds. Because of this remarkable bonding diversity, there are literally millions of known organic compounds.

12.2 HYDROCARBONS

As noted above, hydrocarbon compounds contain only carbon and hydrogen. The major types of hydrocarbons are alkanes, alkenes, alkynes, and aromatic (aryl) compounds. Examples of each are shown in Figure 12.3.

12.2.1 Alkanes

Alkanes, also called *paraffins* or *aliphatic hydrocarbons*, are hydrocarbons in which the C atoms are joined by single covalent bonds (sigma bonds) consisting of two shared electrons (see Chapter 11, Section 11.3). Some examples of alkanes are shown in Figure 12.4. As with other organic compounds, the carbon atoms in alkanes may form straight chains, branched chains, or rings. These three kinds of alkanes are, respectively, *straight-chain alkanes*, *branched-chain alkanes*, and *cycloalkanes*. As shown in Figure 12.3, a typical branched-chain alkane is 2-methylbutane, a volatile, highly flammable liquid. It is a component of gasoline, which may explain why it is commonly found as an air pollutant in urban air. The general molecular formula for straight- and branched-chain alkanes is $C_nH_{2n+2,}$ and that of cyclic alkanes is C_nH_{2n}. The four hydrocarbon molecules in Figure 12.4 contain eight carbon atoms each.

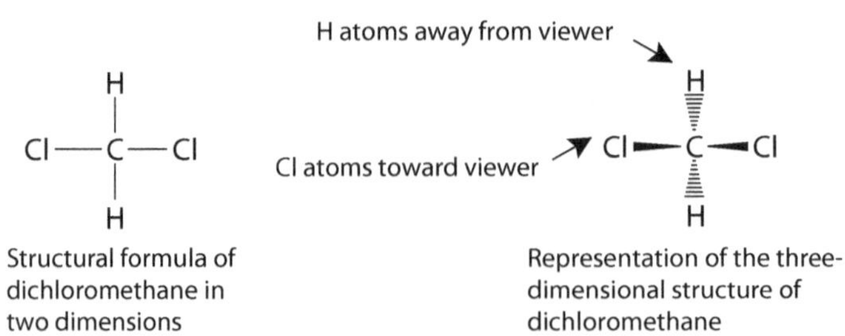

FIGURE 12.2 Structural formulas of dichloromethane, CH_2Cl_2; the formula on the right provides a three-dimensional representation.

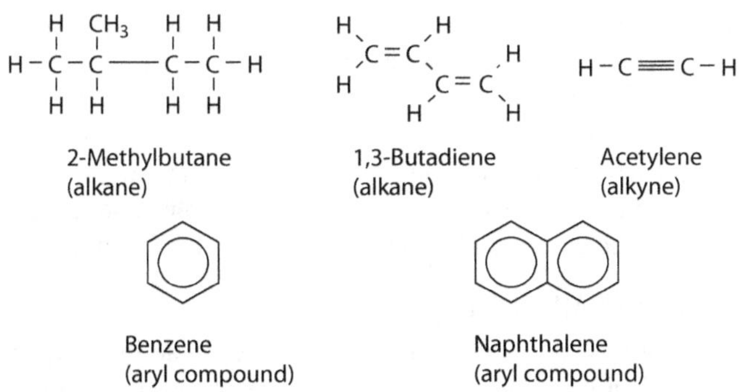

FIGURE 12.3 Examples of major types of hydrocarbons.

In one of the molecules, all of the carbon atoms are in a straight chain and in two they are in branched chains, whereas in a fourth, six of the carbon atoms are in a ring.

12.2.1.1 Formulas of Alkanes

Formulas of organic compounds present information at several different levels of sophistication. *Molecular formulas*, such as that of octane (C_8H_{18}), give the number of each kind of atom in a molecule of a compound. As shown in Figure 12.4, however, the molecular formula of C_8H_{18} may apply to several alkanes, each one of which has

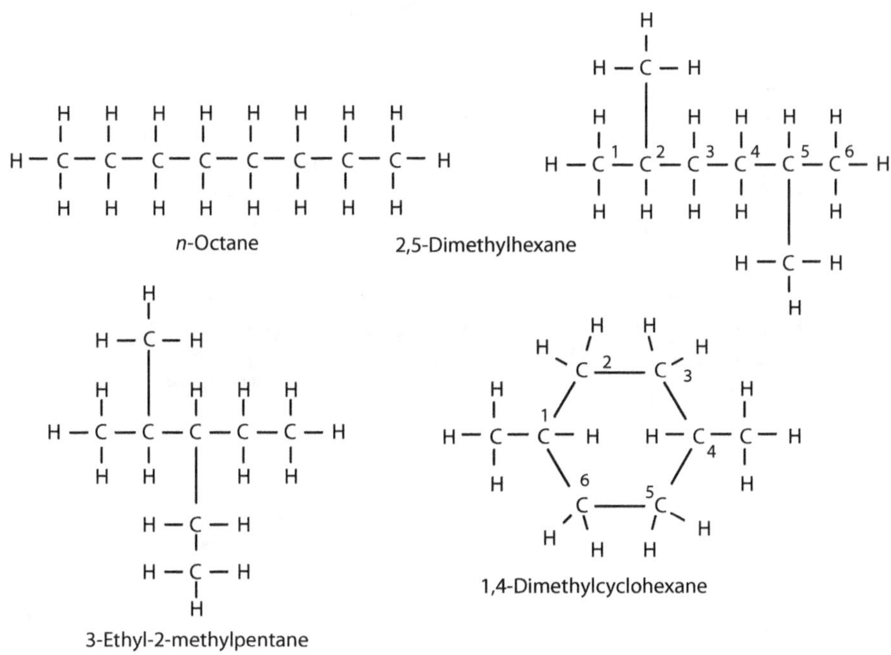

FIGURE 12.4 Structural formulas of four hydrocarbons, each containing eight carbon atoms, that illustrate the structural diversity possible with organic compounds. Numbers used to denote locations of atoms for purposes of naming are shown on two of the compounds.

unique chemical, physical, and toxicological properties. These different compounds are designated by *structural formulas* showing the order in which the atoms in a molecule are arranged. Compounds that have the same molecular but different structural formulas are called *structural isomers*. Of the compounds shown in Figure 12.4, *n*-octane, 2,5-dimethylhexane, and 3-ethyl-2-methylpentane are structural isomers, all having the formula C_8H_{18}, whereas 1,4-dimethylcyclohexane is not a structural isomer of the other three compounds because its molecular formula is C_8H_{16}.

12.2.1.2 Alkanes and Alkyl Groups

Most organic compounds can be derived from alkanes. In addition, many important parts of organic molecules contain one or more alkane groups minus a hydrogen atom bonded as substituents onto the basic organic molecule. As a consequence of these factors, the names of many organic compounds are based upon alkanes, and it is useful to know the names of some of the more common alkanes and substituent groups derived from them, as shown in Table 12.1.

TABLE 12.1
Some Alkanes and Substituent Groups Derived from Them

Alkane | **Substituent Groups Derived from Alkane**

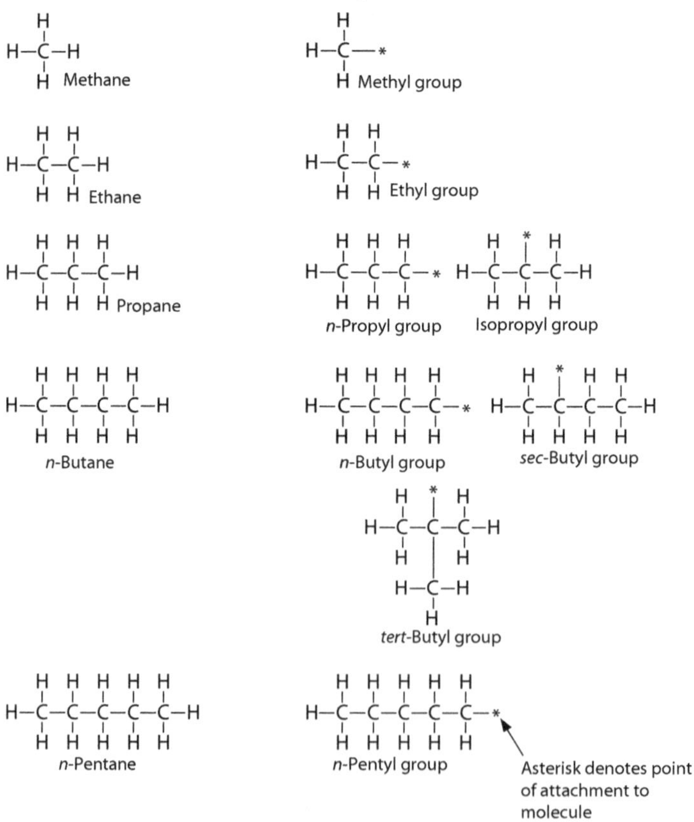

12.2.1.3 Names of Alkanes and Organic Nomenclature

Systematic names, from which the structures of organic molecules can be deduced, have been assigned to all known organic compounds. The more common organic compounds likewise have *common names* that have no structural implications. Although it is not possible to cover organic nomenclature in any detail in this chapter, the basic approach to nomenclature (naming) is presented in the chapter along with some pertinent examples. The simplest approach is to begin with names of alkane hydrocarbons.

Consider the alkanes shown in Figure 12.4. The fact that *n*-octane has no side chains is denoted by "*n*," that it has eight carbon atoms is denoted by "oct," and that it is an alkane is indicated by "ane." The names of compounds with branched chains or atoms other than H or C attached make use of numbers that stand for positions on the longest continuous chain of carbon atoms in the molecule. This convention is illustrated by the second compound in Figure 12.4. It gets the hexane part of the name from the fact that it is an alkane with six carbon atoms in its longest continuous chain ("hex" stands for 6). However, it has a methyl group (CH_3) attached on the second carbon atom of the chain and another on the fifth. Hence, the full systematic name of the compound is 2,5-dimethylhexane, where "di" indicates two methyl groups. In the case of 3-ethyl-2-methylpentane, the longest continuous chain of carbon atoms contains *five* carbon atoms, denoted by *pent*ane, an ethyl group, C_2H_5, is attached to the third carbon atom, and a methyl group on the second carbon atom. The last compound shown in the figure has six carbon atoms in a ring, indicated by the prefix "cyclo," so it is a cyclo*hex*ane compound. Furthermore, the carbon in the ring to which one of the methyl groups is attached is designated by "1" and another methyl group is attached to the fourth carbon atom around the ring. Therefore, the full name of the compound is 1,4-dimethylcyclohexane.

12.2.1.4 Summary of Organic Nomenclature as Applied to Alkanes

Naming relatively simple alkanes is a straightforward process. The basic rules to be followed are the following:

1. The name of the compound is based upon the longest continuous chain of carbon atoms. (The structural formula may be drawn such that this chain is not immediately obvious.)
2. The carbon atoms in the longest continuous chain are numbered sequentially from one end. The end of the chain from which the numbering is started is chosen to give the lower numbers for substituent groups in the final name. For example, the compound

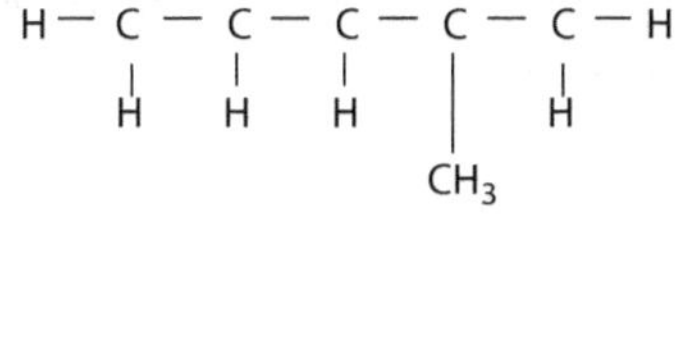

could be named 4-methylpentane (numbering the 5-carbon chain from the left), but should be named 2-methylpentane (numbering the 5-carbon chain from the right).

3. All groups attached to the longest continuous chain are designated by the number of the carbon atoms to which they are attached and by the name of the substituent group ("2-methyl" in the example cited in Step 2 above).
4. A prefix is used to denote multiple substitutions by the same kind of group. This is illustrated by 2,2,3-trimethylpentane

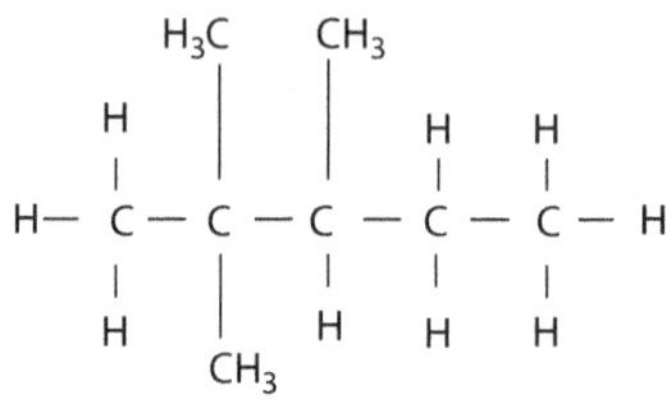

in which the prefix *tri* is used to show that *three* methyl groups are attached to the pentane chain.

5. The complete name is assigned such that it denotes the longest continuous chain of carbon atoms and the name and location on this chain of each substituent group.

12.2.1.5 Reactions of Alkanes

Alkanes contain only C–C and C–H bonds, both of which are relatively strong. For that reason they have little tendency to undergo many kinds of reactions common to some other organic chemicals, such as acid–base reactions or low-temperature oxidation–reduction reactions. However, at elevated temperatures, alkanes readily undergo oxidation, more specifically combustion, with molecular oxygen in air as shown by the following reaction of propane:

$$C_3H_8 + 5O_2 \rightarrow 3CO_2 + 4H_2O + \text{heat} \tag{12.1}$$

Common alkanes are highly flammable and the more volatile lower molecular mass alkanes form explosive mixtures with air. Furthermore, combustion of alkanes in an oxygen-deficient atmosphere or in an automobile engine produces significant quantities of carbon monoxide, CO, a toxic substance that can kill by depriving blood of its oxygen-carrying ability.

In addition to combustion, alkanes undergo *substitution reactions* in which one or more H atoms on an alkane are replaced by atoms of another element. The most common such reaction is the replacement of H by chlorine, to yield *organochlorine* compounds. For example, methane reacts with chlorine to give chloromethane. This reaction begins with the dissociation of molecular chlorine, usually initiated by ultraviolet electromagnetic radiation:

$$Cl_2 + \text{UV energy} \rightarrow Cl^{\bullet} + Cl^{\bullet} \tag{12.2}$$

The $Cl^{\bullet}$ product is a *free radical* species in which the chlorine atom has only seven outer shell electrons, as shown by the Lewis symbol,

$$:\ddot{\underset{\cdot\cdot}{Cl}}\cdot$$

instead of the favored octet of eight outer-shell electrons. In gaining the octet required for chemical stability, the chlorine atom is very reactive. It abstracts a hydrogen from methane,

$$Cl^{\bullet} + CH_4 \rightarrow HCl + CH_3^{\bullet} \qquad (12.3)$$

to yield HCl gas and another reactive species with an unpaired electron, $CH_3^{\bullet}$, called methyl radical. The methyl radical attacks molecular chlorine,

$$CH_3^{\bullet} + Cl_2 \rightarrow CH_3Cl + Cl^{\bullet} \qquad (12.4)$$

to give the chloromethane (CH_3Cl) product and regenerate $Cl^{\bullet}$, which can attack additional methane as shown in Reaction 12.3. The reactive $Cl^{\bullet}$ and $CH_3^{\bullet}$ species continue to cycle through the two preceding reactions.

The reaction sequence shown above illustrates three important aspects of chemistry that are shown to be very important in the discussion of atmospheric chemistry (see Chapter 1, Section 1.4). The first of these is that a reaction can be initiated by a *photochemical process* in which a photon of "light" (electromagnetic radiation, usually in the ultraviolet region) energy produces a reactive species, in this case the $Cl^{\bullet}$ atom. The second point illustrated is the high chemical reactivity of *free radical species* with unpaired electrons and incomplete octets of valence electrons. The third point illustrated is that of *chain reactions*, which can multiply manyfold the effects of a single reaction-initiating event, such as the photochemical dissociation of Cl_2.

12.2.2 Alkenes and Alkynes

Alkenes or olefins are hydrocarbons that have double bonds consisting of four shared electrons. The simplest and most widely manufactured alkene is ethylene,

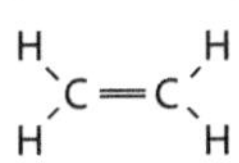

Ethylene (ethene)

used for the production of polyethylene polymer. Another important alkene is 1,3-butadiene (Figure 12.3), widely used in the manufacture of polymers, particularly synthetic rubber. The lighter alkenes, including ethylene and 1,3-butadiene, are highly flammable. Like other gaseous hydrocarbons, they form explosive mixtures with air.

Acetylene (Figure 12.3) is an *alkyne*, a class of hydrocarbons characterized by carbon–carbon triple bonds consisting of six shared electrons. Acetylene is a highly flammable gas that forms dangerously explosive mixtures with air. It is used in large quantities as a chemical raw material. Acetylene is the fuel in oxyacetylene torches used for cutting steel and for various kinds of welding applications.

12.2.2.1 Addition Reactions

The double and triple bonds in alkenes and alkynes have "extra" electrons capable of forming additional bonds. Hence, the carbon atoms attached to these bonds can add atoms without losing any atoms already bonded to them, and the multiple bonds are said to be *unsaturated*. Therefore, alkenes and alkynes both undergo *addition reactions* in which pairs of atoms are added across unsaturated bonds as shown in the reaction of ethylene with hydrogen to give ethane:

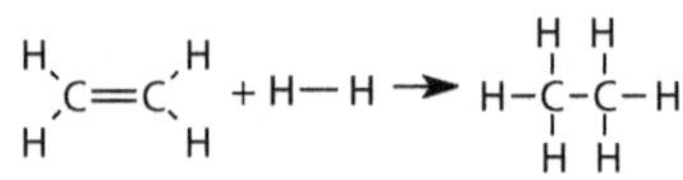

(12.5)

This is an example of a *hydrogenation reaction*, a very common reaction in organic synthesis, food processing (manufacture of hydrogenated oils), and petroleum refining. Another example of an addition reaction is that of HCl gas with acetylene to give vinyl chloride:

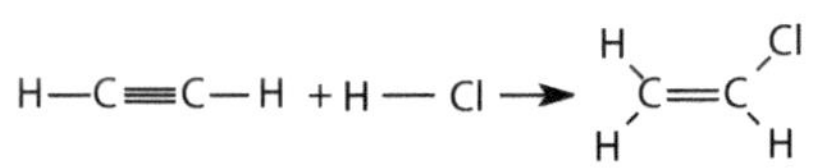

(12.6)

This kind of reaction, which is not possible with alkanes, adds to the chemical and metabolic versatility of compounds containing unsaturated bonds and is a factor contributing to their generally higher toxicities. It makes unsaturated compounds much more reactive, more hazardous to handle in industrial processes, and more active in atmospheric chemical processes such as photochemical smog formation.

12.2.3 ALKENES AND *CIS–TRANS* ISOMERISM

As shown by the two simple compounds in Figure 12.5, the two carbon atoms connected by a double bond in alkenes cannot rotate relative to each other. For this reason, another kind of isomerism, known as *cis–trans* isomerism, is possible for

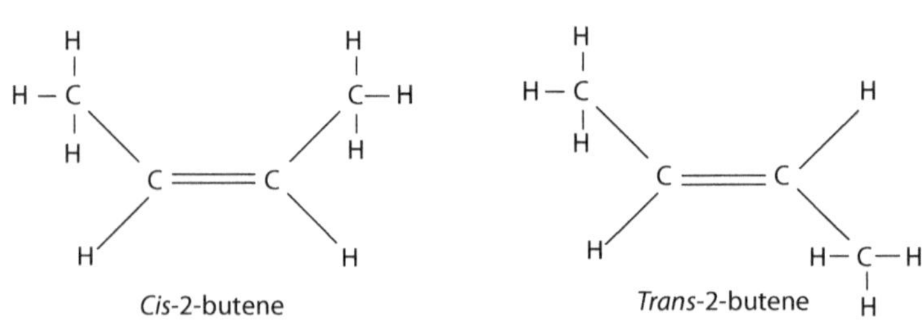

FIGURE 12.5 *Cis* and *trans* isomers of the alkene, 2-butene.

alkenes. *Cis–trans* isomers have different parts of the molecule oriented differently in space, although these parts occur in the same order. Both alkenes illustrated in Figure 12.5 have a molecular formula of C_4H_8. In the case of *cis*-2-butene, the two CH_3 (methyl) groups attached to the C=C carbon atoms are on the same side of the double bond, whereas in *trans*-2-butene they are on opposite sides.

12.2.3.1 Condensed Structural Formulas

To save space, structural formulas are conveniently abbreviated as *condensed structural formulas* such as $CH_3CH(CH_3)CH(C_2H_5)CH_2CH_3$ for 3-ethyl-2-methylpentane, where the CH_3 (methyl) and C_2H_5 (ethyl) groups are placed in parentheses to show that they are branches attached to the longest continuous chain of carbon atoms, which contains five carbon atoms. It is understood that each of the methyl and ethyl groups is attached to the carbon immediately preceding it in the condensed structural formula (methyl attached to the second carbon atom, ethyl to the third).

As illustrated by the examples in Figure 12.6, the structural formulas of organic molecules can be represented in a very compact form by lines and by figures such as hexagons. The ends and intersections of straight line segments in these formulas indicate the locations of carbon atoms. Carbon atoms at the terminal ends of lines are understood to have three H atoms attached, C atoms at the intersections of two lines are understood to have *two* H atoms attached to each, *one* H atom is attached to a carbon represented by the intersection of three lines, and *no* hydrogen atoms are bonded to C atoms where four lines intersect. Other atoms or groups of atoms, such as the Cl atom or OH group that are substituted for H atoms are shown by their symbols attached to a C atom with a line.

12.2.4 Aromatic Hydrocarbons

Benzene (Figure 12.7) is the simplest of a large class of *aromatic* or *aryl hydrocarbons*. Many important aryl compounds have substituent groups containing atoms of elements other than hydrogen and carbon and are called *aromatic compounds* or *aryl compounds*. Most aromatic compounds discussed in this book contain 6-carbon-atom benzene rings as shown for benzene, C_6H_6, in Figure 12.7. Aromatic compounds have ring structures and are held together in part by particularly stable bonds that contain delocalized clouds of the so-called π (pi, pronounced "pie") electrons.

In an oversimplified sense, the structure of benzene can be visualized as resonating between the two equivalent structures shown on the left in Figure 12.7 by the shifting of electrons in chemical bonds to form a hybrid structure. This structure can be shown more simply and accurately by a hexagon with a circle in it.

Aromatic compounds have special characteristics of *aromaticity*, which include a low hydrogen:carbon atomic ratio; C–C bonds that are quite strong and of intermediate length between such bonds in alkanes and those in alkenes; tendency to undergo substitution reactions rather than the addition reactions characteristic of alkenes; and delocalization of π electrons over several carbon atoms. The last phenomenon adds substantial stability to aromatic compounds and is known as *resonance stabilization*.

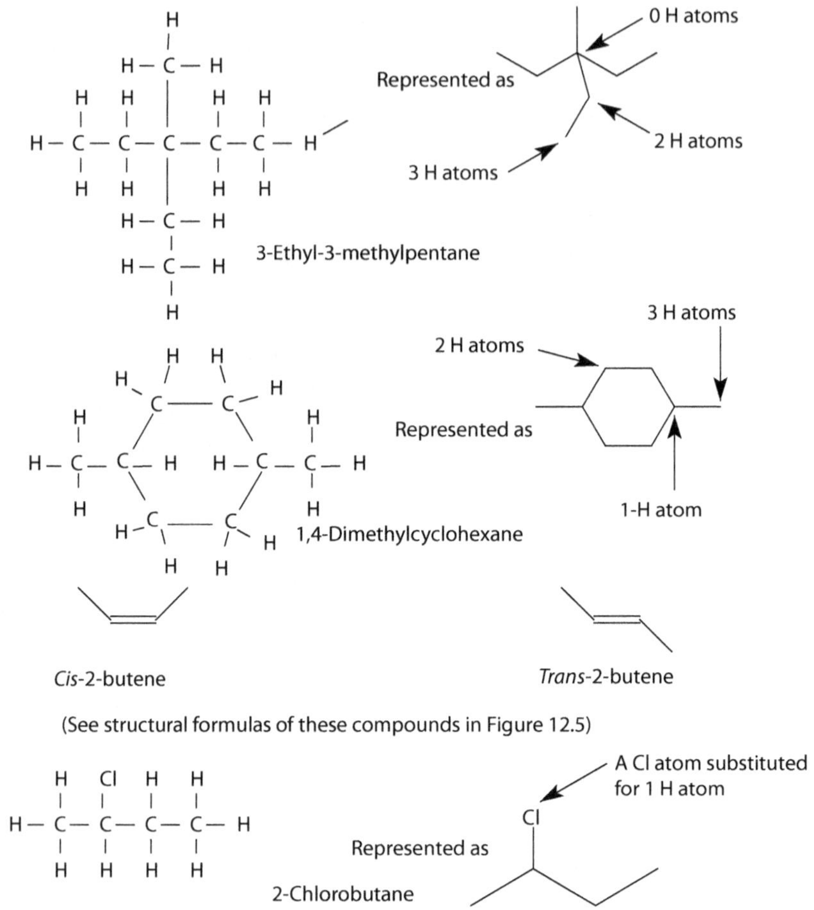

FIGURE 12.6 Representation of structural formulas with lines. A carbon atom is understood to be at each corner and at the end of each line. The numbers of hydrogen atoms attached to carbons at several specific locations are shown with arrows.

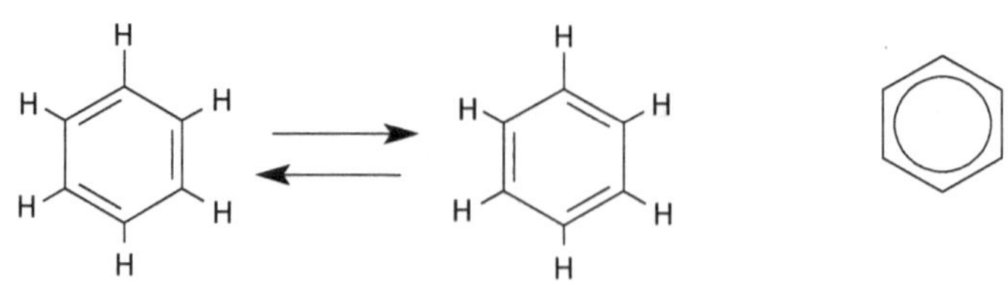

FIGURE 12.7 Representation of the aromatic benzene molecule with two resonance structures (left) and, more accurately, as a hexagon with a circle in it (right). Unless shown by symbols of other atoms, it is understood that a C atom is at each corner and that 1 H atom is bonded to each C atom.

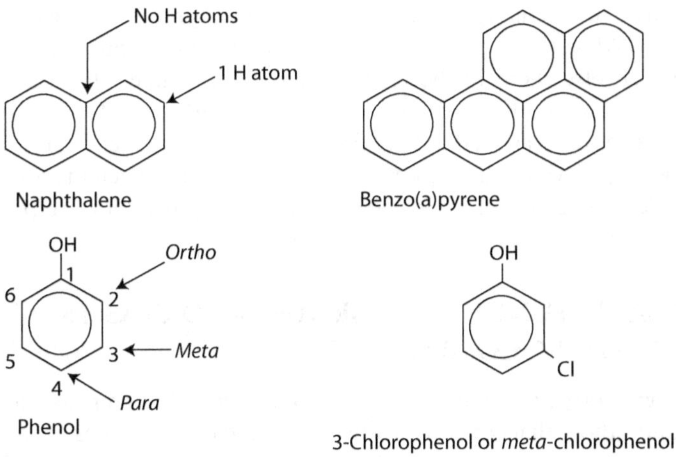

FIGURE 12.8 Aromatic compounds containing fused rings (top) and showing the numbering of carbon atoms for purposes of nomenclature.

Benzene, toluene, and other lighter aromatic compounds were first isolated from hydrocarbon liquids produced as a by-product of coal coking, a process in which the coal is heated in the absence of air to produce a carbonaceous residue used in steelmaking. Now recovered during the refining of crude oil, these aromatic liquid compounds have found widespread use as solvents, in chemical synthesis, and as a fuel. Many toxic substances, environmental pollutants, and hazardous waste compounds, such as benzene, naphthalene, and chlorinated phenols, are aromatic compounds (see Figure 12.8). As shown in Figure 12.8, some arenes, such as naphthalene and the polycyclic aromatic compound, benzo(a)pyrene, contain fused rings.

12.2.4.1 Benzene and Naphthalene

Benzene is a volatile, colorless, highly flammable liquid used to manufacture phenolic and polyester resins, polystyrene plastics, alkylbenzene surfactants, chlorobenzenes, insecticides, and dyes. It is hazardous both for its ignitability and toxicity (exposure to benzene causes blood abnormalities that may develop into leukemia). Naphthalene is the simplest member of a large number of polycyclic (multicyclic) aromatic hydrocarbons having two or more fused rings. It is a volatile white crystalline solid with a characteristic odor and has been used to make mothballs. The most important of the many chemical derivatives made from naphthalene is phthalic anhydride, from which phthalate ester plasticizers are synthesized.

12.2.4.2 Polycyclic Aromatic Hydrocarbons

Benzo(a)pyrene (Figure 12.8) is the most studied of the polycyclic aromatic hydrocarbons (PAHs), which are characterized by condensed ring systems ("chicken wire" structures). These compounds are formed by the incomplete combustion of other hydrocarbons, a process that consumes hydrogen in preference to carbon. The carbon residue is left in the thermodynamically favored condensed aromatic ring system of the PAH compounds.

Because there are so many partial combustion and pyrolysis processes that favor production of PAHs, these compounds are encountered abundantly in the atmosphere, soil, and elsewhere in the environment from sources that include engine exhausts, wood stove smoke, cigarette smoke, and charbroiled food. Coal tars and petroleum residues such as road and roofing asphalt have high levels of PAHs. Some PAH compounds, of which the most widely investigated has been the 5-ringed compound benzo(a)pyrene, are of toxicological concern because they are precursors to cancer-causing metabolites.

12.3 ORGANIC FUNCTIONAL GROUPS AND CLASSES OF ORGANIC COMPOUNDS

The discussion of organic chemistry so far in this chapter has emphasized hydrocarbon compounds, those that contain only hydrogen and carbon. It has been shown that hydrocarbons may exist as alkanes, alkenes, and arenes, depending upon the kinds of bonds between carbon atoms. The presence of elements other than hydrogen and carbon in organic molecules greatly increases the diversity of their chemical properties. As shown in Table 12.2, *functional groups* consist of specific bonding configurations of atoms in organic molecules. Most functional groups contain at least one element other than carbon or hydrogen, although two carbon atoms joined by a double bond (alkenes) or triple bond (alkynes) are likewise considered to be functional groups. Table 12.2 shows some of the major functional groups that determine the nature of organic compounds.

12.3.1 Organooxygen Compounds

The most common types of compounds with oxygen-containing functional groups are epoxides, alcohols, phenols, ethers, aldehydes, ketones, and carboxylic acids. The functional groups characteristic of these compounds are illustrated by the examples of oxygen-containing compounds shown in Figure 12.9.

Ethylene oxide is a moderately to highly toxic, sweet-smelling, colorless, flammable, explosive gas used as a chemical intermediate, sterilant, and fumigant. It is a mutagen and a carcinogen to experimental animals. It is classified as hazardous for both its toxicity and ignitability. *Methanol* is a clear, volatile, flammable liquid alcohol used for chemical synthesis, as a solvent, and as a fuel. It is used as a gasoline additive to reduce emissions of carbon monoxide and other air pollutants. Ingestion of methanol can be fatal, and blindness can result from sublethal doses. *Phenol* is a dangerously toxic aromatic alcohol widely used for chemical synthesis and polymer manufacture. *Methyltertiarybutyl ether*, MTBE, is the ether that became the octane booster of choice to replace tetraethyllead in gasoline. However, its use has now been phased out, largely because of concern over its water pollution potential.

Acrolein is an alkenic aldehyde and a volatile, flammable, highly reactive chemical. It forms explosive peroxides upon prolonged contact with O_2. An extreme lachrimator and strong irritant, acrolein is quite toxic by all routes of exposure. *Acetone* is the lightest of the ketones. Like all ketones, acetone has a carbonyl (C=O) group that is bonded to *two* carbon atoms (i.e., it is somewhere in the middle of a carbon

TABLE 12.2
Examples of Some Important Functional Groups

Type of Functional Group	Example Compound	Structural Formula of Group[a]
Alkene (olefin)	Propene (propylene)	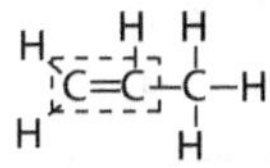
Alkyne	Acetylene	
Alcohol (-OH attached to alkyl group)	2-Propanol	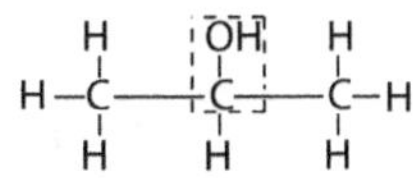
Phenol (-OH attached to aryl group)	Phenol	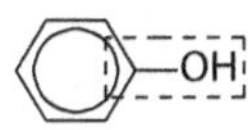
Ketone (When $-\overset{O}{\overset{\|}{C}}-H$ group is on end carbon, compound is an aldehyde)	Acetone	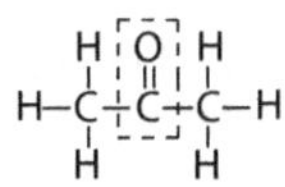
Amine	Methylamine	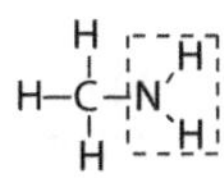
Nitrocompounds	Nitromethane	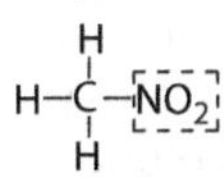
Sulfonic acids	Benzenesulfonic acid	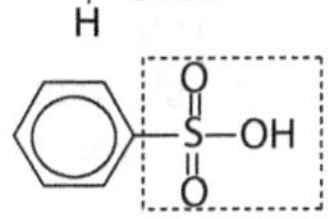
Organohalides	1,1-Dichloro-ethane	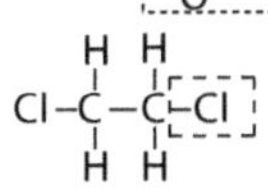

[a] Functional group outlined by dashed line.

atom chain). Acetone is a good solvent and is chemically less reactive than the aldehydes, which all have the functional group

$$-\overset{O}{\overset{\|}{C}}-H$$

in which binding of the C=O to H makes the molecule significantly more reactive. *Propionic acid* is a typical organic carboxylic acid. The $-CO_2H$ group of carboxylic acids may be viewed as the most oxidized functional group (other than peroxides) on an oxygenated organic compound, and carboxylic acids can be synthesized by oxidizing alcohols or aldehydes that have an –OH group or C=O group on an end carbon atom.

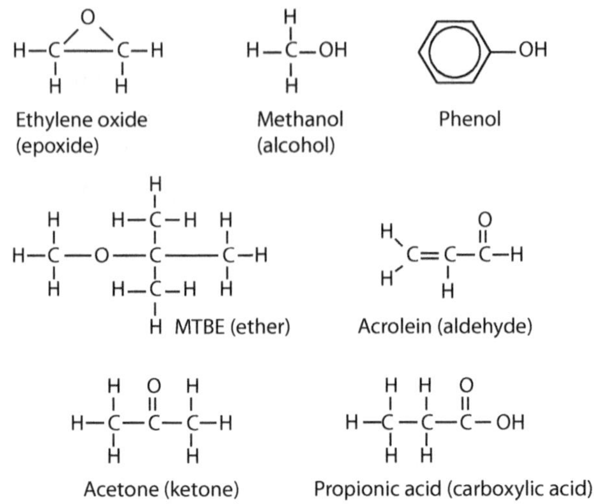

FIGURE 12.9 Examples of oxygen-containing organic compounds that may be significant as wastes, toxic substances, or environmental pollutants.

12.3.2 Organonitrogen Compounds

Figure 12.10 shows examples of three classes of the many kinds of compounds that contain N (amines, nitrosamines, and nitro compounds). Nitrogen occurs in many functional groups in organic compounds, some of which contain nitrogen in ring structures, or along with oxygen.

Methylamine is a colorless, highly flammable gas with a strong odor. It is a severe irritant affecting eyes, skin, and mucous membranes. Methylamine is the simplest of the *amine* compounds, which have the general formula

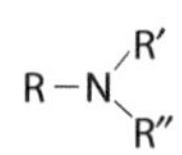

where the Rs are hydrogen or hydrocarbon groups, at least one of which is the latter.

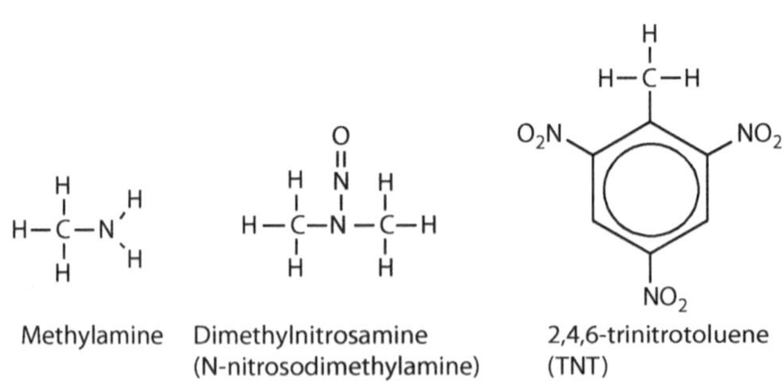

FIGURE 12.10 Examples of organonitrogen that may be significant as wastes, toxic substances, or environmental pollutants.

Dimethylnitrosamine is one of the N-nitroso compounds, all characterized by the N–N=O functional group. It was once widely used as an industrial solvent, but caused liver damage and jaundice in exposed workers. Subsequently, numerous other N-nitroso compounds, many produced as by-products of industrial operations and food and alcoholic beverage processing, were found to be carcinogenic.

Solid *2,4,6-trinitrotoluene* (TNT) has been widely used as a military explosive. TNT is moderately to very toxic and has caused toxic hepatitis or aplastic anemia in exposed individuals, a few of whom have died from its toxic effects. It belongs to the general class of nitro compounds characterized by the presence of $-NO_2$ groups bonded to a hydrocarbon structure.

Some organonitrogen compounds are chelating agents that bind strongly to metal ions. Such chelation of metals may play a role in the solubilization and transport of heavy metal wastes. Prominent among these are salts of the aminocarboxylic acids which, in the acid form, have $-CH_2CO_2H$ groups bonded to nitrogen atoms. An important example of such a compound is the monohydrate of trisodium nitrilotriacetate (NTA):

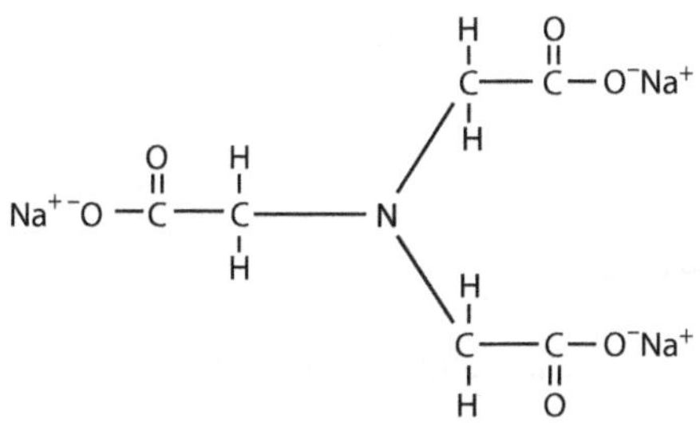

This compound can be used as a substitute for detergent phosphates to bind to calcium ion and make the detergent solution basic. NTA is used in metal plating formulations. It is highly water soluble and quickly eliminated with urine when ingested. It has a low acute toxicity and no chronic effects have been shown for plausible doses. However, concern does exist over its interaction with heavy metals in waste-treatment processes and in the environment.

12.3.3 Organohalide Compounds

Organohalides (Figure 12.11) exhibit a wide range of physical and chemical properties. These compounds consist of halogen-substituted hydrocarbon molecules, each of which contains at least one atom of F, Cl, Br, or I. They may be saturated (*alkyl halides*), unsaturated (*alkenyl halides*), or aromatic (*aromatic halides*). The most widely manufactured organohalide compounds are chlorinated hydrocarbons, many of which are regarded as environmental pollutants or as hazardous wastes.

12.3.3.1 Alkyl Halides

Substitution of halogen atoms for one or more hydrogen atoms on alkanes gives *alkyl halides*, examples of structural formulas are given in Figure 12.11. Most of the

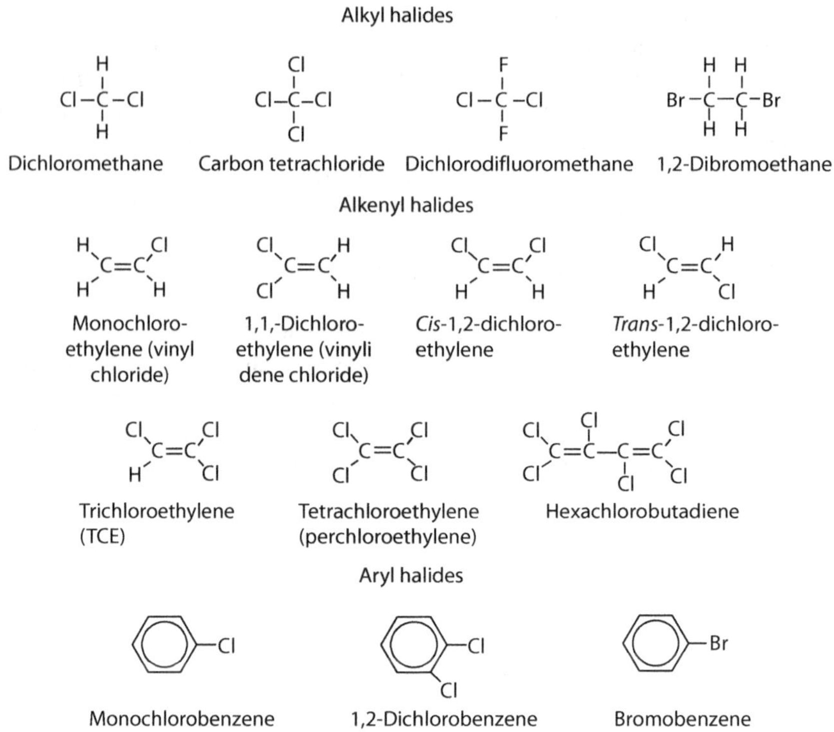

FIGURE 12.11 Examples of organohalide compounds.

commercially important alkyl halides are derivatives of alkanes of low molecular mass. A brief discussion of the uses of the compounds listed in Figure 12.11 is given here to provide an idea of the versatility of the alkyl halides.

Dichloromethane is a volatile liquid with excellent solvent properties for nonpolar organic solutes. It has been used as a solvent for the decaffeination of coffee, in paint strippers, as a blowing agent in urethane polymer manufacture, and to depress vapor pressure in aerosol formulations. Once commonly sold as a solvent and stain remover, highly toxic *carbon tetrachloride* is now largely restricted to uses as a chemical intermediate under controlled conditions. Insecticidal *1,2-dibromoethane* has been consumed in large quantities as a lead scavenger in leaded gasoline and to fumigate soil, grain, and fruit (fumigation with this compound has been discontinued because of toxicological concerns). An effective solvent for resins, gums, and waxes, it serves as a chemical intermediate in the synthesis of some pharmaceutical compounds and dyes.

12.3.3.2 Alkenyl Halides

Viewed as hydrocarbon-substituted derivatives of alkenes, the *alkenyl* or *olefinic organohalides* contain at least one halogen atom and at least one carbon–carbon double bond. The most significant of these, which are materials produced in large quantities for chemical manufacture and other purposes, are the lighter chlorinated compounds, such as those illustrated in Figure 12.11.

Vinyl chloride is consumed in large quantities as a raw material to manufacture pipe, hose, wrapping, and other products fabricated from polyvinylchloride plastic. This highly flammable, volatile, sweet-smelling gas is a known human carcinogen.

There are three possible dichloroethylene compounds, all clear, colorless liquids (Figure 12.11). Vinylidene chloride forms a copolymer with vinyl chloride used in some coatings. The geometrically isomeric 1,2-dichloroethylenes are used as organic synthesis intermediates and as solvents. *Trichloroethylene* is a clear, colorless, nonflammable, volatile liquid. It is an excellent degreasing and drycleaning solvent and has been used as a household solvent and for food extraction (e.g., in decaffeination of coffee). Colorless, nonflammable liquid *tetrachloroethylene* has properties and uses similar to those of trichloroethylene. *Hexachlorobutadiene*, a colorless liquid with an odor somewhat like that of turpentine, is used as a solvent for higher hydrocarbons and elastomers, as a hydraulic fluid, in transformers, and for heat transfer.

12.3.3.3 Aromatic Halides

Aromatic halide derivatives of benzene and toluene have many uses. They are common intermediates and raw materials in chemical synthesis. In addition, they are used as pesticides and raw materials for pesticides manufacture, as solvents, and for a diverse variety of other applications. These widespread uses over many decades have resulted in substantial human exposure and environmental contamination. Three example aromatic halides are shown in Figure 12.11. Monochlorobenzene is a flammable liquid boiling at 132°C. It is used as a solvent, heat transfer fluid, and synthetic reagent. Used as a solvent, 1,2-dichlorobenzene is employed for degreasing hides and wool. It also serves as a synthetic reagent for dye manufacture. Bromobenzene is a liquid boiling at 156°C that is used as a solvent, motor oil additive, and intermediate for organic synthesis.

12.3.3.4 Halogenated Naphthalene and Biphenyl

Two major classes of halogenated aromatic compounds containing two benzene rings are made by the chlorination of naphthalene and biphenyl and have been sold as mixtures with varying degrees of chlorine content. Examples of chlorinated naphthalenes, and polychlorinated biphenyls (PCBs discussed later), are shown in Figure 12.12. The less highly chlorinated of these compounds are liquids, and those with higher chlorine contents are solids. Because of their physical and chemical stabilities and other qualities, these compounds have had many uses, including heat-transfer fluids, hydraulic fluids, and dielectrics. Polybrominated biphenyls (PBBs) have served as flame retardants. However, because chlorinated naphthalenes, PCBs, and PBBs are extremely persistent, their uses have been severely curtailed.

12.3.3.5 Chlorofluorocarbons, Halons, and Hydrogen-Containing Chlorofluorocarbons

Chlorofluorocarbons (CFCs) are volatile 1- and 2-carbon compounds that contain Cl and F bonded to carbon. These extremely stable and nontoxic compounds were once

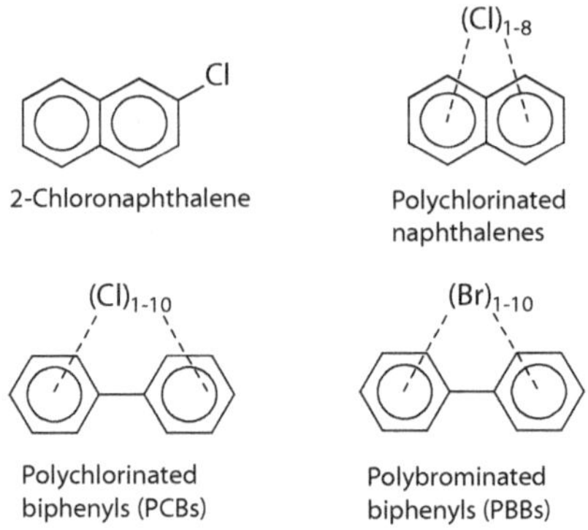

FIGURE 12.12 Halogenated naphthalenes and biphenyls.

widely used in the fabrication of flexible and rigid foams, and as fluids for refrigeration and air conditioning, but have now been essentially phased out because of their potential to cause harm to the stratospheric ozone layer. The most widely manufactured of these compounds in the past were CCl_3F (CFC-11), CCl_2F_2 (CFC-12), $C_2Cl_3F_3$ (CFC-113), $C_2Cl_2F_4$ (CFC-114), and C_2ClF_5 (CFC-115). *Halons* are related compounds that contain bromine and are used in fire extinguisher systems. The most commonly produced commercial halons were $CBrClF_2$ (Halon-1211), $CBrF_3$ (Halon-1301), and $C_2Br_2F_4$ (Halon-2402), where the sequence of numbers denotes the number of carbon, fluorine, chlorine, and bromine atoms, respectively, per molecule. Halons have also been implicated as ozone-destroying gases in the stratosphere and are being phased out. However, finding suitable replacements has been difficult, particularly in the crucial area of aircraft fire extinguishers.

Hydrohalocarbons are hydrogen-containing chlorofluorocarbons (HCFCs) and hydrogen-containing fluorocarbons (HFCs) that are now produced as substitutes for chlorofluorocarbons. These compounds include CH_2FCF_3 (HFC-134a, a substitute for CFC-12 in automobile air conditioners and refrigeration equipment), $CHCl_2CF_3$ (HCFC-123, substitute for CFC-11 in plastic foam-blowing), CH_3CCl_2F (HCFC-141b, substitute for CFC-11 in plastic foam-blowing), and $CHClF_2$ (HCFC-22, air conditioners and manufacture of plastic foam food containers). Because each molecule of these compounds has at least one H–C bond, which is much more readily broken than C–Cl or C–F bonds, the HCFCs do not persist in the atmosphere and pose essentially no threat to the stratospheric ozone layer.

12.3.3.6 Chlorinated Phenols

The chlorinated phenols, particularly *pentachlorophenol* and the trichlorophenol isomers, are significant hazardous wastes. These compounds are biocides that are used to treat wood in order to prevent rot by fungi and to prevent termite infestation. They are toxic, causing liver malfunction and dermatitis. However, contaminant polychlorinated dibenzodioxins ("dioxin") may be responsible for some of the observed effects. Pentachlorophenol and other aromatic halides and aromatic hydrocarbons used as wood preservatives are encountered at many hazardous-waste sites in wastewaters and sludges.

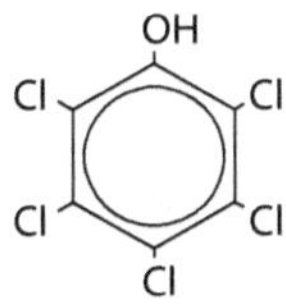

Pentachlorophenol

12.3.4 Organosulfur Compounds

The chemistry of sulfur is similar to but perhaps more diverse than that of oxygen. Whereas, with the exception of peroxides, most chemically combined organic oxygen is in the –2 oxidation state, sulfur occurs in the –2, +4, and +6 oxidation states. Many organosulfur compounds are noted for their "rotten egg" or garlic odors. A number of organosulfur compounds are shown in Figure 12.13.

Substitution of alkyl or aromatic hydrocarbon groups such as phenyl and methyl for H on hydrogen sulfide, H_2S, leads to a number of different organosulfur *thiols* (mercaptans, R–SH) and *sulfides*, also called thioethers (R–S–R). Structural formulas of examples of these compounds are shown in Figure 12.13.

Methanethiol and other lighter alkyl thiols are fairly common air pollutants that have "ultragarlic" odors; both 1- and 2-butanethiol are associated with skunk odor. Gaseous methanethiol is used as an odorant leak-detecting additive for natural gas, propane, and butane; it is also employed as an intermediate in pesticide synthesis. A toxic, irritating volatile liquid with a strong garlic odor, 2-propene-1-thiol (allyl mercaptan) is a typical alkenyl mercaptan. Benzenethiol (phenyl mercaptan) is the simplest of the aromatic thiols. It is a toxic liquid with a severely "repulsive" odor.

Alkyl sulfides or thioethers contain the C–S–C functional group. The lightest of these compounds is dimethyl sulfide, a volatile liquid (bp 38°C) that is moderately toxic by ingestion. It is now known to be a major source of gaseous sulfur entering the atmosphere over the oceans due to its production by marine organisms. Cyclic sulfides contain the C–S–C group in a ring structure. The most common of these compounds is thiophene, a heat-stable liquid (bp 84°C) with a solvent action much like that of benzene, which is used to make pharmaceuticals, dyes, and resins. Its saturated analog is tetrahydrothiophene, or thiophane.

Many important organosulfur compounds also contain nitrogen. One such compound is *thiourea*, the sulfur analog of urea. Its structural formula is shown in Figure 12.13. Thiourea and *phenylthiourea* have been used as rodenticides. Commonly called ANTU, *1-naphthylthiourea* is an excellent rodenticide that is virtually tasteless and has a very high rodent:human toxicity ratio.

Sulfoxides and *sulfones* (Figure 12.13) contain both sulfur and oxygen. *Dimethylsulfoxide* (DMSO) is a liquid used to remove paint and varnish, as a hydraulic fluid, mixed with water as an antifreeze solution, and in pharmaceutical applications as an anti-inflammatory and bacteriostatic agent. *Sulfolane* dissolves both organic and inorganic solutes and is widely used in BTX processing to selectively

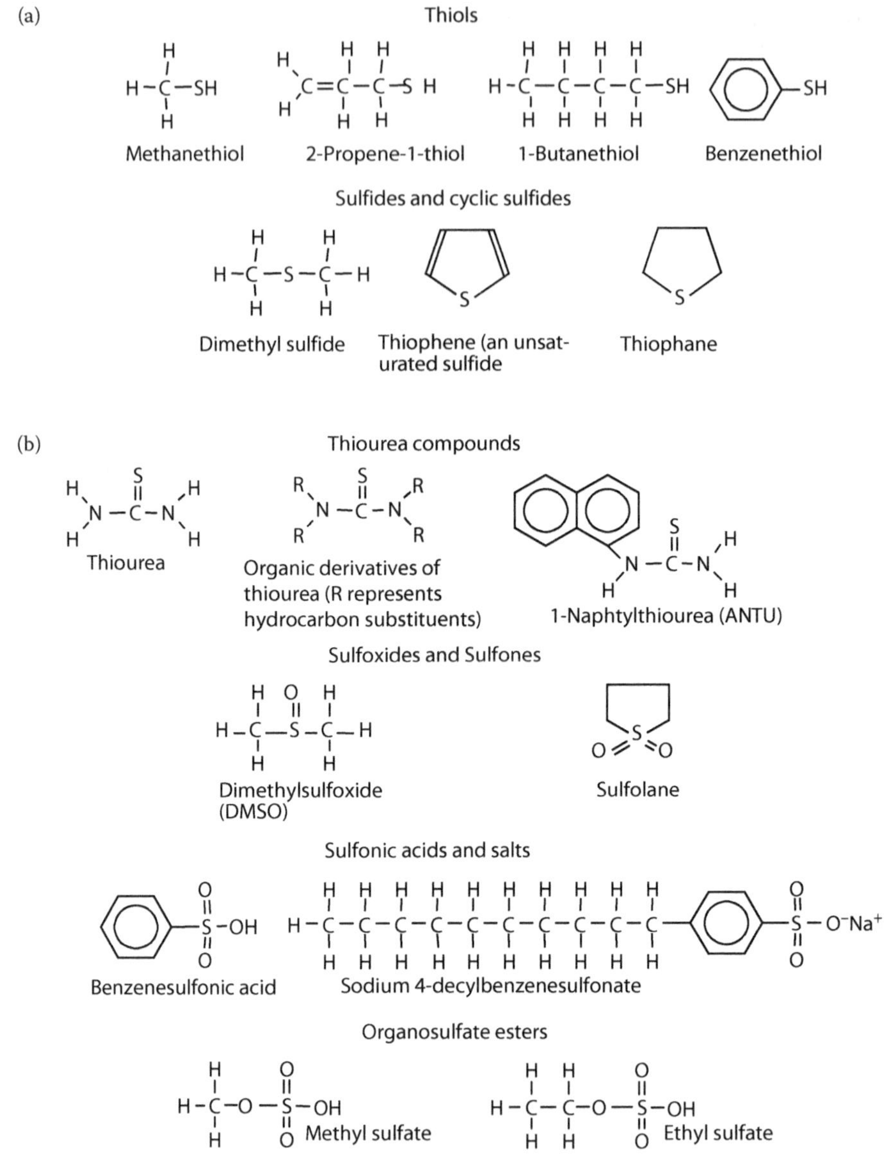

FIGURE 12.13 Examples of organosulfur compounds.

extract benzene, toluene, and xylene from aliphatic hydrocarbons; as the solvent in the Sulfinol process by which thiols and acidic compounds are removed from natural gas; as a solvent for polymerization reactions; and as a polymer plasticizer.

Sulfonic acids and sulfonate salts contain the $-SO_3H$ and $-SO_3^-$ groups, respectively, attached to a hydrocarbon moiety. The structural formula of benzenesulfonic acids and of sodium 4-decylbenzenesulfonate, a biodegradable detergent surfactant, are shown in Figure 12.13. The common sulfonic acids are water-soluble strong acids that lose virtually all ionizable H^+ in aqueous solution. They are used commercially to hydrolyze fat and oil esters to fatty acids and glycerol.

Replacement of 1 H on sulfuric acid, H_2SO_4, with a hydrocarbon group yields an acid ester, and replacement of both yields an ester, examples of which are shown in Figure 12.13. Sulfuric acid esters are used as alkylating agents, which act to attach alkyl groups (such as methyl) to organic molecules in the manufacture of agricultural chemicals, dyes, and drugs. *Methylsulfuric acid* and *ethylsulfuric acid* are oily, water-soluble liquids that are strong irritants to skin, eyes, and mucous tissue.

12.3.5 Organophosphorus Compounds

The first two examples in Figure 12.14, illustrate that the structural formulas of alkyl and aromatic phosphine compounds can be derived by substituting organic groups for the H atoms in phosphine (PH_3), the toxic hydride of phosphorus. *Methylphosphine* is a colorless, reactive gas. Crystalline, solid *triphenylphosphine* has a low reactivity and moderate toxicity when inhaled or ingested.

As shown by the reaction,

$$4C_3H_9P + 26O_2 \rightarrow 12CO_2 + 18H_2O + P_4O_{10} \quad (12.7)$$

combustion of aromatic and alkyl phosphines produces P_4O_{10}, a corrosive, irritant, toxic substance that reacts with moisture in the air to produce droplets of corrosive orthophosphoric acid, H_3PO_4.

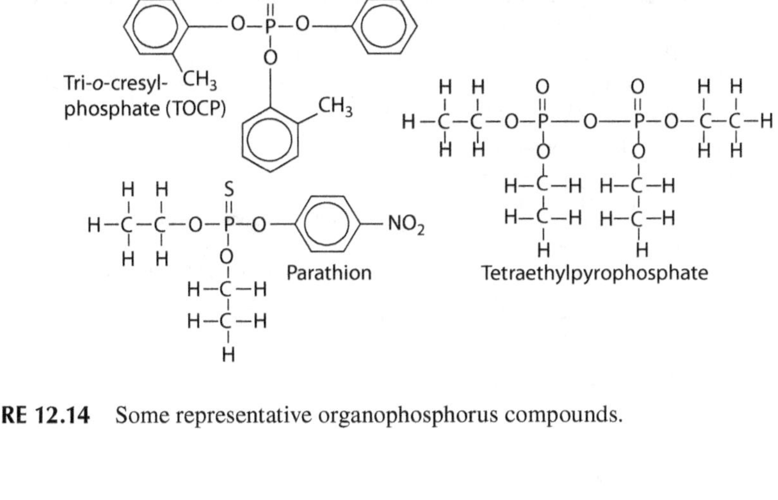

FIGURE 12.14 Some representative organophosphorus compounds.

The structural formulas of three common esters of orthophosphoric acid (H_3PO_4) and an ester of pyrophosphoric acid ($H_4P_2O_6$) are shown in Figure 12.14. Although *trimethylphosphate* is considered to be only moderately toxic, *tri-o-cresyl-phosphate, TOCP*, has a notorious record of poisoning. *Tetraethylpyrophosphate, TEPP*, was developed in Germany during World War II as a substitute for insecticidal nicotine. Although it is a very effective insecticide, its use in that application was of very short duration because, in addition to insects, it kills almost everything else, too.

12.3.5.1 Phosphorothionate Esters

Parathion, shown in Figure 12.14, is an example of *phosphorothionate* esters. These compounds are used as insecticidal acetylcholinesterase inhibitors. They contain the P=S (thiono) group, which increases their insect:mammal toxicity ratios. Since the first organophosphate insecticides were developed in Germany during the 1930s and 1940s, many insecticidal organophosphate compounds have been synthesized. One of the earliest and most successful of these is *parathion, O,O*-diethyl-*O-p*-nitrophenylphosphorothionate (banned from use in the United States in 1991 because of its acute toxicity to humans). From a long-term environmental standpoint, organophosphate insecticides are superior to the organohalide insecticides that they largely displaced because the organophosphates readily undergo biodegradation and do not bioaccumulate.

12.4 SYNTHETIC POLYMERS

The huge polymer manufacture industry is significant to the environment both as a source of environmental pollutants and in the manufacture of materials used to alleviate environmental and waste problems. Synthetic *polymers* are produced when small molecules called *monomers* bond together to form a much smaller number of very large molecules. Many natural products are polymers; for example, cellulose produced by trees and other plants and found in wood, paper, and many other materials, is a polymer of the sugar glucose. Synthetic polymers form the basis of many industries, such as rubber, plastics, and textiles manufacture.

An important example of a polymer is that of polyvinylchloride, shown in Figure 12.15. This polymer is synthesized in large quantities for the manufacture of water

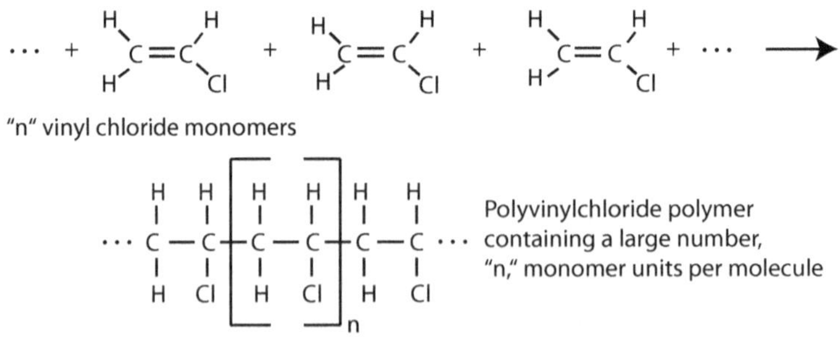

FIGURE 12.15 Polyvinylchloride polymer.

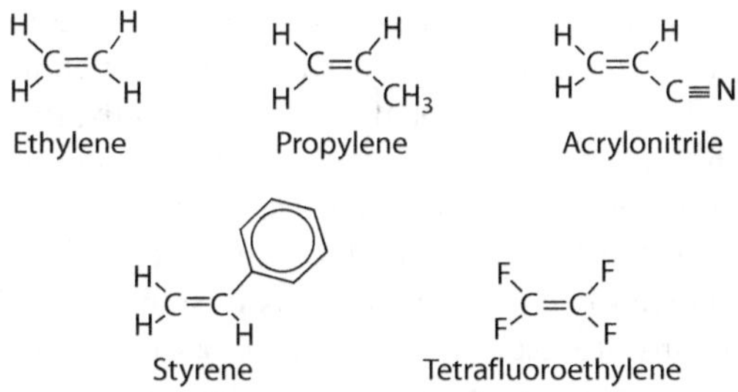

FIGURE 12.16 Monomers from which commonly used polymers are synthesized.

and sewer pipe, water-repellent liners, and other plastic materials. Other major polymers include polyethylene (plastic bags, milk cartons), polypropylene (impact-resistant plastics, indoor–outdoor carpeting), polyacrylonitrile (Orlon, carpets), polystyrene (foam insulation), and polytetrafluoroethylene (Teflon coatings, bearings); the monomers from which these substances are made are shown in Figure 12.16.

Many of the hazards from the polymer industry arise from the monomers used as raw materials. Many monomers are reactive and flammable, with a tendency to form explosive vapor mixtures with air. All have a certain degree of toxicity; vinyl chloride is a known human carcinogen. The combustion of many polymers may result in the evolution of toxic gases, such as hydrogen cyanide (HCN) from polyacrylonitrile, or hydrogen chloride (HCl) from polyvinylchloride. Another hazard presented by plastics results from the presence of *plasticizers* added to provide essential properties such as flexibility. The most widely used plasticizers are phthalates, which are environmentally persistent, resistant to treatment processes, and prone to undergo bioaccumulation.

Polymers have a number of applications in waste treatment and disposal. Waste disposal landfill liners are made from synthetic polymers, as are the fiber filters that remove particulate pollutants from flue gas in baghouses. Membranes used for ultrafiltration and reverse osmosis treatment of water are composed of very thin sheets of synthetic polymers. Organic solutes can be removed from water by sorption

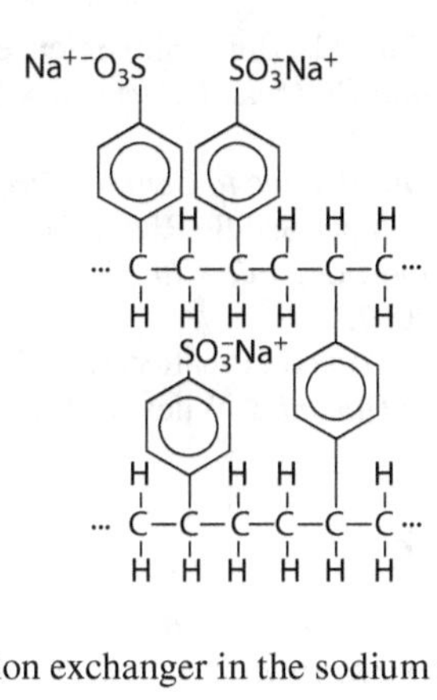

FIGURE 12.17 Polymeric cation exchanger in the sodium form.

onto hydrophobic (water-repelling) organophilic beads of Amberlite XAD resin. Heavy metal pollutants are removed from wastewater by cation exchange resins made of polymers with anionic functional groups. Typically, these resins exchange harmless sodium ion, Na^+, on the solid resin for toxic heavy metal ions in water. Figure 12.17 shows a segment of the polymeric structure of a cation exchange resin in the sodium form. In the treatment of heavy-metal-containing waste solutions, these resins can exchange toxic heavy metal ions in solution, such as Cd^{2+}, for Na^+ ions. Ion exchange resins are used in nuclear reactors to remove traces of metals, some of which may be radioactive, from the water used in the reactor for heat exchange. Ion exchange resins have also been developed in which the ion-exchanging functional group is an iminiodiacetate $\{-N(CH_2CO_2^-)_2\}$ group that has a particularly strong affinity for heavy metals.

SUPPLEMENTARY REFERENCES

Brown, W. H. and C. S. Foote, *Organic Chemistry*, 4th ed., Thomson Brooks/Cole, Belmont, CA, 2005.

Bruice, P. Y., *Organic Chemistry*, 6th ed., Prentice-Hall, Upper Saddle River, NJ, 2010.

Carey, F. A., *Organic Chemistry*, McGraw-Hill, New York, 2007.

Faber, K., *Biotransformations in Organic Chemistry: A Textbook*, 5th ed., Springer Verlag, Berlin, 2004.

McMurry, J., *Fundamentals of Organic Chemistry*, 7th ed., Brooks/Cole/Thomson Learning, Belmont, CA, 2009.

McMurry, J. and M. E. Castellion, *Fundamentals of General, Organic and Biological Chemistry*, 5th ed., Prentice-Hall, Upper Saddle River, NJ, 2007.

McMurry, J. D., S. Ballantine, C. A. Hoeger, and V. E. Peterson, *Fundamentals of General, Organic, and Biological Chemistry*, 6th ed., Prentice-Hall, Upper Saddle River, NJ, 2009.

Raymond, K., *General Organic and Biological Chemistry*, Wiley, Hoboken, NJ, 2010.

Schwarzenbach, R. P., P. M. Gschwend, and D. M. Imboden, *Environmental Organic Chemistry*, 2nd ed., John Wiley & Sons, Hoboken, NJ, 2003.

Seager, S. L. and M. R. Slabaugh, *Chemistry for Today: General, Organic, and Biochemistry*, 6th ed., Thomson Brooks/Cole, Belmont, CA, 2008.

Solomons, T. W. G. and C. B. Fryhle, *Organic Chemistry*, 9th ed., John Wiley & Sons, New York, 2007.

Sorrell, T. N., *Organic Chemistry*, 2nd ed., University Science Books, Sausalito, CA, 2006.

Stoker, H. S., *General, Organic, and Biological Chemistry*, 5th ed., Brooks Cole, Belmont, CA, 2008.

Timberlake, K. C., *Chemistry: An Introduction to General, Organic, and Biological Chemistry*, 10th ed., Pearson Education, Upper Saddle River, NJ, 2009.

Vollhardt, K. P. C. and N. E. Schore, *Organic Chemistry: Structure and Function*, 5th ed., W. H. Freeman, New York, 2007.

Wade, L. G., Jr., *Organic Chemistry*, 7th ed., Prentice-Hall, Upper Saddle River, NJ, 2009.

Winter, A., *Organic Chemistry I for Dummies*, Wiley, Hoboken, NJ, 2005.

QUESTIONS AND PROBLEMS

The use of Internet resources is assumed in answering any of the questions. These would include such things as constants and conversion factors as well as additional information needed to complete an answer.

1. Explain the bonding properties of carbon that make organic chemistry so diverse.
2. Distinguish among alkanes, alkenes, alkynes, and aromatic compounds. To which general class of organic compounds do all belong?
3. In what sense are alkanes saturated? Why are alkenes more reactive than alkanes?
4. Name the compound below:

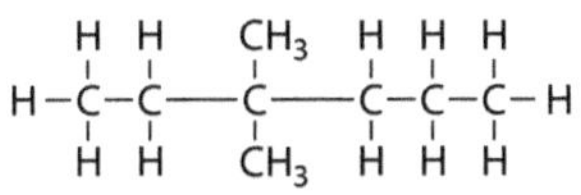

5. What is indicated by "*n*" in a hydrocarbon name?
6. Discuss the chemical reactivity of alkanes. Why are they chemically reactive or unreactive?
7. Discuss the chemical reactivity of alkenes. Why are they chemically reactive or unreactive?
8. What are the characteristics of aromaticity? What are the chemical reactivity characteristics of aromatic compounds?
9. Describe chain reactions, discussing what is meant by free radicals and photochemical processes.
10. Define, with examples, what is meant by isomerism.
11. Describe how the two forms of 1,2-dichloroethylene can be used to illustrate *cis–trans* isomerism.
12. Give the structural formula corresponding to the condensed structural formula of $CH_3CH(C_2H_5)CH(C_2H_5)CH_2CH_3$.
13. Discuss how organic functional groups are used to define classes of organic compounds.
14. Give the functional groups corresponding to (a) alcohols, (b) aldehydes, (c) carboxylic acids, (d) ketones, (e) amines, (f) thiol compounds, and (g) nitro compounds.
15. Give an example compound of each of the following: epoxides, alcohols, phenols, ethers, aldehydes, ketones, and carboxylic acids.
16. Which functional group is characteristic of N-nitroso compounds, and why are these compounds toxicologically significant?
17. Give an example of each of the following: alkyl halides, alkenyl halides, aromatic halides.
18. Give an example compound of a chlorinated naphthalene and of a PCB.
19. What explains the tremendous chemical stability of CFCs? What kinds of compounds are replacing CFCs? Why?
20. How does a thio compound differ from a thioether?
21. How do sulfoxides differ from sulfones?
22. Which inorganic compound is regarded as the parent compound of alkyl and aromatic phosphines? Give an example of each of these.
23. What are organophosphate esters and what is their toxicological significance?
24. Define what is meant by a polymer and give an example of one.

Index

B

C

D

E

Q

R

S

T

U

V

W

Z

For Product Safety Concerns and Information please contact our EU representative GPSR@taylorandfrancis.com
Taylor & Francis Verlag GmbH, Kaufingerstraße 24, 80331 München, Germany

www.ingramcontent.com/pod-product-compliance
Lightning Source LLC
LaVergne TN
LVHW011107300926
849160LV00008B/297
9781439830680